2012 International Conference on Indium Phosphide and Related Materials

(IPRM 2012)

Santa Barbara, California, USA
27-30 August 2012

IEEE Catalog Number: CFP12IIP-PRT
ISBN: 978-1-4673-1725-2

Copyright © 2012 by the Institute of Electrical and Electronic Engineers, Inc
All Rights Reserved

Copyright and Reprint Permissions: Abstracting is permitted with credit to the source. Libraries are permitted to photocopy beyond the limit of U.S. copyright law for private use of patrons those articles in this volume that carry a code at the bottom of the first page, provided the per-copy fee indicated in the code is paid through Copyright Clearance Center, 222 Rosewood Drive, Danvers, MA 01923.

For other copying, reprint or republication permission, write to IEEE Copyrights Manager, IEEE Service Center, 445 Hoes Lane, Piscataway, NJ 08854. All rights reserved.

***This publication is a representation of what appears in the IEEE Digital Libraries. Some format issues inherent in the e-media version may also appear in this print version.

IEEE Catalog Number: CFP12IIP-PRT
ISBN 13:   978-1-4673-1725-2
ISSN: 1092-8669

Additional Copies of This Publication Are Available From:

Curran Associates, Inc
57 Morehouse Lane
Red Hook, NY 12571 USA
Phone: (845) 758-0400
Fax: (845) 758-2633
E-mail: curran@proceedings.com
Web: www.proceedings.com

2012 International Conference on Indium Phosphide and Related Materials (IPRM 2012)

Santa Barbara, California, USA
27-30 August 2012

IEEE Catalog Number: CFP12IIP-POD
ISBN: 978-1-46731-725-2

TABLE OF CONTENTS

Monday, August 27, 2012

PLE   **Plenary Session**

Plenary I   THz Integrated Circuits..1

Plenary IV   Optical Interconnects for Computer-com..5

Mo1C   **Hybrid Integration**

Mo1C.1   Compact InP-based 1 x 2 MMI Splitter on Si Substrate with BCB Wafer Bonding for Membrane Photonic Circuits..8

Mo1C.2   Reflection-Assisted Unidirectional Hybrid-Silicon Microring Lasers..................12

Mo1C.3   Integrated Hybrid III-V/Si Laser and Transmitter...................................16

Mo1C.4   Hybrid InP-Polymer 30 nm Tunable DBR Laser for 10 Gbit/s Direct Modulation in the C-Band 20

Mo1C.5   High-Speed Optical Phased Array Using High-Contrast Grating All-Pass Filters..........22

Mo1D   **Resonant Tunneling Devices**

Mo1D.1   Tunnel Field-Effect Transistor Perspective ....N/A

Mo1D.2   71 mV/dec of Sub-Threshold Slope in Vertical Tunnel Field-Effect Transistors with GaAsSb/InGaAs Heterostructure 25

Mo1D.3   A 1.3 pJ/bit Energy-Efficient Ultra-Low Power On-Off Mode Oscillator Using an InP-based Quantum-Effect Tunneling Device 29

Mo1D.4   Optimized RTD-HBT VCO Design Based on Large Signal Transient Simulations..........32

Mo1D.5   Sensitive High Frequency Envelope Detectors Based on Triple Barrier Resonant Tunneling Diodes...36

MoP   **Joint Poster Session**

MoP.44   Photoluminescence Peak Wavelength Behavior and Luminescent Efficiency of InAs/InGaAsP/InP Quantum Dots Structure....................................40

MoP.45   Lattice Matched and Pseudomorphic InGaAs MOSHEMT with fT of 200GHz 44

MoP.46   Monte Carlo Simulation of InGaAs/Strained-InAs/InGaAs Channel HEMTs Considering Self-Consistent Analysis of 2-Dimensional Electron Gas.......................48

MoP.47   Investigation of GaAs based MOVPE-Grown AlxGa1-xAsyP1-y Strain Compensating Layers 52

MoP.48   Enhancement-Mode Pseudomorphic High-Electron-Mobility Transistor with a Nanoscale Oxidized GaAs Gate...54

MoP.49	Source-Drain Scaling of Ion-Implanted InAs/AlSb HEMTs	57
MoP.50	Electro-Absorption Modulator Chirp Profile Influence on DEML Modulation Scheme at 10 Gb/s	61
MoP.51	Fabrication and DC Characterization of InAs/AlSb Self-Switching Diodes	65
MoP.52	Effect of Temperature on Series Resistance Determination of Au/Polyvinyl Alcohol/n-InP Schottky Structures	69
MoP.53	Estimation of Effective Mass and Subbands in Multiquantum Wells using Polarized Light Irradiation	73
MoP.54	Single-Event Transient Sensitivity to Gate Bias in InAlSb/InAs/AlGaSb High Electron Mobility Transistors	77

Tuesday, August 28, 2012

Tu1D	Optical Nanodevices on Si	N/A
Tu1D.1	III-V Nanowires on Si for Optoelectronics and Solar Applications	N/A
Tu1D.2	Selective Area Heteroepitaxy of InP Nanopyramidal Frusta on Si for Nanophotonics	81
Tu1D.3	Characteristics of InP Nanoneedles Grown on Silicon by Low-Temperature MOCVD	85
Tu1D.4	Silicon-baed Long-Wavelength III-V Quantum-Dot Lasers	88

Tu1E	**High Speed Diode Technologies**	
Tu1E.1	High Efficiency and Broad-Band Operation of Monolithically Integrated W-Band HBV Frequency Tripler	92
Tu1E.2	Lattice-Matched p+-GaAsSb/i-InAlAs/n-InGaAs Zero-Bias Backward Diodes for Millimeter-Wave Detectors and Mixers	95
Tu1E.3	A W-Band InGaAs PIN-MMIC Digital Phase-Shifter Using the Switched Transmission-Line Structure	99
Tu1E.4	Development of a 557 GHz GaAs Monolithic Membrane-Diode Mixer	102
Tu1E.5	Fundamental Oscillation up to 1.31 THz in Thin-Well Resonant Tunneling Diodes	106
Tu1E.6	Design and Fabrication of InGaAs Photodiodes for SWIR Imagers with Low Dark Current	110

Tu2D	**Optical Components**	
Tu2D.1	Integrated Optoelectronic Components for the Transmission and Reception of Polarization- and Phase-Coded Optical Signals	N/A
Tu2D.2	40 Gbit/s Identical Layer InGaAlAs-MQW Electoabsorption-Modulated DFB-Lasers Operating Between 1298 nm and 1311 nm	113
Tu2D.3	Reflective Amplified Modulator Operating at 40 Gb/s up to 85Â°C as Colorless Transceiver for Optical Access Networks	116
Tu2D.4	InP Photonic Integrated Circuit with an AWG-like Design for Optical Beam Steering	120

| Tu2D.5 | Design Fabrication and Preliminary Test Results for a New InGaAsP/InP High-Q Ring Resonator for Gyro Applications | 124 |

Tu2E **Heterogeneous Integration**

Tu2E.1	Double-Layer Stepped Si(100) for III-V-on-Silicon Integration	128
Tu2E.2	Atomic-Plane-Thick Reconstruction Across the Interface During Heteroepitaxial Bonding of InP-Clad Quantum Wells to Si	130
Tu2E.3	Lattice Engineered Substrates	N/A
Tu2E.4	Electrical Conduction Property at InAs/Si(111) Interface by Selective-Area MOVPE	133
Tu2E.5	Interface and Surface Dielectric Anisotropies of GaP/Si(100)	137

Tu3D **Lasers I**

Tu3D.1	Ultralow-Power Nanophotonic Devices Based on Buried-Heterostructure Photonic-Crystal Nanocavities	N/A
Tu3D.2	Metal-Clad Photonic Crystal Membrance Nanolasers	139
Tu3D.3	10-Gbit/s Direct Modulation of Optically Pumped InGaAlAs Multiple-Quantum-Well Photonic-Crystal Nanocavity Laser up to 100°C.	143
Tu3D.4	Multi-Stack Quantum Cascade Laser	147

Tu3E **III-V MOS**

Tu3E.1	Design of High-Current L-Valley GaAs/AlAs$_{0.56}$Sb$_{0.44}$/InP (111)Ultra-Thin-Body nMOSFETs	151
Tu3E.2	High Performance Substitutional-Gate MOSFETs Using MBE Source-Drain Regrowth and Scaled Gate Oxides	155
Tu3E.3	Epitaxy of III-V based Channels on Si and Transistor Integration for 12nm Node CMOS	159
Tu3E.4	Novel Atomic Layer Deposited Thin Film Beryllium Oxide for InGaAs MOS Devices	163
Tu3E.5	Sulfur Cleaning for (100) (111)A and (111)B InGaAs Surfaces with In Content of 0.70 and their Al2O3/InGaAs MOS Interface Properties	167

Tu4A **Late News**

| Tu4A.1 | High-Power InP-based Waveguide Photodiodes and Photodiode Arrays Heterogeneously Integrated on SOI | 171 |

Wednesday, August 29, 2012

We1D **QD Lasers and Technology**

| We1D.1 | Recent Advances in High-Speed Lasers and Amplifiers Based on 1.5 um QD/QDash Material | 173 |

We1D.2	1550nm InAs/InP Quantum Dash Based Directly Modulated Lasers for Next Generation Passive Optical Network	177
We1D.3	20 GHz to 83 GHz Single Section InAs/InP Quantum Dot Mode-Locked Lasers Grown on (001) Misoriented Substrate	181
We1D.4	InAs/InP Quantum Dash Based Mode Locked Lasers for 60 GHz Radio over Fiber Applications	185
We1D.5	MOVPE Growth of Ga(AsBi)/GaAs Quantum Well Structures	N/A

We1E — HBTs

We1E.1	InP HBT with 55-nm-wide Emitter and Relationship Between Emitter Width and Current Density	188
We1E.2	InP/GaInAs DHBT with TiW Emitter Demonstrating fT/fmax ~340/400GHz for 100 Gb/s Circuit Applications	192
We1E.3	Lower Limits to Specific Contact Resistivity	196
We1E.4	Study of the NiGaInAs Alloy as an Ohmic Contact to the p-type Base of InP/GaInAs HBTs	200
We1E.5	Multi-Finger 250nm InP HBTs for 220GHz mm-Wave Power	204
We1E.6	Analysis of InP/GaAsSb DHBT Failure Mechanisms Under Accelerated Aging Tests	208

We2D — Photodiodes

We2D.1	Microwave Photonics	212
We2D.2	Lateral Scalability of Inverted p-down InAlAs/InGaAs Avalanche Photodiode	215
We2D.3	Phase Characterization of Intermodulation Distortion in High-Linearity Photodiodes	219
We2D.4	High Speed AlInGaAs Quantum Well Waveguide Photodiode for Wavelengths Around 2 microns	221
We2D.5	MOCVD Based Zinc Diffusion Process for Planar InP/InGaAs Avalanche Photodiode Fabrication	225

We2E — HEMT Technologies

We2E.1	450 GHz Amplifier MMIC in 50 nm Metamorphic HEMT Technology	229
We2E.2	100nm-Gate InAlAs/InGaAs HEMTs on Plastic Flexible Substrate with High Cut-Off Frequencies	233
We2E.3	High Electron Mobility InAs-Based Heterostructure on Exact (001) Si Using GaSb/GaP Accommodation Layer	N/A
We2E.4	Analysis of Performances of InSb HEMTs Using Quantum-Corrected Monte Carlo Simulation	237
We2E.5	Optimized InP HEMTs for Low Noise at Cryogenic Temperatures	241

We2E.6   Metal-Organic Vapor-Phase Epitaxy Growth of InP-Based HEMT Structures with InGaAs/InAs Composite Channel 245

Thursday, August 30, 2012

Th1D   **Nanowires and Quantum Dots**

Th1D.1   Selective-Area Growth InP-based Nanowires and Their Optical Properties 249

Th1D.2   Single GaAs Nanowire Photovoltaic Devices under Very High Power Illumination 253

Th1D.3   Radial InP/InAsP Quantum Wells with High Arsenic Compositions on Wurtzite-InP Nanowires in the 1.3-Âμm Region 257

Th1D.4   Site-Controlled Growth of InP/InGaP Quantum Dots 261

Th1D.5   Catalyst Design for Native Oxide Based Selective Area InP Nanowire Growth 265

Th2C   **Lasers II**

Th2C.1   High Speed VCSELS for Optical Interconnects 269

Th2C.2   Slotted Tunable Laser with Monolithic Integrated Mode Coupler 273

Th2C.3   C-band Operation of Lateral-Grating-Assisted Lateral Co-Directional Coupler Tunable Laser with High-Mesa Buried Hetero-Structure 277

Th2C.4   Multiple Coherent Outputs from Single Growth Monolithically Integrated Injection Locked Tunable Lasers 281

Th2C.5   Low-Threshold Operation of LCI-Membrance-DFB Lasers with Be-doped GaInAs Contact Layer 285

THz Integrated Circuits using InP HEMT Transistors

Kevin Leong, Gerry Mei, Vesna Radisic, Stephen Sarkozy, William Deal

Northrop Grumman Corporation

Abstract — Over the last few years, operating frequencies of InP HEMT Transistors have pushed above 100 GHz (1 THz). This has allowed electronic circuitry to be realized at frequencies as high as 670 GHz. In particular, Low Noise Amplifiers (LNA), Power Amplifiers (PA), mixers and multipliers have all been implemented. A result of this has been integrated circuit receivers and transmitters operating to frequencies as high as 670 GHz.

Index Terms — Low Noise Amplifier, Power Amplifier, Terahertz

I. INTRODUCTION

In the last few years, the development of > 1 THz f_{MAX} transistor technologies [1] has pushed operating frequencies of amplifiers well into the sub-millimeter wave range. The first demonstrations of sub-millimeter amplification were undertaken at the 340 GHz atmospheric window using InP HEMT [2] and MHEMT [3] technologies. Amplification has now been demonstrated above the 300 GHz sub-millimeter wave threshold, with work in the 460-500 GHz range recently reported including a HEMT amplifier with 11.4 dB packaged gain [4], and noise figure of 11.7 dB [5], and a 16.1 dB gain amplifier measured on-wafer at 460 GHz [6]. In this paper, we provide background into the fundamental developments which now make amplifiers and other types of electronic circuits possible at 670 GHz using 30 nm InP HEMTs.

II. TRANSISTOR TECHNOLOGY

Critical for reaching Terahertz operational frequencies for integrated circuits are transistors with sufficiently high f_{MAX}. This paper describes development of amplifiers targeting 670 GHz. As a rule of thumb, transistor f_{MAX} should be 50-100% higher than the design frequency. Therefore, 1-1.3 THz f_{MAX} transistors are necessary. In this section, 30 nm InP HEMTs are described.

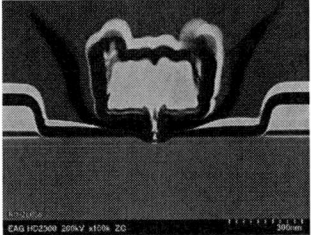

Fig. 1   Scanning Electron Tunneling (SEM) image of 30 nm InP HEMT gate.

An InP-based HEMT was developed as the key enabling technology. The epitaxial wafers were grown in molecular beam epitaxy (MBE) on 3-inch semi-insulating InP (100) substrates. As shown in Fig. 2, the layer structure consists of an n+ InGaAs/InAlAs composite cap for enhanced ohmic contacts, an un-doped InAlAs as Schottky barrier and an InGaAs/InAs composite channel for superior electron transport properties. A Si doping plane was inserted in the Schottky layer to supply electrons for current conduction. A room temperature electron mobility over 15,000 cm2/V•s has been achieved with a sheet charge of 3.3e12 cm2.

$In_{0.6}Ga_{0.4}As/ In_{0.52}Al_{0.48}As$ N+ composite cap
$In_{0.52}Al_{0.48}As$ barrier
$In_{0.52}Al_{0.48}As$ spacer
$In_{0.53}Ga_{0.47}As$
InAs   composite channel
$In_{0.53}Ga_{0.47}As$
$In_{0.52}Al_{0.48}As$ buffer

Fig. 2 Layer profile of the epi wafers.

The device ohmic contact was formed using a non-alloyed metal scheme, enabling an extremely consistent low contact resistance, Rc, typically 0.04 Ω.mm, with the on-wafer standard deviation as low as 3%. The 30nm gate process is another critical element in this technology (Fig. 1). Scaling the size of the gate allowed for a corresponding reduction in the gate capacitance, Cgs, which affects gain for circuits operating at high frequencies. The 30nm gate pattern was defined using electron beam lithography. The self-aligned gate recess was then formed with wet chemical etching, and the T-gate was formed with a refractory Schottky metal stack. The devices were fully passivated with plasma-enhanced chemical vapor deposition SiN. The passivation thickness is 200 Angstroms.

III. SUB-MILLIMETER WAVE AMPLIFICATION

Amplification is a basic circuit function at lower frequencies and is used for small signal and power amplification. Until recently, these functions have been impossible at Terahertz frequencies and these systems have relied on diode based down-conversion and multiplier chains to process small signals and generate power. In this section, we describe several 670 GHz amplifiers and their performance.

Shown in Fig. 3 is a microphotograph of a 10-stage amplifier. The measured S-parameters for the circuit are shown in Fig. 4. The circuit reaches a peak gain of about 30 dB at 670 GHz for a realized transistor gain of approximately 3 dB/stage.

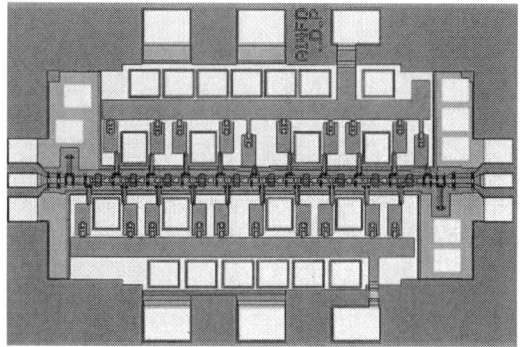

Fig. 3 Microphotograph of 10-Stage 670 GHz low noise amplifier.

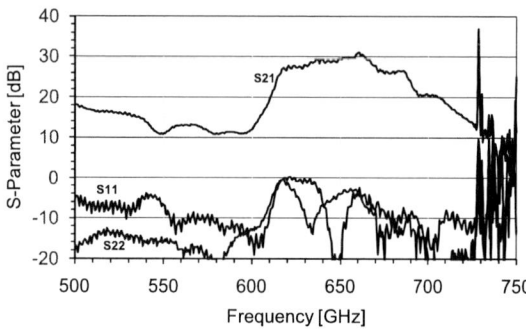

Fig. 4 Measured on-wafer S-Parameters of 10-stage LNA.

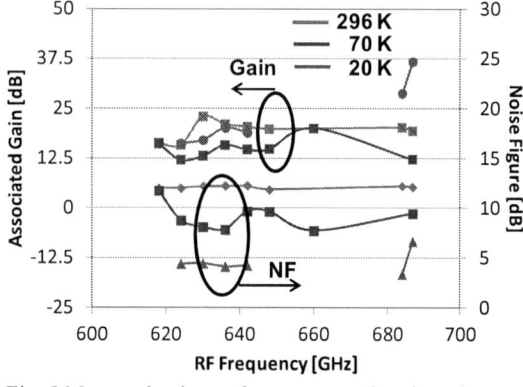

Fig. 5 Measured noise performance as a function of temperature for packaged low noise amplifier

The circuit itself is extremely compact. A typical spacing of only 25 um is present between gate and drain feeds. All drain bias lines are fed to the top and tied together on a single pad and the gate bias lines are routed to the bottom. This means that the entire 10-stage amplifier must operate from a single bias. The amplifier uses 14 um transistors biased at a VDS of 1.2 V and a current density of 450 mA/mm, or ~65 mA for the 10-stage design. To improve input and output match, a single open-circuited stub is used to improve matching. The bright gold rectangles are substrate mode suppression vias to eliminate the parallel plate waveguide mode.

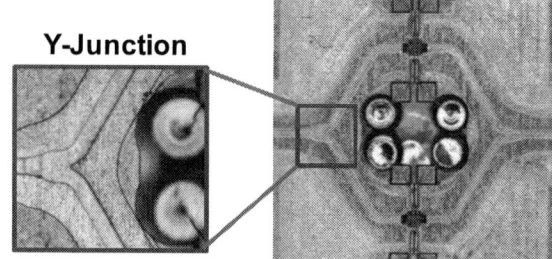

Fig. 6 Photograph of insides of 2-way power combined module operating at 650 GHz.

Noise and power are critical metrics if amplifier technology is to become competitive against mature diode technologies which currently dominate the range from 300-2000 GHz. Both of these parameters have been measured for packaged amplifiers. The noise figure results are shown in Fig. 5. At room temperature (296 K), a packaged noise figure of 12.5 dB is obtained, which is competitive with GaAs Schottky receivers at these frequencies. Lowering the operating temperature to 70 K results in 7-10 dB NF. Further lowering the temperature to 20 K results in noise figures as low as 4 dB. Note that no effort was made to design the circuit for cryogenic operation.

Power amplifiers have also been demonstrated with this technology. A photograph of two chips power combined a package are shown Fig. 6. A two-way power combiner splits the signal, which is then amplified in each arm. A symmetric power combiner is then used to coherently combine the power. Measured small signal S-parameters for the amplifier module are shown in Fig. 7. Small signal gain greater than 5 dB is available from 620-680 GHz, with a peak gain of 12 dB demonstrated at 625 GHz. Shown in Fig. 8 is measured power gain and power from the packaged module at 650 GHz. A peak output power of 3 mW is reached with a power gain of 4

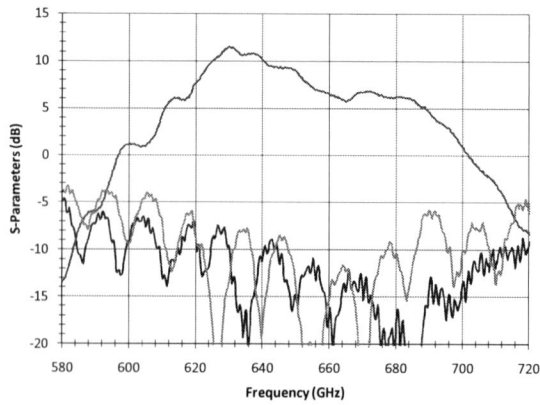

Fig. 7 Measured S-Parameters of PA module

dB. This power level compares favorably to output power generated by diode multiplier chains. We expect output powers to improve as device technology and modeling improve and designs are iterated.

When used in a receiver, this noise and power performance should be acceptable for atmospheric sensing or radioastronomy as both a low noise front-end technology as well as a method for generating Local Oscillator (LO) power for radiometers.

IV. RECEIVER RESULTS

In the prevous section, low noise amplification and power amplification is described at approximately 670 GHz. In addition to amplification, THz HEMTs can be used for other basic circuit functions such as multiplication and mixing. Given the small circuit size due to the short wavelength and compact circuit features, an entire receiver or transmitter can be incorporated onto a single integrated circuit. As frequency

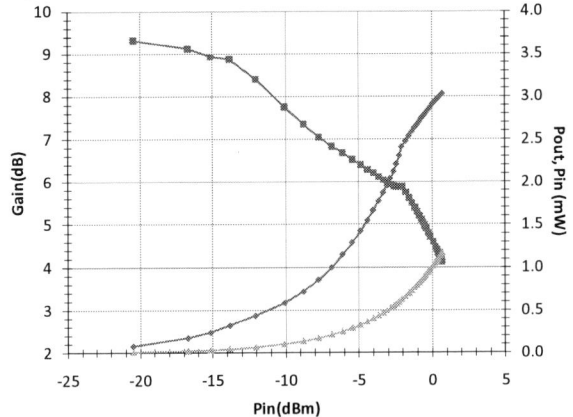

Fig. 3 Measured power and power gain of amplifier module at 650 GHz.

increases, this has significant benefits. First, packaging at submillimeter wave frequencies requires high precision waveguide to convey the signals, which adds significant costs. Traditional submillimeter wave packaging places a single function (mixing or multiplication) in a single package. By integrating functions together, packaging costs are significantly reduced. Moreover, transition losses from the chip to waveguide are quite high at submillimeter wave frequencies. Placing multiple functions on the same chip eliminates these interconnects to improve functionality.

Shown in Fig. 9 is a receiver chip designed for 670 GHz. The input signal is received by a dipole at the left. This dipole electromagnetically couples the signal from the waveguide to the chip. This is an important aspect of integrated circuits operating at submillimeter wave frequencies as the inductive reactance of wirebonds is too high to efficiently transfer signals. The signal is transferred through coplanar waveguide to a 10-stage LNA. This LNA is identical to the chip shown in

Fig. 4, which was designed for integration into this receiver. After amplification, the signal is mixed down to a 70 GHz IF frequency by a sub-harmonic mixer to the right of the LNA. The IF is extracted from the ground-signal-ground pad at the bottom of the chip. The Local Oscillator (LO) multiplier is also included on chip. A 100 GHz LO is applied to the Ground-signal-ground pads on the right of the chip. The LO is tripled to 300 GHz and then amplified by the LO power amplifier to approximately 600 GHz to pump the mixer.

The chip has been packaged and measured for performance. In addition to using the integrated low noise amplifier shown in Fig. 9, we have also test the receiver with one external

Fig. 4 Microphotograph of integrated receiver chip. RF input is to the dipole on left to electromagnetically couple to the integrated circuit.

packaged LNA, and then two external packaged LNA's to help set the receiver noise figure, as shown in Fig. 10. Note that the block title "Macro cell" indicates the integrated reciever. For the packaged amplifiers (with interconnect loss) and on-chip LNA, total gain in front of the mixer is estimated to be 45-50 dB. The results of the measurement is shown in Fig. 11. Note that the blue trace shows the measured receiver by itself (Macro cell). Minimum received noise figure is slightly less than 15 dB and degrades quickly out of the center of the band. After adding an additional LNA, the noise figure improves, indicating that not enough gain was present on the macro cell by itself to set the noise figure. Minimum noise figure is 12.7 dB, and the response is considerably more broadband than the macro cell by itself. However, with an a second additional LNA added, noise figure degraded to more than 16 dB. This result was not expected, and the test was repeated with the two LNA's reversed. Approximately the same result was yielded.

After consideration, the degradation in receiver performance as gain is added is not entirely unexpected. Note from Fig. 6 that the bandwidth is approximately 70 GHz. With the high gain of three LNA's and the large instantaneos bandwidth, we calculate that the noise power over the 70 GHz bandwidth is on the order of 0.1 – 1 mW. In this case, we believe that either the LNA or the mixer is starting to saturate, thus either lowering the LNA gain or increasing the mixer conversion loss. This has the impact of increasing the receiver noise figure.

This result raises several questions about broadband submillimeter wave tranciever architectures. With 10's of GHz of instantaneous RF bandwidth available, submillimeter wave links could transfer massive amounts of data. However, the noise power in the receiver bandwidth will ultimately limit receiver dynamic range. RF filtering should be added to the receiver to band limit the noise power incident on the mixer.

978-1-4673-1725-2/12 $31.00 © 2012 IEEE

This is also a prudent practice for sensitive receivers in radiometers for atmospheric science or radioastronomy.

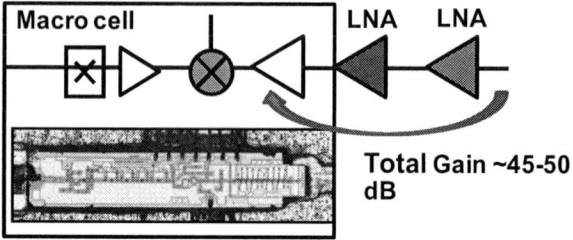

Fig.10 Block diagram illustrating receiver measurements.

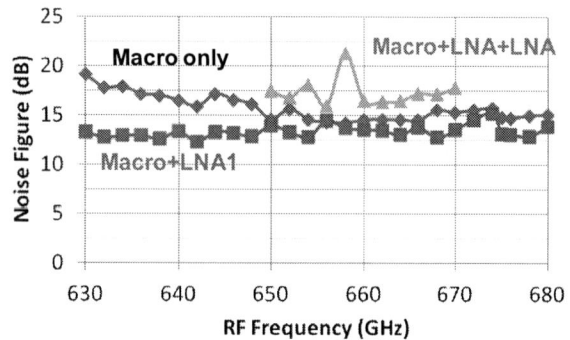

Fig. 11 Measured performance of receiver plus LNA's.

IV. CONCLUSION

In this paper, the basics of a 30 nm InP HEMT is presented. The transistor shows fMAX well above 1 THz, which enables integrated circuits operating to 670 GHz, as is shown in this paper. Low noise amplification and power amplification are presented. The paper concludes with a single chip receiver and measurements of the receiver. The paper shows that integrated circuits at submillimeter wave frequencies are a viable technology with InP HEMT technology.

ACKNOWLEDGEMENTS

This work was supported by the DARPA THz Electronics Program and Army Research Laboratory under the DARPA Contract no. HR0011-09-C-0062. The views, opinions, and/or findings contained in this article/presentation are those of the author/presenter and should not be interpreted as representing the official views or policies, either expressed or implied, of the Defense Advanced Research Projects Agency or the Department of Defense. Approved for Public Release, Distribution Unlimited. The authors would like to thank Dr. John Albrecht of DARPA and Dr. Alfred Hung of ARL and to acknowledge NGAS contributors in HEMT, EBL, MBE, processing, layout, machining, and test groups, ARL THz laboratory for providing test support. The authors thank the Far Infrared Receiver Laboratory at the University of Virginia for developing the WR-1.5 coplanar wafer probes used in this work, as well as Robert Lin and Goutam Chattopadhyay of the Jet Propulsion Laboratory, California Institute of Technology for making noise figure measurements.

REFERENCES

[1] R. Lai, X.B. Mei, W.R. Deal, W. Yoshida, Y.M. Kim, P.H. Liu, J. Lee, J. Uyeda, V. Radisic, M. Lange, T. Gaier, L. Samoska, A. Fung, "sub 50 nm InP HEMT device with Fmax greater than 1 THz," *IEEE 2007 IEDM Conf. Dig.,* pp. 609-611, Dec. 2007.

[2] W.R. Deal, X.B. Mei, V. Radisic, W. Yoshida, P.H. Liu, J. Uyeda, Barsky, T. Gaier, A. Fung, R. Lai, "Demonstration of a S-MMIC LNA with 16-dB gain at 340-GHz," *IEEE 2007 CSIC Conf. Dig.,* pp. 1-4, Oct. 2007.

[3] A. Tessmann, A. Leuther, V. Hurm, H. Massler, M. Zink, M. Kuri, M. Riessle, R. Losch, M. Schlechtweg, O. Ambacher, "A 300 GHz mHEMT amplifier module," *IEEE 2009 IRPM Conf. Dig.,* pp. 196-199, May 2009.

[4] W.R. Deal, K. Leong, V. Radisic, S. Sarkozy, B. Gorospe, J. Lee, P.H. Liu, W. Yoshida, J. Zhou, M. Lange, J. Uyeda, R. Lai, X.B. Mei, "0.48 THz Amplification with InP HEMT Transistors," *IEEE MWCL,* Vol. 19, Issue 5, pp. 289-291, May 2010.

[5] W.R. Deal, "solid-state amplifiers for terahertz electronics," *IEEE MTT-S Int.Symp. Dig.,* pp. 1122-1125, May 2010.

[6] A. Tessmann, A. Leuther, R. Loesch, M. Seelmann-Eggbert, H. Massler, "A metamorphic HEMT S-MMIC amplifier with 16.1 dB gain at 460 GHz", *IEEE 2010 CSIC Conf. Dig.,* pp. 245-248, Oct. 2010.

Optical Interconnects for Computer-com

Marc A. Taubenblatt.

IBM T.J. Watson Research Center, Yorktown Heights, NY

Abstract — Computing Systems are exponentially increasing their dependence on optical interconnects to meet their scaling bandwidth needs. Steady increases in computation density have put pressure on the interconnect infrastructure to keep up. The requirements for these interconnects include a critical set of metrics, that historically have focused on cost, but increasingly consider power and density. Furthermore, reliability (component and data) and latency must be considered as well. Thus the path forward to increasing bandwidth in these systems is becoming an increasingly complex set of trade-offs. This abstract describes the major applications for optical interconnects in computer systems, the relative metrics for these applications and considers new technologies in this context.

Index Terms — optical interconnects; high performance computing, supercomputing.

I. INTRODUCTION

High performance computing systems (HPC), intended to provide vast computational capability, are of steadily growing interest, to provide new computational capabilities, ranging from geophysical data processing to drug discovery to multi-scale modeling to environment and climate modeling. These systems are exponentially increasing their dependence on optical interconnects to meet their scaling BW demands. More recently, the growth of enormous data centers is being driven by factors such as the growing demand for computation and services delivered by the cloud, commercial exploitation of the growing volume of accessible data and increased connectivity and telecommunications BW. These large datacenters require short distance network connectivity that may soon rival those used in HPC systems.

II. TECHNOLOGY TRENDS

Historically the growth trend in computing systems has been fueled by improved transistor speeds, followed by improved chip architectures, including multicore processors, which take advantage of increased transistor density. Their slowing contribution, however, has increased the need for new innovations, such as accelerators (e.g. graphics processing units), and further increased the reliance on continued growth in large scale system parallism to make up the difference. Thus to keep up with computing gains, interconnect BW (bandwidth) requirements continue to scale at all physical levels of the system.

Typical server interconnects include core to core on the same chip, chip to chip (for CPU [central processing unit] to

cache or CPU to CPU), chip to memory on card, CPU to CPU cluster fabric between cards and racks and LAN/WAN links (local/wide area networks) which go beyond the computing or data center [1]. To some degree, clever networking topologies can mitigate BW costs while maintaining reasonable performance, although not without trade-offs. For example, mesh or torus networks require many fewer interconnects, but will need more hops and thus have longer latency to pass data between nodes. However, at each level of packaging hierarchy continued scaling of BW is causing bottlenecks: off chip, off module, off card and rack to rack.

Channel data rates for these off-chip interconnects have been steadily increasing in response, as this has been the best way to improve cost per transported bit, power per transported bit and to meet BW density (wiring, area escape) requirements. However, electrical interconnects become much more difficult to design as data rates begin to exceed 10Gbps, due to frequency dependent losses, crosstalk and frequency resonance effects [2]. Optical interconnects do not suffer such strong signal integrity degradations, and provide additional benefits, including reduced cable bulk, smaller connector size and reduced EMI (Electro-magnetic interference).

Early adoption of parallel optics (space division multiplexing) in HPC systems was primarily driven by the cost competitiveness of optical interconnects compared to the copper cable links they were attempting to displace. A good example of this transition occurred with the IBM ASCI Purple system, which was initially designed for electrical cabling between racks, but later transitioned to optical interconnects. In addition future HPC systems and large scale data centers are now becoming increasingly power constrained, as these systems grow in size and the ability to scale both power and performance of CMOS transistors has fallen off the historical trend in recent years.

Thus optical interconnects for computing systems must simultaneously meet a number of metrics. These include not only cost and bit error rate but now power, cable, connector and module density. Furthermore with the increasing amount of optics used in more critical links, reliability and latency have become increasingly important as well. Thus, the path forward to increasing BW in these systems has become an increasingly complex set of tradeoffs. Figure 1 shows a critical set of six metrics which today's systems must balance. Furthermore, the requirements for these metrics will also vary depending on the system and link within a given system. For example (also shown in Figure 1), a node clustering link in an HPC system will emphasize cost and power, as these become critical factors in very large systems, while an SMP link and a high performance commercial system will more likely emphasize latency and reliability factors.

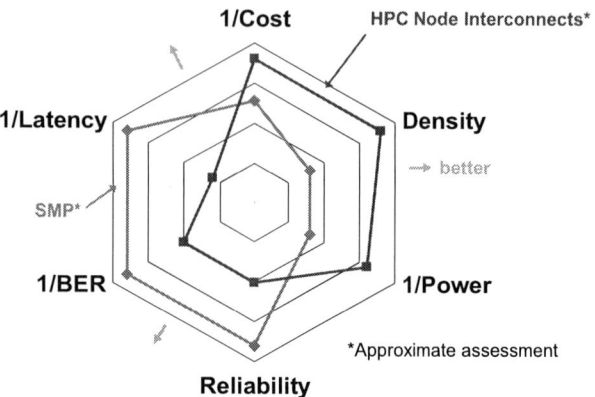

Fig. 1. Critical set of six metrics which today's systems must balance. Requirements will vary with system type and particular link within a system.

Winning technology choices will provide the best tradeoffs to provide higher bandwidth while still meeting all the metrics. For example, cost takedowns by further increasing datarate may be offset by power constraints including the increasing mismatch between optical datarates and on chip clock speeds. Minimization of power in the complete link will simultaneously require very dense optics modules which can be placed close to the CPU or switch chip to reduce the electrical power to and from the optics. Dense optics will benefit from higher datarates as well as tighter pitch optical channels (impacting signal integrity and therefore BER), as well as multichannel methods such as WDM, multi-level coding and multi-core fiber.

III. TECHNOLOGIES OF INTEREST

Historically low cost optical interconnects for data-com and now computer-com have been based on multi-mode fibers and VCSEL technology (Vertical Cavity Surface Emitting Laser). Rack to Rack cluster fabric in particular has made good use of parallel optical modules employing these technologies. This technology will continue to improve, with higher speeds (40Gbps has been shown in lab demonstrations [3]), lower cost, lower power and more compact modules. Density improvements examples are shown in Figures 2 and 3; an ultra-dense high channel count fiber transceiver based on flip chip technology [4] and a multi-core multi-mode fiber transceiver [5].

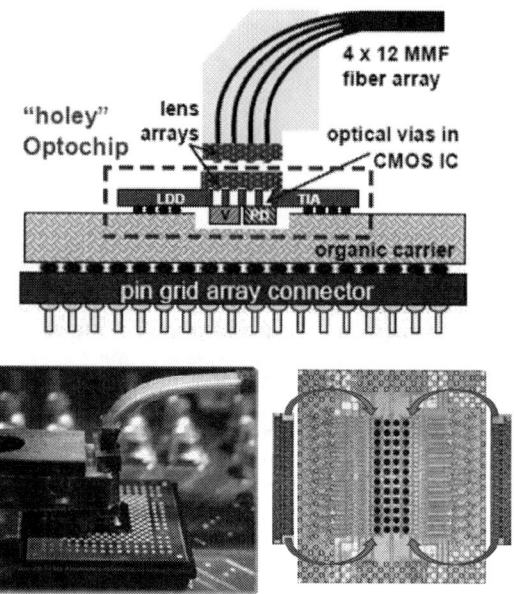

Fig. 2. Holey Optochip, consisting of a CMOS IC containing driver and receiver circuits, with optical vias (holes) etched through the chip.

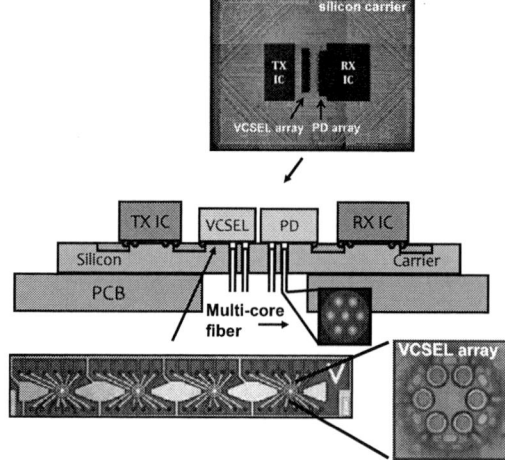

Fig. 3. An optical transceiver utilizing multi-core multi-mode fiber. VCSEL and Photodiode arrays consisting of 4 groups of 6 elements each are coupled to the multicore fiber using a Si carrier with optical vias (holes).

Low power optical links have been demonstrated through use of new circuit and signal processing techniques [6, 7] showing, for example, a few pJ/bit performance at high data rates (<2pJ/bit at 25Gbps for a full optical link).

To make further gains in cost and level of packaging integration, optical printed circuit board technology based on polymer waveguide integration with VCSELS may provide the right combination of low cost manufacturing, module density and semi-customizable integration [8, 9].

Finally, Silicon Photonics, utilizing single mode fiber in combination with DC lasers and silicon based modulators and detectors [10], has made recent progress. This technology may offer the ultimate in integration capability as well as low cost by utilizing low cost CMOS fabs to produce highly integrated assemblies.

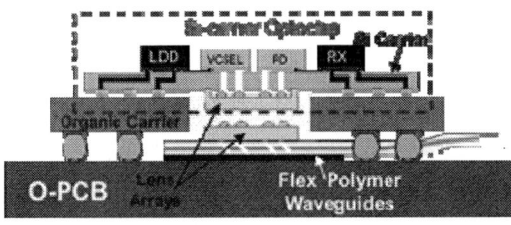

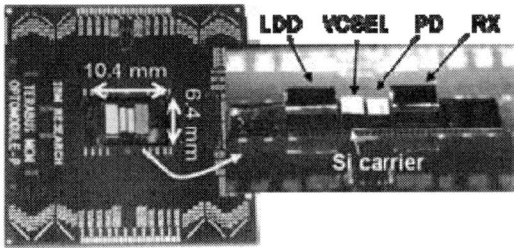

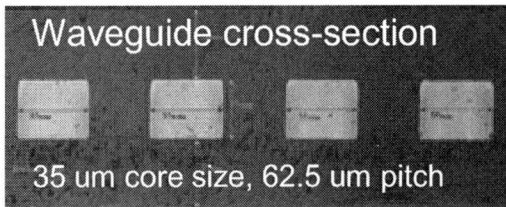

Fig. 4. Optical Printed Circuit board technology. Top: Optical Transceiver and packaging schematic. Middle: Fabricated optical transceiver. Bottom: Cross section of polymer waveguides.

III. CONCLUSIONS

Computer-com is an emerging market with a need for very short links, many <10m, very high aggregate BW links and very high demands for low cost, low power, high reliability and high density. Tightly integrated optics packaging will be required to achieve these goals along with a broad view to optimize technology across system requirements and often between the system providers and component suppliers as well. Winning technology choices will provide the best tradeoffs to provide higher bandwidth while still meeting all the metrics.

ACKNOWLEDGEMENT

The author wishes to thank the many IBM colleagues, too numerous to be named here, who have contributed to this work. Some of this work was partially funded by DARPA.

REFERENCES

[1] A. Benner, M. Ignatowski, J. Kash, D. Kuchta, and M. Ritter, "Exploitation of optical interconnects in future server architectures," IBM J. Res. & Dev., vol. 49, pp. 755-775, 2005.

[2] D. Kam, et al., "Is 25 Gb/s On-Board Signaling Viable?", IEEE Transactions on Advanced Packaging, vol. 32, no. 2, pp. 328 - 344, 2009.

[3] A. V. Rylyakov, C. L. Schow, J. E. Proesel, D. M. Kuchta, C. Baks, N. Y. Li, C. Xie, and K. P. Jackson, "A 40-Gb/s, 850-nm VCSEL-based full optical link," Optical Fiber Communication (OFC) Conference 2012, paper OTh1E, Los Angeles, CA, Mar. 2012.

[4] F. E. Doany et al, "Terabit/sec VCSEL-Based 48-Channel Optical Module Based on Holey CMOS Tansceiver IC", Journal of Lightwave Technology, to be published.

[5] B. G. Lee, et al., "End-to-End Muticore Multimode Fiber Optic Link Operating up to 120 Gb/s," Journal of Lightwave Technology, vol. 30, no. 6, 2012, pp. 886-892).

[6] J. E. Proesel, B. G. Lee, A. V. Rylyakov, C. W. Baks, and C. L. Schow, "Ultra Low Power 10- to 28.5-Gb/s CMOS-Driven VCSEL-Based Optical Links," J. of Optical Comm. And Networking(JOCN), submitted for publication.

[7] A. V. Rylyakov, C. L. Schow, B. G. Lee, F. E. Doany, C. Baks, and J. A. Kash, "Transmitter Pre-Distortion for Simultaneous Improvements in Bit-Rate, Sensitivity, Jitter, and Power Efficiency in 20 Gb/s CMOS-driven VCSEL Links," IEEE J. of Lightw. Technol., vol.30, no.4, pp.399-405, Feb. 2012.

[8] F. E. Doany et al., "Terabit/s-Class 24-Channel Bidirectional Optical Transceiver Module Based on TSV Si Carrier for Board-Level Interconnects," Electronic Components and Technology Conference (ECTC) 2010, pp. 58–65, June 2010.

[9] R. Dangel et al., "Polymer-Waveguide-Based Board-Level Optical Interconnect Technology for Datacom Applications," IEEE Transactions on Advanced Packaging, vol. 41, pp. 759-767, Nov 2008.

[10] Y.A.Vlasov, "Silicon Integrated Nanophotonics: Road from Scientific Explorations to Practical Applications", Conference on Lasers and Electro Optics (CLEO) 2012, San Jose, CA, May 2012.

978-1-4673-1725-2/12 $31.00 © 2012 IEEE

Compact InP-based 1×2 MMI Splitter on Si Substrate with BCB Wafer Bonding for Membrane Photonic Circuits

Jieun Lee[1], Yoshiaki Yamahara[1], Yuki Atsumi[1], Takahiko Shindo[1],
Tomohiro Amemiya[2], Nobuhiko Nishiyama[1], and Shigehisa Arai[1,2]

[1]Department of Electrical and Electronic Engineering, [2]Quantum Nanoelectronics Research Center,
Tokyo Institute of Technology, Meguro-ku, Tokyo 152-8552, Japan
E-Mail lee.j.aj@m.titech.ac.jp

Abstract — Ultra-low power and compact optical interconnects can be realized with III-V-based membrane photonic intergrated circuits (PICs) on Si platform. This study focused design and fabrication of the III-V-based passive components of the PICs, especially, a compact 1×2 multimode interference (MMI) splitter with SiO_2 buried GaInAsP wire structure. The calculated excess loss was as low as 0.28dB in the optimized structure, and the structural tolerance obtained was as low as 0.05dB, when MMI width W_{MMI}, length L_{MMI}, output gap Y were changed in range of ± 50 nm. The footprint of the MMI region was only 1.4×1.5 μm^2. In the spectral range of 1548-1552 nm, the average excess loss measured was around 3dB and the average imbalance between the outputs was 0.5dB allowing the compact size and light splitting function.

Index terms — Multimode ineterference (MMI), III-V-based passive components, membrane photonic integrate circuits(PICs), optcal beam splitters.

I. INTRODUCTION

The processing speed in large-scale integrated cirtuits (LSIs) is being limited by the RC time constant and the power consumption on the global electrical wires [1-2]. As one solution, optical on-chip interconnects have been proposed with the replacement of the copper wiring in Si-LSI [3-4]. To realize them, we proposed the concept of membrane photonic integrated circuits (PICs) such as shown in Fig. 1, which consist of the thin III-V-based PICs with high index-contrast structure of the semiconductor core layer sandwiched by SiO_2 or air claddings [5-6]. The strong optical confinement to the core layer in the membrane structure leads to ultra-compact footprints and low-power dissipation. For integration technique of III-V/Si, a Benzocyclobutene (BCB) adhesive wafer bonding was used to realize the membrane PICs by the way of backend process after the CMOS process [7]. For such III-V PICs, we have reported membrane lasers [8] and photo-detectors [9] as active components. InP-based wire waveguides on the Si substrate was also reported with a propagation loss of 4 dB/cm [10].

For membrane PICs, passive devices for splitting and coupling of the optical signals are necessary. As such a beam splitter, N×N multi-mode interference (MMI) devices have been extensively investigated due to its simple structure and large fabrication tolerance [11-12]. So far, an InP-based 1×2 MMI splitter on Si substrate was reported, but the footprint was rather larger to 2.75×6.64 μm^2 [13].

One of the advantages of this index contrast membrane

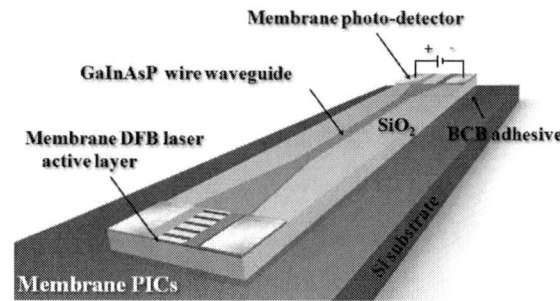

Fig. 1 Schematic representation of a membrane photonic integrated circuits consisting of the membrane DFB laser, the GaInAsP wire waveguide and the membrane photo-detector.

structure is the compactness of components. A 1×2 MMI splitter on Si-on-insulator (SOI) substrate was reported with its compact size (1.8×2.6 μm^2) and low excess loss (0.45dB) structure [14]. Since the index contrast of our InP-based membrane structure is comparable to SOI, similar size InP-based MMI splitter should be realized.

In this study, we focused on design and fabrication of an ultra-compact (to extent of the MMI on SOI) GaInAsP 1×2 MMI splitter with SiO_2 buried wire structure which can be integrated wih the light sources for membrane PICs.

II. DESIGN

A 3dB splitter, 1×2 MMI structure consisting of one input waveguide, one MMI region, and two output waveguides was theoretically investigated, where the GaInAsP 1×2 MMI splitter was assumed to be buried in 1.5-μm-thick SiO_2 upper and bottom claddings on Si substrate by BCB wafer bonding as shown in Fig. 2 (a).

In the calculation, finite difference method (FDM) was used by assuming transverse electric (TE) polarization with the wavelength fixed at 1550 nm. The refractive indices of GaInAsP and SiO_2 were assumed to be 3.35 and 1.45, respectively. While the size of MMI region was designed for compact and low-loss structure, the sizes of the input and the output waveguides were set to be 500-nm-wide and 150-nm-thick, satisfying the singlemode condition at 1550 nm wavelength with the equivalent refractive index n_{eq} 2.0 of the GaInAsP wire waveguide. Figure 2(b) illustrates a top view of the MMI splitter model used for the FDM

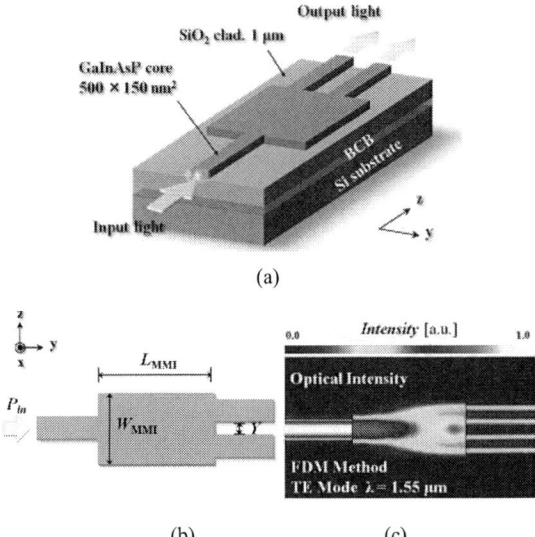

(a)

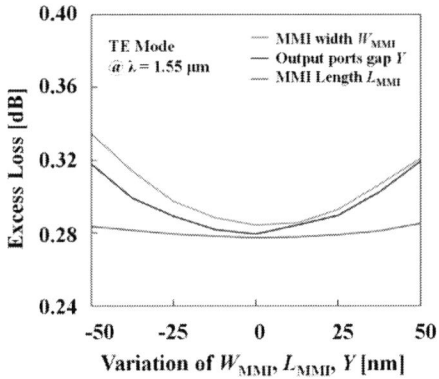

(b)   (c)

Fig. 2 (a) Schematic structure of the MMI splitter; (b) Top view of the FDM calculation model; (c) launched TE polarization optical intensity.

Fig. 3 Estimated structure tolerance by considering excess loss versus MMI width W_{MMI}, output ports gap Y and MMI length L_{MMI}.

simulation. The parameters of MMI region were set as MMI width W_{MMI}, length L_{MMI}, and the gap between output ports Y. The excess loss of 1×2 MMI was calculated from the ratio of total output power of both ports to the input power. By changing the above mentioned parameters, 1×2 MMI structure for compact and low-loss property was investigated. The minimum excess loss of 0.28dB was obtained when the parameters were W_{MMI} = 1.41 μm, L_{MMI} = 1.54 μm, and Y = 0.20 μm. Figure 2(c) shows the calculated optical intensity with TE polarization, which was launched into the input GaInAsP wire waveguide of the MMI splitter with the optimized structure.

Considering fabrication error caused by process variations, the fabrication tolerance was also estimated by changing each parameter in the range of ± 50 nm (which can be regarded as the maximum deviation of our fabrication process by using an electron-beam-lithography followed by

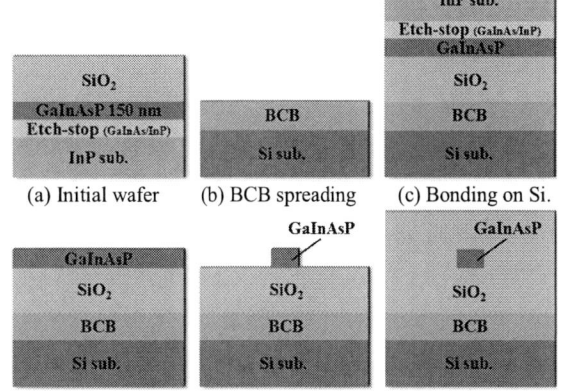

(a) Initial wafer   (b) BCB spreading   (c) Bonding on Si.

(d) Membrane process (e) Core patterning (f) Clad. deposition

Fig. 4 Schematics for fabrication process with GaInAsP membrane structure on Si substrate using BCB adhesive wafer bonding.

a dry etching) from the optimized structure. Figure 3 shows the excess loss as a function of the deviation from the optimum value of each parameter. As can be seen, the excess loss variation was only 0.05dB when the width was changed, and was 0.04dB when output ports gap was varied in the range of ± 50 nm. At the same time, the increased excess loss was as low as 0.01dB during MMI length was also changed in the range of ± 50 nm. The 1×2 MMI splitter size and the excess loss calculated are comparable to the value of the MMI splitter with Si wire waveguide on a SOI substrate, hence compact and low-loss optical elements could be fabricated even in the GaInAsP membrane circuits.

III. FABRICATION

The device fabrication process was performed by the following steps as illustrated in Fig. 4, and comprised of two main steps; a BCB adhesive wafer bonding and a device structure fabrication.

For the BCB bonding process, organo-metallic vapor-phase epitaxy (OMVPE) was used to grow an initial GaInAsP/InP wafer which consists of a 150-nm-thick $Ga_{0.21}In_{0.79}As_{0.46}P_{0.54}$ (λ_g = 1.22 μm) core layer on top of two etch-stop layers (InP/GaInAs). Then, as shown Fig. 4(a), a 1.5-μm-thick SiO_2 bottom cladding layer was deposited on the GaInAsP/InP wafer by plasma-enhanced chemical-vapor deposition (PECVD). Meanwhile, BCB was spin-coated onto the Si wafer, and then, it was thermally pre-cured for its polymerization in N_2 environment at 210°C. Both wafers were bonded with a bonding pressure of 5.9 kPa at 130°C, and completely solidified by hard-curing at 250°C under N_2 atmosphere. With selective wet etching, the InP substrate and two etch-stop layers were removed, then GaInAsP membrane was formed on the Si substrate. Figure 5 shows the GaInAsP membrane of 2 inch wafer formed after BCB bonding and membrane process.

Next, in order to fabricate waveguides, a SiO_2 hard mask

978-1-4673-1725-2/12 $31.00 © 2012 IEEE

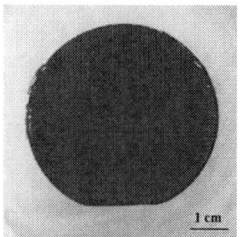

Layer	Thickness [nm]
GaInAsP core	150
SiO₂	1500
BCB	1200
Si sub.	150 μm

Fig. 5 Image of 2-inch GaInAsP membrane surface on a Si substrate after BCB adhesive wafer bonding and selective wet-etching of the InP substrate and etch-stop layers.

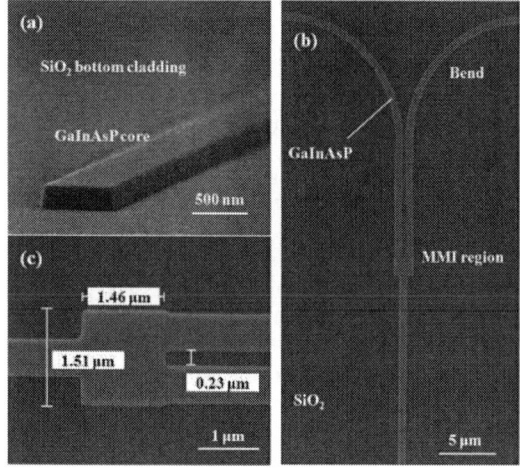

Fig. 6 SEM image of the MMI splitter (a) input waveguide on cross-sectional; (b) the MMI region and (c) enlarged micrograph on top view.

was deposited by PECVD and electron beam resist (ZEP520A/ZEP-C₆₀ double layer) was spun onto the GaInAsP membrane on the Si substrate. After forming the mask pattern using electron beam lithography, the SiO₂ mask was transferred by reactive-ion etching (RIE) using CF₄ gas. And then, the GaInAsP core layer was etched by inductively-coupled plasma RIE (ICP-RIE) using a mixture of CH₄ and H₂ gas. After each etchig process, the surface was cleaned by O₂ plasma ashing, and SiO₂ mask was removed with BHF solution. Finally, to obtain the complete device structure, a 1.5-μm-thick SiO₂ upper cladding was deposited by PECVD. Figure 6 represents the scanning electron microscope (SEM) images of 1×2 MMI splitter fabricated; (a) side view of input portand (b) and (c) top view of device body. The width W_{MMI}, length L_{MMI} and the gap of output ports Y of the fabricated MMI were 1.46 μm, 1.51 μm, and = 0.23 μm, respectively. The width and the thickness of the input and output ports were 0.49 and 0.15 μm, respectively.

IV. MEASUREMENTS

The measurement of fabricated MMI splitter was carried

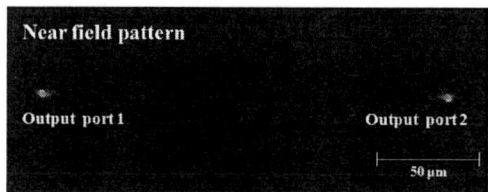

Fig. 7 Near field pattern of MMI splitter at wavelength 1.55 μm.

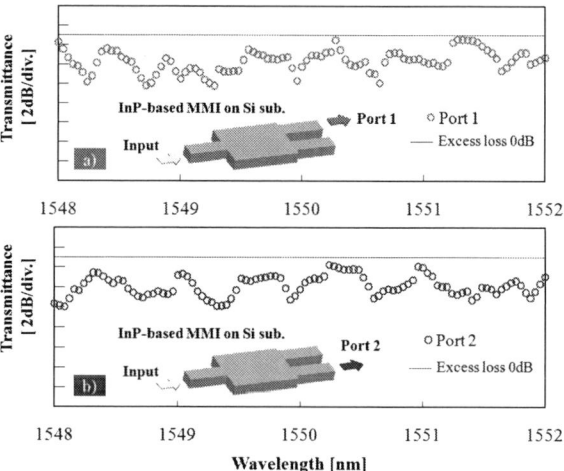

Fig. 8 Transmission of MMI splitter on Si substrate; (a) output port 1 and (b) output port 2.

out in the way that the input and outputs light beams on cleaved facets were coupled to each of the waveguide using spherical-lensed single mode fibers (mode field diameter of ~1.0 μm). Transmission measurement was performed with a wavelength tunable laser. The light transmitted through the device was coupled to the other spherical-lensed single mode fiber and projected on an optical power detector. A polarization controller was used to maintain the transverse electric (TE) incident polarization of the input light from the light source.

Figure 7 indicates the near field pattern on cross-sectional view of upper port and lower port at 1.55 μm wavelength.

Figures 8 shows the transmission of the output port 1 and port 2 of the MMI splitter. The wavelength of input light was swept and the average excess loss was obtained to be 3dB in the range of 1548-1552 nm. The average imbalance, which definded as the power ratio of the output ports was 0.5dB in the spectral range of 1548-1552 nm. The deviation of the excess loss value from the theoretical calculation is considered to be due to the displacement of the device sizes, and imperfect symmetric structure fabricated.

V. CONCLUSION

In conclusion, we designed and fabricated an ultra-compact and low-loss 1×2 MMI splitter with SiO₂ buried GaInAsP wire structure on a Si substrate by BCB adhesive wafer bonding. The excess loss of calculated MMI was as low as 0.28dB in optimized structure, and this structure had a good fabrication tolerance of 0.05dB when W_{MMI}, L_{MMI}, and Y were changed in range of ± 50 nm. The footprint of fabricated 1×2 MMI splitter was as small as $1.51×1.46$ μm^2; in consequence, this device is about 9 times smaller than the reported InP-based 1×2 MMI splitter [13]. The average excess loss was 3dB in C-band and the imbalance between output ports was 0.5dB, but it can still be improved by reduction of its displacement and imperpect summetry of device structure . It is confirmed that compact optical element with low-loss property can be realized even in the GaInAsP membrane circuits.

Acknowledgment

This research was financially supported by the Ministry of Education, Culture, Sports, Science and Technology, Japan (MEXT); the Japan Society for the Promotion of Science (JSPS) under Grants-in-Aid for Scientific Research (#24246061, #22360138, #21226010, #23760305, #10J08973, and #11J08863); by JSPS and the Council for Science and Technology Policy (CSTP) under the FIRST program; and also by the Ministry of Internal Affairs and Communicaions under the SCOPE program.

References

[1] P. Kapur, J. P. Mc Vittie, and K. C. Saraswat, "Technology and reliability constrained future copper interconnects. I. Resistance modeling," *IEEE Trans. Electron Devices*, vol. 49, no. 4, pp. 590-597, Apr. 2002.

[2] P. Kapur, G. Chandra, J. P. Mc Vittie, and K. C. Saraswat, "Technology and reliability constrained future copper interconnects. II. Performance implications," *IEEE Trans. Electron Devices*, vol. 49, no. 4, pp. 598-604, Apr. 2002.

[3] D. A. B. Miller, "Rationale and challenges for optical interconnects to electronic chips," *Proc. IEEE*, vol. 88, no.6, pp. 728-749, Jun. 2000.

[4] M. Haurylau, G. Chen, H. Chen, J. Zhang, N. A. Nelson, D. H. Albonesi, E. G. Friedman, and P. M. Fauchet "On-chip optical interconnect roadmap: challenges and critical directions," *IEEE J. Sel. Top. Quantum Electron*, vol. 12, no. 6, pp. 1699-1705, Sep. 2011.

[5] S. Sakamoto, H. Naitoh, M. Ohtake, Y. Nishimoto, T. Maruyama, N. Nishiyama, and S. Arai, "85°C continuous-wave operation of GaInAsP/InP-membrane buried heterostructure distributed feedback laser with polymer cladding layer," *Jpn. J. Appl. Phys.* vol. 46, no. 47, pp. L1155-L1157, Nov. 2007.

[6] S. Arai, N. Nishiyama, T. Maruyama, and T. Okumura, "GaInAsP/InP membrane lasers for optical interconnects," *IEEE J. Sel. Top. Quantum Electron*, vol. 17, no. 5, pp. 1381-1389, Sep. 2011.

[7] Y. Maeda, J. Lee, Y. Atsumi, N. Nishiyama, and S. Arai, "Uniform BCB bonding process toward low propagation loss," *Proc. Int. Conf. Indium Phosphide and Related Materials*, P08, May 2011.

[8] T. Okumura, T. Koguchi, H. Ito, N. Nishiyama, and S. Arai, "Injection-type GaInAsP/InP membrane buried hetero structure distributed feedback laser with wirelike active

regions," *Appl. Phys. Express*, vol. 4, no. 4, pp. 042101-1-042101-3, Mar. 2011.

[9] T. Okumura, D. Kondo, H. Ito, S. Lee, T. Amemiya, N. Nishiyama, and S. Arai, "Lateral junction waveguide-type photodiode grown on semi-insulating InP substrate," *Jpn. J. Appl. Phys.*, vol.50, no.2 pp. 020206-1-020206-3, Feb. 2011.

[10] J. Lee, Y. Maeda, Y. Atsumi, Y. Takino, N. Nishiyama, and S. Arai, "Low-loss GaInAsP wire waveguide on Si substrate with benzocyclobutene adhesive wafer bonding for membrane photonic circuits," *Jpn. J. Appl. Phys.* vol. 51, no. 4, pp. 042201-1-042201-5, Apr. 2012.

[11] K. Solehmainen, M. Kapulainen, M. Harjanne, and T. Aalto, "Adiabatic and multimode interference couplers on Silicon-on-insulator," *IEEE Photon. Technol. Lett.*, vol. 18, no. 21, pp. 2287-2289, Nov. 2006.

[12] A. Hosseini, D. N. Kwong, Y.Ahang, H. Subbaraman, X. Xu, and R. T. Chen, "1×N Multimode interference beam splitter design techniques for on-chip optical interconnections," *IEEE J. Sel. Top. Quantum Electron*, vol. 17, no. 3, pp. 510-515, June 2011.

[13] F. Bordas, G, Roelkens, R. Zhang, E. J. Geluk, F. Karota, J. J. G. M. Van Der Tol, P. J. Van Veldhoven, R. Nötzel, D. Van Thourhout, R. Baets, and M. K. Smit, "Compact passive devices in InP membrane on silicon," *35th European Conf. on Opt. Commun. (ECOC'09)*, Vienna, Austria, paper 4.2.4, Sept. 2009.

[14] K. Suzuki, H. C. Nguyen, T. Tamanuki, F. Shinobu, Y. Saito, Y. Sakai, and T. Baba, "Slow-light based variable symbol-rate silicon photonics DQPSK receiver," *Opt. Express*. vol. 20, no. 4, pp. 4796-4804, Feb. 2012.

Reflection-Assisted Unidirectional Hybrid Silicon Microring Lasers

Di Liang[1], Sudharsanan Srinivasan[2], David A. Fattal[1], Marco Fiorentino[1],
Zhihong Huang[1], Daryl T. Spencer[2], John E. Bowers[2], Raymond G. Beausoleil[1]

[1] *Intelligence and Infrastructure Lab, HP Labs, 1501 Page Mill Road, Palo Alto, CA, 94304, USA*
[2] *Electrical and Computer Engineering Department, University of California, Santa Barbara, CA, 93106, USA*

Abstract — **We study directional bistability in hybrid silicon microring lasers and demonstrate an unidirectional laser. Unidirectional emission is achieved by integrating a passive reflector that feeds laser emission back into laser cavity to introduce extra unidirectional gain. We show that the length of the passive reflector is a critical parameter in determining the lasing behavior.**

I. INTRODUCTION

Directional bistability, the ability of a laser to operate either in the clock-wise (CW) or counter-clock-wise (CCW)) mode, is a unique characteristic of ring lasers [1]. Bistability has been used to demonstrate optical switching and logic applications [2, 3] in III-V microring lasers. This feature has also been observed in racetrack ring lasers and microring lasers built on a hybrid silicon platform [4, 5].

While useful for some applications, bistability is undesirable for ring lasers used in optical interconnects. Different devices of the same design can lase in different directions or switch lasing direction depending on the injection current and temperature.

In this letter we demonstrate unidirectional lasing in hybrid silicon microring lasers without adding power consumption or chip complexity.

II. HYBRID MICRORING LASER BISTABILITY

Fig. 1 shows the light-current (LI) characteristic of a hybrid microring laser with a 50 μm diameter. Two integrated hybrid silicon photodetectors (inset of Fig. 1(a)) capture CW and CCW laser emission separately. At low injection level (between I_{th} and 33 mA), the laser emits in both CW and CCW modes. At higher injection levels (between 33 and 40 mA), CW and CCW mode degeneracy breaks and the laser operates in the unidirectional bistability regime where one direction dominates in certain injection ranges [1]. The lasing direction can switch in a random fashion (e.g. ~38 mA in Fig 1(a)).

The inset of Fig. 1(b) shows that, for the devices studied here, sidewall roughness causes particularly serious enhancement of intracavity back-reflections. We believe that these back-reflections are primarily responsible for degrading stable unidirectional lasing and triggering frequent lasing direction changes. In Figs. 1(a) and (b) we compare the laser behavior for forward and reverse current sweeps. In contrast with devices reported by other groups [1, 3], our devices do not show any memory (hysteresis) effect, but they display the same pattern of CW and CCW lasing independently of the direction of the current sweep. We attribute this behavior to the sidewall roughness which introduces a strong asymmetric coupling between CW and CCW modes. This coupling, which depends on the lasing wavelength and injection current, favors one mode or the other and therefore triggers the observed unidirectional bistability.

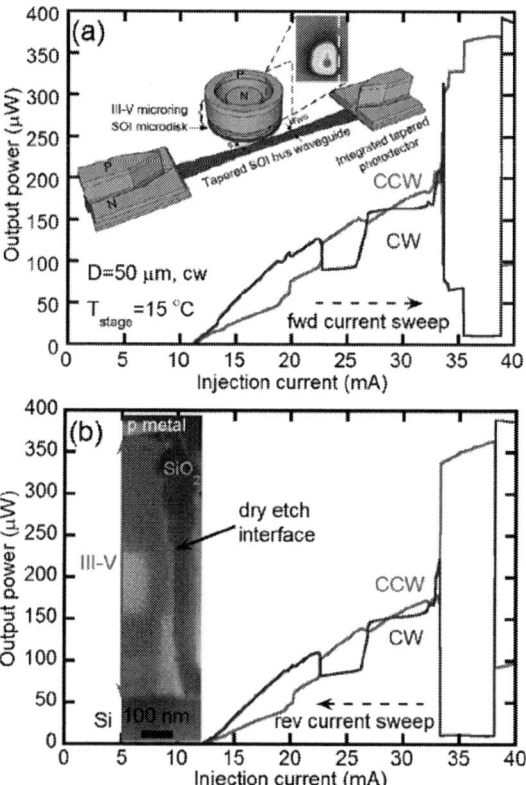

Fig. 1. LI characteristic of a 50 μm diameter hybrid silicon microring laser with (a) increased (fwd) and (b) decreased (rev) injection current sweep. Insets in (a) and (b) are device schematic and SEM cross-section showing the dry etched semiconductor interface, respectively.

III. UNIDIRECTIONAL MICRORING LASER

A number of approaches have been demonstrated to achieve stable unidirectional lasing. "S-shape" ring resonator cavities are designed to introduce asymmetric round trip loss/gain [6]. Optical pulse injection from an external laser or light emitting diode (LED) has also been used to increase the net modal gain in one direction [1, 5, 7]. These approaches introduce additional optical loss or require an external light source, which either degrades laser performance or increases total system complexity and power consumption. Other designs that work for some cavity structures are not applicable to our devices [8].

We fabricated the unidirectional ring laser shown schematically in Fig. 2(a), in which an optical reflector at one end of the bus waveguide induces the laser to emit light toward the other end. In this structure, light emitted in the CW mode is partially coupled back to the CCW mode. The power circled back into the cavity leads to a photon density increase and unidirectional lasing in the desired direction (CCW here). Compared with external injection from another laser this approach does not require additional power and is free of wavelength/mode mismatch. Injection from an on-chip LED is easier to implement. However, because of the LED broad emission spectrum and low output power, it is not very effective in guaranteeing unidirectional operation.

Different types of waveguide reflectors can be used to provide feedback. Figs. 2(b)-(e) show various possible implementations: a cleaved waveguide facet with high reflection coating, a teardrop reflector, a passive ring reflector, and a distributed Bragg reflector (DBR). We use the teardrop design shown in Fig. 2(f) because it is simple to fabricate and has a wide optical bandwidth. Teardrop reflectors have been used as low-loss, high-reflectivity (98%) laser mirrors [9].

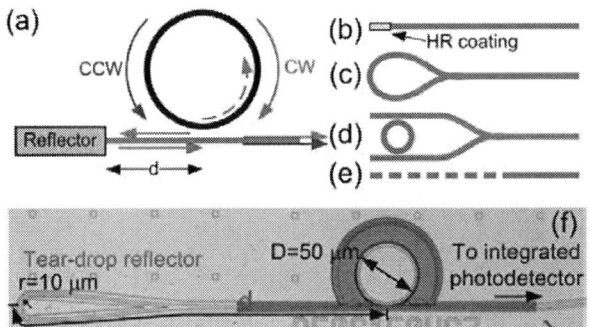

Fig. 2. (a) Schematic of a ring laser with an external reflector integrated on the bus waveguide. (b-e) Schematic of various passive reflector designs, (f) Image of a fabricated hybrid silicon microring laser with tear-drop reflector.

Fig. 3 shows the LI-voltage (LIV) characteristic of 50 μm-diameter lasers. The effective reflector length d (see Fig. 2) for this device was chosen to be 325 μm and 85 μm for Fig. 3(a) and (b), respectively. The inset shows an infrared (IR) image of the unidirectional laser at 30 mA injection current. Fig. 3(a) shows that a LI curve calculated using the laser model developed in Ref. [10] agrees well with the experimental data.. The free-running laser in Fig. 1 and the two unidirectional devices in Fig. 3 all exhibit similar threshold current and output power. This is an indication that most of the injected power in the unidirectional devices goes into the CCW lasing mode. The dips in the LI curves in Fig. 3 are attributed to the interference of the laser CCW mode and

the back-reflected light from teardrop reflector, which is a function of the phase of the reflected light. In our devices, significant lasing in the suppressed mode can co-exist with dominant lasing as seen in the device of Fig. 1 for injection currents above 38 mA. When a reflector is added the light from the CW and CCW light will interfere at the coupler. Depending on the phase of the reflected light constructive or destructive interference will occur. Constructive interference will maximize the laser CCW output. Destructive interference reduces the feedback to the point that the laser goes back to its free-running, bistable operation.

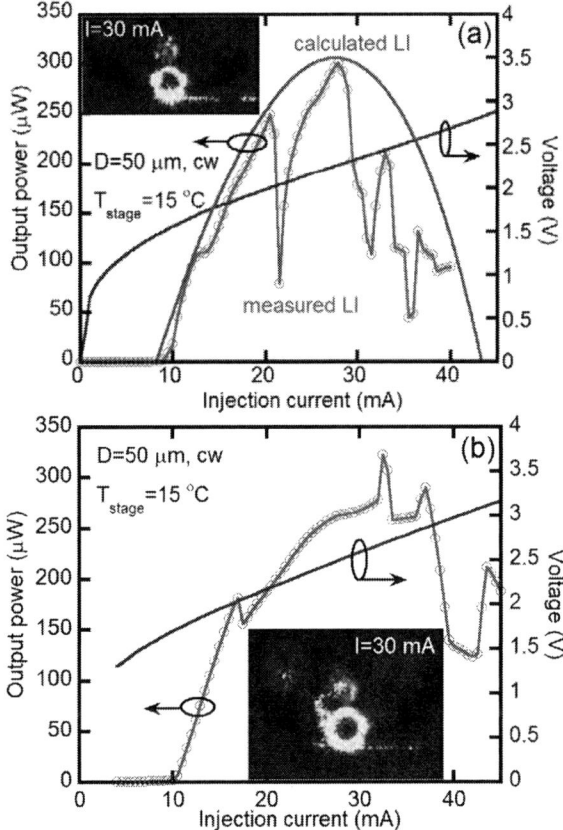

Fig. 3. Simulated and measured LIV characteristic of unidirectional hybrid silicon microring laser with reflector length d=325 μm (a) and 85 μm (b). Inset: IR image of devices unidirectional lasing in CCW mode under 30 mA injection current.

To confirm our understanding of the lasing behavior, we modeled the phase of the back-reflected light. Fig. 4(a) shows that the wavelength shifts as a function of the injected electrical power P_i at a rate of 0.06 nm/mW. In Figs. 4(b) and (c) we overlay the calculated interference intensity with the laser output as a function of P_i. The phase was calculated using a round trip distance of 2d of 650 μm in (b) and 170 μm in (c). The dips in the output correspond to destructive interference between the CCW and reflected light. For regions without dips (e.g. at P_i=59 mW in (b) and P_i=57 mW in (c)) the CCW mode is still favored even in the absence of feedback and no dip is observed. To confirm this we

978-1-4673-1725-2/12 $31.00 © 2012 IEEE   13

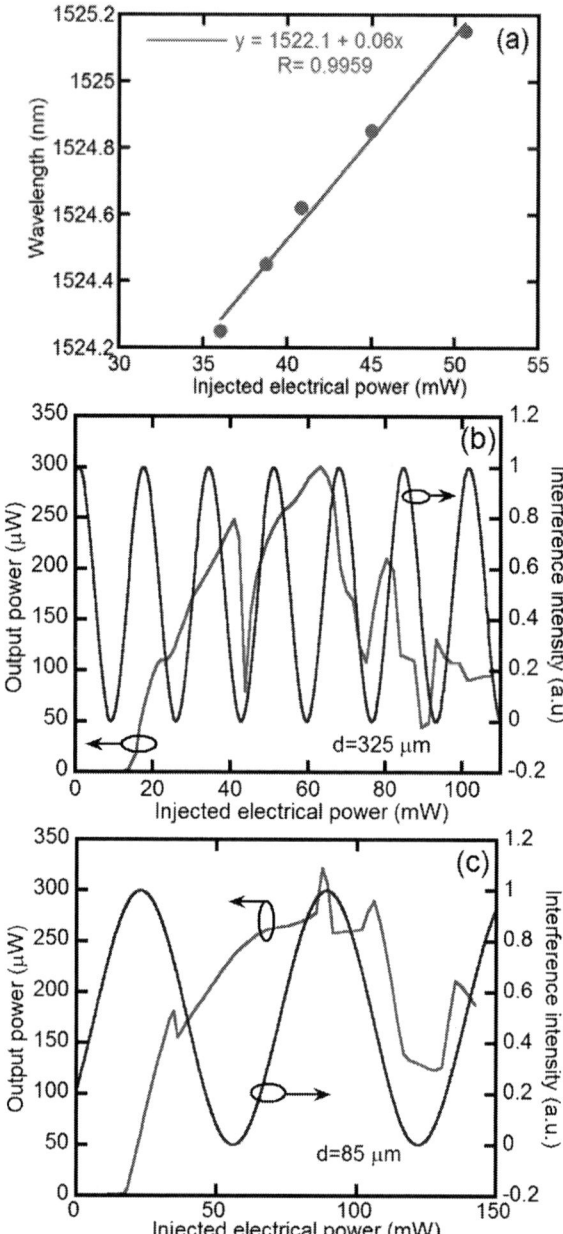

Fig. 4. (a) Lasing wavelength shift vs. injected electrical power. (b) and (c): LI characteristic and simulated interference as a function of injected electrical power for devices shown in Fig. 3 (a) and (b), respectively.

monitored the lasing direction in the bus waveguide using a top down IR camera (insets in Fig. 3) A similar effect has also been observed in heterogeneous micro-disk lasers [11]. The laser power instabilities can be eliminated using several strategies: further decreasing the reflector round-trip distance to reduce the phase shift, a better control of the phase shift through design and fabrication, and reducing device heating to minimize lasing wavelength shift [12]. Reducing the sidewall roughness is also critical not only for enhancing device

performance, but also for minimizing back-reflection to maintain stable unidirectional lasing. The approach we demonstrated here is also applicable to devices with "memory" effects. Because the lasing direction is independent of the external feedback [1] phase these devices would not show power instabilities.

IV. CONCLUSION

In conclusion, we studied bistability in hybrid silicon microring lasers. Excessive back-reflection from rough microring cavity sidewall leads to disappearance of the "memory" effect observed in typical ring lasers. Using feedback from a passive teardrop reflector we have shown that unidirectional lasing can be achieved. This simple design achieves unidirectional operation in ring lasers without additional losses or power consumption. We plan to implement this design into a microring laser array either by adding one reflector to the entire array or one for each individual laser for hybrid coarse wavelength division multiplexing application in our photonic interconnect effort.

REFERENCES

[1] M. Sorel, P. J. R. Laybourn, G. Giuliani, and S. Donati, "Unidirectional bistability in semiconductor waveguide ring lasers," *Applied Physics Letters,* vol. 80, pp. 3051-3053, 2002.

[2] M. T. Hill, H. J. S. Dorren, T. de Vries, X. J. M. Leijtens, J. H. den Besten, B. Smalbrugge, Y.-S. Oei, H. Binsma, G.-D. Khoe, and M. K. Smit, "A fast low-power optical memory based on coupled micro-ring lasers," *Nature,* vol. 432, pp. 206-209, 2004.

[3] L. Liu, R. Kumar, K. Huybrechts, T. Spuesens, G. Roelkens, E.-J. Geluk, T. de Vries, P. Regreny, D. Van Thourhout, R. Baets, and G. Morthier, "An ultra-small, low-power, all-optical flip-flop memory on a silicon chip," *Nature Photonics,* vol. 4, pp. 182-187, 2010.

[4] D. Liang, M. Fiorentino, T. Okumura, H.-H. Chang, D. Spencer, Y.-H. Kuo, A. W. Fang, D. Dai, R. G. Beausoleil, and J. E. Bowers, "Hybrid Silicon (λ=1.5 μm) Microring Lasers and Integrated Photodetectors," *Optics Express,* vol. 17, pp. 20355-20364, 2009.

[5] A. W. Fang, R. Jones, H. Park, O. Cohen, O. Raday, M. J. Paniccia, and J. E. Bowers, "Integrated AlGaInAs-silicon evanescent race track laser and photodetector," *Optics Express,* vol. 15, pp. 2315-2322, 2007.

[6] J. P. Hohimer, G. A. Vawter, and D. C. Craft, "Unidirectional operation in a semiconductor ring diode laser," *Applied Physics Letters,* vol. 62, pp. 1185-1187, 1993.

[7] C. J. Born, S. Yu, M. Sorel, and P. J. R. Laybourn, "Controllable and stable mode selection in a semiconductor ring laser by injection locking," in *Lasers and Electro-Optics, 2003. CLEO '03. Conference on,* 2003, p. 2 pp.

[8] Q. J. Wang, C. Yan, N. Yu, J. Unterhinninghofen, J. Wiersig, C. Pflugl, L. Diehl, T. Edamura, M. Yamanishi, H. Kan, and F. Capasso, "Whispering-gallery mode resonators for highly unidirectional laser action," *Proceedings of the National Academy of Sciences,* vol. USA107, pp. 22407-22412, 2010.

[9] Y. Zheng, D. K.-T. Ng, Y. Wei, W. Yadong, Y. Huang, Y. Tu, C.-W. Lee, B. Liu, and S.-T. Ho, "Electrically pumped heterogeneously integrated Si/III-V evanescent lasers with micro-loop mirror reflector," *Applied Physics Letters,* vol. 99, p. 011103, 2011.

[10] D. Liang, M. Fiorentino, S. Srinivasan, S. T. Todd, G. Kurczveil, J. E. Bowers, and R. G. Beausoleil, "Optimization of Hybrid Silicon Microring Lasers," *IEEE Photonics Journal,* vol. 3, pp. 580-587, 2011.

[11] T. Spuesens, F. Mandorlo, P. Rojo-Romeo, P. Regreny, N. Olivier, J. M. Fedeli, and D. Van Thourhout, "Compact Integration of Optical Sources and Detectors on SOI for Optical Interconnects Fabricated in a 200 mm CMOS Pilot Line," *Journal of Lightwave Technology,* vol. 30, pp. 1764-1770, 2012.

[12] D. Liang, S. Srinivasan, M. Fiorentino, G. Kurczveil, J. E. Bowers, and R. G. Beausoleil, "A Metal Thermal Shunt Design for Hybrid Silicon Microring Laser " in *IEEE Optical Interconnects Conference.* paper TuD2 Santa Fe, NW, USA, 2012.

Integrated Hybrid III-V/Si Laser and Transmitter

G.-H. Duan[1], C. Jany[1], A. Le Liepvre[1], M. Lamponi[1], A.Accard[1], F. Poingt[1], D. Make[1], F. Lelarge[1],
S. Messaoudene[2], D. Bordel[2], and J.-M. Fedeli[2], S. Keyvaninia[3], G. Roelkens[3], D. Van Thourhout[3], D. J. Thomson[4],
F. Y. Gardes[4] and G. T. Reed[4]

[1]III-V Lab, a joint lab of 'Alcatel-Lucent Bell Labs France', 'Thales Research and Technology' and 'CEA Leti',
Campus Polytechnique, 1, Avenue A. Fresnel, 91767 Palaiseau cedex, France.
[2]CEA LETI, Minatec, 17 rue des Martyrs, F-38054 GRENOBLE cedex 9, France.
[3]Photonics Research Group, INTEC, Ghent University-IMEC, Sint-Pietersnieuwstraat 41, B-9000 Ghent, Belgium
[4]School of Electronics and Computer Science, University of Southampton, Southampton, United Kingdom.

Abstract — This paper reports on recent advances on integrated hybrid InP/SOI lasers and transmitters. Based on a molecular wafer bonding technique, we develop hybrid III-V/Si lasers exhibiting new features: narrow III-V waveguide width of less than 3 µm, tapered III-V and silicon waveguides for mode transfer. These new features lead to good laser performances: a lasing threshold as low as 30mA and an output power of more than 10 mW at room temperature in continuous wave operation regime from a single facet. Continuous wave lasing up to 70 C is obtained. Moreover, hybrid III-V/Si lasers, integrating two intra-cavity ring resonators, are fabricated. Such lasers achieve a thermal tuning range of 45 nm, with a side mode suppression ratio higher than 40 dB. More recently we demonstrate a tunable transmitter, integrating a hybrid III-V/Si laser fabricated by wafer bonding and a silicon Mach-Zehnder modulator. The integrated transmitter exhibits 9 nm wavelength tunability by heating an intra-cavity ring resonator, high extinction ratio from 6 to 10 dB, and excellent bit-error-rate performance at 10 Gb/s.

Index Terms — Hybrid integrated circuits, silicon laser, silicon-on-insulator (SOI) technology, adiabatic taper.

I. INTRODUCTION

Silicon photonics is drawing increasing attention due to the promise of fabricating low-cost, compact circuits that integrate photonic and microelectronic elements [1]. It can address a wide range of applications from short distance data communication to long haul optical transmission. Today, practical Si-based light sources are still missing, despite the recent demonstration of an electrically pumped germanium laser [2]. This situation has driven research to the heterogeneous integration of III-V semiconductors on silicon. In order to densely integrate the III-V semiconductors with the silicon waveguide circuits, mainly DVS-BCB adhesive wafer bonding and molecular bonding techniques are used and are actively reported in state-of-the-art hybrid lasers [3-7]. In these approaches, unstructured InP dies are bonded, epitaxial layers down, on a SOI waveguide circuit wafer, after which the InP growth substrate is removed and the III-V epitaxial film is processed. The design space for hybrid InP/SOI lasers is large and in particular the coupling between the optical mode in the top active III-V waveguide and that in the bottom passive SOI waveguide plays an important role.

This paper reports on recent advances on integrated hybrid InP/SOI lasers and transmitters [8-11]. Based on a molecular wafer bonding technique, we developed hybrid III-V/Si lasers exhibiting new features. For instance, III-V waveguides have a narrow width of less than 3 µm, reducing the power consumption of the devices. In order to make the mode coupling efficient, both the III-V waveguide and silicon waveguide are tapered, with a tip width for the III-V waveguide down to 300 nm for some devices. These new features lead to good laser performances. Moreover, hybrid III-V/SI lasers, integrating two intra-cavity ring resonators, are fabricated. Such lasers achieve a thermal tuning range of 45 nm, with side mode suppression ratio higher than 40 dB. More recently we demonstrate a tunable transmitter, integrating a hybrid III-V/Si laser fabricated by wafer bonding and a silicon Mach-Zehnder modulator. The integrated transmitter exhibits 9 nm wavelength tunability by heating an intra-cavity ring resonator, high extinction ratio from 6 to 10 dB, and excellent bit-error-rate performance at 10 Gb/s.

II. III-V/SI LASER STRUCTURE AND FABRICATION

We developed a new type of hybrid III-V/Si lasers, as shown in Fig. 1. The laser structure can be divided into three parts. In the center of the device the optical mode is confined to the narrow III-V waveguide, which provides the gain. At both sides of this section there is a coupling region that couples light from the III-V waveguide to the underlying silicon waveguide. After the coupling region the light is guided by a silicon waveguide without III-V on top.

The III-V region consists of a p-InGaAs contact layer, a p-InP cladding layer, InGaAsP quantum wells surrounded by two InGaAsP separate confinement heterostructure (SCH) layers, and an n-InP layer. The SOI substrate (200mm wafer manufactured by SOITEC) is composed of a mono-crystalline silicon layer on top of a 2μm thick buried oxide layer on a silicon substrate. The silicon rib waveguides have a height of 400nm, an etch depth of 180nm and a width of 1μm. The III-V epitaxial layers are transferred to the patterned SOI wafer through DVS-BCB adhesive bonding. The bonding layer thickness is around 80nm.

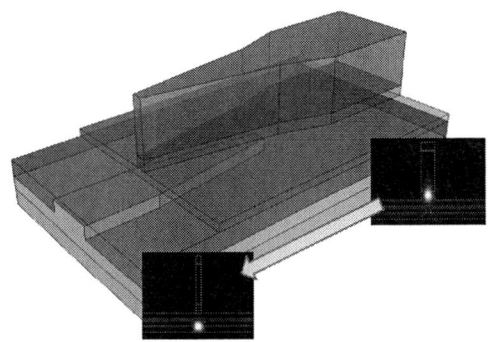

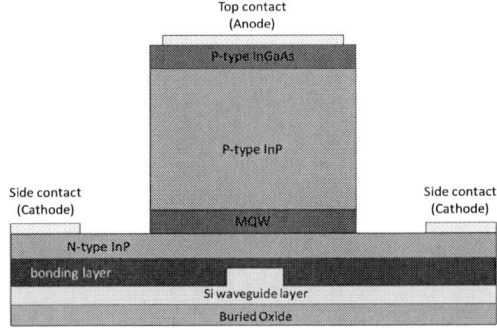

Figure 1 (up) 3D view of the coupling structure of the hybrid laser with representative mode profiles in two cross-sections, (down) the detailed cross-sectional view of the center of the hybrid laser

The active waveguide structure and the guided mode profile are illustrated in Fig. 1 (up). The straight active waveguide has a typical width of 1.7μm. The mode profiles are calculated using the mode matching method. The calculated optical confinement in the MQW is typically around 10%. To achieve index matching between the two waveguides a deep ridge III-V waveguide is used in the double taper region. As shown in Fig. 1(up), a double taper structure is used to allow the efficient coupling of light from the III-V waveguide to the silicon waveguide [8,9]. In the double taper region, the silicon waveguide tapers from 350nm to 1μm, while the III-V waveguide taper has two linear parts: the first part of 30μm

long with a decrease of width from 1.7 to 1.0 μm, and the second part of 150 μm long starting with a III-V waveguide width of 1.0 μm and ending at 500nm.

After wafer bonding and InP substrate removal, an SiO₂ hard mask was defined using 248nm UV lithography. ICP etching was used to etch through the InGaAs layer and partly through the InP p-doped layer. The InP p-doped layer etching was completed by chemical selective etching. The MQW layer was etched by CH_4:H_2 RIE. The active waveguide is encapsulated with DVS-BCB. A Ti/Pt/Au alloy was used for metallisation on both p and n sides. Finally, the III-V/silicon wafer is cleaved, to form a Fabry-Perot laser cavity.

The devices are mounted on a temperature controller set to different temperatures. An example of the L-I curves is shown in figure 2 for a hybrid laser with a III-V waveguide width of 1.7μm. The length of the straight III-V waveguide section is 490μm. The coupling section length is 230μm. The overall cavity length is 1700μm, including the passive silicon waveguide. The device has a threshold current of 35mA and a maximum single facet continuous wave output power of 5mW at 20°C. At 60°C the laser still exhibits an output power of more than 1mW. The series resistance is 5 Ohms, while the slope efficiency is 0.043 W/A.

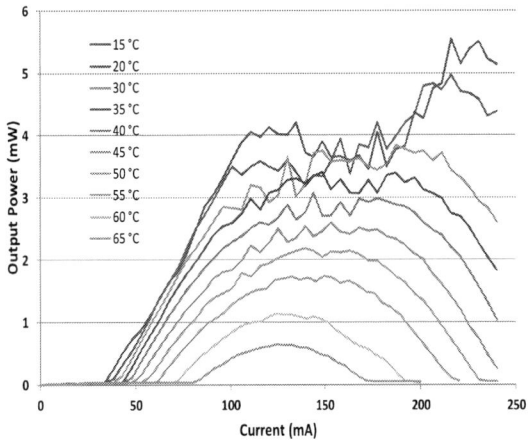

Figure 2: Continuous wave single facet laser power as a function of drive current at different stage temperatures.

III. WIDELY TUNABLE HYBRID III-V/SI LASERS

The tunable laser, as schematically shown on figure 3 (up), consists of an InP based amplification section, tapers for the modal transfer between III-V and Si waveguides, two ring resonators (RRs) for single mode selection, metal heaters on top of the rings for the thermal wavelength tuning and Bragg gratings providing reflection and output fibre coupling [10].

978-1-4673-1725-2/12 $31.00 © 2012 IEEE

The straight III-V waveguide has a width of 1.7 μm and a length of 500 μm. Both silicon and III-V waveguides are tapered over a length of 200 μm to allow an efficient adiabatic mode transfer between those two waveguides [10]. In the silicon sections, ring resonators 1 (R1) and 2 (R2) have free spectral range (FSR) of 650 and 590 GHz, respectively. The slight difference between these two values allows taking advantage of the Vernier effect for the wavelength tuning. Moreover, the bandwidth of the double ring filter is designed to select only one Fabry-Perot mode of the cavity. The Bragg reflectors are made by partially etching the silicon waveguide. The two Bragg reflectors, each with a pitch of 290 nm and 60 periods, are designed to have a reflectivity of more than 90%, and a 3 dB bandwidth larger than 100 nm. Figure 3 (bottom) shows a picture of the fabricated tunable lasers [10].

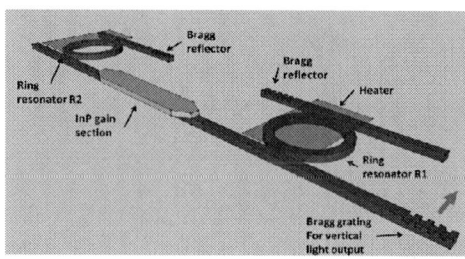

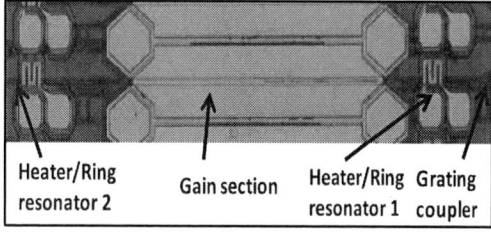

Figure 3: Schematic view (up) and picture (down) of the widely tunable single mode hybrid laser

The lasers are tested on wafers with vertical Bragg gratings coupling the output light into a cleaved single mode fiber. The coupling losses were measured to be around 10 dB. At 20°, the laser has a threshold current of 21mA. Figure 4 shows the super-imposed laser emission spectra by changing heating power levels to the two RRs. On the backgrounds of those spectra curves, one can observe transmission peaks created by R2 and the transmission dips created by R1. With less than 400mW of combined power in both heaters, a high SMSR (>40dB) wavelength range over 45nm is achieved. For wavelength setting, both ring power must be adjusted so that one transmission peak of R1 matches one of R2 at a desired wavelength. The wavelength tuning range is currently limited by the too large difference in the FSR between the two RRs.

An optimized design should allow covering the whole gain bandwidth of the III-V active material.

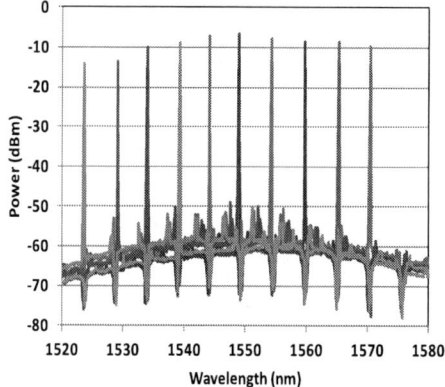

Figure 4 Super-imposed laser spectra of the tunable laser

IV. TUNABLE HYBRID III-V/SI TRANSMITTER

Figure 5 shows a schematic view (up) and a picture (down) of an Integrated Tunable laser – Mach Zehnder Modulator (ITLMZ) [11]. The ITLMZ chip consists of a single mode hybrid III-V/silicon laser, a silicon MZM and an optical output coupler. The single-mode hybrid laser includes an InP waveguide providing light amplification, and a ring resonator (RR) allowing single mode operation. Two Bragg reflectors are etched on silicon waveguides in order to close the laser cavity. The MZM allows modulation of the output light emitted by the hybrid laser [12].

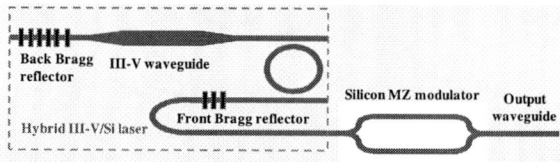

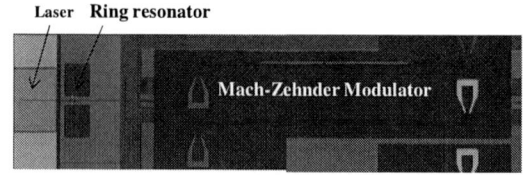

Figure 5. Schematic view (up), and picture (down) of the ITLMZ chip

The output of the ITLMZ chip is coupled to a lensed fiber, amplified by an erbium doped fiber amplifier and then filtered out. One arm of the MZM is modulated with a voltage swing of around 7 V, at 10 Gb/s using a pseudo-random binary sequence (PRBS). The BER measurement is performed for 8 different wavelengths by changing the power dissipated in the RR heater. Fig. 6 (left) shows the BER curves for all the wavelengths and also a reference curve for a directly modulated laser, measured using a high sensitivity receiver including an avalanche photodiode. The PRBS length is 27-1, limited by the photo-receiver used in this experiment. Fig. 4 (right) shows the corresponding eye diagram for all those channels, independent of the length of PRBS in the range from 2^7-1 to 2^{31}-1. The ER of all those wavelengths varies from 6 to 10 dB, while the ER for the reference is only 4 dB. One can see from Fig. 4 (left) that all channels have better BER performance than the reference for received power levels lower than -25 dBm, due to the higher ER of the ITLMZ compared to that of the reference. For power levels higher than -25 dBm, channels λ2, λ3, λ4 and λ5 behave slightly better than the reference, achieving error free operation with BER < 10-9. Other channels have minimum BER between 10-7 and 10-8, mainly limited by the optical signal to noise ratio (OSNR) due to the high coupling losses between the ITLMZ output waveguide and the used lensed fiber. The power level difference to achieve the same BER among all channels is around 4 dB, explained by the difference in OSNR and the achieved ER among those channels. Finally the smaller slopes for all wavelength channels compared to that of the reference in the BER curves is attributed to their lower OSNR

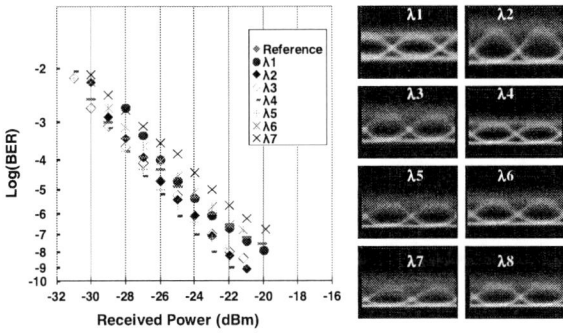

Figure 4 Bit error rate (left) and corresponding eye diagrams for different wavelengths (right)

V. CONCLUSION

This paper reports on recent advances on hybrid III-V/Si lasers and transmitters. The performance of hybrid III-V/Si lasers approaches that of monolithically integrated InP lasers. Moreover, high quality silicon waveguides offer new design possibilities to make for example high performance and robust tunable lasers. We demonstrate also the integrated

transmitter by incorporating hybrid tunable III-V/Si lasers and silicon modulators. We believe that the integrated hybrid III-V/Si transmitters will have important applications for WDM transmission for both short range and long haul optical transmission.

Acknowledgments: This work was supported in part by the European Community's Seventh Framework Program (FP7/2007-2013) under grant agreement n° 224312 HELIOS (pHotonics Electronics functional Integration on CMOS), and by the French Agence Nationale pour la Recherche VERSO program under the grant agreement MICROS.

REFERENCES

[1] C. Gunn, "CMOS photonics for high-speed interconnects", IEEE Proceedings of Computer Science, 2006.

[2] R E. Camacho-Aguilera, *et al,* "An electrically pumped germanium laser", Optics Express, Vol. 20 Issue 10, pp.11316-11320 , 2012

[3] A.W. Fang, *et al.*, "Hybrid silicon evanescent devices", Materials Today, Volume 10, Issues 7-8, p. 28-35, 2007.

[4] H. Park, *et al.,* "Device and Integration Technology forSilicon Photonic Transm", IEEE JOURNAL OF SELECTED TOPICS IN QUANTUM ELECTRONICS, VOL. 17, NO. 3, pp. 671-688, 2011.

[5] S. Stankovic, *et al.*, "1310nm evanescent hybrid III-V/Si laser based on DVS-BCB bonding", Integrated Photonics Research, Silicon and Nano-Photonics (IPR), Canada, p.IWC3, 2011

[6] G. Roelkens, et al. , "III-V/Si photonics by die-to-wafer bonding" , Materials Today, Volume 10, Issues 7-8, p.36-43, 2007.

[7] B. Ben Bakir, *et al.,* "Electrically driven hybrid Si/III-V lasers based on adiabatic mode transformers ", Proceedings of the SPIE Conference Photonics Europe, Bruxelles, 2010.

[8] M. Lamponi, *et al.* ,"Heterogeneously integrated InP/SOI laser using double tapered single-mode waveguides through InP die to SOI wafer bonding", Group Four Photonics Topical Meeting, Beijing, China Sept. 2010

[9] M. Lamponi, *et al.,* "Low-threshold heterogeneously integrated InP/SOI laser with a double adiabatic taper coupler", IEEE Photonics Technology Letters, Volume: 24, Page(s): 76 – 78, 2012.

[10] A. Le Liepvre, *et al.,* "Widely Wavelength Tunable Hybrid III-V/Silicon Laser with 45 nm Tuning Range Fabricated Using a Wafer Bonding Technique", To be presented at Group Four Photonics Topical Meeting, San Diego, Aug. 2012

[11] G-.H. Duan, *et al.,* "10 Gb/s Integrated Tunable Hybrid III-V/Si Laser and Silicon Mach-Zehnder Modulator", To be presented at European Conference on Optical Communication, Amsterdam, Sept. 2012.

[12] D. J. Thomson, *et al.,* "50Gbit/s Silicon Optical Modulator" IEEE Photonics Technology Letters, vol. 24, n 4, pp. 234 - 236, 2012

Hybrid InP-Polymer 30 nm tunable DBR Laser for 10 Gbit/s direct Modulation in the C-Band

Holger Klein[a], Christoph Wagner[a], Walter Brinker[a], Francisco Soares[a],
David de Felipe[a], Ziyang Zhang[a], Crispin Zawadzki[a], Norbert Keil[a] and Martin Moehrle[a]

[a]Fraunhofer Institute for Telecommunications, Heinrich-Hertz-Institute,
Einsteinufer 37, 10587 Berlin, Germany

Abstract — A C-Band hybrid InP-polymer tunable DBR laser for direct modulation is presented. The DBR tunes 34 nm with output powers larger 8 mW and SMSRs above 40 dB. 10 Gbit/s operation shows open eye diagrams.

I. INTRODUCTION

Over the past years, wavelength-division-multiplexing passive optical networks (WDM-PON) for fiber-to-the-x (FTTx) applications have attracted great attention. The demand for cost-effective and yet high-performance tunable laser sources has increased ever since. Hybrid-integrated tunable waveguide grating lasers comprising an active gain element and a thermo-optically tuned polymer Bragg reflector have been proposed [1,2].

We have developed a novel version of such a hybrid laser consisting of a curved-stripe (CS) InP-based gain chip and a waveguide integrated thermo-optically tunable polymer Bragg grating using HHI's PolyBoard integration platform [3]. A tuning range of more than 30nm is achieved using a buried heater geometry. Small-signal 3-dB bandwidths over 10 GHz are demonstrated, and large-signal 10 Gbit/s direct modulation shows clear open eye diagrams.

II. DEVICE STRUCTURE

A schematic of the developed polymer waveguide grating laser (PWGL) structure is shown in Fig.1. An InP-based gain chip with a high-reflection (HR) coated rear facet and an anti-reflection (AR) coated front facet acts as optical gain medium while the polymer waveguide grating is used as wavelength selective feedback mirror. The buried-heterostructure (BH) -type curved-stripe gain chip comprises a MOVPE grown InGaAsP multi-quantum well active region with a PL wavelength of 1570 nm and pn-InP blocking layers. Chip lengths between 250 μm and 400 μm length have been realized where all chips have a 9° angled facet to avoid undesired reflections at the InP-polymer interface in the laser cavity. To obtain high mode overlap between the InP gain chip and the polymer waveguide (i.e. low coupling loss), the gain chip waveguide is laterally tapered down to about 0.4 μm at the front facet. The optical far field is <20° (FWHM) in

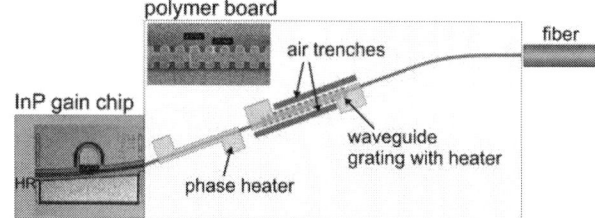

Fig. 1.   Schematic of the hybrid InP-Polymer DBR laser.

vertical and lateral direction. To enable high speed modulation a low parasitic capacitance BH laser structure was realized.

Wavelength shifting of the Bragg grating is achieved through a heater electrode in the vicinity of the polymer waveguide that thermo-optically changes the effective refractive index of the waveguide. The waveguide is fabricated as a buried structure with an index contrast of Δn=0.02. The wavelength selective mirror is realized with variable lengths between 700 μm and 1250μm using 3rd-order sidewall-corrugated Bragg gratings in the waveguide layer without the need for an additional lithographic and etching step. An additional phase control heater (cf. Fig. 1) with 600 μm length is added for precise tuning of the wavelength and the side-mode suppression ratio (SMSR).

III. DC PERFORMANCE

To perform DC and RF characterization a PWGL has been assembled on a Si-submount using a 300 μm long InP gain chip a polymer waveguide with a 700 μm long grating and a 600 μm phase section. Fig. 2 shows the optical output powers vs. gain currents of the hybrid polymer waveguide grating laser. The light was butt-coupled into a cleaved fiber at the right output port (cf. Fig. 1). The PI-curves were measured at 20°C operation temperature using an optical power meter. The laser threshold currents are 8.0 mA, 8.5 mA and 10.0 mA, the optical output powers (100mA) are 9.7 mW, 9.1 mW and 8.3 mW for heater powers of 0 mW, 40 mW and 80 mW respectively.

The optical spectra for grating heater powers of 0 mW, 30 mW, 50 mW and 80 mW are shown in Fig. 3. A gain current of 100 mA has been used at a temperature of 20°C. A

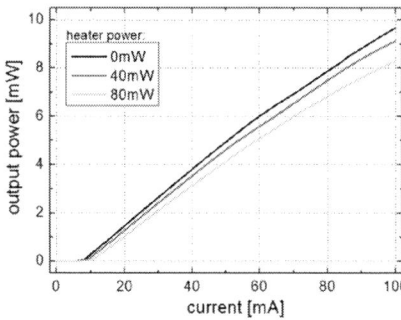

Fig. 2. Power vs. injection current characteristics of the hybrid InP-polymer laser for different heater powers. T=20°C.

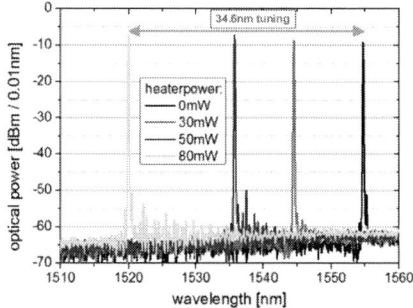

Fig. 3. Wavelength spectra of the hybrid InP-polymer laser with side mode suppression ratios (SMSR) > 40 dB. I_{gain}=100 mA, T=20°C.

maximum wavelength tuning from 1554.6 nm to 1520.0 nm has been realized by applying a maximum heating power of 80 mW. Single-mode operation of the PWGL is achieved with side mode suppression ratios (SMSR) exceeding 40 dB across the full tuning range.

IV. RF PERFORMANCE

The small signal electro-optic response of the hybrid InP-polymer laser has been measured from 100 MHz up to a frequency of 18 GHz. Fig. 4 shows the measured response functions with 3dB-bandwidths of 9.0 GHz, 9.7 GHz and 10.0 GHz for tuning heater powers of 0 mW, 40 mW and 80 mW, respectively. The gain current was set to 80 mA and the temperature was kept constant at 20°C. No additional heater power was applied to the phase section during the small signal characterization.

Large signal measurements were performed at 10 Gbit/s using an NRZ PRBS 2^{31}-1 bit sequence and a gain current of 80mA at a temperature of 20°C. The clear open eye diagrams measured at tuning heater powers of 0 mW, 40 mW and 80 mW are shown in Fig. 5. Dynamic extinction ratios of 7.17 dB, 7.12 dB and 7.23 dB are obtained for wavelengths of 1554.6 nm, 1540.6 nm and 1520.0 nm, respectively. The phase tuning heater electrode has been used to achieve linear PI-curves w/o any mode jumps during large signal operation. The applied phase section heater powers in Fig. 5a, 5b and 5c

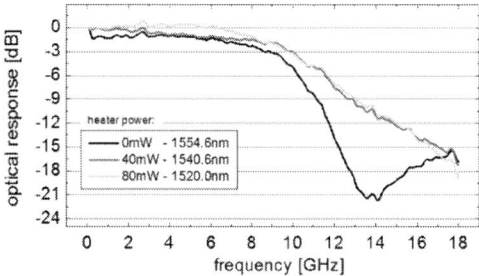

Fig. 4. Electro-optic small signal responses of the hybrid polymer laser at three different wavelengths. I_{gain}=80 mA, T=20°C

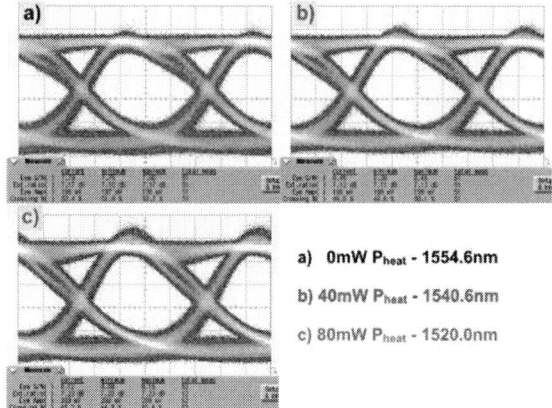

Fig. 5 Large signal eye diagrams of the hybrid polymer laser at three different wavelengths. PRBS 2^{31}-1. I_{gain}=80 mA, T=20°C.

have been 1.5 mW, 3.15 mW and 0 mW, respectively.

V. CONCLUSION

We successfully developed and fabricated a hybrid tunable directly-modulated DBR laser consisting of an InP-based gain chip and a polymer waveguide integrated thermo-optically tunable Bragg-grating. A tuning range of 34.6 nm with optical output powers of more than 8 mW and SMSR values above 40 dB have been demonstrated. The 3dB-bandwidth exceeds 9 GHz and clear open eye diagrams for 10 Gbit/s NRZ modulation have been shown.

REFERENCES

[1] G. Jeong. M. Y. Park. C. Y. Kim, S.-H. Cho, W. Lee, G. Jeong, and B. W. Kim, "Over 26 nm Wavelength Tunable External Cavity Laser Based on Polymer Waveguide Platforms for WDM Access Network", Photon Techn. Lett., vol. 18 no. 20, pp 2102-2104, 2006.

[2] K.-H. Yoon, S.H. Oh, K. S. Kim, O-K. Kwon, and D. K. Oh, "2.5-Gb/s hybridly-integrated tunable external cavity laser using a superluminiscent diode and a polymer Bragg reflector", Optics Express, vol. 18 no. 6, pp 5556-5561, March 2010.

[3] N. Keil, C. Zawadzki, Z. Zhang, J. Wang, N. Mettbach, N. Grote, and M. Schell, "Polymer PLC as an Optical Integration Bench," in Proc. OFC'11 (Los Angeles, CA, USA), paper OWM1, 2011.

High-Speed Optical Phased Array
Using High-Contrast Grating All-Pass Filters

Weijian Yang[1], Tianbo Sun[1], Yi Rao[1], Mischa Megens[2], Trevor Chan[2], Byung-Wook Yoo[1], David A. Horsley[2], Ming C. Wu[1], and Connie J. Chang-Hasnain[1, *]

[1] *Department of Electrical Engineering and Computer Sciences, University of California at Berkeley, Berkeley, California 94720, United States*

[2] *Department of Mechanical and Aerospace Engineering, University of California at Davis, Davis, California 95616, United States*

[*]*Email:* cch@eecs.berkeley.edu

Abstract — **A novel optical phased array is experimentally demonstrated with high speed (0.577 MHz) beam steering, which consists of 8x8 tunable 1550-nm all-pass filters with ultrathin high-contrast grating as the micro-electro-mechanically actuated reflector for fast tuning.**

Index Terms — **Beam steering, gratings, metamaterials, microelectromechanical devices, optical devices, optical phased arrays.**

I. INTRODUCTION

Optical phased array for free-space beam steering has been of intense research interests for a wide range of applications, e.g. imaging, display, chemical-bio sensing, precision targeting, surveillance, etc. Chip-scale optical phased array is desirable due to their integration capability, small footprint, and low power consumption. Several phase tuning mechanisms have been demonstrated using electro-mechanical [1], electro-optic [2], thermo-optic effect [3], etc. However, most of them are relatively slow at a few kHz to tens of kHz. Tuning in the order of MHz is critical to enable the advanced application in light detection and ranging (LIDAR) as well as optical circuit switching.

In this paper, we experimentally demonstrate a novel 8x8 optical phased array with high-speed micro-electro-mechanical actuation. Each array element is a tunable all-pass filter (APF) with high-contrast near-wavelength grating (HCG) [4, 5] as the top reflector and distributed Bragg reflector (DBR) as the bottom reflector. An HCG is a single layer of high-index material fully surrounded by low-index material with the period less than one wavelength, which can be designed to have broadband surface-normal reflection. An all-pass filter is an asymmetric Fabry-Perot (FP) etalon with carefully designed top and bottom reflectivities such that, when tuned across the FP resonance, the reflected light experiences continuous phase change approaching 2π without significant change of reflectivity. The APF enables a high efficiency phase tuning with small actuation distance. The ultrathin light-weight HCG further ensures a high tuning speed.

II. HCG ALL-PASS FILTER OPTICAL PHASED ARRAY

A. HCG All-Pass Filter Design and Characterization

Figure 1 shows the schematics of an individual array element. The HCG is fabricated on a p-$Al_{0.6}Ga_{0.4}As$ epitaxial layer, which is on top of a sacrificial layer and 22 pairs of GaAs/$Al_{0.9}Ga_{0.1}As$ n-DBR. With the sacrificial layer subsequently etched, the suspended HCG and bottom DBR forms a tunable Fabry-Perot cavity. We design the HCG period (Λ), bar width (s) and thickness (t_g) to be 1150 nm, 700 nm and 450 nm respectively, such that its reflectivity is ~90%. The incident light polarization is TE, i.e. electrical field along the HCG bars. The static cavity length is 700 nm, corresponding to a resonance wavelength ~1550 nm. The optical phased array is composed of 8x8 individual pixels. Each HCG mirror size is 20 μm by 20 μm; the pitch is ~28.5μm. Figure 2 shows the scanning electron microscope (SEM) image of the fabricated device.

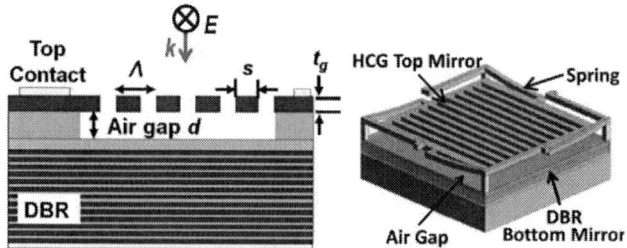

Fig. 1   Schematic of an individual pixel of the optical phased array. The $Al_{0.6}Ga_{0.4}As$ HCG and 22 pairs of GaAs/$Al_{0.9}Ga_{0.1}As$ DBR serve as the top and bottom reflector of the Fabry-Perot etalon. The incident light's polarization is parallel to the grating bar. Λ, HCG period; s, grating bar width; t_g, HCG thickness; d, air gap between HCG and DBR.

978-1-4673-1725-2/12 $31.00 © 2012 IEEE

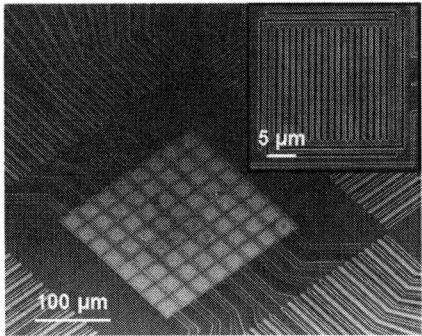

Fig. 2 SEM image of the fabricated device. The HCG array consists of 8x8 pixels. Groups of four pixels are electrically connected to increase the fill-factor. Inset, zoomed-in image of an individual pixel.

The HCG can be actuated by applying a reverse electrical bias between the HCG and DBR. This changes the cavity length and thus the reflection phase. Figure 3 shows the reflection phase of an individual etalon versus applied voltages, measured with an interferometer. A total of ~1.75 π phase change is achieved with a very small actuation voltage, corresponding to an actuation distance of the HCG ~10 nm (calculated from Fig. 4). This is expected for the all-pass filters with high quality factor (Q).

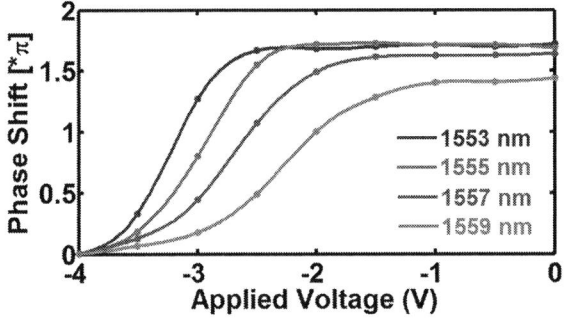

Fig. 3 Reflection phase shift versus applied voltage on an individual pixel of the phased array. ~1.75 π phase shift is achieved with small actuation voltage range. As the incident wavelength increases, the resonance cavity length increases, and thus less voltage is needed to actuate the HCG to reach the resonance cavity length. This all-pass filter design enables a small actuation distance and thus fast tuning speed. The dots on the curves are the experimental sampling points. The curves are the fitting results using smoothing-spline algorithm.

By tuning the cavity resonance wavelength through different applied voltages, this all-pass filter can operate in a large wavelength range. Figure 4 shows the reflection spectrum of the all-pass filter. A large reflection dip was observed, indicating a sharp resonance wavelength and thus a high quality factor. Based on the reflection spectrum of all 64 pixels, Q value is estimated to be 170~260. The HCG reflectivity is back calculated to be 95%~98%, and the bottom DBR reflectivity is 99.3%±0.3%. The HCG reflectivity is higher than the design value of 90% due to an inadvertent inaccuracy in electron beam lithography and etching process.

Nevertheless, the Q value is in good agreement with the calculated value based on actual HCG dimensions measured by SEM.

The resonance wavelength can be extracted for each applied voltage on Fig. 4. As the reverse-bias voltage amplitude increases, the cavity length decreases, resulting in a blue shift of the resonance wavelength. Based on the round trip phase condition of the cavity resonance, the actuation distance of the HCG can be calculated for each applied voltage. At an applied voltage of -4V, the actuation distance is ~10 nm.

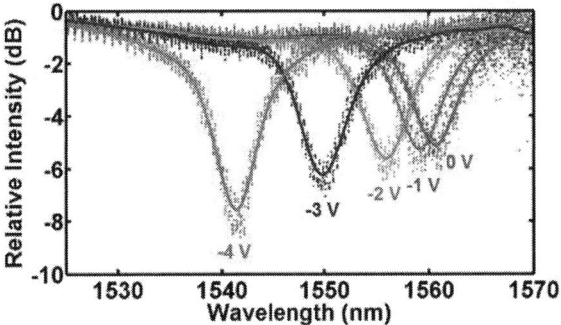

Fig. 4 Reflection spectrum of an individual pixel of the phased array for different applied voltages. The large and sharp resonance dip indicates a high quality factor. As the reverse-bias voltage amplitude increases, the cavity length decreases, resulting in a blue shift of the resonance wavelength. The resonance wavelength can be extracted for each applied voltage. Based on the round trip phase condition of the cavity resonance, the actuation distance of the HCG can be calculated for each applied voltage. With the reflection spectrum of all 64 pixels, Q value is estimated to be 170~260. The dots on the curves are the experimental sampling points. The curves are the fitting results using smoothing-spline algorithm.

B. Beam Steering Experiment

The desired reflection phase front of the HCG array can be created by individually applying different voltages to different pixels; the reflected beam can thus be steered. The HCG bar is 45° aligned to the input beam's polarization, and thus the HCG sees both TE and TM polarization. Only TE polarization can be steered. TM polarization does not "see" the cavity effect due to a very low HCG reflectivity (~30%). Its reflection phase does not change with applied voltage, and it contributes to the non-steered beam. The reflection beam then passes through a polarizer, 90° oriented to the incident light's polarization. This eliminates the background reflection. The steered beam's angle is further amplified by a lens pair. Figure 5 shows the far field intensity distribution of the steered beam, in good agreement with the simulation results. The energy of TM polarization is subtracted from the 0th order beam. Since only a very small actuation distance is needed for large phase shift, the beam steering bandwidth can be estimated by the mechanical resonance frequency f_r of the HCG actuation, measured by laser Doppler velocimetry. f_r is ~ 0.39 MHz for the current array, thanks to the light weight of the HCG mirror

(~0.67 ng). A shorter spring design has yielded an f_r ~0.577 MHz, shown in Fig. 6. Bandwidth in excess of 1 MHz can be achieved by the optimization of the micro-electro-mechanical design.

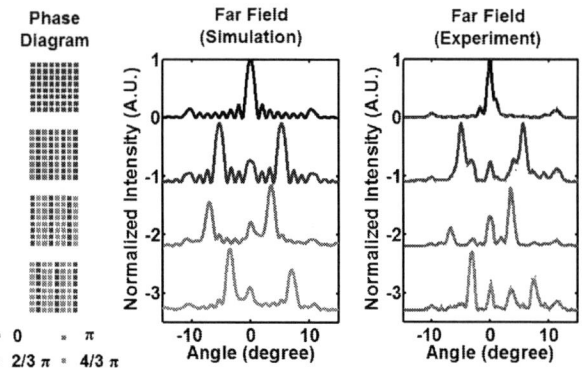

Fig. 5 Beam steering experiment. The experimental results are well-matched with simulation results. The discrepancy is due to some non-uniformity of the HCG array. The TM polarized light experiences a very low Q etalon, and thus its reflection phase does not change with applied voltage. It contributes to the 0th order beam. The TM light's energy is thus subtracted from the 0th order beam.

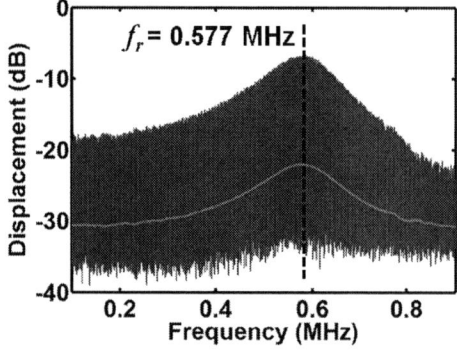

Fig. 6 Mechanical resonance f_r measured by laser Doppler velocimetry. f_r can be increased by optimization of the micro-electro-mechanical design of the individual pixel.

III. SUMMARY

We experimentally demonstrate a novel high speed optical phased array for beam steering based on the micro-electro-mechanically actuated HCG all-pass filter array. A large optical phase tuning (~1.75 π) is achieved by a small actuation distance (~10 nm) of the ultrathin lightweight HCG mirror. This enables a high efficiency and high speed (0.577 MHz) beam steering. Optimization of the HCG all-pass filter and the micro-electro-mechanical structure will further improve the beam steering quality and increase its operation bandwidth.

ACKNOWLEDGEMENT

We thank the support of DARPA SWEEPER Program (No. HR0011-10-2-0002) and a Guggenheim Fellowship (CCH).

REFERENCES

[1] U. Krishnamoorthy, K. Li, K. Yu, D. Lee, J. P. Heritage, and O. Solgaard, "Dual-mode micromirrors for optical phased array applications," *Sens. Actuators A, Phys.*, vol. 97-98, pp. 21-26, April 2002.

[2] Q. W. Song, X.-M. Wang, R. Bussjager, and J. Osman, "Electro-optic beam-steering device based on a lanthanum-modified lead zirconate titanate ceramic wafer," *App. Opt.*, vol. 35, no. 17, pp. 3155-3162, June 1996.

[3] J. K. Doylend, M. J. R. Heck, J. T. Bovington, J. D. Peters, L. A. Coldren, and J. E. Bowers, "Two-dimensional free-space beam steering with an optical phased array on silicon-on-insulator," *Opt. Express*, vol. 19, no. 22, pp. 21595-21604, October 2011.

[4] C. J. Chang-Hasnain, Y. Zhou, M. C. Y. Huang, C. Chase, "High-contrast grating VCSELs," *IEEE J. Sel. Topics Quantum Electron.*, vol. 15, no. 3, pp. 869-878, May/June 2009.

[5] C. J. Chang-Hasnain, "High-contrast gratings as a new platform for integrated optoelectronics," *Semicond. Sci. Technol.*, vol. 26, no.1, 014043, January 2011.

71 mV/dec of Sub-threshold Slope in Vertical Tunnel Field-effect Transistors with GaAsSb/InGaAs Heterostructure

Motohiko FUJIMATSU[1], Hisashi SAITO[1], and Yasuyuki MIYAMOTO[1]
[1]*Department of Physical Electronics, Tokyo Institute of Technology,*
2-12-1-S9-2, O-okayama, Meguro-ku, Tokyo 152-8552, Japan

Abstract — We fabricated a vertical tunnel field-effect transistor (TFET) with a GaAsSb/InGaAs heterojunction using a 5-nm-thick Al₂O₃ dielectric. The 26-nm width of the narrow channel mesa structure was confirmed using citric acid solution. The minimum sub-threshold slope (SS) was 71 mV/dec. On the basis of our simulated and experimental results, the SS was estimated to be 54 mV/dec for an effective oxide thickness (EOT) of 1 nm.

Index Terms — Tunnel FET, vertical transistor, GaAsSb, type-II heterojunction, sub-threshold slope, Al₂O₃.

I. INTRODUCTION

Tunnel field-effect transistors (TFETs) are expected to be used as low-power switching devices, because the filtering of high-energy electrons by the inter-band tunneling mechanism ensures a steep sub-threshold slope (SS) [1]. We previously reported the fabrication of a vertical TFET with a GaAsSb/InGaAs heterostructure [2] using the InP/InGaAs vertical MISFET process established in our laboratory [3]. In this study, we reduced the thickness of the Al₂O₃ dielectric to 5 nm and confirmed the reduction in the SS. The minimum SS was 71 mV/dec. To the best of our knowledge, this is the lowest value for a TFET with a type-II heterojunction [4].

II. SIMULATION

Figure 1 shows the layer structure of the TFET used for the simulation. The source region was p-GaAs$_{0.51}$Sb$_{0.49}$ (3×10^{19} cm^{-3}), the channel region was i-In$_{0.53}$Ga$_{0.47}$As and the drain region was n- In$_{0.53}$Ga$_{0.47}$As (1×10^{19} cm^{-3}). The type-II heterojunction was formed between p-GaAsSb and i-InGaAs. Because of the limitations of the fabrication process, the thickness of each layer was 40 nm and the channel width was 20 nm. The band diagram was obtained with Silvaco's ATLAS simulator. The dependence of the energy band on EOT is shown in Fig. 2. When the EOT was 1 nm, the tunneling distance was shorter near the heterojunction. The band-to-band tunneling (BTBT) probability was calculated using a triangle barrier model as shown in Figs. 3(a)–(b) [5]. The probability is given as

$$T(E) \propto \exp\left(-\frac{4\sqrt{2m^*}E_{g,\text{eff}}^{\frac{3}{2}}}{3q\hbar F(E)}\right)$$

where m* is the effective mass, E$_{g,\text{eff}}$ is the effective band gap, q is the elementary electric charge, $\hbar$ is the Dirac constant, and

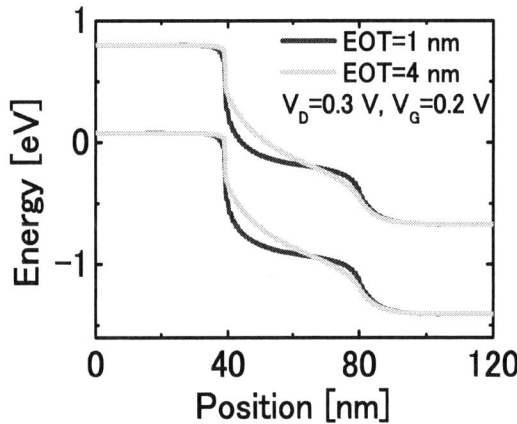

Fig. 2 Simulated band diagrams.

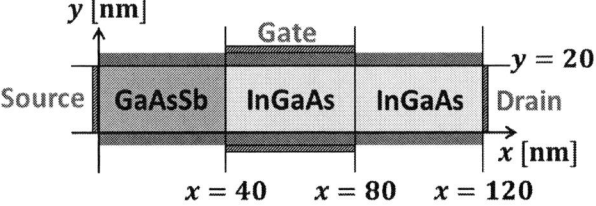

Fig. 1 Schematic image of simulated TFET.

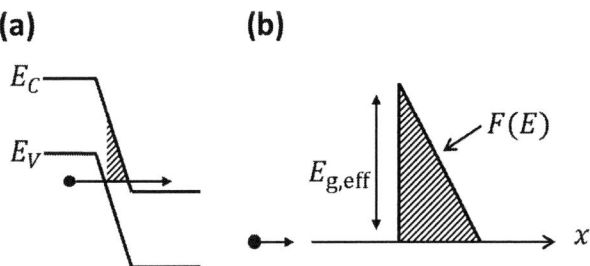

Figs. 3(a)–(b) Schematic image of tunneling electron and triangle barrier model.

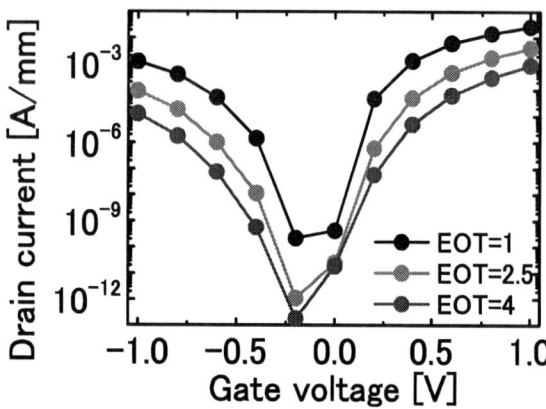

Fig. 4 Simulated transfer characteristics.

F(E) is the electric field at energy E. If the electrons were not scattered during BTBT, then, we can calculate the BTBT current using the Tsu-Esaki formula [6]. The BTBT current is given as

$$J = \frac{qm^*k_BT}{2\pi^2\hbar^3}\int_{E_{Cn}}^{E_{Vp}} T(E)\left[\frac{1+\exp\left(-\dfrac{E+E_{Fp}}{k_BT}\right)}{1+\exp\left(-\dfrac{E+E_{Fn}-qV+E_{Fp}}{k_BT}\right)}\right]dE$$

where k_B is the Boltzmann constant, T is the room temperature, E_{Cn} is the conduction band energy in the drain region, E_{Vp} is the valence band energy in the source region, E_{Fp} is the Fermi energy in the source region, E_{Fn} is the Fermi energy in the drain region, and V is the gate voltage. Figure 4 shows the simulated transfer characteristics at a drain voltage V_D of 0.3 V. When the EOT is 1, 2.5 and 4 nm, the SS is 39, 44, and 56 mV/dec. These values are lower than the conventional limit for MOSFETs at room temperature, which is typically 60 mV/dec.

III. EXPERIMENT

Figure 5 shows the schematic image of the fabricated TFET. The type-II heterojunction was formed between p-GaAsSb (40 nm, 3×10^{19} cm^{-3}) and i-InGaAs (40 nm). The drain region was InGaAs (40 nm, 1×10^{19} cm^{-3}). A 5-nm-thick layer of Al$_2$O$_3$ (EOT = 2.3 nm) was used as the gate dielectric. The step of the fabrication process of the narrow mesa structure are as follows. First, a 150-nm-thick TiW drain electrode was deposited by sputtering, and a 5-um-long Cr line was formed on the surface of the TiW electrode by electron beam evaporation. This Cr line was used as an etching mask for the CF$_4$ reactive ion etching (RIE) of the TiW layer. Figure 6(a) shows the schematic image of a narrow drain electrode after

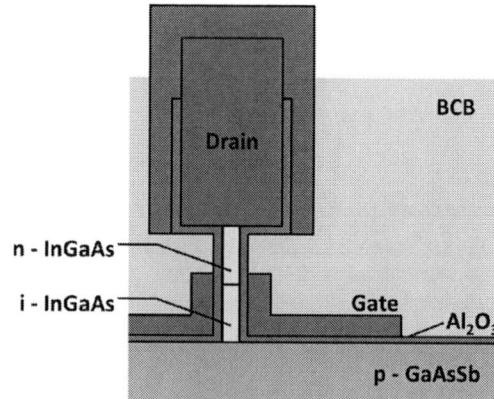

Fig. 5 Schematic image of fabricated TFET.

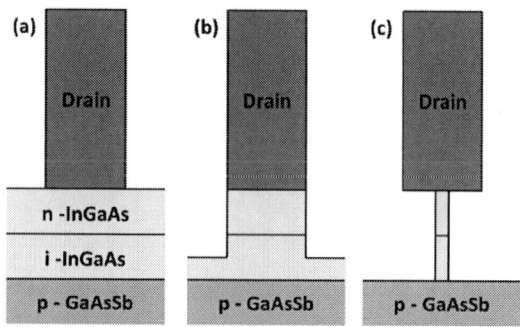

Fig. 6(a)Drain electrode formed by RIE. (b) Channel mesa structure formed by ICP-RIE. (c) Narrow mesa structure obtained after wet etching.

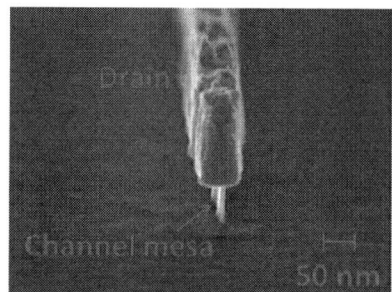

Fig. 7 SEM image of the fabricated mesa structure.

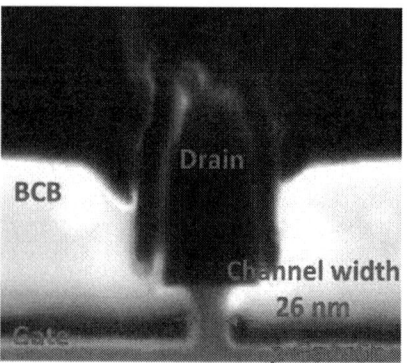

Fig. 8 Cross-sectional TEM image of fabricated TFET.

the RIE process. Then, a narrow channel mesa structure was formed by CH_4/H_2 inductively coupled plasma reactive ion etching (ICP-RIE) and anisotropic etching using a citric acid solution [7]. Figure 6(b) shows the schematic image of a mesa structure after the ICP-RIE process. Figure 6(c) shows the schematic image and Fig. 7 shows the SEM image of a narrow channel structure after using citric acid solution. We can fabricate narrow mesa structures because the etching-speed of citric acid in the <011> direction is 0.5 nm/s. Figure 8 shows the cross-sectional TEM image of the fabricated TFET. The formation of a 26-nm-wide channel mesa structure was confirmed. Figure 9 shows the transfer characteristics measured at room temperature. The off-state was improved in comparison with the previous work [2], which had a 25-nm-wide channel TFET with an EOT of 3.8 nm. The SS at a drain voltage of 0.25 V was reduced from 130 mV/dec to 85 mV/dec, the on/off current ratio was improved to 1.7×10^4 from 1.5×10^4, and the drain-induced barrier lowering (DIBL) was reduced from 239 mV/V to 200 mV/V. Figure 10 shows the output characteristics. We observed that the sub-threshold characteristics shows improvement with a reduction in the drain voltage. As shown in Fig. 11, an SS of 71 mV/dec was obtained at a drain voltage of 0.05 V.

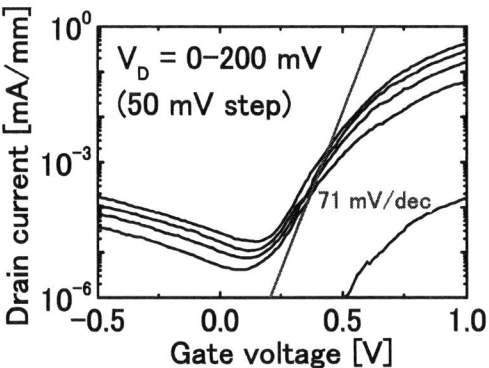

Fig. 11 Transfer characteristics at low drain voltage.

IV. DISCUSSION

We evaluated the SS using the gate stack model shown in Fig. 12. In this model,

$$SS_m = SS_{cal}\left(1 + \frac{C_{it}}{C_{ox}}\right)$$

where SS_m is the measured SS, SS_{cal} is the simulated SS, C_{it} (= qD_{it}) is the interface trap capacitance, and C_{ox} is the effective oxide capacitance. As shown in Table I, D_{it} was estimated to be 7.4×10^{12} eV^{-1}cm^{-2} when the EOT was 3.8 nm, and it was 8.6×10^{12} eV^{-1}cm^{-2} when the EOT was 2.3 nm. When the EOT was 1 nm, the SS_m was estimated to be 54 mV/dec by using the average of these D_{it} values. Additionally, by using the reported D_{it} of 1.7×10^{12} eV^{-1}cm^{-2} for the D_{it} in Al_2O_3/InGaAs interface [8], we calculated SS_m to be 42 mV/dec when the EOT was 1 nm [9].

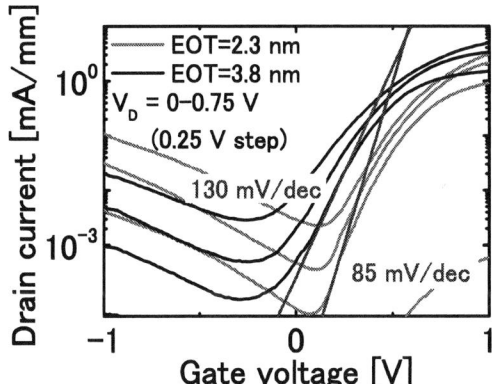

Fig. 9 Comparison of transfer characteristics.

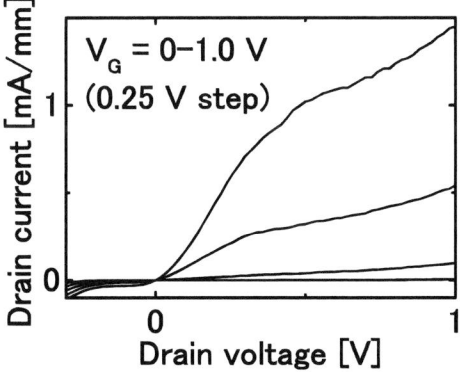

Fig. 10 Output characteristics.

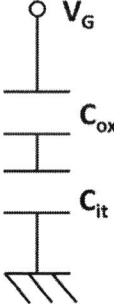

Fig. 12 Model of gate stack structure.

V. CONCLUSION

A vertical TFET with a GaAsSb/InGaAs heterostructure fabricated using a 5-nm-thick Al_2O_3 dielectric was shown to have a minimum SS of 71 mV/dec, an on/off current ratio of 1.7×10^4, and DIBL of 200 mV/dec. The EOT scaling showed improvement in the off-state. However, the characteristics degraded ownig to the interface trap. We estimated the SS to be 54 mV/dec when EOT was 1 nm, even if the characteristics degraded by the interface trap.

ACKNOWLEDGEMENTS

This work was supported by a Grant-in Aid for scientific research by MEXT/JSPS and SCOPE by MIC.

REFERENCES

[1] Q. Zhang, W. Zhao, and A. Seabaugh, "Low-Subthreshold-Swing Tunnel Transistors", *IEEE Electron Device Lett.*, vol. 27, p.297, 2006.

[2] M. Fujimatsu, H. Saito, and Y. Miyamoto, "GaAsSb/InGaAs Vertical Tunnel FET with a 25 nm-wide Channel Mesa Structure", *SSDM 2011*, A-4-1.

[3] H. Saito, Y. Miyamoto, and K. Furuya, "Fabrication of Vertical InGaAs Channel Metal-Insulator-Semiconductor Field Effect Transistor with a 15-nm-wide Mesa Structure and Drain Current Density of 7 MA/cm^2," *Appl. Phys. Express*, vol. 3, 084101, 2010.

[4] D. Mohata, S. Mookerjea, A. Agrawal, Y. Li, T. Mayer, V. Narayanan, A. Liu, D. Loubychev, J. Fastenau, and S. Datta, "Experiment Staggered-Source and N+ Pocket-Doped Channel III-V Tunnel Field-Effect Transistors and Their Scalabilities", *Appl. Phys. Express*, vol.4, 024105, 2011.

[5] S. M. Sze and K. K. Ng, "Physics of Semiconductor Devices", 3rd. ed., Wiley Interscience, p.423, 2007.

[6] R. Tsu and L. Esaki, "Tunneling in a Finite Superlattice", *Applied Physics Letters*, vol. 22 (1973) 562.

[7] G. C. DeSalvo, R. Kaspi, and C. A. Bozada, "Citric Acid Etching of $GaAs_{1-x}Sb_x$, $Al_{0.5}Ga_{0.5}Ab$, and InAs for Heterostructure Device Fabrication", *J. Electrochem. Soc.*, vol.141, p.3526, 1994.

[8] A. D. Carter, W. J. Mitchell, B. J. Thibeault, J. J. M. Law, and M. J. W. Rodwell, "Al_2O_3 Growth on (100) In0.53Ga0.47As Initiated by Cyclic Trimethylaluminum and Hydrogen Plasma Exposures", *Appl. Phys. Express*, vol.4, 091102, 2011.

[9] R. Suzuki, N. Taoka, M. Yokoyama, S. Lee, S. H. Kim, T. Hoshii, T. Yasuda, W. Jevasuwan, T. Maeda, O. Ichikawa, N. Fukuhara, M. Hata, M. Takenaka, and S. Takagi, "1-nm-capacitance-equivalent-thickness $HfO_2/Al_2O_3/InGaAs$ Metal-oxide-semiconductor structure with Low Interface Trap Density and Low Gate Leakage Current Density", *Applied Physics Letters.*, vol. 100, 132906, 2012.

TABLE I
ESTIMATED D_{it} from EOT and SS.

EOT [nm]	SS_{cal} [mV/dec]	SS_m [mV/dec]	D_{it} [eV^{-1}cm^{-2}]
3.8	56	130	7.4×10^{12}
2.3	44	85	8.6×10^{12}
1.0	39	(54)	(8.0×10^{12})

A 1.3 pJ/bit Energy-Efficient Ultra-Low Power On-off mode Oscillator Using an InP-based Quantum-effect Tunneling Device

Jooseok Lee, Jongwon Lee, Jaehong Park and Kyounghoon Yang.

Department of Electrical Engineering
Korea Advanced Institute of Science and Technology (KAIST)
373-1, Guseong-Dong, Yuseong-Gu, Daejeon, Republic of Korea.
leejooseok@kaist.ac.kr

Abstract — A low power on-off mode oscillator is demonstrated by using an InP-based RTD. In order to achieve the low power operation, the NDC (negative differential conductance) characteristic at a low voltage of the RTD is used for RF signal generation. The implemented RTD-based oscillator by using an InP-based RTD/HBT MMIC technology shows low power consumption of 1.3 mW at an oscillation frequency of 5.8 GHz. The RTD-based oscillator operates in an on-off mode with a high data rate of 1 Gb/s. The obtained energy efficiency of 1.3 pJ/bit is found to be the best reported up to date.

Index Terms — microwave oscillator, negative differential resistance circuit, resonant tunneling diode.

I. INTRODUCTION

Due to the excellent NDC characteristics and strong I-V nonlinearity, the InP-based RTD (resonant tunneling diode) has attracted a great deal of interests in microwave applications. In the previous works, we demonstrated various RTD-based oscillators with extremely low power consumption [1]-[3]. The extremely low power performance of the RTD-based oscillators is attributed to the excellent inherent NDC characteristic of the RTD achieved at a low applied voltage. Furthermore, because of the high-speed switching capability due to the fast quantum resonant tunneling phenomena and the related small parasitic capacitance [4], the on-off time of the RTD-based oscillator is very fast. Therefore, the RTD-based oscillator allows the high data-rate operation as an on-off mode oscillator for OOK-type transmitter applications.

In this work, we demonstrate a low power RTD-based on-off mode oscillator, which is an essential component of the wireless OOK transmitter applications. The proposed on-off mode oscillator has been implemented by using an RTD/HBT MMIC (monolithic microwave integrated circuit) technology.

II. OPERATION PRINCIPLE AND DEVICE TECHNOLOGY

Fig. 1 shows the circuit schematic diagram of the proposed RTD-based on-off mode oscillator, which consists of an RTD oscillator core and an output buffer stage. The previously reported design theories of the tunneling-diode based oscillator [1], [5]-[6] are considered to satisfy the oscillation condition and to achieve the fast start-up of the RTD oscillator.

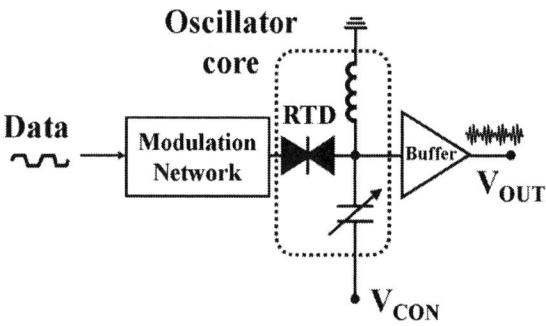

Fig.1. Schematic diagram of the RTD-based on-off mode oscillator.

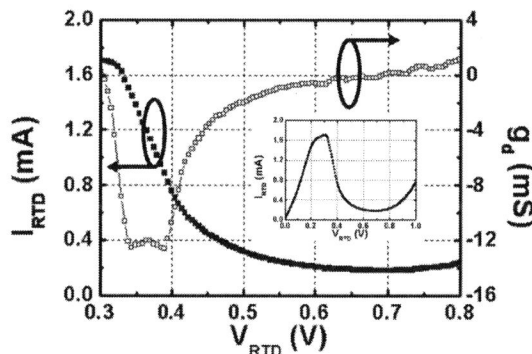

Fig.2. Measured dc I–V and differential conductance (g_d) characteristics of the fabricated RTD at the NDC region at room temperature (inset: the full I-V characteristic).

The emitter follower buffer is also used to isolate the RTD oscillator core from the 50 Ω output load for measurement. The RTD is used as the negative conductance cell in the one-port oscillator. To achieve the on-off mode operation of the oscillator, the bias current of the RTD is modulated by a control signal applied at the Data node. When the voltage of the Data node is 0.4 V, the RTD is biased in the negative differential conductance region as shown in Fig. 2. Consequently, the differential conductance of the RTD is

978-1-4673-1725-2/12 $31.00 © 2012 IEEE   29

TABLE I
PERFORMANCE COMPARISON WITH OTHER LOW-POWER OOK TRANSMITTER ICS OPERATING AT THE ISM BAND.

	[7]	[8]	This work
Technology	180 nm CMOS	250 nm CMOS	**1.5 μm RTD/HBT**
f_{center} (GHz)	2.46	5.73	**5.8**
Data Rate (Mbps)	136	100	**1000**
Power Consumption (mW)	3	5	**1.3**
Energy efficiency (pJ/bit)	22	50	**1.3**

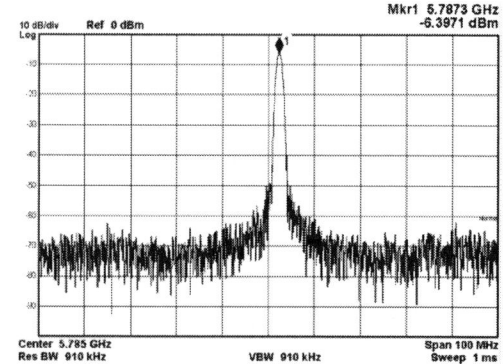

Fig. 3. Measured output spectrum of the fabricated RTD oscillator

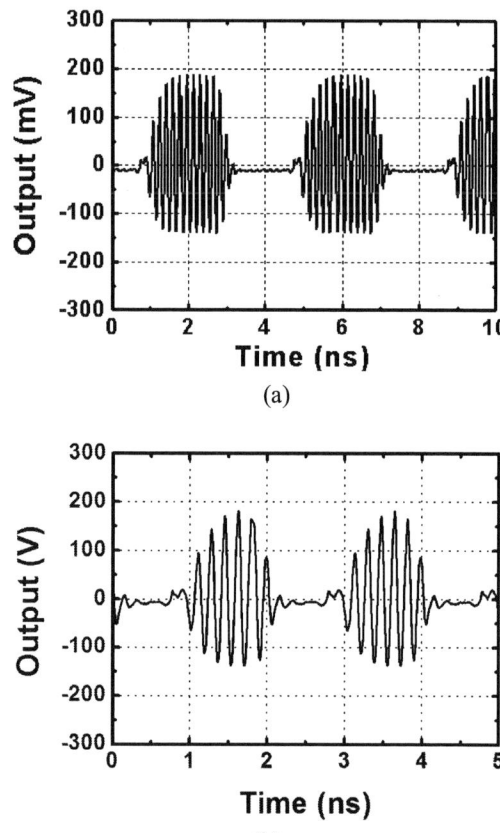

Fig. 4. Measured time-domain waveforms at the output of the fabricated RTD on-off mode oscillator with an on-off modulation data rate of (a) 500 Mb/s and (b) 1 Gb/s.

negative and the oscillator starts the oscillation in the on-mode. On the other hand, when the voltage at the Data node is 0 V, the RTD is biased in the positive conductance region and the oscillator stops the oscillation in the off mode condition.

An InP-based RTD/HBT MMIC technology has been used to fabricate the proposed RTD oscillator. The double-barrier quantum-well structure of RTD, stacked on the HBT layers, is composed of AlAs/InGaAs/InAs/InGaAs/AlAs layers. The detailed layer structure and the fabrication process have been described elsewhere [1]. The measured dc characteristics of the fabricated RTD with an emitter area of $2.0 \times 1.0 \ \mu m^2$ are shown in Fig. 2. The differential conductance (g_d) of $\partial I/\partial V$ obtained from the dc I−V characteristics is also shown. The RTD shows a peak voltage (V_P) of 0.31 V and a peak current (I_P) of 1.7 mA with a relatively high PVCR of 9.4 at room temperature. The RTD exhibits the negative conductance characteristics in a voltage range from 0.31 to 0.70 V. The fabricated $1.5 \times 4.0 \ \mu m^2$ HBT shows a maximum dc current gain of 50, and the maximum f_T and f_{Max} are 100 GHz and 100 GHz, respectively. As for the LC resonator, the inductance of the used spiral inductor is 1.96 nH. The varactor diode has been fabricated using the base-collector junction of the InP-based HBT with a junction size of $28 \times 28 \ \mu m^2$.

III. MEASUREMENT AND DISCUSSION

In order to confirm the fundamental performances of the fabricated RTD oscillator, the fully on-mode operation of the oscillator was first measured on wafer using a N9030A PXA signal analyzer with a system impedance of 50 Ω at room temperature. As shown in Fig. 3, the output spectrum of the oscillator shows an oscillation frequency of 5.8 GHz. The RF power of -6.4 dBm was obtained at the output port of the oscillator, under a bias condition of RTD current of 750 μA at

a supply voltage of 0.40 V. The corresponding core power consumption of the oscillator was 0.3 mW excluding the output buffer. The total power consumption of the RTD oscillator including the emitter follower output buffer was measured to be 1.3 mW.

In order to characterize the on-off mode operation, the time-domain measurement for the fabricated RTD oscillator was conducted. The NRZ (non return to zero)-mode digital input data signal was generated by using a pulse pattern generator (Anritsu MP1763B). The input voltages for the on and off states of the input patterns are 0.4 V and 0 V, respectively. Fig. 4 shows the measured time-domain waveforms from the output of the RTD on-off mode oscillator at OOK data rates of 500 Mb/s and 1 Gb/s. The obtained high data rate performance is due to the inherent high speed switching capability of the RTD as well as the high NDC arising from the strong quantum-effect-related I-V nonlinearity of the RTD. To evaluate the performances of the on-off mode oscillator for OOK transmitter applications, the widely used FOM (figure of merit) for the energy efficiency is used, which is given by FOM (energy efficiency) = dc power consumption / data rate. The energy efficiency in a unit of Joule/bit represents the average amount of energy required to transmit a single bit of data. The obtained FOM from the fabricated on-off mode oscillator in this work is 1.3 pJ/bit including the power consumption of the buffer stage.

Table I shows the performance comparison of the fabricated on-off mode RTD oscillator with other low-power OOK transmitter ICs reported in the ISM-band frequency range. The results show that, compared to the conventional transistor-based OOK transmitters, the fabricated proposed RTD on-off mode oscillator operates at extremely low power consumption with high energy efficiency. The extremely low power consumption with the excellent energy efficiency, which is at least 10 times better than the other transistor-based approaches, indicates that the quantum-effect RTD oscillator is very favorable for the low-power short-range wireless transmitters, especially such as biomedical applications [9].

IV. CONCLUSION

A low power RTD on-off mode oscillator for OOK transmitter applications has been proposed and demonstrated based on an InP RTD/HBT MMIC technology. The fabricated oscillator showed extremely low power consumption of 1.3

mW at an oscillation frequency of 5.8 GHz. In addition, the high energy efficiency of 1.3 pJ/bit at a 1 Gb/s data rate was achieved. The extremely low power consumption with the excellent energy efficiency, which is at least 10 times better than the other transistor-based approaches, indicates that the quantum-effect RTD oscillator is very promising for the low-power short-range wireless transmitters.

REFERENCES

[1] S. Choi, Y. Jeong, and K.Yang, "14 GHz InP-based RTDMMIC VCOs with ultra-low dc power consumption," in *Proc. IEEE Int. Conf. Indium Phosphide Related Mat.*, 2006, pp. 439–441.

[2] Y. Jeong, S. Choi, and K. Yang, "A Sub-100 µW Ku-band RTD VCO for extremely low power applications," *IEEE Microw. Wireless Compon. Lett.*, vol. 19, no. 9, pp. 569–571, Sep. 2009.

[3] Y. Jeong, S. Choi, and K. Yang, "Novel antiphase-coupled RTD microwave oscillator operating at extremely low DC-power Consumption," *IEEE Trans. on Nanotech.*, vol. 9, No. 3, pp. 338–341, May. 2010.

[4] N. Shimizu, T. Nagatsuma, T. Waho, M. Shinagawa, M. Yaita, and M. Yamamoto, "InGaAs/AlAs resonant tunneling diodes with switching time of 1.5 ps," *Electron. Lett.*, vol. 31, no. 19, pp. 1695–1697, Sep. 1995.

[5] N. Deparis, C. Loyez, N. Rolland, and P. A. Rolland, "UWB in Millimeter Wave Band With Pulsed ILO," *IEEE Trans. Circuit Syst. II*, vol. 55, no. 4, pp. 339–343, Apr. 2008.

[6] M. Egard, M. A¨ rlelid, E. Lind, G. Astromskas, and L.-E. Wernersson, "20 Ghz wavelet generator using a gated tunnel diode," *IEEE Microw. Wireless Compon. Lett.*, vol. 19, no. 6, pp. 386 –388, Jun. 2009.

[7] J. Jung, S. Zhu, P. Liu, Y. E. Chen, and D. Heo, "22-pJ/bit energy efficient 2.4-GHz implantable OOK transmitter for wireless biotelemetry systems: In vitro experiments using rat skin-mimic," *IEEE Trans. Microw. Theory Tech.*, vol. 58, no. 12, pp. 4102–4111, Dec. 2010.

[8] P. Upadhyaya, M. Rajashekharaiah, D. Heo, D.M. Rector, and Yi-Jan Emery Chen, "100 Mbs OOK transmitter with low power and low phase noise LC VCO for neurosensory application," in *IEEE Southeast Conf. Proceedings.*, pp.75–78. Apr. 2005.

[9] E. Y. Chow, C.-L. Yang, A. Chlebowski, S. Moon, W. J. Chappell, and P. P. Irazoqui, "Implantable wireless telemetry boards for In Vivo transocular transmission," *IEEE Trans. Microw. Theory Tech.*, vol. 56, no. 12, pp. 3200–3208, Dec. 2008.

Optimized RTD-HBT VCO Design Based On Large Signal Transient Simulations

Benjamin Münstermann, Anselme Tchegho, Gregor Keller and F.-J. Tegude

Solid-State Electronics Department, Center for Nanointegration, University of Duisburg Essen,
Lotharstr. 55, D-47057 Duisburg, Germany

Abstract — This paper presents an optimized RTD-HBT single ended voltage controlled oscillator topology with increased tuning range and improved frequency stability at 20 GHz oscillation frequency. By connecting the RTD to the emitter of an HBT, a larger voltage swing at the parallel resonator compared to the direct connection can be achieved. In addition the RTD-capacitance influence on the oscillation frequency can be suppressed efficiently. Transient assisted harmonic balance simulations promise an increased oscillation power by 2 dB and a doubled tuning range of about 3.2 GHz compared to the conventional RTD-HBT circuits.

Index Terms — resonant tunneling diodes, MMIC, voltage controlled oscillators, phase noise.

I. INTRODUCTION

Resonant tunneling diodes are excellent active devices for oscillators especially at very high frequencies. The built-in negative differential resistance is useable up to 1.1 THz as demonstrated in [2]. Another very attractive application can be found in microwave VCOs with very low phase noise and power consumption [1, 3]. By combining the RTD with HBT technology, frequency tuning can be achieved by using reverse biased base collector junctions. Additionally buffer stages can be integrated to drive a 50 Ohm load. This way low phase noise and further excellent figures of merit have been achieved [3].

In the conventional circuit, depicted in Fig. 1a, the parallel resonator consisting of a spiral inductance and the pn-varactor diode is directly connected to the resonant tunneling diode, which is biased in the negative differential region. The RTD-device capacitance is strongly voltage dependent in this region [5] which leads to unwanted frequency changes even with bias drifts of only few tens of mV. The RTD-size is restricted by the target tuning range and the stability criterion introduced by Kidner [4], therefore the maximum available power is directly limited by the RTD. In this paper we propose an alternative approach to use the negative resistance of the RTD in the emitter branch of the HBT to reduce frequency detuning by the RTD quantum capacitance and improve the loaded Q of the resonator for large signal conditions.

II. DEVICE MODELS AND TECHNOLOGY PARAMETERS

All device models are fitted to measurement data of manufactured test devices on InP substrates (Fig. 2). The DC-characteristics have been measured with a Keithley 4200 characterisation system. The rf-properties of the devices have been investigated by performing s-parameter measurements with a 8510C network analyzer.

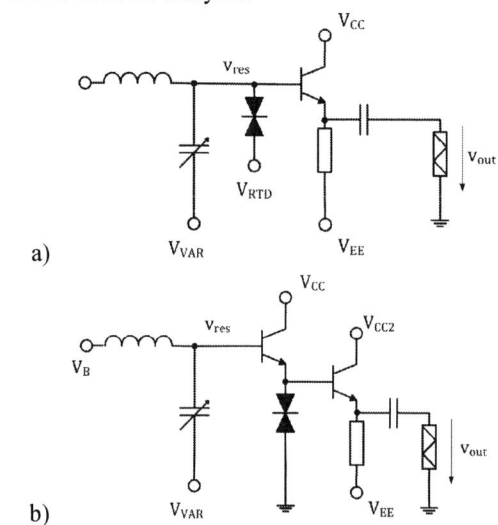

Fig. 1.   a) RTD circuit used in [3] b) proposed circuit

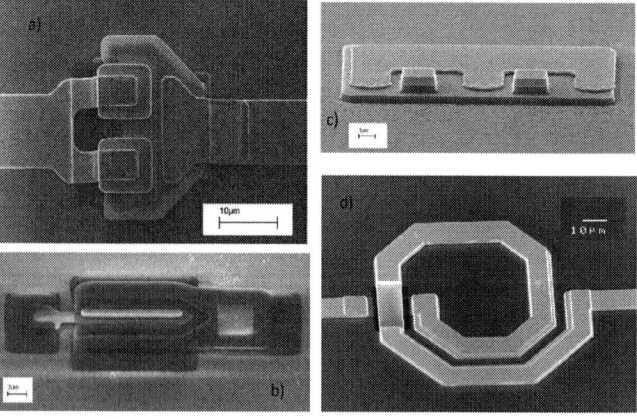

Fig. 2.   SEM-pictures of the manufactured test devices
a) pn-varactor diode   b) triple mesa SHBT
c) multiple mesa RTD   d) spiral inductor

The RTD multiple mesa devices with 1 µm² top contact areas have shown good scalability in current density and bias dependent device capacitance. The peak current density of $1.9\,\mathrm{mA/\mu m^2}$ is reached at $0.46\,\mathrm{V}$ and the peak to valley current ratio is 8. The pn-varactor diodes offer a capacitance of $0.5\,\mathrm{fF}\,/\,0.9\,\mathrm{fF}$ per μm^2 and quality factor of $70\,/\,52$ at $3\,\mathrm{V}\,/\,0\,\mathrm{V}$ bias voltage. The spiral inductance with $0.31\,\mathrm{nH}$ exhibit a quality factor of 25 at the center frequency of $20\,\mathrm{GHz}$. The HBT with $1\,\mu m \times 10\,\mu m$ emitter size has a dc-current gain of 42 and $150\,\mathrm{GHz}$, $230\,\mathrm{GHz}$ f_T,f_{max} at collector current density of $1.4\,\mathrm{mA/\mu m^2}$.

The RTD model consists of a series resistor r_s, a voltage controlled current source parallel to a voltage dependent capacitance as described in [6]. For enabling an accurate representation of the RTD-I-V characteristics and capacitance variations we used the equations of Yan [8] (Fig.3). The resistance r_s is determined by s-parameter measurements at voltage biases near the peak or the valley voltage using the extraction method of [7]. The HBT are modeled with the help of the AgilentHBT model and fitted to gummel plots, diode characteristics, output characteristics and s-parameter data up to $50\,\mathrm{GHz}$ (Fig.4). The varactor diodes, represented by pn-junction diode models, were fitted with respect to capacitance and quality factor of the measured test elements in the target frequency range. The spiral inductor is modeled with a frequency dependent quality factor to achieve good emulation of the measured resonator losses.

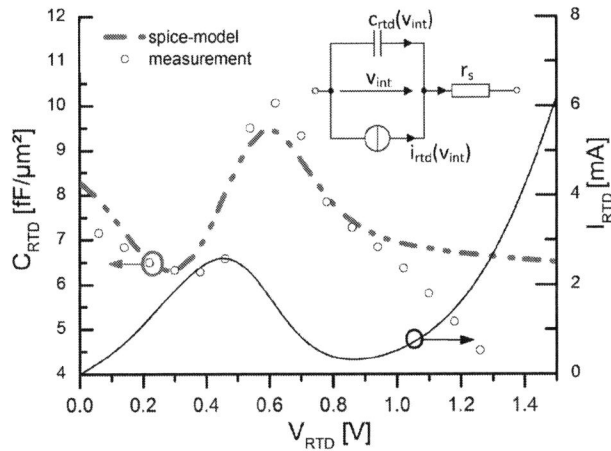

Fig. 3.   Voltage dependent RTD-capacitance extracted from s-parameter measurements of $2\,\mu m^2$ RTD

III. SIMULATION

Transient assisted harmonic balance simulations have been performed using the "Agilent Advanced Design System" software to design and optimize the circuits for the target frequency of $20\,\mathrm{GHz}$ using the same spiral inductance and scalable varactor diode model. The designs were examined in terms of losses related to phase noise, available output power, tuning range and robustness against technological uncertainties like deviating RTD-size. By investigating the transient RTD currents and voltages of the two circuit topologies in Fig. 5 the

advantage of the improved RTD to resonator coupling is obvious.

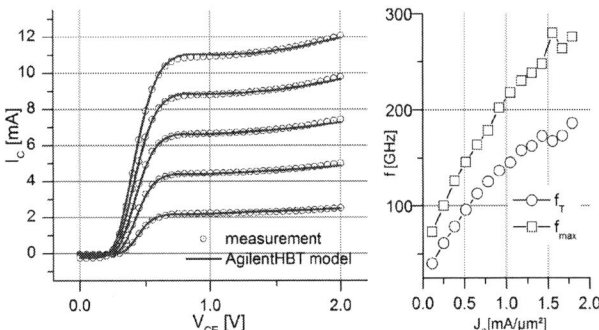

Fig. 4.   Output characteristic and f_T,f_{max}, extracted from S-parameter data of a fabricated $1 \times 10\mu m^2$-SHBT

The RTD-impedance seen by the resonator is transformed by the current gain of the HBT. This includes a reduced negative conductance on the one hand, on the other hand the negative effect of the RTD device capacitance and the additional load conductance originating from the RTD outside the NDC region is reduced significantly. Additionally, in circuit configuration a) the base emitter-junction conductance reduces the loaded Q of the VCO when biased with high base current levels, so high frequency current gain of the HBT is limited by the fact that the collector current density is limited.

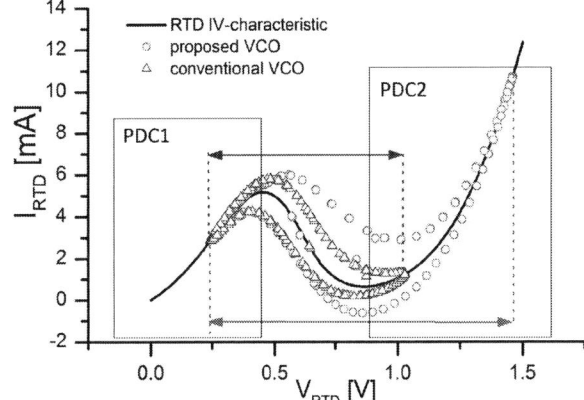

Fig. 5.   Simulated current and voltage swing of the $2\,\mu m^2$ RTD in both circuit configurations

The voltage swing at the resonator can be used as an indicator for the energy stored in the LC-tank and should be maximized to concentrate the signal power at the oscillation frequency and suppress phase noise. In configuration a) this voltage swing is equal to the voltage swing at the RTD. For high voltage levels the positive differential conductance of the RTD (PDC1 and PDC2 in Fig.2) limits the voltage swing to $800\,\mathrm{mV}$. In the proposed configuration b) the RTD impedance is coupled more efficiently to the resonator resulting in an improved loaded Q. The oscillation amplitude at the RTD is increased by 40%, the voltage swing at the resonator is more than 100% larger than in the conventional RTD-HBT circuit

978-1-4673-1725-2/12 $31.00 © 2012 IEEE

(Fig. 5). The HBT-RTD combination also allows to drive the buffer transistor with more current to achieve a good high frequency power amplification of the signal.

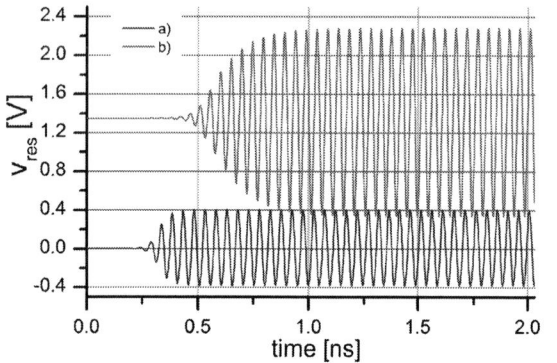

Fig. 6. Simulated resonator voltage v_{res} for both VCO configurations

A set of simulations have been made to see the influence of the RTD-parameters to the circuit performance. In Fig. 7 one can compare the available output power at the oscillation frequency versus rtd-area, which is scaling peak- and valley currents and device capacitance. Both oscillators show increasing output power with growing rtd sizes. Oscillator a) seems to be more stable in output power compared to the proposed version. This can be explained by the RTD-voltage swing limitation, given by the highly conductive PDR1 and PDR2 regions, so the additional power generated in NDR is consumed partially in the RTD.

This situation is more relaxed in oscillator b) so the oscillation power is more sensitive to the RTD-size. Since RTD-size is limited, because of the stability criteria [4], oscillation power can only be scaled to a certain extend. In case of the new topology, for a fixed RTD-size and accordingly negative conductance a 2 dB increase in output power could be achieved without compromising bias stability.

RTD-mesa definition in the submicron range are subject to uncertainties in lithographie and vertical etching steps. Therefore a robust VCO design should be robust against the technological variations in device capacitance. To evaluate the sensitivity of the two circuits to the RTD-mesa definition error the oscillation frequency is plotted versus the tuning voltage and the RTD-size in Fig 8. It can be observed that an area uncertainty directly influences the center frequency of the conventional RTD oscillator, on the other hand output signal frequencies of the proposed circuit are nearly independent of the area size. Due to the reduced influence of the RTD-capacitance range the varactor-area can be designed larger resulting in an improved tuning range of up to 3 GHz for the proposed VCO.

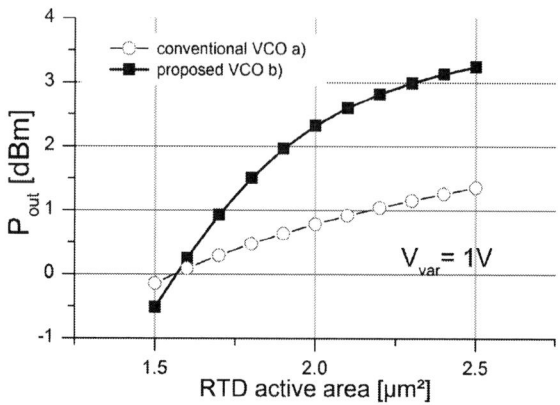

Fig. 7. Simulated output power with a 50 Ohm load versus RTD active area of the two circuit designs

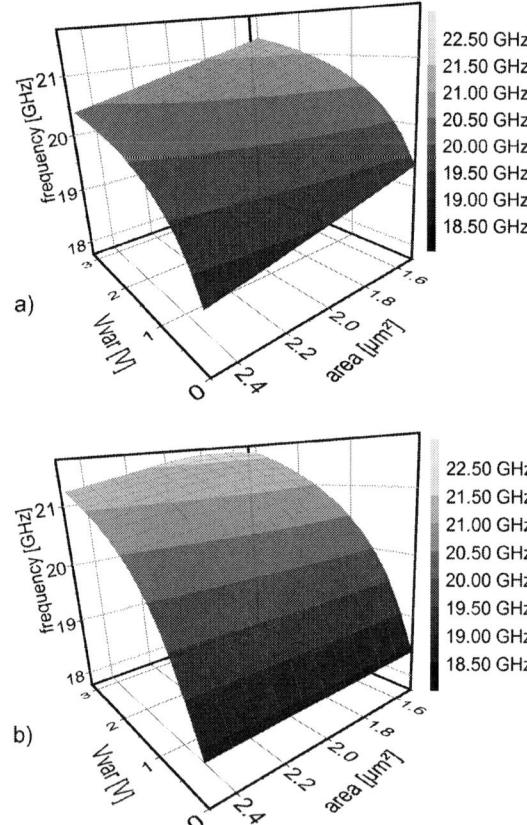

Fig. 8. Simulated oscillation frequency versus tuning voltage V_{var} and RTD active area for VCO-configuration a),b)

V. CONCLUSION

A novel circuit topology using the RTD in the emitter branch of an HBT to create oscillations in a parallel LC-resonator has been simulated based on measured device data and revealed advantages in frequency tuning and robustness against RTD technology variatons. By using the additional transistor it is possible to significantly improve signal power, whereby biasing instabilities in case of high current RTDs can be avoided. The increased amount of energy stored in the high Q LC tank should lead to further improvement of the excellent low phase noise performance of RTD HBT VCOs, which is going to be validated experimentally soon.

REFERENCES

[1] Sunkyu Choi, Yongsik Jeong, and Kyounghoon Yang. "14 Ghz InP-based RTD MMIC VCOs with Ultra Low DC Power Consumption". In *Indium Phosphide and Related Materials Conference Proceedings, 2006 International Conference on*, pages 439 –441, 2006.

[2] Michael Feiginov, Cezary Sydlo, Oleg Cojocari, and Peter Meissner. "Resonant-tunnelling-diode oscillators operating at frequencies above 1.1 Thz". *Applied Physics Letters*, 99(23):233506 –233506–3, 2011.

[3] Yongsik Jeong, Sunkyu Choi, and Kyounghoon Yang. "Development of sub-100uW microwave RTD VCOs. In *Nanotechnology (IEEE-NANO), 2010 10th IEEE Conference on*, pages 214 –216, 2010.

[4] C. Kidner, I. Mehdi, J.R. East, and G.I. Haddad. "Power and stability limitations of resonant tunneling diodes". *Microwave Theory and Techniques, IEEE Transactions on*, 38(7):864 –872, 1990.

[5] R. Lake and Junjie Yang. "A physics based model for the RTD quantum capacitance". *Electron Devices, IEEE Transactions on*, 50(3):785 – 789, 2003.

[6] A. Matiss, A. Poloczek, W. Brockerhoff, W. Prost, and F.-J. Tegude. "Large-signal analysis and ac modelling of sub micron resonant tunnelling diodes". In *Microwave Integrated Circuit Conference, 2007. EuMIC 2007. European*, pages 207 –210, 2007.

[7] A. Tchegho, B. Muenstermann, C. Gutsche, A. Poloczek, K. Blekker, W. Prost, and F.J. Tegude. "Scalable high-current density RTDs with low series resistance". In *Indium Phosphide Related Materials (IPRM), 2010 International Conference on*, pages 1 –4, 2010.

[8] Zhixin Yan and M.J. Deen. "New RTD large-signal DC model suitable for pSpice". *Computer-Aided Design of Integrated Circuits and Systems, IEEE Transactions on*, 14(2):167 –172, 1995.

Sensitive high frequency envelope detectors based on triple barrier resonant tunneling diodes

Gregor Keller, Anselme Tchegho, Benjamin Münstermann, Werner Prost, and Franz-Josef Tegude.

Solid-State Electronics Department, Center for Nanointegration, University of Duisburg Essen, Lotharstr. 55, D-47057 Duisburg, Germany

Abstract — InP-based resonant tunneling diodes with symmetrical I/V-characteristics have shown their excellent high frequency performance for THz signal generation. We present a modification with an additional third barrier to create an unsymmetrical I/V-characteristic. With their large current densities and low capacitances these devices are promising candidates for zero bias high frequency envelope detectors. Based on simulations two layer stacks are grown by MBE technology. The fabricated devices were measured at dc- and high frequencies. First measurement results for the short circuit responsivity are discussed.

Index Terms — Indium phosphide, Rectifiers, Resonant tunneling devices, Triple barrier RTD

I. INTRODUCTION

RF-Signal detectors have been important devices since the early days of radio technology. These devices require strong asymmetry to allow efficient rectification. Resonant tunneling diodes (RTD) have demonstrated their excellent high frequency performance in THz signal generation [1]. The latest improvements are mainly driven by an increase of current density up to 1000 kA/cm² [2]. Based on our RTDs with current densities up to 500 kA/cm², the approach of resonant tunneling diodes with 3 Barriers is investigated [3, 4, 5]. By design of the epitaxial layer structure of the second well and the third barrier, an asymmetry in layer structure leads to an asymmetric current-voltage characteristic.

II. OPTIMIZATION OF THE DEVICES

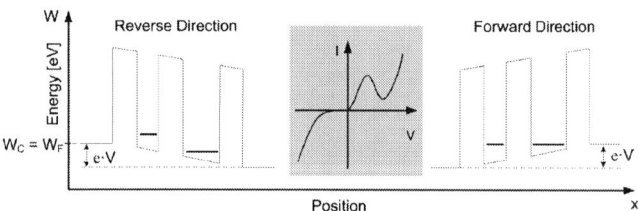

Fig. 1.   Reverse and forward current characteristic

The typical conduction band configuration for a triple barrier resonant tunneling diode (TBRTD) is presented in figure 1. Here the asymmetry is created by a thicker second quantum well, leading to a lower discrete energy level compared to the first one. In reverse direction this difference causes a blocking effect. For larger applied voltages the parasitic thermionic current dominates the characteristic. In forward direction the resonant tunneling through the device is

possible and the device behavior is analog to the well-known RTD.

To optimize the devices for operation as sensitive envelope detector, numerical device simulation based on Green's function have been performed. Critical parameters for the optimization are the width of the second well and the material and thickness of the third barrier. Here it is important to provide low energy levels in the well to allow low voltage operation of the device. The third barrier must be designed for an optimum between high current density and low thermionic current through the device.

III. LAYER STRUCTURE

Table 1 Layer sequence of the fabricated TBRTD. The material X is AlAs for sample A and InAlAs for sample B.

function	Material	thickness [nm]	doping [10^{18}cm-3]
Contact layer	In$_{0.7}$Ga$_{0.3}$As	8	37.4
Contact layer	InGaAs	100	37.4
Grading layer	InGaAs	50	37.4 to 1
Spacer	InGaAs	1.17	
Barrier	X	1.7	
Well, smoothing	InGaAs	1.17	
Well	InAs	2.42	
Well, smoothing	InGaAs	1.17	
Barrier	AlAs	1.7	n.i.d.
Well, smoothing	InGaAs	1.17	
Well	InAs	1.21	
Well, smoothing	InGaAs	1.17	
Barrier	AlAs	1.7	
Spacer	InGaAs	1.17	
Contact layer	InGaAs	10	1
Grading layer	InGaAs	50	1 to 37.4
Contact layer	InGaAs	300	37.4
Substrate	s.i. InP		

The investigated double quantum well devices are InAs/InGaAs/AlAs and/or InAlAs-heterostructure grown by molecular beam epitaxy on semi-insulating InP-substrate. Wide band gap materials (AlAs and/or InAlAs) are used to realize barriers. The InGaAs/InAs/InGaAs-quantum wells are sandwiched between 1.7 nm barrier layers [5, 6]. In contrast to the InP-based RTD with double AlAs-barriers presented in [5], a second quantum well with thicker InAs-layer were

978-1-4673-1725-2/12 $31.00 © 2012 IEEE

added, to ensure an alignment or misalignment of two quantum levels under forward or reverse voltage bias respectively (fig. 1). The designed layer stack is presented in Table 1. AlAs layer was used for the barriers of sample A. The lattice matched InGaAs layers sandwiched between well and barrier material was used to smooth the surface above the lattice mismatched layers. For sample B, the lattice matched InAlAs was used for the third barrier. To provide good ohmic contacts to the device and therefore low intrinsic series resistance, heavily Si-doped InGaAs contact layers with doping concentration up to 3.74E+19 cm^{-3} were used.

IV. TECHNOLOGY

Fig. 2.   REM picture of a realized 4*2.5*2.5 device

The processing of TBRTD was carried out using optical lithography, dry and wet etching, and Ti/Pt/Au lift-off technology. Ni is used to protect the metallization during dry etching. First the TBRTD-anodes are patterned with positive photoresist. The metallization was done under vacuum in an evaporation system. Ni was added on the top of this metallization as a mask for ICP-RIE dry etching. After the first etching process, the lower interconnection layer was metallized by a similar process. A wet chemistry etching of the lower InGaAs interconnection layer provides an electrical separation of the devices on the wafer. The air bridge contacts to the top and lower electrodes are connected to an on wafer GSG measurement pad configuration. Figure 2 presents a fabricated device with four active mesa areas, to provide low parasitic capacitances and low contact resistances through the lower contact layer.

V. DC MEASUREMENT RESULTS

The current-voltage characteristics of the fabricated devices were measured on-wafer with a semiconductor parameter analyzer. The results are presented in figure 3. For both samples they show characteristics with good agreement to the theory. In forward direction a fast increasing current can be observed, up to the peak current density, equivalent to single quantum well RTD devices. With the low quasi discrete levels in the quantum well, induced by the InAs material, they show low peak voltages. This is an advantage for rectification of small signals without the need of biasing, because small voltages in forward direction lead to a fast increase in current density. In backward direction good current blocking can be observed for small voltages. For voltages larger than 0.5 V for sample A and 0.4 V for sample B the increasing of the thermionic current can be observed. Even with three barriers the devices show high current densities comparable to RTDs. The values at peak voltage can be found in table 2.

Table 2 Peak currents of the devices (300K)

device	J_{peak} [kA/cm^2]	V_{peak} [V]	PVCR
Sample A	80	0.4	1.6
Sample B	230	0.46	3

The peak to valley current ratio (PVCR) in forward direction also shows competitive values for a three barrier device at room temperature, representing a good quality of the used epitaxial layers.

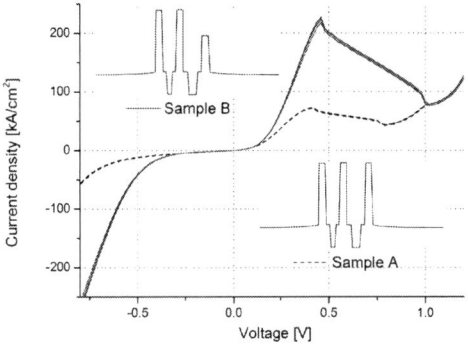

Fig. 3.   Current voltage characteristics of measured devices (300K)

The differences between the samples are caused by the material used for the third barrier. Here the current density can be improved by the use of InAlAs for sample B. This leads to higher tunnel probability through the last barrier. This improvement comes at the cost of faster increasing the thermionic current in reverse direction, but only for larger voltages. So for rectification of small voltages there is no significant influence. To investigate the rectification of the devices based on the current voltage characteristics, the rectification factor G can be calculated with the knowledge of the forward ($J_F(V_{pos})$) and backward ($J_B(V_{neg})$) current density.

$$G = \frac{J_F(Vpos)}{J_B(Vneg)} \ , \ |V_{pos}| = |V_{neg}| \qquad (1)$$

Results of this calculation can be found in figure 4. Sample A provide a maximum rectification ratio of 10. For the second sample this value is improved up to 20 at a slightly lower voltage. The main task to improve these values is the suppression of the thermionic current in backward direction. In contrast to devices based on Schottky junctions that are mainly defined by the parameters of the used materials [7] the triple barrier rtd approach provides several parameters for

978-1-4673-1725-2/12 $31.00 © 2012 IEEE

optimization. Additional to the degree of freedom, by the material used, the doping profile in the contact layers and the thickness of the barriers and wells can be adjusted.

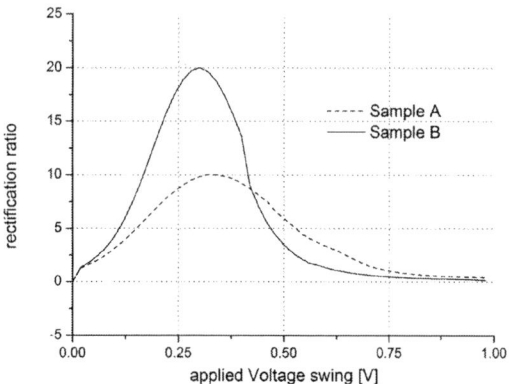

Fig. 4. Comparison of the rectification ratio of both samples at room temperature

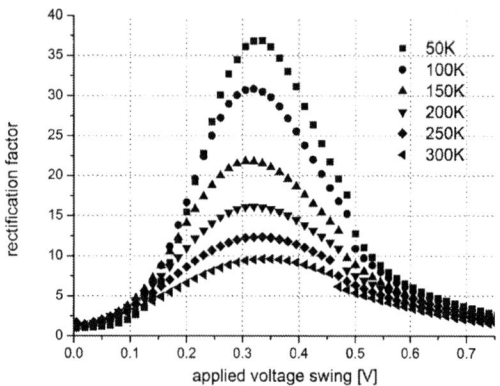

Fig. 4. Rectification ratio of sample A at low temperature

To investigate the expected improvements by suppressing the thermionic current component at low temperatures measurements down to 50 K are performed with sample A. As expected from theory, the current caused by tunneling processes is not significantly influenced by changes in temperature. This leads to improved rectification factors for lower temperature. The values increase up to 37 at 50 K for the investigated sample.

VI. RF MEASUREMENT RESULTS

To analyze the high frequency behavior of the samples, measurements are performed with a vector network analyzer for frequencies up to 50 GHz. All measured data were deembedded by using open and short structures on the investigated wafers. For the analysis of the devices the small signal equivalent circuit, presented in the inset of figure 5 is used. In a first step measurements at peak current densities are performed. These were used to calculate the series resistance

R_s of the device like presented in [8]. For sample A, this leads to a value of 5 Ohm for a device with an active area of 8 μm^2. With this knowledge the bias dependent elements R_{diff} and C_{RTD} can be calculated. Extracted values for an 8 μm device are presented in figure 5.

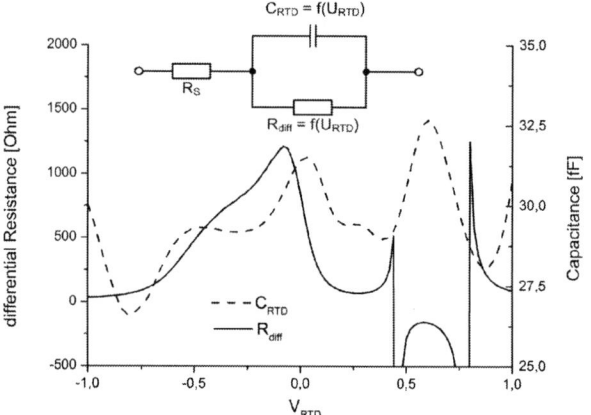

Fig. 5. Differential resistance and capacitance of a 8 μm^2 device on sample A

The capacitance of the device is one of the key elements for providing rectification for signals at high frequencies. To get an impression in comparison with other possible elements the capacitance for a defined current density is calculated. The capacitance per current density is 0.39 fF/kA for sample A and 0.14 fF/kA for sample B. The speed-index s is calculated by equation 2.

$$\Delta V * C = \Delta I * \Delta t \Rightarrow s = \frac{\Delta t}{\Delta V} = \frac{C}{\Delta I} \left[\frac{ps}{V}\right] \qquad (2)$$

For the investigated samples the following table shows the results with fast switching speeds, up to 0.14 ps/V for sample B.

Table 3 Speed index of the devices (300K)

Device	J_{peak} [kA/cm^2]	C_{peak} / kA [fF/kA]	S [ps/V]
Sample A	80	0.39	0.39
Sample B	230	0.14	0.14

To optimize the capacitance the undoped spacer layer and the design of the graded contact layer provide possibilities to improve the device performance by lowering the capacitance.

VII. FIRST RESPONSIVITY MEASUREMENTS

Short circuit responsivity is evaluated in first measurements, with a bias tee and an rf-signal source for frequencies up to 20 GHz. The dc-path is short circuited and the current was measured. The short circuit responsivity measured on sample B was nearly constant over frequency up to 20 GHz.

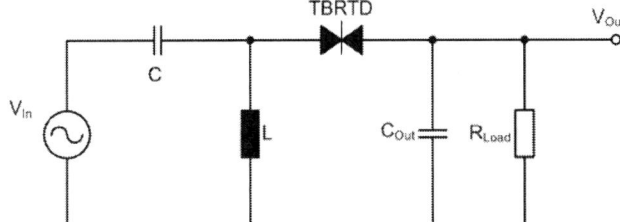

Fig. 4.   Simple detector circuit with a bias tee

Neglecting the losses by cables and coplanar probe, 2.5 A/W can be measured for a 6.25 µm device with -5 dBm input power. This value is factor 5 smaller than theoretical values presented for a 100 µm² Schottky diode [7]. The measured value could be improved by matching the device to the high frequency source, and by including a correction for the losses of the measurement setup.

VIII. CONCLUSION

In this work we present a triple barrier tunneling diode for zero bias rectification. The nonlinearity is based on an unsymmetrical design of the two quantum wells. While the second well is designed with a thicker InAs layer, the first discrete energy level has significant lower energy. This leads to a blocking behavior for negative voltages. The fabricated elements show high current densities up to 240 kA/cm² and low capacitances down to 0.14 fF/kA. With this the devices are good candidates for high frequency rectification, perhaps up to the THz frequency range.

REFERENCES

[1]   M. Feiginov, C. Sydlo, O. Cojocari, and P. Meissner, "Resonant-tunneling-diode oscillators operating at frequencies above 1.1 THz", Applied Physics Letters, vol 99, no. 23, December 2011.

[2]   A. Teranishi, K. Shizuno, S. Suzuki, M. Asada, H. Sugiyama, and H. Yokoyama, "Fundamental oscillation up to 1.08 THz in resonant tunneling diodes with high-indium-composition transit layers for reduction of transit delay", IEICE Electronics Express, Vol. 9, No. 5, pp. 385-390, 2012

[3]   A. Tchegho, B. Muenstermann, C. Gutsche, A. Polozek, K. Blekker, W. Prost, and F.J. Tegude, "Scalable High-Current density RTDs with low series resistance", IPRM 2010.

[4]   T. Nakagawa, H. Imamoto, T. Kojima, and K. Ohta, "Observation of resonant tunneling in AlGaAs/GaAs triple barrier diodes", Applied Physics Letters, vol 49, no. 73, 1986

[5]   S. Takahagi, H. Shin-ya, K. Asakawa, M. Saito, M. Suhara, "Equivalent Circuit Model of Triple-Barrier Resonant Tunneling Diodes Monolithically Integrated with Bow-Tie Antennas and Analysis of Rectification Properties towards Ultra Wideband Teraherz Detections", Japanese Journal of Apploed Physics, January 2011

[6]   A. Tchegho, B. Muenstermann, R. Geitmann, O. Benner, K. Blekker, W. Prost, and F.J. Tegude, "High performance submicron RTD design for mm-Wave Oscillator Applications", IPRM 2011.

[7]   A. C. Young, J. D. Zimmermann, E. R. Brown, and A. C. Gossard, "Semimetal-semiconductor rectifiers for sensitive room-temperature microwave detectors" Applied Physics Letters, vol. 87, no. , pp. 163506, 2005.

[8]   A. Matiss et al. "Large-signal analysis and AC modeling of sub-micron resonant tunneling diodes", Microwave Integrated Circuit Conference, 2007. EuMIC 2007. European, vol., no., pp. 207-210, 8-10 Oct. 2007

Photoluminescence peak wavelength behavior and luminescent efficiency of InAs/InGaAsP/InP quantum dots structure

Rie Sato, Ayako Fukuda, Tomomi Suzuki, and Hajime Imai

Faculty of Science, Japan Women's University
2-8-1 Mejirodai, Bunkyo-ku, Tokyo, 112-8681 JAPAN
m1236003sr@gr.jwu.ac.jp

Abstract — **We measured the photoluminescence (PL) spectra of InAs/InGaAsP/InP quantum dots structures. We examined the PL peak wavelength and the luminescent efficiency. The polarization of the excitation light was changed between the p-polarization and the s-polarization. We analyzed the incident angle dependence of the PL peak shift and that of the luminescent efficiency for the each polarization. From the results, we estimated that the temperature change due to the phonon emission affected the PLspectra for the p-polarization.**

Index Terms — **InGaAs/InP quantum well structure, InAs/InGaAsP/InP quantum dots structure, Photoluminescence**

I. INTRODUCTION

The quantum structures are effective to improve the characteristics of the optical semiconductor devices especially laser diodes. We studied the photoluminescence(PL) spectra of quantum dots structures. We have already reported that the phenomenon of the short wavelength shift of the PL spectra peak according to the increase in the excitation intensity was seemed to be explained by the band filling effect. We considered that this phenomenon was caused by the spread of the quantum level due to the variation in the quantum dot sizes. We here analyzed the incident angle dependence and temperature dependence on the PL spectra peak shift and the luminescent efficiency .

II. SAMPLES AND EXPERIMENTAL SET-UP

Fig.1. shows the experimental set-up. We used the YAG laser as the excitation light whose the wavelength and the intensity were 1064nm and 1W, respectively .The grating type spectrometer [Jobin Yvon, Triax 320], and the PbS photodiode were used to perform the PL spectra measurement. The polarization of the excitation light was set at p-polarization or s-polarization.The measurement was performed by changing the incident angle of the excitation light from 30 degrees to 70 degrees for each polarization . The temperature of the samples were adjusted at 40 ℃ by using the peltier cooler. The excitation light intensity was changed by using the ND filters :10% , 50% set between the excitation light source and the sample.

We measured the absorption coefficient of the samples for the YAG laser light to calculate how much the intensity was absorbed in the sample for each incident angle.

The samples are InAs/InGaAsP/InP quantum dots structures; the structure of quantum dots was oval about 30nm ×40nm×6nm. The substrate was InP and the spacer layer was InGaAsP. Samples were multi layered structures ; InAs quantum dot layers consisted of 7 layers. The packing density of quantum dots was about 40%.

Fig.2 shows the schematic structure of samples. We used different structures's samples as shown in Table 1. The $\Delta a/a$ represented the parameter of the lattice mismatch ; a is the lattice constant of the InP substrate and Δa is the difference in the lattice constant between the InP substrate and the InGaAsP.

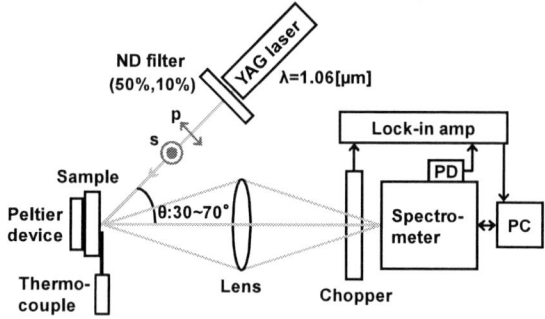

Fig.1. Experimental set-up

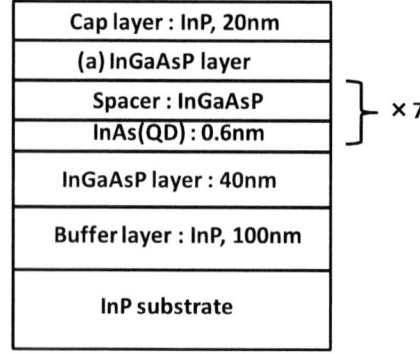

Fig.2. Schematic structure of samples

Table 1. Parameters of samples

	(a)	Spacer	Δa/a[%]
Sample#1	20nm	20nm	0
Sample#2			-0.4
Sample#3			-0.8
Sample#4	10nm	30nm	0
Sample#5			-0.4
Sample#6	20nm	40nm	0
Sample#7		20nm	-0.6

III. RESULT AND DISCUSSIONS

Fig.3 shows the PL peak spectra against the temperature . The temperature were at 20℃, 40℃ and 70℃. The excitation intensity and the incident angle were constant. The dashed lines showed in there figures the temperature dependence of the band gap energy of InAs. From the results , the PL spectra peak wavelength for both of the polarization shifted gradually to longer wavelength side with the same slope and the temperature dependence of Eg for InAs.

Fig4 shows the luminescent efficiency against the temperature change. The measurement conditions were the same on Fig.3. The luminescent efficiency decreased for both polarization when the temperature rose. We considered that the nonradiation recombination were increased with the increase with temperature.

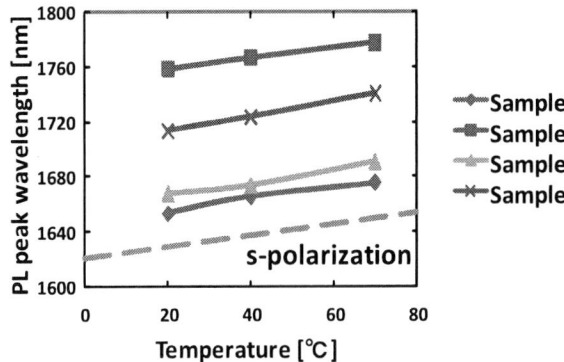

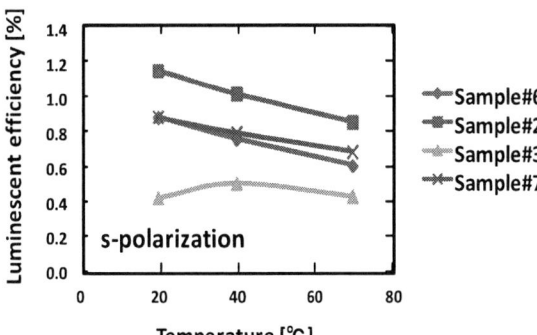

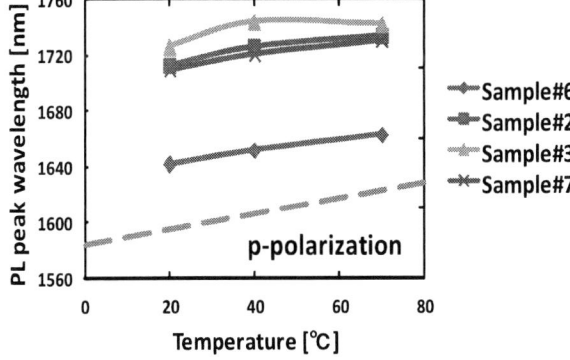

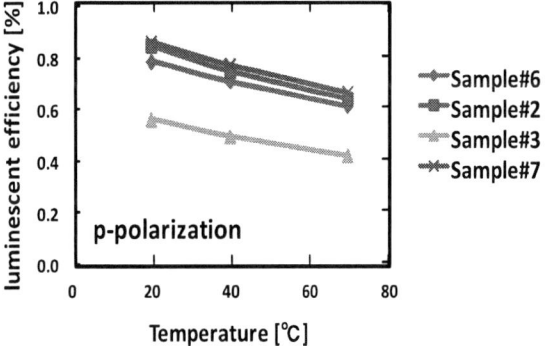

Fig.3. λ_{peak} versus temperature change

Fig.4. The luminescent efficiency versus temperature change

978-1-4673-1725-2/12 $31.00 © 2012 IEEE

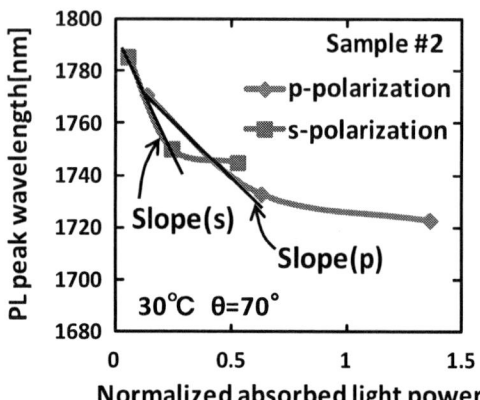

Fig.5. λ_{peak} versus normalized absorbed light power

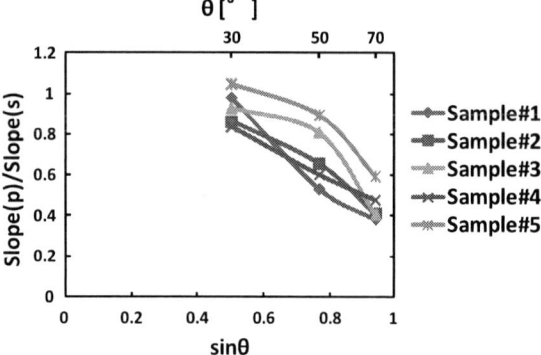

Fig.6. The ratio of slope versus incident angle θ

Fig.7 shows the luminescent efficiency against the incident angle for the polarization and the p-polarization. In the luminescent efficiency of the s-polarization , the dependence on was very slight . The s-polarization always had only the TE component. Therefore we considered that the polarization component of the excitation light absorbed in the semiconductor was always constant whenever the incident angle was changed. On the other hand the luminescent efficiency of the p-polarization decreased. For the p-polarization , when the incident angle increased , TM ($\sin\theta$) component increased. We considered that the phonon emission was increased with the increase in the TM component. Therefore we estimated that the luminescent efficiency was decreased with the increase with the incident angle.

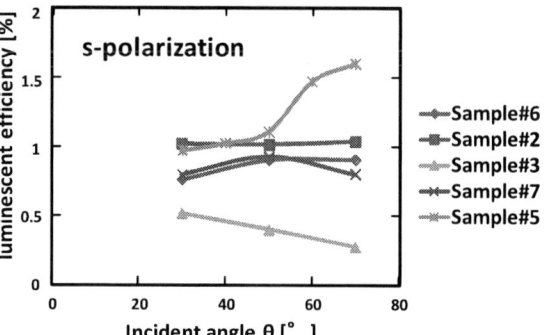

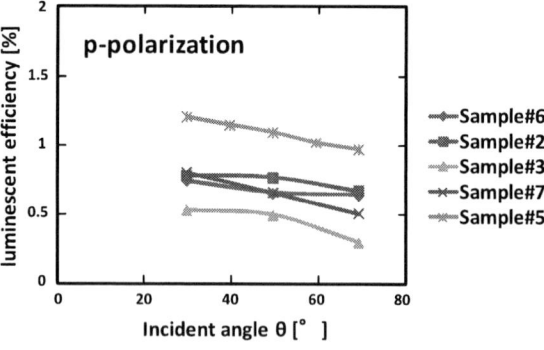

Fig.7. The luminescent efficiency versus the incident angle

Fig.5. shows the change in the PL peak wavelength against the absorption power of the excitation light for the p-polarization or s- polarization. The measurement was performed at 30 ℃ and the incident angle of 70 degree. The PL peak initially shifted to shorter as the increase of the excitation light power and then saturated.

The slope of the PL peak shift was calculated for each polarization and the relationship between the ratio of these slopes and $\sin\theta$ was shown in Fig.6 ; θ is the incident angle. The ratio of the slope decreases with the increase in $\sin\theta$. We assumed that the p-polarization light was divided into TE ($\cos\theta$) component and TM ($\sin\theta$) component at the incident angle θ. We estimated that TE component contributed to the photon emission and TM component contributed mainly to the phonon emission. When $\sin\theta$ (TM component) increased, the temperature rose due to the phonon emission. Therefore we considered that this temperature rise was competing with the phenomena like the band filling , and that the short wavelength shift for the p-polarization was less steeper than that of the s-polarization.

978-1-4673-1725-2/12 $31.00 © 2012 IEEE   42

IV. CONCLUSION

We measured the PL spectra of InAs/InGaAsP/InP quantum dots structures. We examined the temperature dependence and the incident angle dependence of the PL peak shift and the luminescent efficiency.

The PL peak wavelength shift to the long wavelength was observed for the temperature rise.

When the temperature rise , the luminescent efficiency decreased regardless of the polarization . We considered this reason that the contribution of the non-radiation process was increased by raising the temperature.

The PL peak shifted to shorter with the increase of the excitation light. The slope of the p-polarization was less steeper than that of the s-polarization. We considered that this reason was due to temperature rise by phonon emission caused to TM ($\sin\theta$) component of p-polarization and that competed with the phenomena like the bandfilling.

The luminescent efficiency of the p-polarization decreased as the incident angle of the excitation light increased. As TM ($\sin\theta$) component increased with the increase in the incident angle, we considered the luminescent efficiency was decreased by the temperature rise due to the phonon emission.

We estimated the temperature rise due to the phonon emission took the important effect to the PL spectra and luminescent efficiency for the p-polarization.

V. REFERENCES

[1] H. Imai and A. Motomura, presented at 2007 InternationalConference on Indium Phosphide and Related Materials, PA4, pp. 103-106(2007)

[2] H. Imai, M. Esaki, and Y. Saito, presented at 2009 InternationalConference on Indium Phosphide and Related Materials, WP3(2009)

[3] M. Esaki, N. Inaba, A. Fukuda and H. Imai, presented at 2010 InternationalConference on Indium Phosphide and Related Materials, WeP5(2010)

[4] M. Esaki, A. Fukuda and H. Imai, presented at 2010 in Japanese The Society of Applied Physics, 16p-NB-15(2010)

[5] A. Fukuda, M. Esaki, M. Akimoto and H. Imai, presented at 2011 InternationalConference on Indium Phosphide and Related Materials, (2011)

Lattice matched and Pseudomorphic InGaAs MOSHEMT with f_T of 200GHz

J.J. Mo, N. Wichmann, Y. Roelens, M. Zaknoune, L. Desplanque, X. Wallart, S. Bollaert

IEMN, UMR-CNRS 8520, UNIVERSITY OF LILLE 1, AVENUE POINCARE BP-60069, 59652 VILLENEUVE D'ASCQ, FRANCE

Abstract — We present lattice matched (LM) $In_{0.53}Ga_{0.47}As$ MOSHEMT and pseudomorphic (PM) $In_{0.75}Ga_{0.25}As$ MOSHEMT with gate-first process as well as their related post process annealing (PPA) effects in this paper. PM $In_{0.75}Ga_{0.25}As$ MOSHEMT with 7nm $In_{0.75}Ga_{0.25}As$ inserted channel promotes higher DC and RF performances due to higher electron mobility with higher indium content of the channel. MOSHEMT structure is preferred to MOSFET structure since the channel layer is moved away from the oxide/semiconductor interface by using of an $In_{0.52}Al_{0.48}As$ barrier layer between channel and Al_2O_3 oxide. We obtain a high cut-off frequency f_T of 200GHz for a 100nm-gate-length device of PM $In_{0.75}Ga_{0.25}As$ structure, which is 20GHz higher than the LM $In_{0.53}Ga_{0.47}As$ MOSHEMT. The PPA process at 400°C for 1 minute in forming gas N_2H_2 brings no difference to the DC and RF performances, showing that the detrimental effect of the interface defects is attenuated by using buried channel.

Index Terms — MOSFET III-V, ALD Al_2O_3, interface defect, post process annealing, pseudomorphic structure.

I. INTRODUCTION

MOSFETs with III-V materials have attracted more and more attention as Si-based CMOS technology reaches its scale and performance limits. Even though acceptable interface defects density $D_{it} \sim 5*10^{11}$ [1-4] have been achieved with different oxide deposition and surface treatment methods, MOSFETs III-V with surface channel remain as a challenge for even higher performances. Instead of working on the surface property improvements for surface channel structure, MOSHEMT structure is employed here with a barrier layer inserted between the channel InGaAs and the oxide Al_2O_3. As the channel is moved away from the interface, higher performances can be contemplated with less interface defects influence. In this work, we report on the fabrication of PM $In_{0.75}Ga_{0.25}As$ MOSHEMT with referenced LM $In_{0.53}Ga_{0.47}As$ MOSHEMT and $In_{0.53}Ga_{0.47}As$ MOSFET on parallel. Both two types of MOSHEMTs were capped by $In_{0.53}Ga_{0.47}As/In_{0.52}Al_{0.48}As$ barrier with larger band gap ($In_{0.53}Ga_{0.47}As$ was used on top of $In_{0.52}Al_{0.48}As$ to avoid $In_{0.52}Al_{0.48}As$ oxidation). PPA effects have also been studied on InGaAs MOSHEMT, and it shows no notable improvements considering the ease of interface defects influence thanks to the buried channel.

II. FABRICATION

The PM $In_{0.52}Al_{0.48}As/In_{0.75}Ga_{0.25}As$ MOSHEMT and referenced LM $In_{0.52}Al_{0.48}As/In_{0.53}Ga_{0.47}As$ and $In_{0.53}Ga_{0.47}As$ MOSFET were fabricated with the same process with respective expitaxial structures shown in Fig.1. They started with an 8nm Al_2O_3 deposition, realized by atomic layer deposition (ALD) at 300°C after NH_4OH surface treatment during 1 minute, and followed by post deposition annealing (PDA) at 500°C for 1 minute in vacuum. Transistor gate was defined at first using refractory metal Tantalum by reactive ion etching (RIE) patterning, serving as an implantation mask. Then self-aligned source/drain (S/D) implantation was realized with Si dose of $10^{14}/cm^2$ at 20keV, followed by an activation process as rapid thermal annealing (RTA) at 600°C for 20 seconds. S/D metallization was performed by lift-off process using stack metal contact Ti/Pt/Au, followed by a thermal annealing at 400°C to achieve a good ohmic contact. From these implantation and annealing conditions, low contact resistance R_c=0.08Ωmm and low sheet resistance R_{sh}=137Ω/sh were achieved by TLM (Transmission Line Model) measurements. Mesa was etched using H_3PO_4 acid. Eventually, a coplanar air-bridge structure was used to make the multi-finger structure to reduce the total gate resistance and limits the parasitic capacitance as well for favorable RF performances as

$$R_{g,Total} = \frac{R_g \cdot W}{3n^2} \qquad (1)$$

With R_g is the single gate resistance, W is the gate width and n is the gate finger number. Increasing the gate finger number n is more effective to reduce the total gate resistance as compared with enlarging the gate width W. PPA process was carried out at 400°C for 1minute in forming gas N_2H_2 on one sample to be compared with the as-fabricated one to investigate the annealing effect.

III. CHARACTERIZATION AND DISCUSSION

The DC and RF characterizations of InGaAs MOSHEMT and MOSFET were carried out at room temperature. Fig.2 and 3 show the output and transfer characteristics of gate-first

978-1-4673-1725-2/12 $31.00 © 2012 IEEE

processed PM $In_{0.75}Ga_{0.25}As$ MOSHEMT with a gate length L_g of 100nm. An I_{dmax}= 325mA/mm and g_m=210mS/mm at V_{ds}=1V were obtained with a related subthreshold swing SS=210mV/dec at V_{ds}=0.2V. Compared with the LM $In_{0.53}Ga_{0.47}As$ MOSHEMT in Fig.4, it can be observed that PM MOSHEMT has a two times higher output current I_d and a better charge command g_m. This can be explained that higher indium content channel promotes higher charge mobility and inversion efficiency [5,6]. Furthermore, it could be also due to the larger band offset between channel and barrier (ΔEc=0.67eV for $In_{0.75}Ga_{0.25}As$ versus ΔEc=0.47eV for $In_{0.53}Ga_{0.47}As$) with induced better charge confinement and less charge splitting into the barrier. Meanwhile, the obtained DC performances of MOSHEMT are much higher than the parallel $In_{0.53}Ga_{0.47}As$ surface channel MOSFET (Fig.5), where the dielectric oxide and interface quality play an important role. While For MOSHEMT with buried $In_{0.53}Ga_{0.47}As$ or $In_{0.75}Ga_{0.25}As$ channel, the channel was sufficiently kept away from the problematic interface to reduce the interaction between electrons in the channel and defects at the interface, leading to a higher mobility. We also observed that there is a V_{TH} shift towards smaller voltage for MOSHEMT structure compared to MOSFET, which could be due to different interface dipoles arrangements. For MOSFET structure, the oxide was realized by ex-situ ALD, leading to relatively poor channel/oxide interface with randomly fluctuated dipoles, while for MOSHEMT structure, since the barrier layer was deposited directly by in-situ MBE, the barrier/channel interface has therefore a good quality with ideally ordered dipoles, bringing a negative V_{TH} shift [7].

A (45MHz-67GHz) Agilent Network Analyzer has been used to perform on-wafer RF measurements of the InGaAs MOSHEMT and MOSFET. Fig.6 shows the short circuit current gain $|H_{21}|^2$ and unilateral mason's gain U of the PM and LM InGaAs MOSHEMT and InGaAs MOSFET with gate length of 100nm. It is to be noted that an 'open' and 'short' like de-embedding has been used to obtain S-parameters of the devices [8]. By extrapolation of -20dB/dec, we obtained a cut-off frequency f_T of 200GHz and a maximum oscillation frequency f_{MAX} of 50GHz for PM MOSHEMT, which are both higher than LM MOSHEMT and MOSFET. The f_T we obtained is among the highest values to author's knowledge for $In_xGa_{1-x}As$ MOSFET/MOSHEMT at inversion mode. The relatively poor f_{MAX} is partly due to the high gate resistance R_g coming from the high resistivity of Tantalum. Compared to $In_{0.53}Ga_{0.47}As$ surface channel MOSFET we fabricated in parallel, the f_T we obtained here for MOSHEMT is 3 times higher than that of MOSFET (Fig.6). This can be attributed to the higher channel mobility and lower interface scattering events thanks to the buried channel configuration.

The PPA effects have also been studied by comparing one sample annealed at 400°C for 1minute in forming gas N_2H_2 with an as-fabricated one. As can be observed from Fig.7 and 8, there is a little threshold voltage V_{TH} shift towards higher voltage after PPA. We can deduce that PPA can improve the oxide quality by reducing the fixed charges situated in the oxide and the interface defects at the semiconductor/oxide interface. The PPA effect is also confirmed by $In_{0.53}Ga_{0.47}As$ MOS capacitor measurements (Fig.9 and 10), where before PPA, there is large frequency dispersion at accumulation region and prominent stretch-out phenomenon. While after PPA, the stretch-out phenomenon is reduced with lower frequency dispersion at accumulation region, indicating that the interface defect density has been reduced while there are still some fixed charges in the oxide, with their frequency response depending on their distance from the interface. Furthermore, there is a V_{TH} shift towards higher voltage for MOS capacitors, which confirms us that the interface defects are dominated by donor-type ones.

IV. SUMMARY AND CONCLUSIONS

In this paper, we realized PM $In_{0.75}Ga_{0.25}As$ MOSHEMT, LM $In_{0.53}Ga_{0.47}As$ MOSHEMT and $In_{0.53}Ga_{0.47}As$ MOSFET on parallel with gate-first process. PM $In_{0.75}Ga_{0.25}As$ MOSHEMT shows higher DC and RF performances with I_{dmax}=325mA/mm, g_{mmax}=210mS/mm at V_{ds}=1V and f_T=200GHz for 100nm-gate-length. The PPA effects have also been studied showing no prominent DC, RF performance improvements owing to less interface interference by using buried channel.

Pseudomorphic $In_{0.75}Ga_{0.25}As$ MOSHEMT	Lattice matched $In_{0.53}Ga_{0.45}As$ MOSHEMT	$In_{0.53}Ga_{0.45}As$ MOSFET
Al_2O_3 4nm	Al_2O_3 4nm	Al_2O_3 4nm
$In_{0.53}Ga_{0.47}As$ NID 4nm	$In_{0.53}Ga_{0.47}As$ NID 4nm	$In_{0.53}Ga_{0.47}As$ Na=10^{17}/cm^3 300nm
$In_{0.52}Al_{0.48}As$ NID 4nm	$In_{0.52}Al_{0.48}As$ NID 4nm	$In_{0.53}Ga_{0.47}As$ Na=10^{19}/cm^3 500nm
$In_{0.75}Ga_{0.25}As$ NID 7nm	$In_{0.53}Ga_{0.47}As$ Na=10^{17}/cm^3 300nm	InP Substrate
$In_{0.53}Ga_{0.47}As$ Na=10^{17}/cm^3 300nm	$In_{0.53}Ga_{0.47}As$ Na=10^{19}/cm^3 500nm	
$In_{0.53}Ga_{0.47}As$ Na=10^{19}/cm^3 500nm	InP Substrate	
InP Substrate		

Fig.1. Epitaxial structures of pseudomorphic $In_{0.75}Ga_{0.25}As$ MOSHEMT, lattice matched $In_{0.53}Ga_{0.47}As$ MOSHEMT and $In_{0.53}Ga_{0.47}As$ MOSFET.

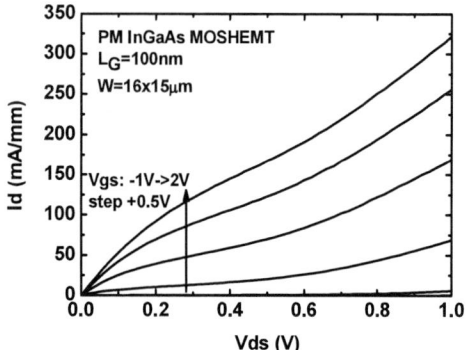

Fig.2. Output characteristics of pseudomorphic $In_{0.75}Ga_{0.25}As$ MOSHEMT with L_g=100nm.

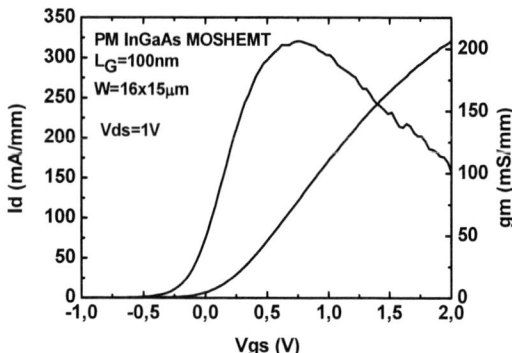

Fig.3. Transfer characteristics of pseudomorphic $In_{0.75}Ga_{0.25}As$ MOSHEMT with L_g=100nm.

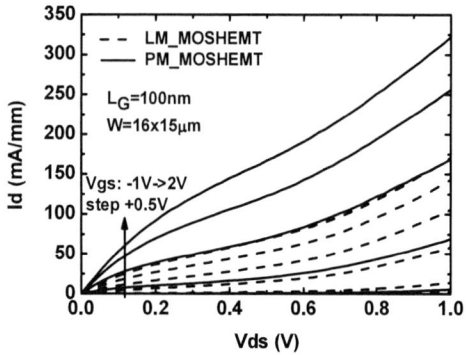

Fig.4. Comparison of output characteristics of pseudomorphic $In_{0.75}Ga_{0.25}As$ MOSHEMT and lattice matched $In_{0.53}Ga_{0.47}As$ MOSHEMT with L_g=100nm.

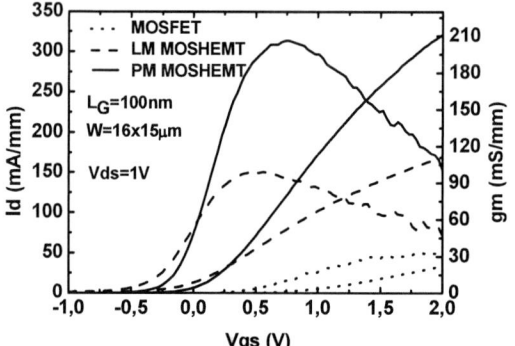

Fig.5. Comparison of transfer characteristics of pseudomorphic $In_{0.75}Ga_{0.25}As$ MOSHEMT, lattice matched $In_{0.53}Ga_{0.47}As$ MOSHEMT and $In_{0.53}Ga_{0.47}As$ MOSFET with L_g=100nm.

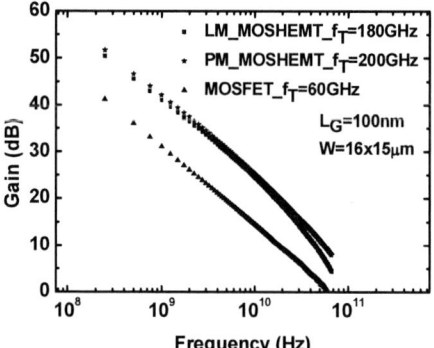

Fig.6. Comparison of $|H_{21}|^2$ and related f_T characteristics between pseudomorphic $In_{0.75}Ga_{0.25}As$ MOSHEMT, lattice matched $In_{0.53}Ga_{0.47}As$ MOSHEMT and $In_{0.53}Ga_{0.47}As$ MOSFET with L_g=100nm.

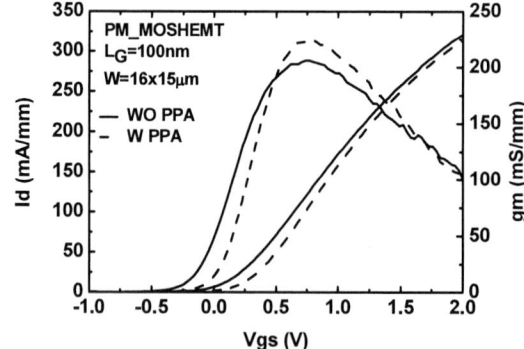

Fig.7. Output characteristics of pseudomorphic $In_{0.75}Ga_{0.25}As$ MOSHEMT with L_g=100nm before (solid curve) and after PPA (dashed curve).

978-1-4673-1725-2/12 $31.00 © 2012 IEEE

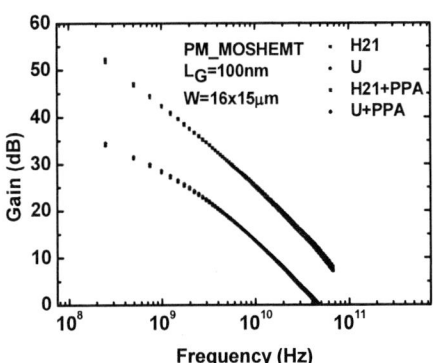

Fig.8. Transfer characteristics of pseudomorphic $In_{0.75}Ga_{0.25}As$ MOSHEMT with L_g=100nm before and after PPA.

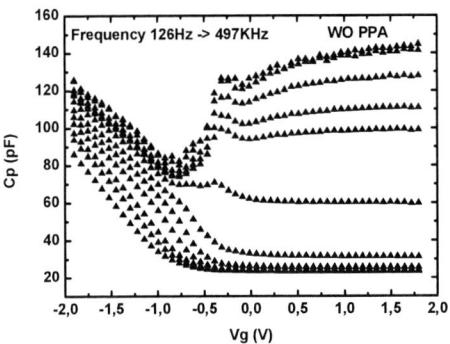

Fig.9. *C-V* evolutions with varied frequencies from 126Hz to 497kHz of InGaAs MOS capacitor before PPA.

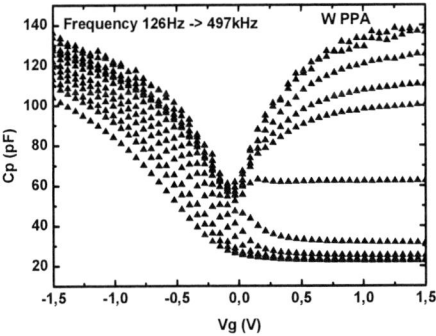

Fig.10. *C-V* evolutions with varied frequencies from 126Hz to 497kHz of InGaAs MOS capacitor after PPA.

V. ACKNOWLEDGEMENT

This work was financially supported by ANR (Agence Nationale de la Recherche) of France as a MOS35 project. N°contract: ANR-08-NANO-022. This work was realized with the cooperation of CEA-LETI in France for the Al_2O_3 deposition.

REFERENCES

[1] Y.-T. Chen, Y. Wang, F. Xue, F. Zhou, et J. C. Lee, *IEEE Electron Device Lett.*, vol. 32, p. 1531–1533, (2011).

[2] Hock-Chun Chin, Xinke Liu, Xiao Gong, et Yee-Chia Yeo, *IEEE Trans. Electron Devices*, vol. 57, p. 973–979, (2010).

[3] J. Huang, N. Goel, H. Zhao, C. Y. Kang, K. S. Min, G. Bersuker, S. Oktyabrsky, C. K. Gaspe, M. B. Santos, P. Majhi, P. D. Kirsch, H.-H. Tseng, J. C. Lee, et R. Jammy, *IEEE International Electron Devices Meeting (IEDM)*, p. 1–4, (2009).

[4] H. C. Chiu, L. T. Tung, Y. H. Chang, Y. J. Lee, C. C. Chang, J. Kwo, et M. Hong, *Appl. Phys. Lett.*, vol. 93, p. 202903, (2008).

[5] Y.-C. Wu, E. Y. Chang, Y.-C. Lin, C.-C. Kei, M. K. Hudait, M. Radosavljevic, Y.-Y. Wong, C.-T. Chang, J.-C. Huang, et S.-H. Tang, *Solid-State Electron.*, vol. 54, p. 37–41, (2010).

[6] H. Zhao, Y.-T. Chen, J. H. Yum, Y. Wang, N. Goel, et J. C. Lee, *Appl. Phys. Lett.*, vol. 94, p. 193502–193502–3, (2009).

[7] Y. Urabe, N. Miyata, H. Ishii, T. Itatani, T. Maeda, T. Yasuda, H. Yamada, N. Fukuhara, M. Hata, M. Yokoyama, N. Taoka, M. Takenaka, et S. Takagi, *IEEE International Electron Devices Meeting (IEDM)*, p. 6.5.1-6.5.4, (2010).

[8] P. J. van Wijnen, H. R. Claessen, and E. A. Wolsheimer, A new straightforward calibration and correction procedure for "on wafer" high frequency S -parameter measurements (45 MHz-18 GHz), 1987.

Monte Carlo Simulation of InGaAs/Strained-InAs/InGaAs Channel HEMTs Considering Self-Consistent Analysis of 2-Dimensional Electron Gas

Akira Endoh[1,2], Issei Watanabe[1], and Takashi Mimura[1,2]

[1]*National Institute of Information and Communications Technology (NICT),*
4-2-1 Nukui-kitamachi, Koganei, Tokyo 184-8795, Japan
[2]*Fujitsu Laboratories Ltd., 10-1 Morinosato-Wakamiya, Atsugi, Kanagawa 243-0197, Japan*

Abstract — We calculated the unstrained and the strained band structures of InAs and carried out Monte Carlo simulation of InGaAs/strained-InAs/InGaAs composite channel high electron mobility transistors (HEMTs) considering 2-dimensional electron gas (2DEG) self-consistent analysis by solving Schrödinger and Poisson equations. With considering the effect of 2DEG, the drain-source current I_{ds} decreases. However, the negative threshold voltage shift due to the short-channel effects is not affected by considering 2DEG. The threshold voltage shift occurs in the region $L_g/d <$ ~5 (L_g: gate length, d: sum of the barrier and channel layer thicknesses). At $L_g = 20$ nm, the simulated cutoff frequency f_T values were 943 GHz without 2DEG and 813 GHz with 2DEG. The trend of the f_T values with L_g reflects that of the electron velocities mainly.

Index Terms — Band structure, cutoff frequency, f_T, HEMTs, Monte Carlo simulation, strained InAs, threshold voltage shift, 2-dimensional electron gas.

I. INTRODUCTION

InP-based $In_{0.52}Al_{0.48}As/In_xGa_{1-x}As$ ($x \geq 0.53$) high electron mobility transistors (HEMTs) [1] are one of the most promising devices for future ultrahigh-speed applications. To achieve higher-speed operations, an InGaAs/strained-InAs/InGaAs composite layer is used as a channel [2]. In our previous work [3], we reported Monte Carlo (MC) simulation results of InGaAs/InAs/InGaAs composite channel HEMTs without considering strain and quantum confinement effects in InAs layer.

In this work, we calculated the band structures of the unstrained and the strained InAs and carried out MC simulation of InGaAs/strained-InAs/InGaAs composite channel HEMTs considering 2-dimensional electron gas (2DEG) self-consistent analysis by solving Schrödinger and Poisson equations. We examined the effect of the gate length L_g on DC and RF performances in the HEMTs with and without considering the effect of 2DEG.

II. BAND STRUCTURES OF UNSTRAINED AND STRAINED InAs

The unstrained and the strained band structures of InAs were calculated by using all-electron full-potential linearized augmented-plane-wave (FLAPW) method in the local density

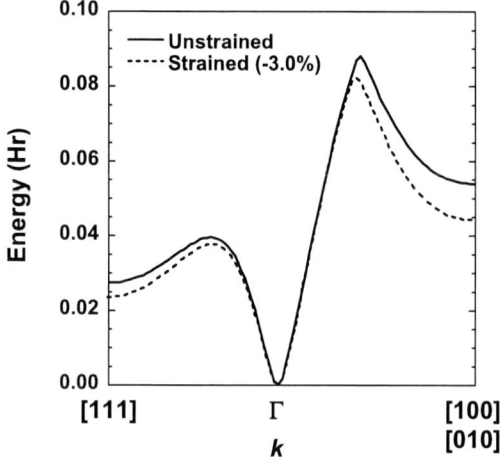

Fig. 1. Conduction band structures of unstrained and strained InAs.

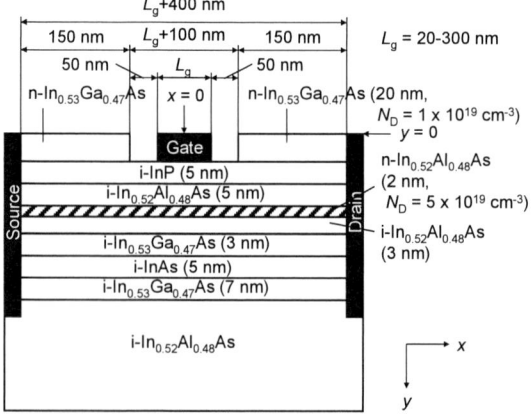

Fig. 2. Schematic cross-sectional model structure of InGaAs/strained-InAs/InGaAs composite channel HEMT.

978-1-4673-1725-2/12 $31.00 © 2012 IEEE

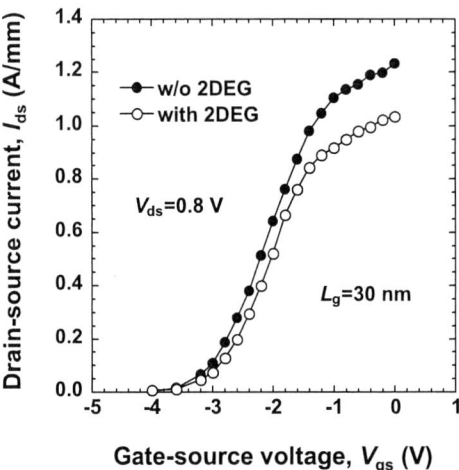

Fig. 3. Drain-source current vs gate-source voltage (I_{ds}-V_{gs}) characteristics of 30-nm-gate HEMTs with and without 2DEG. The drain-source voltage V_{ds} is 0.8 V.

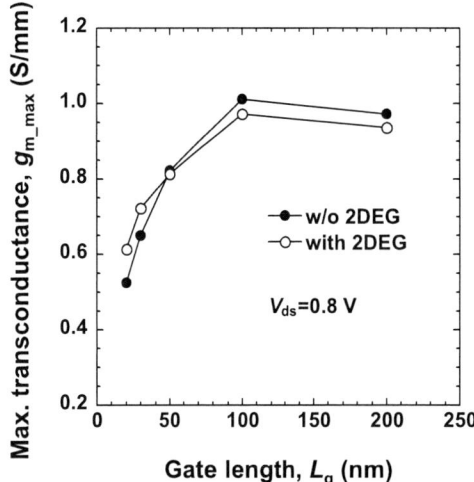

Fig. 4. Gate length L_g dependence of the maximum transconductance g_{m_max} under a drain-source voltage V_{ds} of 0.8 V.

approximation (LDA). The computational code used was originally developed by Kodama, Hamada, and Yanase [4]. Figure 1 shows the conduction band structures of the unstrained and the strained InAs. By applying -3% (compressive) strain, the increase of the electron effective mass in the Γ valley and the decrease of the Γ-L and the Γ-X valley energy separations were observed. We obtained the electron effective mass in the Γ-valley along the [100] ([010]) direction to be $0.032m_0$ for the unstrained InAs and $0.038m_0$ for the -3% (compressive) strained InAs with considering nonparabolicity, where m_0 is electron rest mass. The electron effective mass in the Γ valley for the unstrained InAs is almost same as that calculated by the empirical pseudopotential method ($0.031m_0$) [5, 6]. On the other hand, the electron effective mass for the -3% (compressive) strained InAs is lighter than that by the pseudopotential method ($0.043m_0$) [6].

III. MONTE CARLO SIMULATION

MC simulations were carried out at a lattice temperature of 300 K by using the program, "COSMOS," developed by Mizuho Information & Research Institute, Inc. [7] Figure 2 shows a schematic cross-sectional model structure of the composite channel HEMT used in the present MC simulations. We used a three-valley model (Γ, L, X) with nonparabolicity for the conduction band structures of InAs, $In_{0.53}Ga_{0.47}As$, $In_{0.52}Al_{0.48}As$, and InP. $In_{0.53}Ga_{0.47}As$ and $In_{0.52}Al_{0.48}As$ layers are lattice-matched to InP layer. InAs layer is compressively strained. We used the unstrained and the strained band structures of InAs obtained in Sec. II. The parameters of AlAs, GaAs, and InP were taken from the literature [5]. We considered the 2DEG in the channel layer by solving the 1-

dimensional Schrödinger equation at various x-positions. The electron scattering mechanisms [8, 9] considered were polar optical phonon scattering, non-polar optical phonon scattering, acoustic phonon scattering, inter-valley phonon scattering, and ionized impurity scattering. We did not consider impact ionization in the present simulations since the previous MC simulation results [10] show that the influence of impact ionization is relatively small for InGaAs/strained-InAs/InGaAs channel HEMT in the drain-source voltage V_{ds} range between 0 and 0.8 V. Sub-band scattering was considered for 2DEG. Dirichlet boundary conditions were applied to all metal-semiconductor interfaces, and Neumann boundary conditions (the zero normal derivative of the potential) were applied to other surfaces. The potential was calculated by the finite difference method. The time step was set to 0.5 fs. The gate length L_g was varied from 20 to 300 nm.

IV. RESULTS AND DISCUSSION

A. DC Performances

Figure 3 shows the drain-source current vs gate-source voltage (I_{ds}-V_{gs}) characteristics of a 30-nm-gate HEMT with and without considering the effect of 2DEG under a drain-source voltage V_{ds} of 0.8 V. The I_{ds} with 2DEG is lower than that without 2DEG. With considering the effect of 2DEG, the reductions of electron velocity and electron density in the channel occur due to the quantum levels in the channel. Figure 4 shows the gate length L_g dependence of the maximum transconductance g_{m_max} under a V_{ds} of 0.8 V. The g_{m_max} increases with decreasing L_g in the region $L_g > \sim 100$ nm, and decreases in the region $L_g < \sim 100$ nm. The decrease of g_{m_max}

978-1-4673-1725-2/12 $31.00 © 2012 IEEE

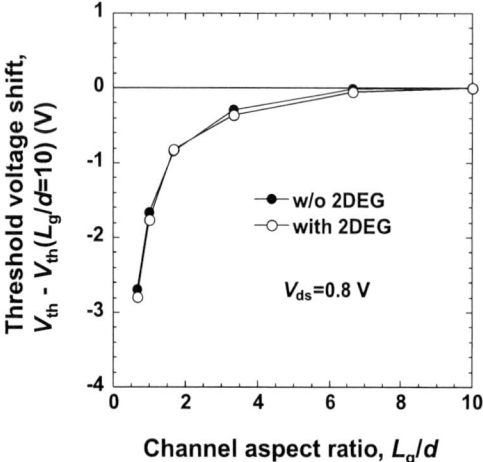

Fig. 5. Channel aspect ratio L_g/d dependence of the threshold voltage shift under a drain-source voltage V_{ds} of 0.8 V.

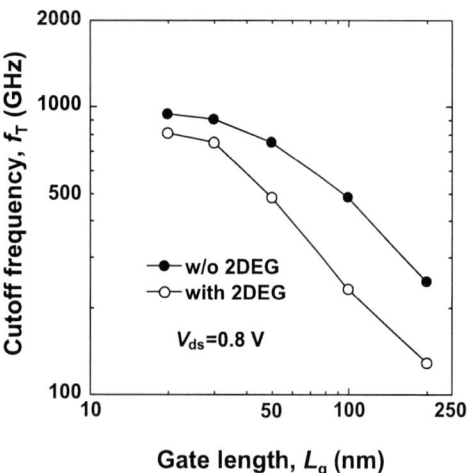

Fig. 6. Gate length L_g dependence of the cutoff frequency f_T under a drain-source voltage V_{ds} of 0.8 V.

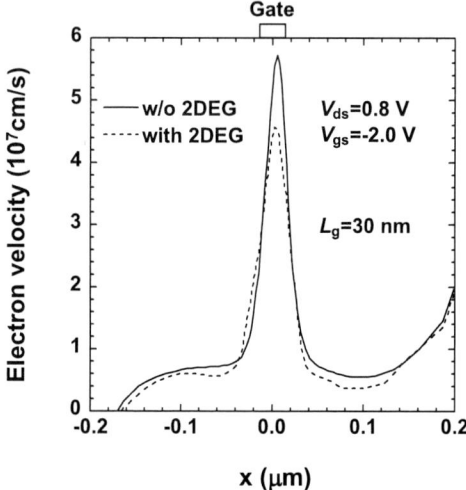

Fig. 7. Electron velocity profiles in the channel layer for 30-nm-gate HEMTs with and without 2DEG. The drain-source voltage V_{ds} is 0.8 V, and the gate-source voltage V_{gs} is -2.0 V.

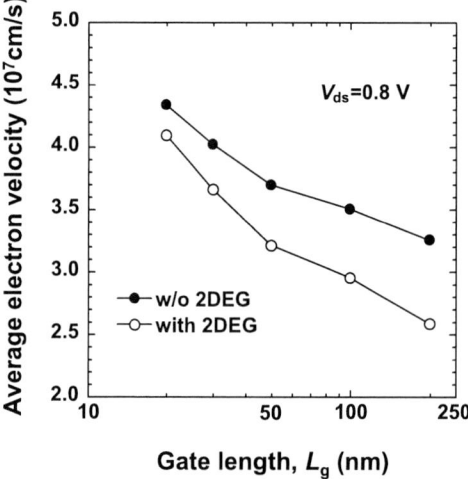

Fig. 8. Gate length L_g dependence of the average electron velocity in the channel layer under the gate electrode. The drain-source voltage V_{ds} is 0.8 V.

in the region $L_g < \sim 100$ nm results from the short-channel effects [11, 12].

To clarify the short-channel effects on performances further, we obtained the threshold voltage shift in the I_{ds}-V_{gs} curves. Figure 5 shows the channel aspect ratio L_g/d dependence of the threshold voltage shift. In this work, d was defined as the sum of the barrier and channel layer thicknesses. The threshold voltage shift was defined as the difference from the threshold voltage V_{th} at $L_g/d = 10$. The negative threshold voltage shift due to the short-channel effects occurs in the region $L_g/d < \sim 5$. This phenomenon is similar to those for AlGaAs/GaAs [11] and InAlAs/InGaAs [12] HEMTs. The

difference between with and without 2DEG is very small. Thus, the same scaling rule is valid for the InGaAs/strained-InAs/InGaAs composite channel HEMTs.

B. Cutoff frequency

The cutoff frequency f_T values were calculated by the method presented by Kwon and Pavlidis [13]. Figure 6 shows the gate length L_g dependence of the cutoff frequency f_T under a V_{ds} of 0.8 V. The f_T increases with decreasing L_g in the present simulation range. The f_T's without 2DEG are higher than those with 2DEG. At an L_g of 20 nm, f_T without 2DEG reaches 943 GHz. On the other hand, f_T with 2DEG reaches 813 GHz. Thus, very high f_T might be obtained by reducing L_g to around 20 nm.

To understand the trend of the cutoff frequency f_T further, we obtained the electron velocity in the channel layer. Figure 7 shows the electron velocity profiles in the channel layer of the 30-nm-gate HEMTs with and without considering the effect of 2DEG under a V_{ds} of 0.8 V and a V_{gs} of -2.0 V. There is a velocity overshoot under the gate. In the almost whole region, the electron velocity without 2DEG is higher than that with 2DEG. The peak electron velocities are about 5.7×10^7 cm/s without 2DEG and about 4.6×10^7 cm/s with 2DEG. Figure 8 shows the gate length L_g dependence of the average electron velocity in the channel layer under the gate electrode. The V_{ds} is 0.8 V, and the V_{gs} values are the voltages with the g_{m_max} values. The average electron velocity increases with decreasing L_g in the simulation range. The increase of average electron velocity results from the increase of the electric field under the gate electrode. On the other hand, the average velocity without 2DEG is higher than that with 2DEG. The trend of the f_T values with L_g reflects that of the electron velocities mainly.

V. SUMMARY

In summary, we calculated the unstrained and the strained band structures of InAs and carried out MC simulation of InGaAs/strained-InAs/InGaAs composite channel HEMTs considering the 2DEG self-consistent analysis by solving Schrödinger and Poisson equations to examine the effect of L_g on DC and RF performances. The I_{ds} with 2DEG is lower than that without 2DEG due to the reductions of electron velocity and electron density. However, the negative threshold voltage shift due to the short-channel effects is not affected by considering 2DEG. The trend of the threshold voltage shift is similar to those of AlGaAs/GaAs and InAlAs/InGaAs HEMTs, i.e. the threshold voltage shift occurs in the channel aspect region $L_g/d <$ ~5. We also obtained the cutoff frequency f_T values. The f_T values at $L_g = 20$ nm are 943 GHz without 2DEG and 813 GHz with 2DEG. The trend of the f_T values with L_g reflects that of the electron velocities mainly.

ACKNOWLEDGMENT

This work was supported in part by "The research and development project for the expansion of radio spectrum resources" of the Ministry of Internal Affairs and Communications, Japan (Multi-tens gigabit wireless communication technology at subterahertz frequencies).

REFERENCES

[1] D.-H. Kim, B. Brar, and J. A. del Alamo, "f_T = 688 GHz and f_{max} = 800 GHz in L_g = 40 nm $In_{0.7}Ga_{0.3}$As MHEMTs with g_{m_max} > 2.7 mS/μm," *IEDM Tech. Dig.*, no. 13.6, pp. 319-322, December 2011.

[2] D.-H. Kim, and J. A. del Alamo, "30-nm InAs PHEMTs With f_T = 644 GHz and f_{max} = 681 GHz," *IEEE Electron Device Lett.*, vol. 31, no. 8, pp. 806-808, August 2010.

[3] A. Endoh, I. Watanabe, N. Hirose, T. Mimura, and T. Matsui, "Monte Carlo Simulations of Electron Transport in Nanogate $In_{0.7}Ga_{0.3}$As/InAs/$In_{0.7}Ga_{0.3}$As Composite Channel HEMTs," *Abstracts of 37th Int. Symp. Comp. Semicon.*, no. MoP28, p. 30, May 2010.

[4] M. Umekawa, N. Hamada, A. Kodama, and Y. Moritomo, "Electronic Structure of $RbMnFe(CN)_6$: Ground State," *J. Phys. Soc. Jpn.*, vol. 73, no. 2, pp. 430-433, February 2004. The code MIZUHO/ABCAP is supplied from Mizuho Information & Research Institute, Inc.

[5] M. V. Fischetti, "Monte Carlo Simulation of Transport in Technologically Significant Semiconductors of the Diamond and Zinc-Blende Structures — Part I: Homogeneous Transport," *IEEE Trans. Electron Devices*, vol. 38, no. 3, pp. 634-649, March 1991.

[6] H. Nishino, I. Kawahira, F. Machida, S. Hara, and H. I. Fujishiro, "Monte Carlo Study of Strain Effect on High Field Electron Transport in InAs and InSb," *Proceedings of 22nd Int. Conf. Indium Phosphide Related Mat.*, no. WeB2-6, pp. 156-159, June 2010.

[7] http://www.mizuho-ir.co.jp/solution/research/semiconductor/ devicemeister/montecarlo/index.html

[8] K. Tomizawa, *Semiconductor Device Simulation — Visualization of the Operation of Submicron Semiconductor Devices by Simulation —*, Tokyo: Corona, 1996.

[9] C. Jacoboni, and P. Lugli, *The Monte Carlo Method for Semiconductor Device Simulation*, Heidelberg: Springer, 1989.

[10] F. Machida, H. Nishino, J. Sato, H. Watanabe, S. Hara, and H. I. Fujishiro, "Strain Effects on Performances in InAs HEMTs," *Proceedings of 23rd Int. Conf. Indium Phosphide Related Mat.*, no. Th-8.2.2, pp. 437-440, May 2011.

[11] Y. Awano, M. Kosugi, K. Kosemura, T. Mimura, and M. Abe, "Short-Channel Effects in Subquarter-Mocrometer-Gate HEMT's: Simulation and Experiment," *IEEE Trans. Electron Devices*, vol. 36, no. 10, pp. 2260-2266, October 1989.

[12] A. Endoh, Y. Yamashita, K. Shinohara, K. Higashiwaki, K. Hikosaka, T. Mimura, S. Hiyamizu, and T. Matsui, "Fabrication Technology and Device Performance of Sub-50-nm-Gate InP-Based High Electron Mobility Transistors," *Jpn. J. Appl. Phys.*, vol. 41, part 1, no. 2B, pp. 1094-1098, February 2002.

[13] Y. Kwon, and D. Pavlidis, "Delay Time Analysis of Submicron InP-Based HEMT's," *IEEE Trans. Electron Devices*, vol. 43, no. 2, pp. 228-237, February 1996.

Investigation of GaAs based MOVPE-grown $Al_xGa_{1-x}As_yP_{1-y}$ strain compensating layers

A. Maaßdorf*, Anatol Lochmann**[2] and M. Weyers*

*Ferdinand-Braun-Institut, Leibniz-Institut fuer Hoechstfrequenztechnik
Gustav-Kirchhoff-Straße 4, D-12489 Berlin, Germany
**LayTec AG, Seesener Straße 10-13, D-10709 Berlin, Germany

Abstract — **GaAs-based edge-emitting diode lasers designed for the near-infrared spectral region usually contain waveguide and cladding layers consisting of $Al_xGa_{1-x}As$. $Al_xGa_{1-x}As$-on-GaAs is known to be almost perfectly lattice matched. We have grown $Al_xGa_{1-x}As_yP_{1-y}/Al_xGa_{1-x}As$ test samples targeting a partial compensation of the room temperature wafer bow by incorporating up to 4% phosphorus in $Al_{0.85}GaAs$. The proposed strain compensation scheme has been applied in complete laser devices by partially replacing $Al_{0.85}Ga_{0.15}As$ with $Al_{0.85}Ga_{0.15}As_yP_{1-y}$. The results will be discussed.**

Index Terms — **Metalorganic vapor phase epitaxy, high resolution X-ray diffraction, laser diodes, in-situ curvature measurements.**

GaAs-based edge-emitting diode lasers designed for the near-infrared spectral region usually contain waveguide and cladding layers consisting of $Al_xGa_{1-x}As$. $Al_xGa_{1-x}As$-on-GaAs is known to be almost perfectly matched. While this is true for typical growth temperatures of 700°C-800°C the different thermal expansion coefficients of GaAs and AlAs cause convex wafer bow at room-temperature [1], [2]. For thicker laser structures with a higher average Al content this wafer bow can reach critical values making subsequent wafer processing impossible. Adding phosphorus (P) to $Al_xGa_{1-x}As$ i.e. by growing $Al_xGa_{1-x}As_yP_{1-y}$ this wafer bow can be reduced.

We have grown $Al_xGa_{1-x}As_yP_{1-y}/Al_xGa_{1-x}As$ test samples targeting a partial compensation of the room temperature wafer bow by incorporating up to 4% phosphorus in $Al_{0.85}GaAs$.

Samples were grown using a 5 4" planetary MOVPE reactor (AIX2400 G3) that is equipped with a EpiCurve®TT AR sensor. The investigated test samples are composed of a 500 nm $Al_{0.85}GaAs$ layer, a 150 nm $Al_{0.85}GaAs_yP_{1-y}$ layer and a 20 nm GaAs cap layer. XRD analysis allows to precisely determine the amount of incorporated phosphorus, assuming the added phosphorus is not affecting the Al incorporation efficiency.

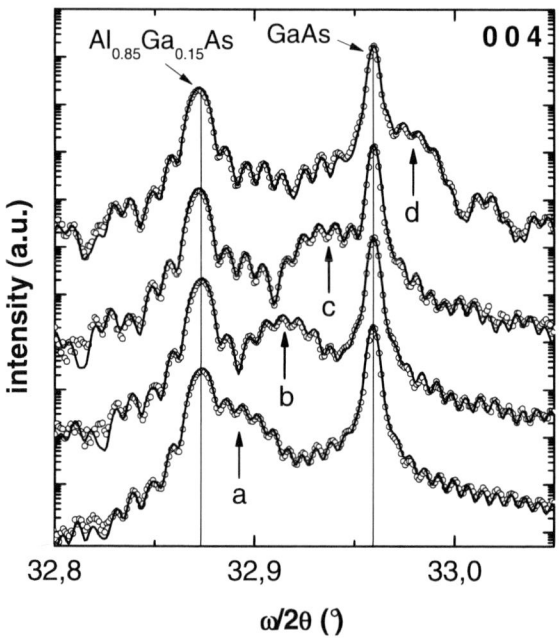

Fig. 1. Measured and simulated XRD rocking curves for different $Al_{0.85}Ga_{0.15}As/Al_{0.85}Ga_{0.15}As_yP_{1-y}$ test samples. Vertical arrows indicate the $Al_{0.85}Ga_{0.15}As_yP_{1-y}$ peak position.

Fig. 1 shows measured and simulated rocking curves of test samples (a)-(d). It can be seen that to obtain perfectly lattice matched $Al_{0.85}GaAs_yP_{1-y}$ layer on GaAs at room temperature, 3-4% phosphorus are required.

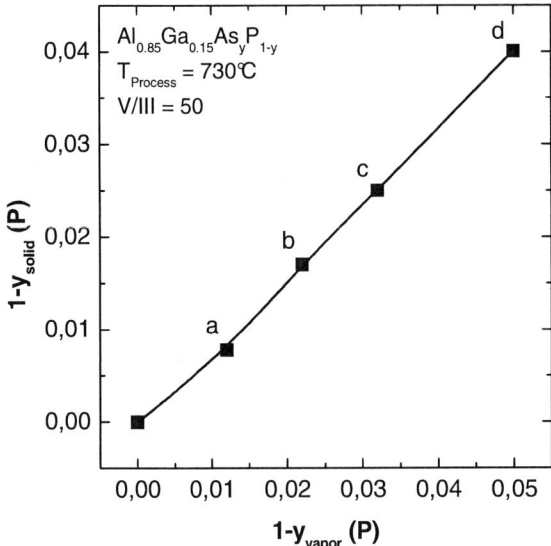

Fig. 2. Observed relation between incorporated phosphorus (1-y_{solid}) and vapor phase composition (1-y_{vapor}).

Fig. 2 shows the corresponding almost linear relation between solid and vapor phase composition for these samples.

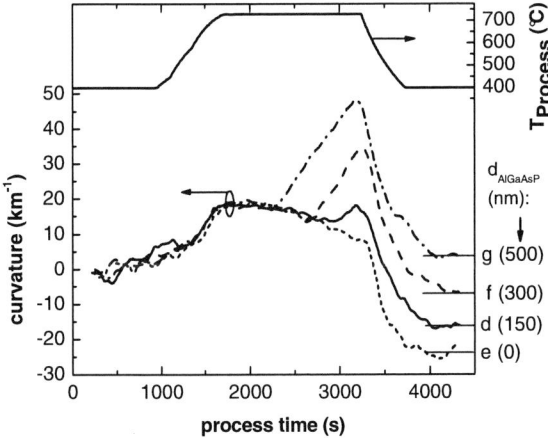

Fig. 3. In-situ curvature transients of samples with different $Al_{0.85}Ga_{0.15}As_{0.96}P_{0.04}$ layer thickness.

Fig. 3 shows the curvature transient of sample (d) from Fig. 1, in comparison to three other test samples with the same $Al_xGa_{1-x}As_yP_{1-y}$ composition and layer thicknesses of 0 nm (i.e. AlGaAs) (e), 300 nm (f) and 500 nm (g) on top of a lower AlGaAs layer. The thickness of both layers together was kept constant at 650 nm. It can be seen that the $Al_{0.85}GaAs_yP_{1-y}$ layer induces positive bow, offsetting the bow introduced

during cooling down. In case of sample (g) this leads to zero bow at room temperature.

The proposed strain compensation scheme has been applied in complete laser devices by partially replacing $Al_{0.85}Ga_{0.15}As$ with $Al_{0.85}Ga_{0.15}As_yP_{1-y}$. The results will be discussed.

REFERENCES

[1] Sadao Adachi, "Properties of Aluminum Gallium Arsenide", INSPEC 1993, No. 7, P43
[2] S. Novikova, Sov. Phys. - Solid State 3 (1961) 129

ENHANCEMENT-MODE PSEUDOMORPHIC HIGH-ELECTRON-MOBILITY TRANSISTOR WITH A NANOSCALE OXIDIZED GaAs GATE

Kuan-Wei Lee[1,*], Hsien-Cheng Lin[2], and Yeong-Her Wang[2,**]

[1] Department of Electronic Engineering, I-Shou University, Kaohsiung 840, Taiwan
*E-mail: kwlee@isu.edu.tw

[2] Institute of Microelectronics, Department of Electrical Engineering, Advanced Optoelectronic Technology Center, National Cheng-Kung University, Tainan 701, Taiwan
**E-mail: yhw@ee.ncku.edu.tw

Abstract

This study examines the feasibility of an enhancement-mode (E-mode) pseudomorphic high-electron-mobility transistor (PHEMT) with a nanoscale oxidized GaAs gate using liquid phase oxidation (LPO). Using LPO, the threshold voltage (V_{th}) can be shifted positively. Results indicated a reduced leakage current, a higher breakdown voltage, and an improved subthreshold swing compared to those of the Schottky-gate PHEMT. Therefore, LPO did not degrade the DC performance of the device.

I. Introduction

Enhancement-mode (E-mode) pseudomorphic high-electron-mobility transistors (PHEMTs) have attracted significant attention as power devices for single-supply operation. Because they require only a positive voltage to apply to the gate, E-mode-based circuits are easier and less complex for circuit design and construction. Therefore, circuit designers want to further the development of E-mode devices. However, few attempts have been made in high-performance E-mode PHEMTs because of certain substantial problems. One of these issues is the critically reduced gate-to-channel distance necessary for achieving a positive threshold voltage (V_{th}). An additional issue is the increased source resistance caused by the region surrounding the gate periphery being depleted. The buried-gate approach [1-2], or double-recess process [3-4], has been reported to effectively address these issues. However, traditional E-mode PHEMTs exhibiting several tenths of a volt forward bias limit from the Schottky barrier heights can still suffer from high gate leakage. This issue can be overcome by using oxide film as an insulator between the two-dimensional electron gas (2DEG) channel and the gate electrode. In addition, although E-mode PHEMTs have performed well, to the best of our knowledge, research is scant on E-mode metal-oxide-semiconductor pseudomorphic high-electron-mobility transistors (MOS-PHEMTs).

In this study, the V_{th} shift originating from liquid phase oxidation (LPO) [5, 6] was successfully applied to the fabrication of E-mode MOS-PHEMT. LPO is a conversion method that differs from the oxide deposited on (i.e., upward) the semiconductor. LPO occurs through the in-diffusion of oxygen at the oxide-semiconductor interface; therefore, a fresh interface region can be achieved because the original semiconductor surface contaminants are ended up on the oxide surface. The MOS structure created using the LPO method not only passivates the semiconductor surface, but also shortens the gate-to-channel distance needed to achieve a positive V_{th}.

II. Experimental

The selective oxidation process is schematically illustrated in Fig. 1. First, the photoresist (PR) was coated on the GaAs cap layer, and the pattern of which was designed using the photolithographic processes. Thereafter, the sample was transferred into the growth solution (initial pH = 5.0) at 50 °C. An oxide layer composed of Ga_2O_3 and As_2O_3 can only be grown on a bare GaAs surface that is not covered by PR. Following the removal of the PR by using acetone, the final selectively oxidized structure was obtained. The oxide film can be etched using diluted HF solution (HF:H_2O = 1:200), and the GaAs seemed to be consumed because of the loss of oxide species.

The proposed MOS-PHEMT structures, shown in Fig. 2, were grown by metal-organic chemical vapor deposition (MOCVD) on semi-insulating GaAs substrates. For the AlGaAs/InGaAs MOS-PHEMT structure, the buffer layer consists of 100 nm undoped GaAs, in addition to 250 nm undoped AlGaAs and 60 nm GaAs. Thereafter, a 10 nm AlGaAs donor layer with a silicon doping density of 1.5×10^{18} cm^{-3} and a 2 nm undoped AlGaAs spacer layer were grown on the buffer layer, followed by a 14 nm undoped $In_{0.15}Ga_{0.85}As$ channel layer, a 2 nm undoped InGaP spacer layer, a 20 nm InGaP donor layer with a silicon doping, and a 50 nm GaAs ohmic layer. Hall measurements revealed that the electron mobility was 6472 cm^2/Vs and that the electron sheet density was 1.44×10^{12} cm^{-2} at room temperature.

For the MOS-PHEMT fabrication, an active layer was defined by mesa wet etching, and the Au/Ge/Ni ohmic metal was annealed rapidly. The gate recess was etched by H_3PO_4:H_2O_2:H_2O (= 1:1:30). Following the etching of the partial capping layer, an LPO growth solution (initial pH = 5.0) was used to generate the oxidized GaAs as a gate insulator at 50

°C. This process converted the GaAs layer into the native oxide film, as shown in Fig. 2. Finally, the $1 \times 100 \mu m^2$ gate pattern was defined with Au. The details of the LPO growth solution and oxidation system were reported earlier in reference [5].

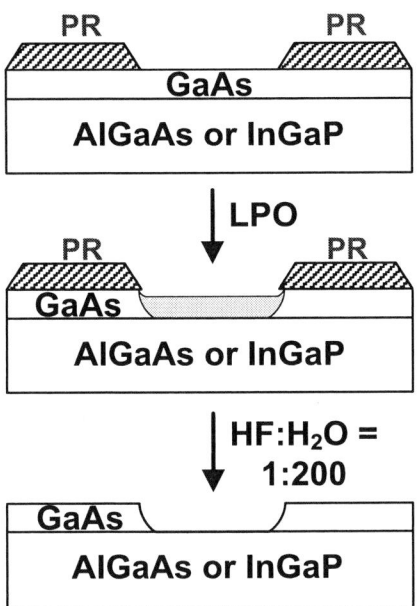

Fig. 1. Schematic cross section of selective oxidation on GaAs using PR as a mask.

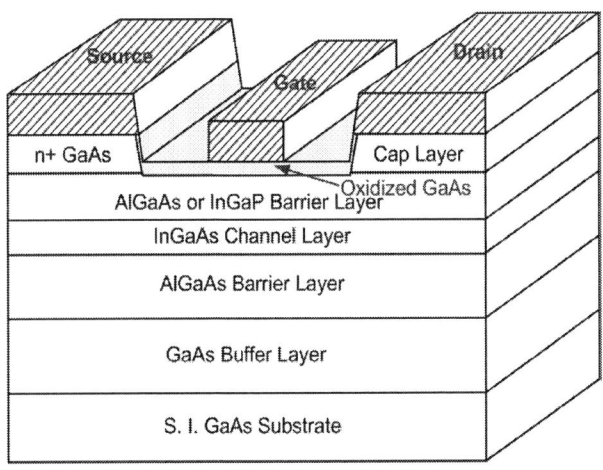

Fig. 2. Structure of the proposed E-mode MOS-PHEMT.

III. Results and Discussion

Figure 3 shows the cross-sectional view of the MOS (i.e., metal/oxidized GaAs/Schottky layer) structure using a transmission electron microscope (TEM). A 5-nm-thick oxidized GaAs layer was clearly observed. With help from the LPO, the V_{th} shifted positively because of the reduced gate-to-channel distance, forming the E-mode MOS-PHEMT.

Fig. 4 shows the measured *I-V* characteristics for the AlGaAs/InGaAs E-mode MOS-PHEMT. A V_{th} of 0.3 V was determined as the gate-to-source voltage intercept of the linear extrapolation of the drain current at the point of peak transconductance (g_m). A maximum gate voltage of 2.5 V was larger than that of conventional Schottky-gate PHEMTs because of its higher energy barrier at the gate-Schottky layer interface, which enhanced the current driving capability. The maximum drain current density was 93 mA/mm at V_{GS} = 2.5 V and V_{DS} = 7 V.

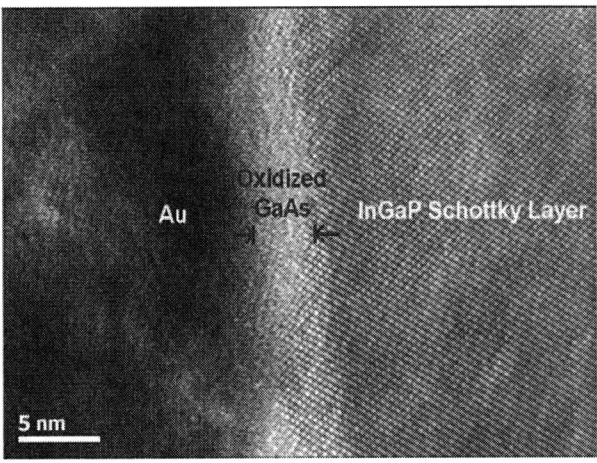

Fig. 3. Cross-sectional TEM image of the MOS (metal/oxidized GaAs/Schottky layer) structure.

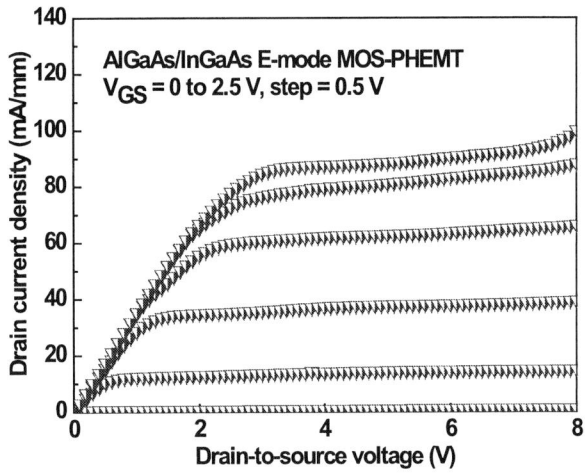

Fig. 4. *I-V* characteristics for the 1 μm gate-length AlGaAs/InGaAs E-mode MOS-PHEMT.

As shown in Fig. 5, g_m was 171 (140) mS/mm and the gate voltage swing was approximately 0.5 (0.5) V at V_{DS} = 5 (4) V for the InGaP/InGaAs MOS-PHEMT. The gate voltage swing in this study was comparable to that of traditional E-mode InGaP/InGaAs PHEMTs [7, 8]. The oxide film provided an improved breakdown voltage for the gate leakage current of the MOS-PHEMT, and was supported by the two-terminal (i.e., floating source) gate-to-drain diode characteristics [9, 10]. For the InGaP/InGaAs MOS-PHEMT, the gate leakage current density was suppressed by approximately three orders of magnitude, and the corresponding reverse gate-to-drain breakdown voltage (BV_{GD}) was -29.3 V. The turn-on voltage (V_{on}) (0.90 V) was higher than that of the referenced Schottky-gate PHEMT (0.75 V). V_{on} and BV_{GD} were defined as the voltage at which the gate current reached 1 mA/mm [11, 12]. An increased V_{on} typically accompanies an improved gate voltage swing. This trend was consistent with the results shown in Fig. 4. The smaller gate leakage current of the InGaP/InGaAs (or AlGaAs/InGaAs) MOS-PHEMT was caused by the MOS structure and the elimination of sidewall leakage paths that were directly passivated during LPO. The subthreshold swing (SS) of the E-mode MOS-PHEMT was better than that of the referenced case. The improved SS was associated with the suppressed leakage characteristics of the MOS-PHEMT. Therefore, the oxidized GaAs by LPO did not degrade the DC performance of the device.

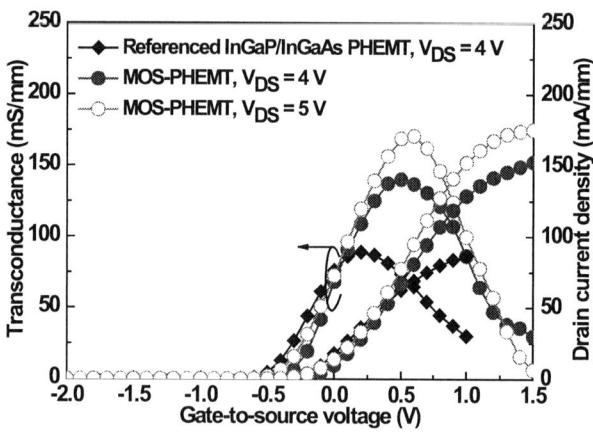

Fig. 5. The g_m and drain current density versus gate-to-source voltage for the InGaP/InGaAs PHEMT with and without LPO.

IV. Conclusions

Results demonstrated the feasibility of an oxidized GaAs gate on E-mode MOS-PHEMTs prepared by LPO. The LPO method not only passivated the semiconductor surface, but also shortened the gate-to- channel distance, achieving a positive threshold voltage. In addition, the device had a higher breakdown voltage, a lower gate leakage current, and an improved subthreshold swing compared to the referenced Schottky-gate PHEMT, making the proposed low-temperature and low-cost LPO suitable for power applications.

Acknowledgements

This study was supported in part by the National Science Council of Taiwan under the Contracts Nos. NSC 99-2221-E-214-067, NSC 98-2221-E-214-061, 982C12, and MOE Program for Promoting Academic Excellence of Universities under the Grant No. D98-3600.

References

[1] Y. Takanashi, M. Hirano, and T. Sugeta, "Control of threshold voltage of AlGaAs/GaAs 2DEG FET's through heat treatment," *IEEE Electron Device Lett.*, vol. EDL-5, pp. 241-243, 1984.

[2] K. J. Chen, T. Enoki, K. Maezawa, K. Arai, and M. Yamamoto, "High-performance InP-based enhancement-mode HEMT's using non-alloyed ohmic contacts and Pt-based buried-gate technologies," *IEEE Trans. Electron Devices*, vol. 43, pp. 252-257, 1996.

[3] K. Ohmuro, H. I. Fujishiro, M. Itoh, H. Nakamura, and S. Nishi, "Enhancement-mode pseudomorphic inverted HEMT for low noise amplifier," *IEEE Trans. Microw. Theory Tech.*, vol. 39, pp. 1995-2000, 1991.

[4] Y. Okamoto, K. Matsunaga, and M. Kuzuhara, "Enhancement-mode buried gate InGaP/AlGaAs/InGaAs heterojunction FETs fabricated by selective wet etching," *Electron. Lett.*, vol. 31, pp. 2216-2218, 1995.

[5] H. H. Wang, C. J. Huang, Y. H. Wang, and M. P. Houng, "Liquid phase chemical-enhanced oxidation for GaAs operated near room temperature," *Jpn. J. Appl. Phys.*, vol. 37, pp. L67-L70, 1998.

[6] J. Y. Wu, H. H. Wang, Y. H. Wang, and M. P. Houng, "A GaAs MOSFET with a liquid phase oxidized gate," *IEEE Electron Device Lett.*, vol. 20, pp. 18-20, 1999.

[7] H. Y. Tu, T. H. Chou, Y. S. Lin, H. C. Chiu, P. Y. Chen, W. C. Wu, and S. S. Lu, "DC and RF characteristics of E-mode $Ga_{0.51}In_{0.49}P$-$In_{0.15}Ga_{0.85}As$ pseudomorphic HEMTs," *IEEE Electron Device Lett.*, vol. 24, pp.132-134, 2003.

[8] L. H. Chu, E. Y. Chang, S. H. Chen, Y. C. Lien, and C. Y. Chang, "2 V-operated InGaP-AlGaAs-InGaAs enhancement-mode pseudomorphic HEMT," *IEEE Electron Device Lett.*, vol. 26, pp. 53-55, 2005.

[9] J. B. Shealy, T. Y. Liu, M. A. Thompson, R. G. Wilson, L. D. Nguyen, and U. K. Mishra, "High threshold uniformity, millimeter-wave p⁺-GaInAs/n-AlInAs/GaInAs JHEMT's," *IEEE Electron Device Lett.*, vol. 16, pp. 560-562, 1995.

[10] R. Menozzi, "Off-state breakdown of GaAs PHEMTs review and new data," *IEEE Trans. Device and Materials Reliability*, vol. 4, pp. 54-62, 2004.

[11] M. Matloubian, L. M. Jelloian, A. S. Brown, L. D. Nguyen, L. E. Larson, M. J. Delaney, M. A. Thompson, R. A. Rhodes, and J. E. Pence, "V-Band high-efficiency high-power AlInAs/GaInAs/InP HEMT's," *IEEE Trans. Microwave Theory and Techniques*, vol. 41, pp. 2206-2210, 1993.

[12] M. Zaknoune, B. Bonte, C. Gaquiere, Y. Cordier, Y. Druelle, D. Théron, and Y. Crosnier, "InAlAs/InGaAs metamorphic HEMT with high current density and high breakdown voltage," *IEEE Electron Device Lett.*, vol. 19, pp. 345-347, 1998.

Source-Drain Scaling of Ion-Implanted InAs/AlSb HEMTs

G. Moschetti[1*], P.-Å. Nilsson[1], A. Hallén[2], L. Desplanque[3], X. Wallart[3] and J. Grahn[1].

[1]Department of Microtechnology and Nanoscience, Chalmers University of Technology,
SE-41296 Göteborg, Sweden
[2]Department of Microelectronics and Applied Physics, Royal Institute of Technology (KTH),
SE-16440 Kista, Sweden
[3]Institute of Electronics, Microelectronics and Nanotechnology, IEMN/CNRS UMR 8520,
University of Lille 1, 59652 Villeneuve d'Ascq, France
[*]giuseppe.moschetti@chalmers.se

Abstract — We report on the lateral scaling of true planar InAs/AlSb high electron mobility transistors (HEMTs) based on ion implantation for device isolation. When reducing the source drain distance, d_{sd}, from 2.5 μm to 1 μm, the HEMTs showed up to 56% higher maximum drain current, 23% higher peak transconductance and f_T of 185 GHz (+32%). A trade-off in the lateral scaling is needed due to increased gate leakage current and pinch-off degradation for d_{sd} below 1.5 μm. The ability to withstand oxidation of the InAs/AlSb heterostructure makes the planar technology based on ion implantation extremely promising for MMIC integration of InAs/AlSb HEMTs.

Index Terms — InAs/AlSb HEMT, ion implantation, oxidation resistant, lateral scaling, low-power, MMIC.

I. INTRODUCTION

The InAs material is characterized by the combination of low bandgap (0.36 eV), high electron mobility and high peak electron velocity [1]. This makes the InAs/AlSb HEMT a promising device candidate for high speed portable and space communication systems where ultra-low power consumption is a key parameter [2-4].

The extremely high oxidation tendency of the AlSb material results in a challenging fabrication process for devices based on traditional mesa isolation technology [5]. A shallow mesa process, where the isolation etch is stopped in the more stable AlGaSb alloy, allows improved stability against oxidation at the expenses of poorer electrical isolation [6]. An air-bridge-gate technology [7] can instead offer improved electrical isolation but at the cost of exposure of the deep AlSb mesa walls to air and of difficult integration in monolithic microwave integrated circuits (MMICs).

We have recently demonstrated that ion implantation can be used for device isolation in InAs/AlSb heterostructures [8]. This technology enables fully planar HEMTs, thus particularly suitable for MMIC applications. Furthermore, very high stability against oxidation is achieved since no AlSb layers are ever exposed to the environment. In this work we have further developed the fabrication process of true planar InAs/AlSb HEMTs based on Ar[+] ion implantation by investigating the device performance upon lateral scaling of the source-to-drain distance.

II. HEMT TECHNOLOGY

The HEMT layers are grown on two-inch semi-insulating InP wafers by molecular beam epitaxy (MBE). The heterostructure design is similar to that reported in [8]. Hall measurements performed at room temperature shows electron mobility and carrier concentration of 17,300 cm^2/Vs and 2.5×10^{12} cm^{-2}, respectively.

The fabrication process starts with implantation mask definition by the use of S1813 photoresist. Thereafter, the heterostructure is bombarded outside the active regions of the HEMTs with Ar[+] ions at a dose and energy of 2×10^{15} cm^{-2} and 100 keV, respectively. After the ion implantation the isolating area shows a sheet resistance of 60 kΩ/sq, which increases to 6 MΩ/sq after annealing at 370 °C; see Fig. 1.

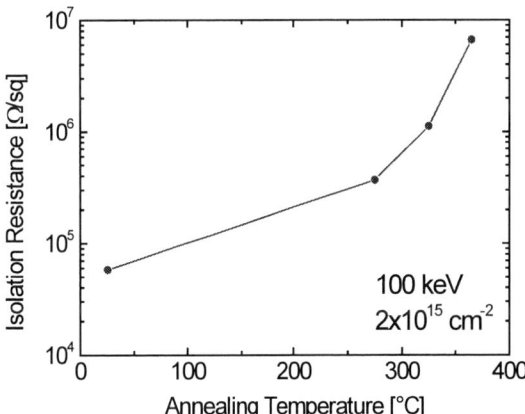

Fig. 1. Isolation sheet resistance versus annealing temperature for implantation energy and dose of 100 keV and 2×10^{15} cm^{-2}, respectively. The first data point represents a non-annealed sample.

Subsequently, the ohmic contacts pattern is defined using Electron beam (E-beam) lithography. Four different designs of source-drain distances d_{sd}, ranging from 1 to 2.5 μm, are defined and a Pd/Pt/Au metal stack is deposited. The contact resistance R_c is as low as 0.05 Ω·mm after a thermal annealing at 275 °C. Ti/Pt/Au T-gates with a footprint of 110 nm are also defined by E-beam lithography and oriented along the [110] direction in order to maximize the HEMT performance [9]. In Fig. 2, a scanning transmission electron microscopy (STEM) of the gate foot and of the underlying HEMT structure is shown. Finally, Ti/Au probing pads are fabricated using photolithography.

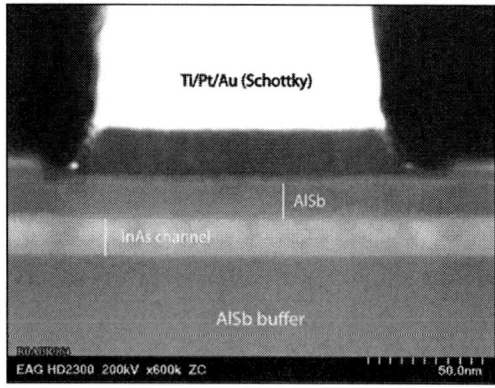

Fig. 2. STEM micrograph of the gate foot as well as of the underlying epitaxial layers.

III. DC CHARACTERIZATION

The DC measurements are performed on-wafer using coplanar GSG waveguide probes and an HP4156B parameter analyzer. The maximum drain-source voltage V_{DS} is limited to 0.3 V in order to avoid degradation of the HEMTs caused by the impact ionization effect occurring at higher biases.

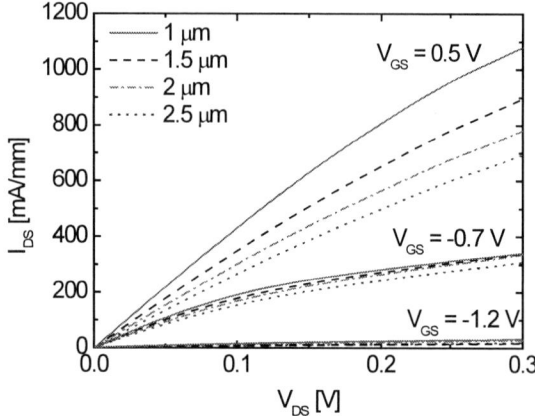

Fig. 3. I_{DS} versus V_{DS} for ion-implanted InAs/AlSb HEMTs with d_{sd} ranging from 1 μm to 2.5 μm.

Fig. 3 shows the output characteristics of Ar$^+$ implanted InAs/AlSb HEMTs with different d_{sd}. When d_{sd} is scaled from 2.5 μm to 2 μm, 1.5 μm and 1 μm, the maximum drain-source current I_{DS}, at V_{DS} of 0.3 V, increases by 12%, 29% and 56%, respectively.

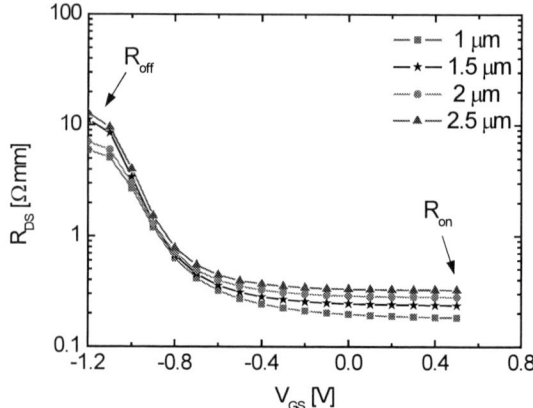

Fig. 4. R_{DS} as a function of V_{GS}. The on-resistance is extracted at $V_{GS} = 0.5$ V.

As shown in Fig. 4, the on-resistance R_{on}, given by the access resistance R_{DS} at high gate voltage, scales with d_{sd}, ranging from 0.32 Ω·mm down to 0.18 Ω·mm when d_{sd} is scaled from 2.5 μm to 1 μm. Moreover, the off-resistance R_{off} is degraded when d_{sd} is reduced.

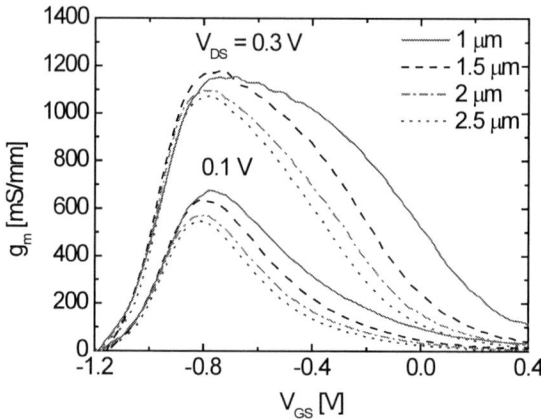

Fig. 5. g_m as a function of V_{GS} for V_{DS} biases of 0.1 V and 0.3 V.

When scaling d_{sd} from 2.5 μm to 1 μm, the transconductance g_m peak at $V_{DS} = 0.1$ V is enhanced from 547 mS/mm to 672 mS/mm (+23%), as can be seen in Fig. 5. Moreover, the g_m peak becomes wider with scaling. A wider g_m peak leads to lower values of the first and second order derivatives g_m' and g_m'', thus to improved second and third order intercept point IP2 and IP3.

Fig. 6 shows the subthreshold current I_{DS} and the gate current leakage I_G for the four different designs. Both, the non pinch-off I_{DS} and I_G are increased when reducing d_{sd} possibly due to the higher lateral leakage between gate and source/drain contacts. The device with d_{sd} of 2.5 μm shows however a higher I_G, compared to the 1 µm and 1.5 µm devices, probably due to a not perfectly uniform depth of the recess etch beneath the gate fingers.

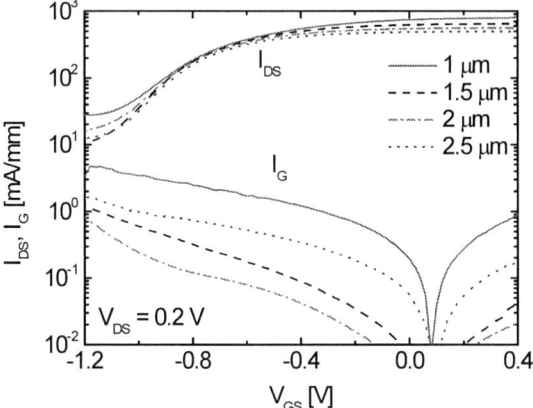

Fig. 6. Subthreshold and gate current characteristics at V_{DS} of 0.1 V and 0.3 V.

The I_G becomes even more severe because of the lateral diffusion of the ohmic metal shown in the STEM of Fig. 7. A similar diffusion of the ohmic metal has been observed in [5]. The subthreshold slope, S, is degraded from 248 mV/dec up to 334 mV/dec when d_{sd} is scaled from 2.5 µm to 1 µm.

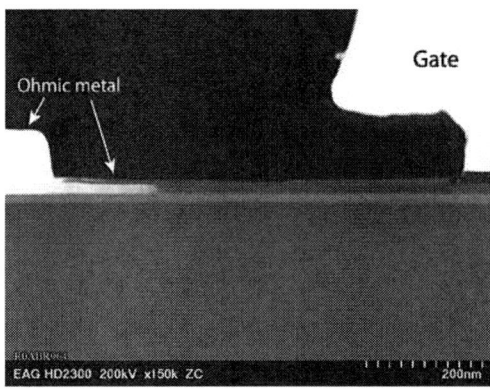

Fig. 7. STEM micrograph of the ohmic metal, showing its lateral diffusion.

Moreover, as can be seen in Fig. 8, the drain induced barrier lowering (*DIBL*) increases from 92 mV/V up to 170 mV/V (+85%) when d_{sd} is scaled from 2.5 µm to 1 µm. Therefore, a tradeoff in the DC performance of the HEMT should be taken into account when reducing d_{sd}.

IV. RF CHARACTERIZATION

S-parameters were measured using an Agilent E8361A vector network analyzer in the range between 0.1 GHz and 67 GHz. Fig. 9 shows the unity-current gain $|h_{21}|^2$ and the

Mason's gain U at $V_{DS} = 0.2$ V. When scaling d_{sd}, $|h_{21}|^2$ is enhanced leading to a f_T increase from 140 GHz up to 185 GHz (+32%) at the shortest distance of 1 µm.

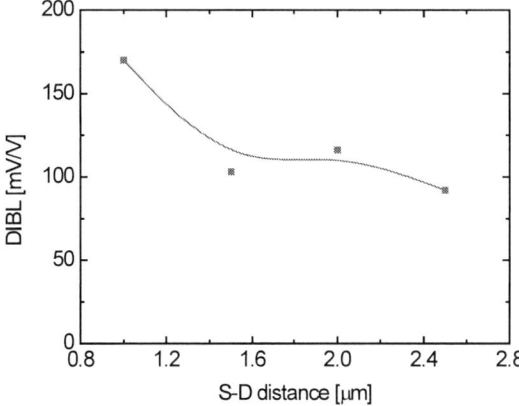

Fig. 8. DIBL as a function of source-drain distance of ion-implanted InAs/AlSb HEMTs.

A different behavior is however observed for U. At low frequency ($\sim$ 10 GHz) it is around 11% higher when scaling from 2.5 to 2 µm. When reducing d_{sd} to 1.5 µm and 1 µm, U is 13% and 18% lower, respectively.

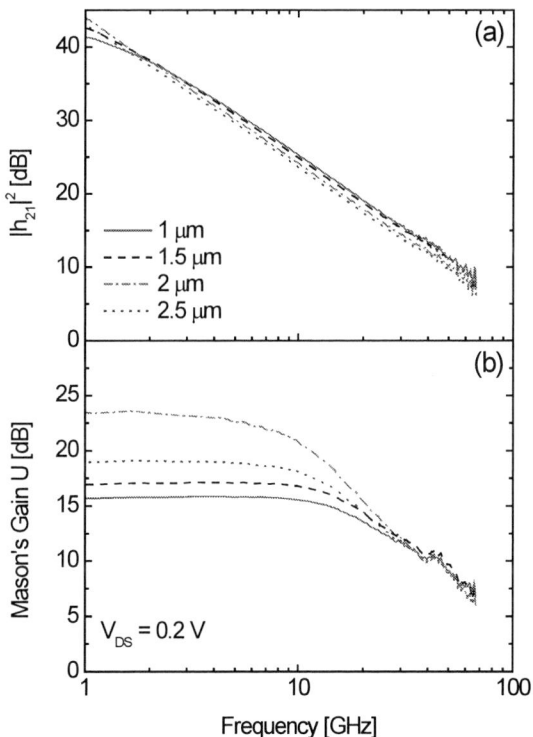

Fig. 9. (a) $|h_{21}|^2$ and (b) U versus frequency at V_{DS} of 0.2 V for d_{sd} ranging from 1 to 2.5 µm.

The reduction of U in the low frequency range can be related to the higher I_G occurring for short d_{sd} values shown in Fig. 6. The values of U and $|h_{21}|^2$ at 10 GHz and for a drain bias of 0.2 V are plotted as a function of the source-drain distance in Fig. 10. In this plot it becomes clearer that a large reduction of d_{sd} leads to a tradeoff in the RF performance of the HEMT as well.

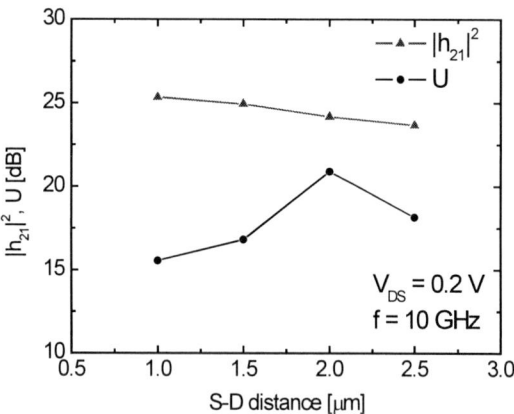

Fig. 10. $|h_{21}|^2$ and U at 10 GHz and 0.2 V as a function of the source-drain distance.

V. CONCLUSIONS

The lateral scaling of true planar InAs/AlSb HEMTs based on the ion-implantation for device isolation is here investigated. Upon scaling, the HEMTs showed improved g_{ds}, higher g_m peak and I_{DS} as well as enhanced f_T. The results also suggest that a tradeoff in the source-drain distance with respect to I_G, pinch-off and Mason's gain has to be taken into account. The advantage of the planar process combined with the DC and RF performance demonstrated here, shows the high potential of this device technology for ultra-low power Sb-based MMICs.

ACKNOWLEDGEMENT

Gösta Widman from Uppsala University, Sweden, is acknowledged for having performed the ion implantation of the measured samples.

REFERENCES

[1] B.R. Bennett, R. Magno, J.B. Boos, W. Kruppa, and M.G. Ancona, "Antimonide-based compound semiconductors for electronic devices: A review," *Solid-State Electron.*, vol. 49, no. 12, pp. 1875-1895, Dec. 2005.

[2] B.Y. Ma, J. Bergman, P. Chen, J.B. Hacker, G. Sullivan, G. Nagy, and B. Brar, "InAs/AlSb HEMT and its application to ultra-low-power wideband high-gain low-noise amplifiers," *IEEE Trans. Microw. Theory Tech.*, vol. 54, no. 12, pp. 4448-4455, Dec. 2006.

[3] W.R. Deal, R. Tsai, M.D. Lange, J.B. Boos, B.R. Bennett, and A. Gutierrez, "A Low Power/Low Noise MMIC Amplifier for Phased-Array Applications using InAs/AlSb HEMT," *IEEE Int. Microw. Symp.*, pp. 2051-2054, June 2006.

[4] G. Moschetti, N. Wadefalk, P.-Å. Nilsson, M. Abbasi, L. Desplanque, X. Wallart and J. Grahn, "Cryogenic InAs/AlSb HEMT Wideband Low-Noise IF Amplifier for Ultra-Low-Power Applications," *IEEE Microw. Wireless Comp. Lett.*, vol. 22, no. 3, pp. 144-146, Mar. 2012.

[5] Y.C. Chou, J.M. Yang, M.D. Lange, S.S. Tsui, D.L. Leung, C.H. Lin, M. Wojtowicz, and A.K. Oki, "Degradation mechanisms of 0.1 μm AlSb/InAs HEMTS for ultralow-power applications," *IEEE Int. Rel. Phys. Symp.*, pp. 436-440, Apr. 2008.

[6] J.B. Boos, B.V. Shanabrook, D. Park, J.L. Davis, H.B. Dietrich, and W. Kruppa, "Impact ionisation in high-output-conductance region of 0.5 μm AlSb/InAs HEMTs," *Electronics Letters*, vol. 29, no. 21, pp. 1888-1890, Oct. 1993.

[7] A. Noudéviwa, Y. Roelens, F. Danneville, A. Olivier, N. Wichmann, N. Waldhoff, S. Lepilliet, G. Dambrine, L. Desplanque, X. Wallart, G. Moschetti, J. Grahn, and S. Bollaert, "Sb-HEMT: Toward 100-mV Cryogenic Electronics," *IEEE Trans. Electron Devices*, vol. 57, no. 8, pp. 1903-1909, Aug. 2010.

[8] G. Moschetti, P.-Å. Nilsson, A. Hallén, L. Desplanque, X. Wallart and J. Grahn, "Planar InAs/AlSb HEMTs with ion-implanted isolation," *IEEE Electron Device Lett.*, vol. 33, no. 4, pp. 510-512, Mar. 2012.

[9] G. Moschetti, H. Zhao, P.-Å. Nilsson, S. Wang, A. Kalabukhov, G. Dambrine, S. Bollaert, L. Desplanque, X. Wallart, and J. Grahn, "Anisotropic transport properties in InAs/AlSb heterostructures," *Appl. Phys. Lett.*, vol. 97, no. 24, pp. 243510, Dec. 2010.

Electroabsorption Modulator Chirp Profile Influence on D-EML Modulation Scheme at 10 Gb/s

T. Anfray[1], C. Aupetit-Berthelemot[1], K. Kechaou[2], D. Erasme[2], G. Aubin[3], C. Kazmierski[4] and P. Chanclou[5]

[1]XLIM Dpt. C2S2 UMR CNRS 7252/University of Limoges, 16 Rue Atlantis, 87068 Limoges, France
[2]Institute TELECOM, TELECOM ParisTech, CNRS LTCI, 46 Rue Barrault, 75634 Paris, France
[3]Laboratory for Photonics and Nanostructures/CNRS, Route de Nozay, 91460 Marcoussis, France
[4]III-V Lab, Route de Nozay, 91460 Marcoussis, France
[5]Orange Labs, 2 Avenue Pierre Marzin, 22307 Lannion, France

thomas.anfray@ensil.unilim.fr, christelle.aupetit@ensil.unilim.fr

Abstract—The Dual Electroabsorption Modulated Laser (D-EML) is a transmitter technology specially design for dual drive. It is composed of an integrated Distributed Feed Back (DFB) laser and an Electro-Absorption Modulator (EAM) with dual independent modulation access. By controlling independently laser and modulator modulation amplitudes, it's possible to adjust the chirp of this optical source and thus increase the transmission distance far beyond the chromatic dispersion limit. We develop a D-EML model based on experimental characterizations in order to study benefits of dual modulation scheme for optical systems at 10 Gb/s and above. In this paper, we propose to analyze the impact of the EAM chirp profile on the performances.

Index Terms—Optical Communications; Optical Sources.

I. INTRODUCTION

Gigabit Passive Optical Networks (GPON) are being deployed by operators in several countries and are among the fastest access technologies currently available. Such systems play a key role in delivering high-bandwidth services to a diverse range of users. With the rise of new content-rich services and the increasing demand for bandwidth-intensive service such as High Definition Television, File Sharing, Video On Demand, Cloud Computing Services, *etc.*, it is expected that the required bandwidth will steadily increase. It is expected that at least 1 Gb/s will be proposed in few years. Another important factor to take into account is the increasing number of customers which want to benefit from these services. For these reasons, operators continue to develop and standardize new technologies for Next Generation Access (NGA) networks so as to support the increased data rates and customers. Currently GPON and its update at 10 Gb/s (XGPON) is based on Time Division Multiplexing / Time Division Multiple Access and Non Return to Zero (NRZ) modulation format. Data rate transmissions at 10 Gb/s and above suffer from fiber chromatic dispersion which combined to the chirp of the optical source induced Inter Symbol Interferences (ISI) limiting the transmission distance and/or bit rate for a given Quality of Service.

There are many solutions to counter chromatic dispersion such as Electronic or Optical Dispersion Compensation, Preemphasis, Equalization, *etc.* Some other techniques focus on

the optical sources in order to make them efficient for high bit rates and/or high distances by reducing there modulated spectrum. This is the case of Chirp Managed Laser (CML) for example [1], [2]. CML require external Frequency Modulation-Amplitude Modulation (FM-AM) converters precisely tuned and stabilized on an optical carrier making them difficult to implement in access network. More recently some results have been reported with the use of an integrated Electroabsorption Modulated Laser - Semiconductor Optical Amplifier (EML-SOA) [3]. With such a component, the SOA negative chirp can compensate EAM positive chirp and produce pre-chirping effect leading to higher transmission distance and/or bit rate. Kim Hoon *et al.* proposed to separate completely FM and AM by using a simultaneous modulation of a laser and an external modulator integrated on the same chip [4]. J. Petit *et al.* confirmed experimentally the concept of dual modulation applied to an integrated D-EML at 10 Gb/s [5]. More recently we reported experimental results of an error-free transmission at 20 Gb/s over 40 km [6].

In this paper, we propose to demonstrate by simulation that D-EML associated with Direct Detection without any kind of chromatic dispersion compensation can improve transmission distances at 10 Gb/s. More in depth, we studied the impact of EAM chirp profile on transmission performances. First, D-EML was characterized and the experimental results have been used as benchmarks to validate the developed model. Next, system simulations have been carried out and compared to experimental transmission tests. One goal of this study is to provide a help in order to optimize the modulation schemes and component requirements.

II. COMPONENT DESCRIPTION AND OPERATION PRINCIPLE

A. Component Description

The D-EML is a transmitter technology specifically designed for dual drive. The device is a monolithically integrated EAM and DFB laser that emits at 1534 nm. The integration technology is based on AlGaInAs-QW (Quantum Well) material for its large electronic confinement providing enhanced

978-1-4673-1725-2/12 $31.00 © 2012 IEEE

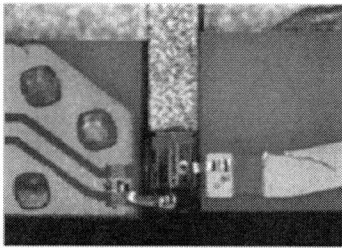

Fig. 1. D-EML chip on submount showing EAM and laser access ceramics respectively on the left and on the right (chip size 0.25 mm×0.5 mm).

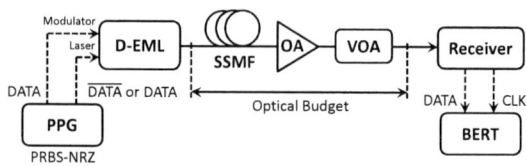

Fig. 2. Synoptic of the link for system simulation with D-EML model at 10 Gb/s.

electro-absorption properties and reduced thermal carrier leakage. The same active layer is used for both laser and modulator section and their gaps were engineered by Selective Area Growth. The component waveguide is selectively buried with a tandem layer of semi-insulating InP separately doped with Fe and Ru. The semi-insulating buried structure assures low EAM capacitance and low thermal resistance of the laser. This is obtained by Fe doped InP due to its high resistivity. Ru doping was introduced to prevent Zn-Fe interdiffusion and to provide InP single optical index confinement for all waveguide [7]. The diodes are completed by a standard p-layer doped confinement and contact structure. Large p-type regrowth surface favors a low series resistance of both component sections in order to enhance their RC bandwidth limits. EAM section is 75 μm long and may be modulated up to 40 GHz. DFB laser section is 470 μm long and is designed to improve the FM modulation properties versus standard EML version (device photograph is shown in Fig. 1). Its AM electro-optical bandwidth is around 11 GHz allowing NRZ modulation at 20 Gb/s. The coupled optical power in the single mode fiber is around 4 dBm.

With such a design, it's possible to separate completely optical FM and AM by applying digital or analog modulation respectively on the high FM-index laser section and on the high AM-index modulator section. Residual AM produced by the laser section is negligible compared to AM produced by modulator.

B. Operation Principle

J. Binder and U. Kohn explained in [8] that it is possible to generate SSB-LC (Single Side Band Large Carrier) modulation from a FM coupled with an AM. They stated the one better condition on AM/FM proportion for a sinus modulation and extended this condition to digital NRZ intensity modulations in order to obtain SSB-LC signal. This is equation (1).

$$\Delta f_{pp} = \frac{B}{2} \cdot \frac{ER - 1}{ER + 1} \qquad (1)$$

ER is the linear optical Extinction Ratio as measured on the eye diagram, Δf_{pp} is the peak-to-peak optical frequency deviation between marks and spaces and B is the bit rate of the digital signal. Knowing the FM efficiency of the laser, expressed for example in GHz/mA, it is possible to adjust the modulation current of the laser to produce target optical frequency deviation that corresponds to a given extinction ratio

and bit rate. This implies that AM has to be managed by another component, such as EAM in the case of the D-EML, in order to preserve this given extinction ratio.

III. SIMULATION SETUP AND RESULTS

A. Simulation Setup

Setup for network simulation including D-EML is based on a Point-to-Point link whose elements are depicted on Fig. 2. The Pulse Pattern Generator (PPG) feeds the D-EML with a Pseudo-Random Binary Sequence (PRBS) coded in NRZ. The PPG takes into account random Time Jitter (TJ_{RMS} is 1% of bit time and TJ_{pp} is 20% of bit time) and thermal noise (N_{th} is 10^{-11} A/$\sqrt{\text{Hz}}$). At the output of D-EML model, there is a Standard Single Mode Fiber (SSMF) which includes non-linear effects. The linear attenuation and the chromatic dispersion are respectively set to 0.2 dB/km and 16 ps/nm/km at 1.55 μm. The output power of D-EML model is around 10 dBm what allows to suppose that non-linear effects can be neglected. After the SSMF an Optical Amplifier (OA) has been added because the linear attenuation of the fiber becomes limiting for long distances. The optical gain is 10 dB and the added noise is very low. A Variable Optical Attenuator (VOA) is used after the optical amplifier for simulating optical losses. Receiver part is composed of a PIN photodiode which includes thermal noise (N_{th} is $1.41 \cdot 10^{-11}$ A/$\sqrt{\text{Hz}}$) and shot noise, an electrical pre-amplifier of 10 dB gain (thermal noise is included in the photodiode one) and a 2nd order Butterworth electrical low-pass filter with a band pass at 75% of bit rate. After receiver, a Bit Error Rate Tester (BERT) is added for measuring the transmission performances. The Optical Budget is defined as the sum of optical losses, due to the linear attenuation of the fiber and the variable optical attenuator, minus optical gains, due to the optical amplifier.

Dual modulation consists in sending DATA signal on the EAM and DATA or inverted DATA signal on the laser. Four cases were investigated to prove benefits of dual modulation principle at 10 Gb/s. (a) is the case of pure AM with $ER \approx 16$ dB. (b), (c) and (d) are the cases of mixed AM/FM with respectively $ER \approx 16$ dB and $\Delta f_{pp} \approx 5$ GHz, $ER \approx 9.5$ dB and $\Delta f_{pp} \approx 4$ GHz and finally $ER \approx 6$ dB and $\Delta f_{pp} \approx 3$ GHz. All Δf_{pp} verify (1).

In the simulations, the D-EML is modeled by a data sheet laser model including RIN (Relative Intensity Noise), linewidth, driver transconductance, dynamic and adiabatic chirp, slope efficiency, etc. and an EAM model with a voltage-dependent transmission function and alpha-factor. Although

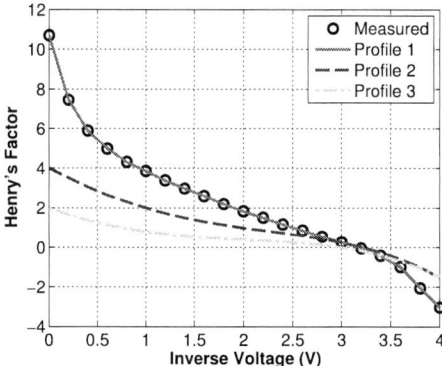

Fig. 3. The EAM voltage-dependent chirp profiles studied. EAM model includes only transient chirp also known as Henry's Factor.

the laser model takes into account transient and adiabatic chirp, the last one is much more important than the first one in order to match with the D-EML component behavior. The EAM model includes only transient chirp because it is widely predominant effect. The EAM voltage-dependent chirp profiles studied are shown in Fig. 3.

The first model that we implemented (profile 1) is a fitting of a measured transient chirp versus bias voltage onto a D-EML chip. We proposed two other profile models (profiles 2 and 3) by reducing positive and negative chirp respectively for low and high reverse voltage. By that way, we wanted to know the influence of high positive and negative chirp on transmission performances based on dual modulation scheme.

The optical frequency deviation generated by the D-EML model is the sum of the laser adiabatic and transient chirp contribution and the modulator transient chirp contribution. For each studied case, only three parameters are adjusted to control the D-EML model output signal in order to match to the SSB conditions (1). On the one hand, the optical frequency deviation is partially controlled by means of laser modulation current and on the other hand, the extinction ratio is preserved by adjusting the modulator modulation and bias voltages. For example, in the case (b), the laser modulation current is 5 mA peak-to-peak around bias at 90 mA and the EAM modulation voltage is 1.7 V peak-to-peak around bias at 2.35 V.

B. Simulation Results

Our study is based on power penalty criterion express in dB which is the difference between the Received Optical Power (ROP) at X km of transmission and in Back-to-Back (BtB) configuration for a Bit Error Rate (BER) at 10^{-3}. Thus we plotted in Fig. 4 power penalty versus transmission distance for the three transient chirp profiles at 10 Gb/s for the four cases explained in part III-A in DATA/DATA configuration.

Let us consider the chirp profile introducing the lowest transient chirp, the profile 3. The cases of dual modulation (b), (c) and (d) allow to reach longer transmission distances than the case of single modulation (a) due to their narrowest modulated optical spectra. In time domain, the adiabatic chirp

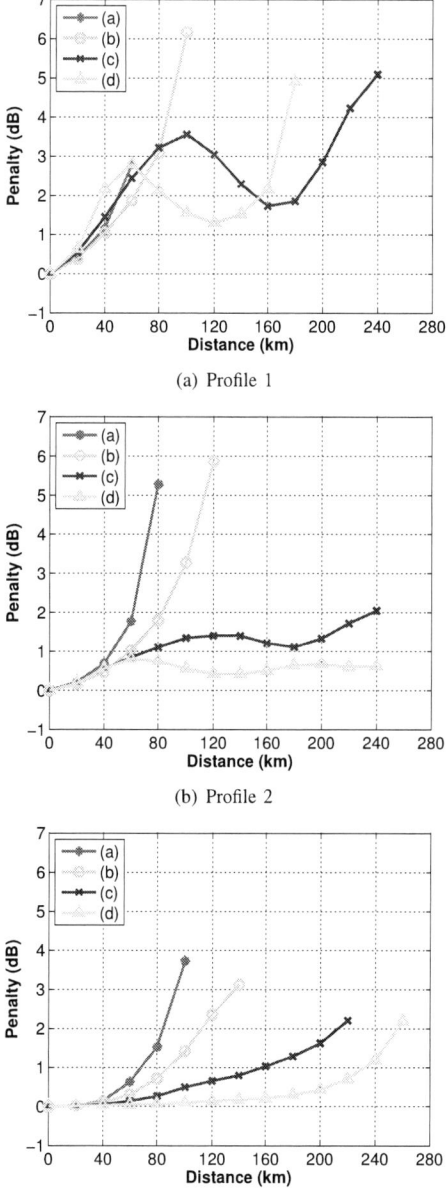

(a) Profile 1

(b) Profile 2

(c) Profile 3

Fig. 4. Calculated power penalty for a BER at 10^{-3} versus distance for the three chirp profiles and the four cases described in the part III-A.

of the laser contributes to compress 0-1-0 transitions leading in a better eye opening after transmission. Decreasing extinction ratio from case (b) to (d) results in an increase of transmission distances. This is explained by the requirement of low AM and FM indexes to form a narrower SSB spectrum as it is depicted in Fig. 5. Compared to single modulation (a), the dual modulation (c) with an extinction ratio around 10 dB results in an increase of about 120 km for a 2 dB power penalty.

Fig. 5 depicts the optical spectrum at the output of the D-EML model for each case in the EAM chirp profile 3 configuration. This chirp profile was chosen for its low tran-

978-1-4673-1725-2/12 $31.00 © 2012 IEEE

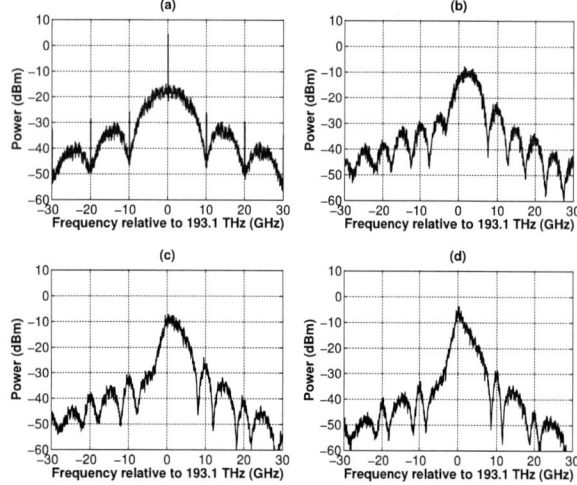

Fig. 5. Optical spectra for the cases (a), (b), (c) and (d) after D-EML model at 10 Gb/s in DATA/DATA configuration with the chirp profile 3.

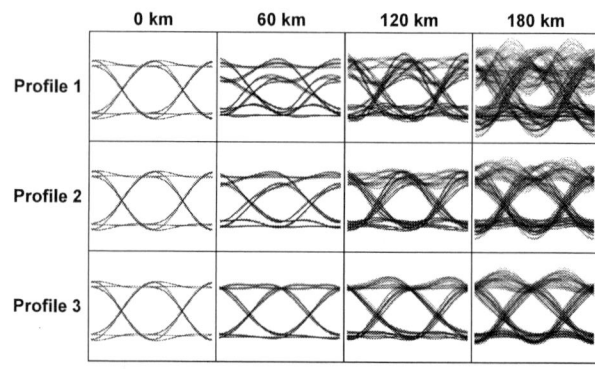

Fig. 6. Electrical eye diagrams after receiver for the case (d) in the three profile configuration for different transmission distances.

sient chirp in order to completely fill the SSB requirement. Compared to the case of pure AM (a), the cases of mixed AM/FM (b), (c) and (d) allow spectrum sculpting by reducing one of the side band (Lower Side Band in DATA/DATA configuration) in the signal spectral content and lead therefore to SSB spectra. Note that the sign of the data (DATA or inverted DATA on the laser input) determines which sideband disappears. Thus adiabatic chirp leading to blue-shift or red-shift reduces equally the spectral width. The narrower spectrum obtained is less penalized by the chromatic dispersion during its propagation through the fiber and the transmission distance can be enhanced by this way.

In the case of single modulation (a), increasing positive and negative transient chirp results in a significant decrease of transmission distance for a 2 dB power penalty. By comparing profiles 1 and 3 in Fig. 4, we observe a distance decrease of 30 km. The case of dual modulation (b) is equally affected when more transient chirp is introduced. For the cases of dual modulation (c) and (d), the introduction of higher transient chirp is more penalizing because of their lower extinction ratio. We note that some long transmission distances permit better performances compared to other shorter. For example in the case (d) for the profile 1, 120 km distance presents better performances than 60 km distance. Indeed, the transient chirp induces additional distortions of the temporal power-signal as it is depicted in eye diagrams of Fig. 6 for the case (d). These distortions reduce more or less eye opening depending on transmission distance leading to fluctuations in performances. On the other hand, the transient chirp induces wider SSB spectrum at D-EML model output.

Fig. 6 shows the electrical eye diagrams after receiver for the case (d) for different transmission distances and the three chirp profiles. As it is shown on this figure, the more the transient chirp is high, the less the eye opening is high because of distortions induced by transient chirp.

IV. CONCLUSION

Firstly, this paper proves, with component characterization-based simulations, that dual modulation scheme using the D-EML model can improve transmission distances far beyond chromatic dispersion limit. Moreover, we show that the EAM transient chirp profile is a key parameter and it is necessary to reduce it for increasing transmission distances without any additive dispersion compensation device.

ACKNOWLEDGMENT

The authors are grateful to the MODULE project from the French national initiative ANR-VERSO program, Systematic Paris-region and Elopsys Limousin-region competitiveness clusters for financial support.

REFERENCES

[1] Y. Matsui, D. Mahgerefteh, X. Zheng, C. Liao, Z. Fan, K. McCallion, and P. Tayebati, "Chirp-Managed Directly Modulated Laser (CML)," *IEEE Photonics Technology Letters*, vol. 18, no. 2, pp. 385–387, 2006.
[2] D. Mahgerefteh, Y. Matsui, X. Zheng, and K. McCallion, "Chirp Managed Laser and Applications," *IEEE Journal of Selected Topics in Quantum Electronics*, vol. 16, no. 5, pp. 1126–1139, 2010.
[3] M. Ngo, H. Nguyen, C. Gosset, D. Erasme, Q. Deniel, and N. Genay, "Transmission Performance of Chirp-Controlled Signal Emitted by Electroabsorption Modulated Laser Integrated with a Semiconductor Optical Amplifier," in *Optical Fiber Communication Conference*, 2012, OW4F.6.
[4] H. Kim, S. Kim, H. Lee, S. Hwang, and Y. Oh, "A Novel Way to Improve the Dispersion-Limited Transmission Distance of Electroabsorption Modulated Lasers," *IEEE Photonics Technology Letters*, vol. 18, no. 8, pp. 947–949, 2006.
[5] J. Petit, D. Erasme, C. Kazmierski, C. Jany, J. Decobert, F. Alexandre, N. Dupuis, and R. Gabet, "Enhanced 10-Gb/s NRZ Transmission Distance Using Dual Modulation of an Integrated Electro-Absorption Modulated Laser Transmitter," in *Optical Fiber Communication Conference*, 2009, OThG2.
[6] K. Kechaou, T. Anfray, K. Merghem, C. Aupetit-Berthelemot, G. Aubin, C. Kazmierski, C. Jany, P. Chanclou, and D. Erasme, "First Demonstration of Dispersion Limit Improvement at 20 Gb/s with a Dual Electroabsorption Modulated Laser," in *Optical Fiber Communication Conference*, 2012, OTh3F.1.
[7] O. Patard, F. Alexandre, F. Martin, J. Decobert, N. Lagay, R. Guillamet, and C. Kazmierski, "REAM-SOA Integrating Ru Doped InP Current Blocking Layer," in *Indium Phosphide and Related Materials Conference*, 2011, Post-deadline Paper.
[8] J. Binder and U. Kohn, "10 Gb/s-Dispersion Optimized Transmission at 1.55 μm Wavelength on Standard Single Mode Fiber," *IEEE Photonics Technology Letters*, vol. 6, no. 4, pp. 558–560, 1994.

Fabrication and DC characterization of InAs/AlSb Self-Switching Diodes

Andreas Westlund, Giuseppe Moschetti, Huan Zhao, Per-Åke Nilsson, Jan Grahn

Department of Microtechnology and Nanoscience (MC2), Chalmers University of Technology, SE 412 96, Göteborg, Sweden

[1#]andreas.westlund@chalmers.se

Abstract — Fabrication and DC measurements of an InAs/AlSb self-switching diode (SSD), aimed for THz detection, is presented. An SSD with a channel width of 160 nm and a trench width of 240 nm was designed and fabricated in a process using an *in situ* passivation procedure of the oxidation-sensitive trench. Rectifying behavior was observed in the I-V characteristics. The device performance was relatively stable over a period of three months.

I. INTRODUCTION

The self-switching diode (SSD) is a unipolar rectifying nanodiode [1] with potential for detection applications in the THz range. Recent room-temperature measurements for GaAs SSDs showed detection capability at 1.5 THz [2].

InAs/AlSb is a promising material system for room temperature THz applications due to its narrow band gap and high velocity electrons [3]. Compared to GaAs, the InAs material exhibits about three times greater carrier concentration and 2.5 times higher electron mobility. Such properties suggest that InAs may allow even higher operating frequencies of the SSD. Monte Carlo simulations have shown that InAs/AlSb SSDs can operate well above 2 THz [4].

The SSD operation is schematically described in Fig. 1. A 2DEG flows across a narrow channel made by vertical etching of trenches. In equilibrium, surface charges cause a lateral depletion around the trenches,. In reverse bias, the depletion pinches the channel, whereas in forward bias, electrons are free to flow [1].

A large obstacle for effective processing of InAs/AlSb SSDs is the AlSb material used in the buffer layer of the epitaxial heterostructure. The AlSb is strongly prone to oxidation and must therefore be protected throughout the fabrication. Otherwise, the device will quickly degrade [5].

In this work, we have designed and fabricated InAs/AlSb SSDs showing rectifying DC behavior. Similar work has been done using wet etching [6]. In this work devices were defined using dry etching to minimize oxidation. The full fabrication process and I-V results is presented.

II. DESIGN AND FABRICATION

All fabricated SSDs had five channels in parallel to increase current levels. Different designs of channels were

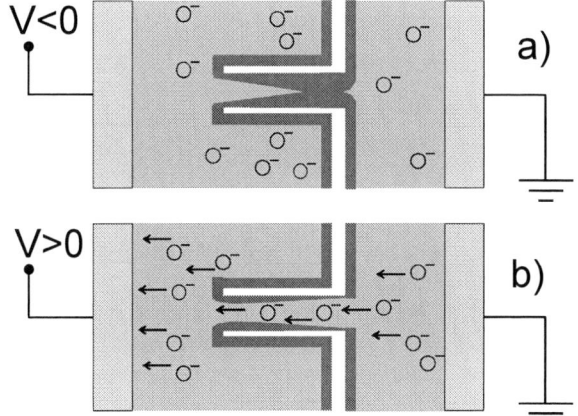

Fig. 1. The working principle of the SSD (top view). Dark blue is lateral depletion. (a) Reverse and (b) forward bias.

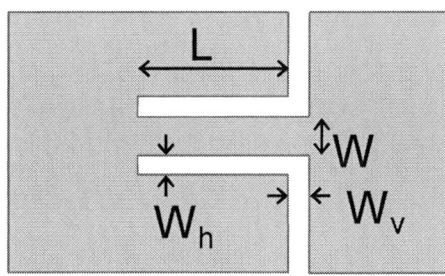

Fig. 2. Definition of diode dimensions of the SSD. All fabricated designs included five channels in parallel.

tested. For definitions, see Fig. 2. Designs tested were W in the range 250 - 750 nm, W_h in the range 150-250 nm, W_v in the range 200-1000 nm and L in the range 1-10 μm.

The epitaxial InAs/AlSb heterostructure was grown by molecular beam epitaxy on a GaAs bulk; See Fig. 3. The channel was pure InAs, grown on top of an AlSb buffer with an intermediate layer of AlGaSb. In contrast to AlSb, AlGaSb is much more oxidation resistant, although more conductive.

Room-temperature Hall measurements gave an electron concentration and mobility of 1.8×10^{12} cm^{-2} and 18,000 cm^2/Vs, respectively. The sheet resistance was 194 Ω/square.

A process overview is given in Fig. 4. First, ohmic contacts of Pd/Pt/Au are deposited by evaporation (step 1). The

InAs	2×10^{19} cm^{-3}	40 Å
In$_{0.52}$Al$_{0.48}$As		40 Å
Barrier	AlSb	70 Å
Doping layer	2 mono-layers InAs	
Doping plane	Si=3×10^{12} cm^{-2}	
Doping layer	2 mono-layers InAs	
Spacer	AlSb	60 Å
Channel	InAs	150 Å
Buffer	AlSb	500 Å
Metamorphic Al$_{0.8}$Ga$_{0.2}$Sb		2500 Å
Metamorphic buffer AlSb		7500 Å
GaAs (001) SI		

Fig. 3. Epitaxial structure for the SSD grown by molecular beam epitaxy.

contacts were annealed at 275°C. Second, the trench pattern was defined in resist with e-beam lithography (step 2). A trench was then formed in the semiconductor by dry-etching with a Cl$_2$:Ar inductively-coupled plasma reactive ion-etch process (step 3). Etching was stopped in the Al$_{0.8}$Ga$_{0.2}$Sb buffer layer, thereby minimizing the exposed area of AlSb in the device.

To electrically passivate the inside of the trench and protect it from oxidation, an approximately 25 nm thick layer of silicon nitride was deposited by plasma-enhanced chemical vapor deposition at room temperature, see Fig. 4 (step 4). In order to minimize oxidation of the exposed AlSb, the deposition was done *in situ*. Following deposition, excess passivation was removed in a lift-off procedure, (step 5). After the trench step, a pad layer of gold was deposited by evaporation (step 6).

The last step in the process the mesa definition, see Fig. 4 (step 7-9). Wet mesa etching is known to cause considerable sidewall- oxidation [5], thus a dry method was chosen also for the mesa etch. All areas except for the active device and metallization region were etched down to the substrate with a Cl:Ar inductively coupled reactive ion plasma etch. This deep mesa etch was necessary to avoid oxidation of the buffer between mesas. The mesa sidewalls were left unpassivated. However, for this process this was acceptable since the major problem of oxidation was in the trench.

A top view of a finished 5-channel SSD using scanning electron microscopy (SEM) is shown in Fig. 5. Measured W from the SEM picture was found to be 160 nm after compensating for the 25 nm thick silicon nitride covering the sidewalls of the channel. W_v and W_h were measured to 1 μm and 240 nm, respectively.

Also seen in Fig. 5 is the protective effect of the silicon nitride, most clearly observed by comparing the inside of the trench with the rough non-passivated mesa edge.

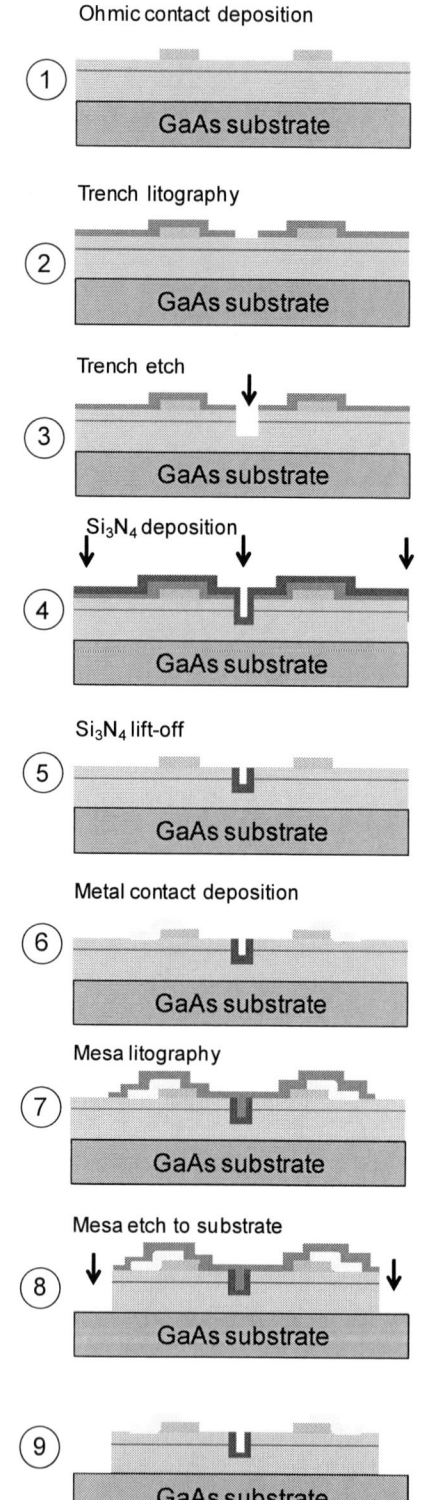

Fig. 4. Process flow of InAs SSD fabrication. (1) Ohmic contacts fabrication, (2) Trench resist patterning, (3) Trench Cl:Ar RIE/ICP etch, (4) *In situ* PECVD silicon nitride deposition, (5) Silicon nitride lift-off, (6) Metal contact fabrication, (7) Mesa lithography, (8) Deep Cl:Ar RIE/ICP mesa etch to substrate, (9) Final device.

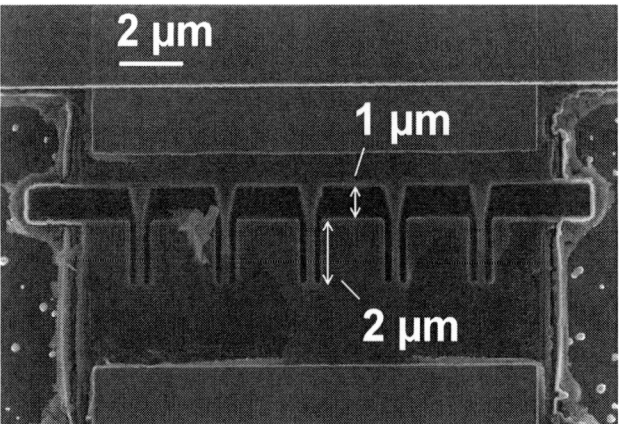

Fig. 5. SEM image of a fabricated five-channel InAs/AlSb SSD. Channel width was 150 nm and trench width 240 nm. The isolating trench was 1 μm wide.

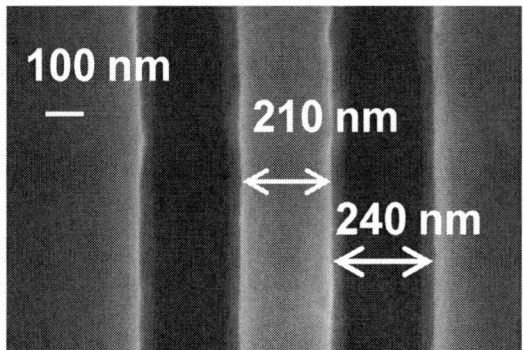

Fig. 6. Magnification of a channel in the device shown in Fig. 5. Channels and trenches appear well-defined. The channel is covered by 25 nm of silicon nitride, meaning the physical width of the channel is 160 nm.

III. RESULTS AND DISCUSSION

The DC performance of the SSDs was measured just after fabrication and after three months. I-V-characteristics for W=250 nm, W_h=200 nm, W_v=1000 nm and L=2 μm are plotted in Fig. 7. The SSD exhibits a clear knee in the I-V. For +1 V bias, the current is 4.5 μA. There is a significant diode leakage of 2 μA for -1 V, but the rectifying action is clear.

For other designs, asymmetric I-V was observed but none as distinct as in Fig. 7. Notably, the device exhibiting rectifying behavior was designed for the smallest W.

After three months of storage in a nitrogen ambience, the SSD exhibited current levels of +5.5 μA and -4 μA at +1V and -1 V, respectively; See Fig. 5. A doubling in diode reverse leakage was thus seen and a somewhat increased forward current. The region 0-0.8 V was relatively unaffected by the three months storage. The observed differences may be due to a slow degradation of the devices in air. Ion implantation has recently been proven to combine high electrical isolation and large resistance against oxidation for

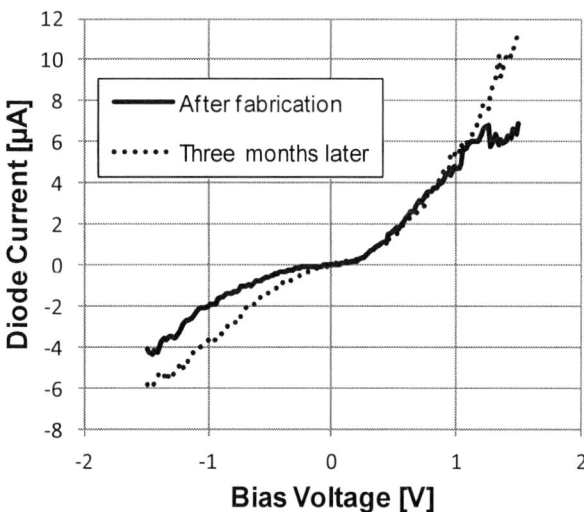

Fig. 7. IV characteristics measured at room temperature of a fabricated five-channel SSD just after fabrication and three months later. In the region of maximum curvature, 0-0.8V, the output current is unaffected.

AlSb materials and is a promising option for producing stable mesas [7].

Some observations were made regarding the differential resistance of the device in Fig. 7. A total series resistance of 140 kΩ is observed at 0.7 V bias. For detection applications, zero voltage bias is preferred due to the low low-frequency noise associated with zero bias current [8]. At 0 V bias, the resistance of the presented InAs SSD is approximately 1.2 MΩ. In a detector application where the SSD must be matched to a lower impedance, this high impedance will result in inefficient power transfer [2]. In the reported InGaAs SSD detection experiment at 1.5 THz [2], a four-channel SSD with an impedance of 150 kΩ was used. To reduce the impedance of the InAs SSD, the channel should be made wider, allowing a zero-threshold SSD [2].

IV. CONCLUSIONS

An InAs/AlSb SSD has been designed, fabricated and measured. Rectification was observed in a device with 160 nm wide channel. The DC properties of the SSD was measured after three months of storage in nitrogen ambience and found to be relatively stable.

V. ACKNOWLEDGEMENTS

The authors wish to acknowledge the financial support of this work through the project ROOTHz, a part of the Future and Emerging Technologies (FET) programme within the Seventh Framework Programme for Research of the European Commission, under FET-Open grant number 243845.

REFERENCES

[1] A.M. Song, M. Missous, P. Omling, A.R. Peaker, L. Samuelson, and W. Seifert, "Unidirectional electron flow in a nanometer-scale semiconductor channel: A self-switching device," *Applied Physics Letters,* vol. 83, pp. 1881-1883, September 2003.

[2] C. Balocco, S.R. Kasjoo, X.F. Lu, L.Q. Zhang, Y. Alimi, S. Winner, and A.M. Song, "Room-temperature operation of a unipolar nanodiode at terahertz frequencies," *Applied Physics Letters*, vol. 98, 223501, May 2011.

[3] B.R. Bennett, R. Magno, J.B. Boos, W. Kruppa, and M.G. Ancona, "Antimonide-based compound semiconductors for electronic devices : A review," *Solid-State Electronics.*, vol. 49, pp. 1875-1895, November 2005.

[4] I. Iniguez-de-la-Torre, H. Rodilla, J. Mateos, D. Pardo, A.M. Song, and T. Gonzalez, "Terahertz tunable detection in self-switching diodes based on high mobility semiconductors: InGaAs, InAs and InSb," *Journal of Physics.: Conf. Ser.* 193, 012082, 2009.

[5] A. Olivier, T. Gehin, L. Desplanque, X. Wallart, Y. Roelens, G. Dambrine, A. Cappy, S. Bollaert, E. Lefebvre, M. Malmkvist, and J. Grahn, "AlSb/InAs HEMTs on InP substrate using wet and dry etching for mesa isolation," *IEEE IPRM '08*, pp. 316-318, 2008.

[6] L.Q. Zhang, C. Balocco, Y. Alimi, H. Zhao, A. Westlund, G. Moschetti, and P. Nilsson, "Novel Unipolar Nanodiodes in InAs / AlGaSb Heterostructure Fabricated by Wet Chemical Etching," *UK Semiconductors*, 2011.

[7] G. Moschetti, P.-A. Nilsson, A. Hallen, L. Desplanque, X. Wallart, and J. Grahn, "Planar InAs/AlSb HEMTs With Ion-Implanted Isolation," *Electron Device Letters*, vol. 33, pp. 510-512, April 2012.

[8] C. Balocco, S.R. Kasjoo, L.Q. Zhang, Y. Alimi, and A.M. Song, "Low-frequency noise of unipolar nanorectifiers," *Applied Physics Letters*, vol. 99, 113511, September 2011.

Effect of temperature on series resistance determination of Au/polyvinyl alcohol/n-InP Schottky structures

M. Siva Pratap Reddy[1*], Hee-Sung Kang[1], Dong-Seok Kim[1], Young-Woo Jo[1], Chul-Ho Won[1], Ryun-Hwi Kim[1], Kyu-Il Jang[1], Chandrashekhar C.H[1], Jung-Hee Lee[1] and V. Rajagopal Reddy[2]

[1]School of Electrical Engineering & Computer Science, Kyungpook National University, Daegu, Korea

[2]Department of Physics, Sri Venkateswara University, Tirupati, India

Abstract — The current-voltage (I-V) characteristics of Au/polyvinyl alcohol (PVA)/n-InP Schottky diode have been measured at temperature range 175-425 K. It is found that the series resistance (R_S) values of Au/PVA/n-InP Schottky diode estimated from Cheung's and Norde's methods, are strongly temperature dependent. The values of barrier height and R_S have very different especially towards to the lower temperatures. This is attributed to non-ideal I-V characteristics of the MIS structure and non-pure thermionic emission (TE) mechanism due to the low temperature effects.

Index Terms — Au/PVA/n-InP Schottky diode, barrier height, ideality factor, series resistance, thermionic emission.

I. INTRODUCTION

Developing organic-inorganic heterojunction is a promising technology due to several advantages of organic materials such as flexibility, light weight, low cost, easy manufacturing and versatile applicability. The electrical properties of the metal-semiconductor can be usefully modified by inserting organic layer in metal and semiconductor. Such type of heterojunction fabrication and the extraction of the diode parameters is an attractive research field for photovoltaic devices [1].

II. EXPERIMENTAL TECHNIQUE

The Schottky diodes are prepared using a one side polished n-InP wafer with a donor concentration of $4.9\text{-}5.0 \times 10^{15}$ cm^{-3}. The wafer is cleaned with warm organic solvents like trichloroethylene, acetone and methanol by means of ultrasonic agitation for the duration of 5 min each to remove contaminants, followed by rinsing in deionized (DI) water and drying in N_2 flow. Then wafer is etched with HF (49%) and H_2O (1:10) to remove the native oxides from the substrate. Before forming the organic layer to the n-InP substrate, the ohmic contact is made by evaporating indium on the rough side of the n-InP. The contacts are then annealed at 350 °C for 1 min in N_2 ambient. The polyvinyl alcohol (PVA) film is deposited on n-type InP by spin-coating technique. The thickness of the film so obtained is in the range of 10-20 nm across the full substrate surface used is uniform. Finally top metal dots with a diameter of 0.7 mm are deposited of Au (50

nm) by e-beam evaporation system. The temperature dependence of the I-V characteristics of the Au/PVA/n-InP Schottky diodes are measured in the temperature range 175-425 K in steps of 25 K using a Keithley 2400 and DLS-83D spectrometer. The temperature is controlled with an accuracy of ± 1 K using temperature controller DLS-83D-1 cryostat.

III. RESULTS AND DISCUSSION

The forward and reverse bias I-V characteristics of Au/PVA/n-InP Schottky diodes in the range of 175-425 K by the steps of 25 K are shown in Fig. 1.

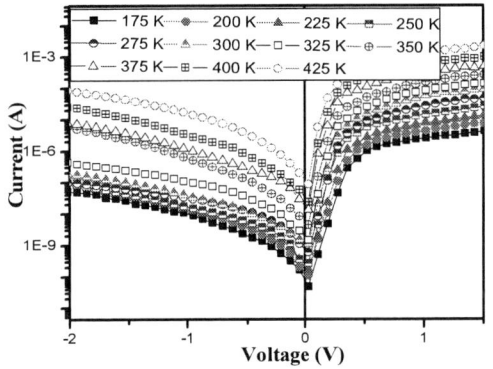

Fig. 1.   The current-voltage (I-V) characteristics of Au/PVA/n-InP Schottky barrier diode in the temperature range of 175-425K

The current-voltage characteristics of the Schottky structure can be described by the following relation [1].

$$I = I_o \exp\left(\frac{qV}{nkT}\right)\left[1 - \exp\left(-\frac{qV}{kT}\right)\right] \qquad (1)$$

$$I_o = AA^{*}T^{2}\exp\left(-\frac{q\Phi_{bo}}{kT}\right) \qquad (2)$$

where I_o is the saturation density, q is the electron charge, V is the applied voltage, T is the absolute temperature, k is the Boltzmann constant, Φ_{bo} is the Schottky barrier height (SBH), A is the diode area and A^{*} is the Richardson constant for n-InP (A^{*}=9.4 A/cm^{2}K^2). The ideality factor of Au/PVA/n-InP Schottky diode from Eq. (1) is contained in the slope of

978-1-4673-1725-2/12 $31.00 © 2012 IEEE

straight line region of the forward-bias characteristics of I-V through the relation.

$$n = \left(\frac{q}{kT} \right) \left[\frac{dV}{d \ln I} \right] \quad (3)$$

where n is a measure of conformity of the diode to pure thermionic emission (TE). Calculations showed that the SBH (Φ_{bo}) and ideality factor n values of the Au/PVA/n-InP Schottky diodes are 0.48 eV and 3.59 at 175 K, and 0.93 eV and 1.31 at 425 K, respectively.

As can seen in Fig. 2 the zero-bias barrier height increases with increasing temperature, whereas the ideality factor decreases with increasing temperature. This is indicating that the current transport deviates from the ideal TE model. Because the current transport across the MIS interface is a temperature-activated process electrons at low temperatures are able to surmount the lower barriers [2-4]. Therefore, the current transport will be dominated by the current flowing through the patches of lower SBH, leading to a larger ideality factor. Larger ideality factors are attributed to secondary mechanisms at the interface such as lateral inhomogeneous distribution of BH which may be created by interface defects.

The series resistance (R_S) is a crucial parameter of Schottky diode. The resistance of the Schottky diode is the sum of total resistance value of the resistors in series and resistance in semiconductor device in the direction of current flow. The diode parameters as the Φ_{bo}, n and R_S were also achieved using a method developed by Cheung's [5]. The forward bias I-V characteristics due to thermionic emission of a Schottky contact with R_S can be expressed as the Cheung's function and given by

$$\frac{dV}{d(\ln I)} = IR_S + \frac{nkT}{q} \quad (4)$$

$$H(I) = V - \frac{nkT}{q} \ln\left(\frac{I}{AA^*T^2} \right) \quad (5)$$

$$H(I) = IR_S + n\Phi_{bo} \quad (6)$$

From forward I-V characteristics of Au/PVA/n-InP Schottky diode show a straight line in the downward curvature region. In Fig. 3(a) and (b), dV/dlnI versus I and H(I) versus I plots are presented at different temperatures for Au/PVA/n-InP Schottky diode. Equation (4) should give straight line for the data of down curvature region in the forward bias I-V. Thus the values of R_S and nkT/q have been obtained from the slope and y-axis intercepts of the dV/dlnI versus I plots from Fig. 3(a), respectively. As function of temperature, it is evident that the plot of H(I) versus I from Fig. 3(b) gives a straight line with the y-axis intercept equal to $n\Phi_{bo}$ and the slope parameter (R_S) can be used to check the consistency of Cheung's function. However, it can clearly be seen that there is a comparative variation between the values of the ideality factor obtained from the downward curvature region of the forward bias I-V plots and from the linear regions of the same characteristics. This difference can be attributed to the existence of effects such as the R_S and the bias dependence of the SBH.

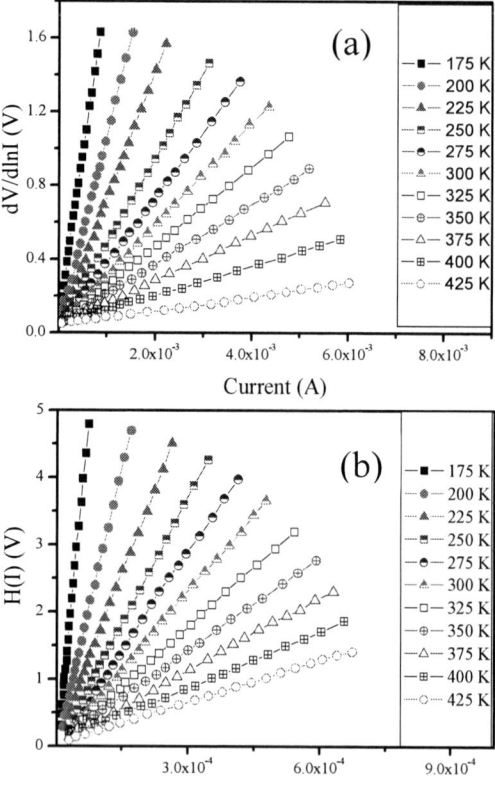

Fig. 3. Cheung's functions of (a) plots of dV/dlnI vesus I, and (b) H(I) verus I for the Au/PVA/n-InP Schottky diode at various temperatures.

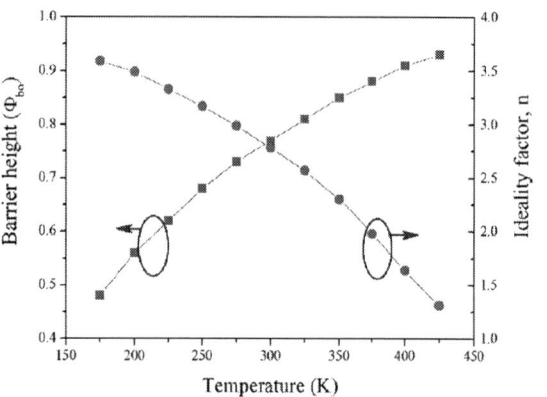

Fig. 2. The variation in the zero-bias barrier height and ideality factor with temperature for Au/PVA/n-InP Schottky diode.

To determine the value of R_S using an alternative method, proposed by Norde [6]. Using this method, F(V) is defined as

$$F(V) = \frac{V}{2} - \frac{kT}{q}\ln\left[\frac{I(V)}{AA^*T^2}\right] \quad (7)$$

where I(V) is the current obtained from the I-V curve. A plot of F(V) versus V for the diode at different temperatures is shown in Fig. 4. From the plots of F(V) versus V, the value of the BH of the diode can be determined as follows

$$\Phi_{bo}^{Norde} = F(V_{min}) + \frac{V_{min}}{2} - \frac{kT}{q} \quad (8)$$

where $F(V_{min})$ is the minimum value of F(V) and V_{min} is the corresponding voltage. For real contacts (n>1), the R_S can be expressed as

$$R_S = \frac{(2-n)kT}{qI_{min}} \quad (9)$$

where I_{min} is the value of the forward current at the voltage V_{min} where the function F(V) exhibits a minimum. The values of R_S obtained from the equation (9), which is decrease with an increase in the temperature and the values are given in Table I. As seen in Table I, the values of the Φ_{bo} and R_S obtained from the Norde's method have been slightly changed with increasing temperature. It can be said that, especially, the values of R_S obtained from both methods are slightly different from each other. In most cases, the parameters obtained from the Cheung's method and the Norde's methods are not in agreement with each other. From the F(V) versus V plots, the values of the Φ_{bo} and R_S have been determined. In general, the values obtained from the Norde's method are higher than the estimated from the Cheung's method. Because, Norde's method is applied to the full forward bias I-V characteristics whereas, the Cheung's method is only applied to the non-linear region (high voltage regime) of the semi-logarithmic forward bias I-V characteristics of the junctions.

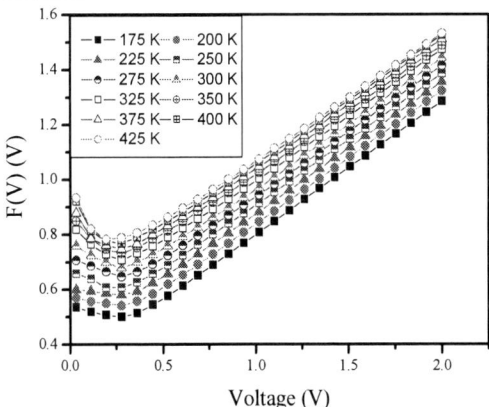

Fig. 4. F(V) versus V of Au/PVA/n-InP Schottky structure as function of temperature.

TABLE I

TYPICAL PARAMETERS OBTAINED FROM I-V MEASUREMENTS

T (K)	I-V		Cheung's				Norde's	
	Φ_{bo} (eV)	n	dV/dlnI		H(I)		R_S (Ω)	Φ_{bo} (eV)
			R_S (Ω)	n	R_S (Ω)	Φ_{bo} (eV)		
175	0.48	3.59	656	3.89	668	0.51	984	0.62
200	0.56	3.49	567	3.64	581	0.58	849	0.66
225	0.62	3.33	484	3.51	503	0.65	742	0.70
250	0.68	3.17	417	3.36	422	0.70	655	0.73
275	0.73	2.99	362	3.19	379	0.75	570	0.76
300	0.77	2.78	315	2.97	326	0.79	504	0.79
325	0.81	2.57	268	2.81	279	0.83	443	0.82
350	0.85	2.30	243	2.63	253	0.87	396	0.84
375	0.88	1.98	209	2.38	218	0.89	351	0.86
400	0.91	1.64	177	2.01	185	0.92	322	0.87
425	0.93	1.31	152	1.75	159	0.94	298	0.88

IV. CONCLUSIONS

In this work, the electrical properties of Au/PVA/n-InP Schottky structure are measured in range between 175 and 425 K at 25 K intervals. The zero-bias barrier height and ideality factor have been obtained from forward bias I-V characteristics assuming TE mechanism show strong temperature dependence. It is revealed that the R_S values estimated from Cheung's and Norde's functions decreases as the temperature increases. It is seen that the electrical properties of Au/PVA/n-InP Schottky structure are remarkably changed with temperature.

ACKNOWLEDGEMENTS

This work was supported by 2008 Brain Korea 21 (BK21), the National Research Foundation of Korea (NRF) grant founded by the Korea government (MEST) (No. 2011-0016222, 2012-0000627), R&D program of MKE/KETEP (2011101050017B, 'Development of high efficiency GaN power device for an power grid inverter system', and WCU (World Class University) program through the Korea Science and Engineering Foundation founded by the Ministry of Education, Science and Technology (R33-10055), and the IT R&D program of MKE/KEIT (10038766, Energy Efficient Power Semiconductor Technology for Next Generation Data Center).

REFERENCES

[1] S.M. Sze, *Physics of Semiconductors*, New York: Wiley, 1981.

[2] J. H. Werner, and H. H. Guttler, "Barrier inhomogeneities at Schottky contacts," *J. Appl. Phys.*, vol. 69, no. 3, pp. 1522-1533, February 1991.

[3] A. Gumus, A. Turut, and N. Yalcin, "Temperature dependent barrier characteristics of CrNiCo alloy Schottky contacts on n-

type molecular-beam epitaxy GaAs," *J. Appl. Phys.*, vol. 91, no. 1, pp. 245-250, October 2001.

[4] S. Karatas, S. Altinal, A. Turut, and A. Ozmen, "Temperature dependence of characteristic parameters of the H-terminated Sn/p-Si (100) Schottky contacts," *Appl. Surf. Sci.*, vol. 217, no. 1, pp. 250-260, March 2003.

[5] S.K. Cheung, and N.W. Cheung, "Extraction of Schottky diode parameters from forward current-voltage characteristics," *Appl. Phys. Lett.*, vol. 49, no. 2, pp. 85-87, May 1986.

[6] H. Norde, "A modified forward I-V plot for Schottky diodes with high series resistance," *J. Appl. Phys.*, vol. 50, no. 7, pp. 5052-5053, February 1979.

ESTIMATION OF EFFECTIVE MASS AND SUBBANDS IN MULTI-QUANTUM WELLS USING POLARIZED LIGHT IRRADIATION

K. Tanaka,[a] N. Happo,[a] M. Fujiwara,[a] N. Kotera[b].

[a] Hiroshima City University, 3-4-1 Ozuka-Higashi, Asa-Minami-Ku, Hiroshima, 731-3194, JAPAN

[b] Kyushu Institute of Technology, 680-4 kawazu, Iizuka, 820-8502, JAPAN

Abstract — Nonparabolic nature of optical interband transitions of multi-quantum wells structure is important for related device designs. The optical interband transitions of $In_{0.53}Ga_{0.47}As/In_{0.52}Al_{0.48}As$ MQWs including 20-nm-wide well clearly appeared on photocurrent spectra with polarized light source. The nonparabolic nature of conduction subbands was decided from experimental interband transitions. The electron effective mass smoothly increased from 0.041 m_0 to 0.07 m_0 in the conduction quantum well.

Index Terms — Photocurrent, Transverse-Electric Mode, Transverse-Magnetic Mode.

I. INTRODUCTION

Nonparabolic nature of optical interband transitions of multi-quantum wells (MQWs) structure is important for related device designs [1-2]. The optical interband transitions of $In_{0.53}Ga_{0.47}As/In_{0.52}Al_{0.48}As$ MQWs including 20-nm-wide well clearly appeared on photocurrent spectra with polarized light source. The nonparabolic nature of subbands was decided from experimental interband transitions.

II. EXPERIMENTAL

Our specimen was made by an MBE technique on a (100) surface of each n-type InP substrate. Undoped $In_{0.53}Ga_{0.47}As/In_{0.52}Al_{0.48}As$ MQWs were in succession to an n-InAlAs buffer layer on an n-type InP substrate. Width of the InGaAs quantum wells was 20 nm. Number of quantum wells was 16. Width of the InAlAs barriers was 10 nm. The heavily doped p-InGaAs layer, which was used as an electrode in contact with a metal probing needle on the photocurrent measurement, was on top of such a p-i-n junction. The photocurrent was measured when dc reverse bias voltages were applied to the p-i-n junction and some electric field was intentionally generated within the quantum well layers under infra-red irradiation. In the such 20 nm wide quantum well which it was hard to distinguish of light hall subbands became clear. The infra-red light of Transverse-Electric Mode(TE) and Transverse-Magnetic Mode(TM) that was polarized with a polarizing plate was irradiated on the specimen with the photocurrent measurement.

III. RESULTS AND DISCUSSIONS

The electron effective mass was discussed with a result of making the experimental electron eigen-energies to fit the envelope function model in the effective mass approximation.

A. Photocurrent Spectra

Photocurrent spectra of the MQWs were measured between 293 K and 9.3 K. The polarized light was radiated with incidence angle, 55°. These spectra are normalized by a photocurrent spectrum of a similar p-i-n junction structure including an undoped InGaAs layer of 500 nm instead of the MQWs in Fig. 1. Steplike structures of the spectra which consist of edges and sequent plateaus are believed to be due to the two-dimensional density of states. Exciton peaks of interband transitions caused by heavy hole between the conduction subband and the valence subband must be bigger than the ones caused by the light hole, because of the excitonic transition probably. Series of these steps are assigned to be allowed optical transitions having same quantum number, n, in conduction and valence subbands. Peaks originated from heavy and light holes are respectively shown in Hn and Ln.

This is started from an eminent edge of band gap of the n-InP substrate at 0.74 eV and ended to a big rise of band gap of the n-InP substrate at 1.26 eV. This big rise appears at same energy position in other spectra. A notch between 1.20 and 1.28 eV is a systematic noise for the TE spectra. Transition types were determined from the two-dimensional spectral shape and comparison of TM and TE spectral shape. The energies of the seven allowed transitions of heavy hole, E_{H1}, E_{H2}, E_{H3}, E_{H4}, E_{H5}, E_{H6} and E_{H7} are 0.74, 0.78, 0.85, 0.95, 1.05, 1.19 and 1.33 eV as well. Light hole transitions, E_{L1}, E_{L2}, E_{L3}, E_{L4} and E_{L5} are 0.76, 0.83, 0.96, 1.13 and 1.28 eV. When the big rise of band gap of the n-InP substrate at 1.26 eV is represented in the Fig. 1, the peaks of transitions appear to be smaller. Hence, H7 and L5 that are higher the band gap of the n-InP are not included in the display range.

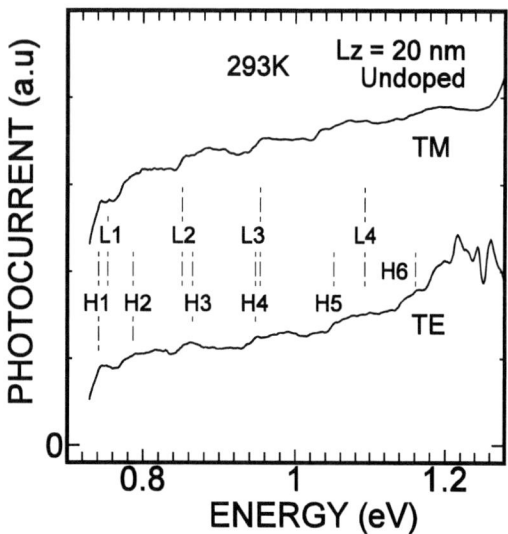

Fig. 1 Photocurrent spectra measured at 293 K with polarized light. Allowed transitions caused by heavy hole and light hole, Hn and Ln, are marked with broken lines.

Photocurrent spectra of the MQWs were measured at 9.3 K, too, as shown in Fig. 2. The energies of the seven allowed transitions of heavy hole, E_{H1}, E_{H2}, E_{H3}, E_{H4}, E_{H5}, E_{H6} and E_{H7}, are 0.81, 0.84, 0.92, 1.01, 1.12, 1.26 and 1.38 eV as well. Light hole transitions, E_{L1}, E_{L2}, E_{L3}, E_{L4} and E_{L5} are 0.82, 0.89, 1.02, 1.18 and 1.36 eV. Transition energies at 9.3 K are about 66 eV higher than ones at 293 K. These transitions are located at higher position by the band gap energy additional.

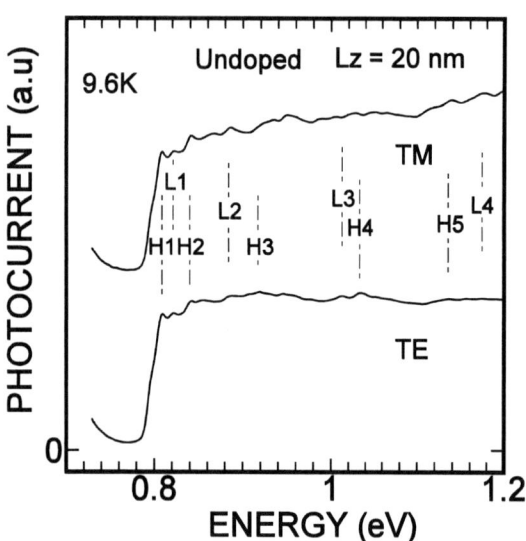

Fig. 2 Photocurrent spectra measured at 9.3 K. Allowed transitions are located at higher energy position by the band gap energy additional comparing with 293 K spectrum.

The peaks of the transitions originated from light hole subbands are emphasized on the TM spectrum compared with the TE spectrum. If it takes these differences at each temperature, the peak due to the heavy holes and light holes will be emphasized. The transitions due to the heavy holes are shown as downward peaks, and the transitions due to the light holes are shown as upward peaks in Fig. 3. For 9.3 K spectrum, the energies of the seven allowed transitions of heavy hole, E_{H1}, E_{H2}, E_{H3}, E_{H4}, E_{H5}, E_{H6} and E_{H7}, are shown as downward peaks at 0.81, 0.84, 092, 1.01, 1.12, 1.26 and 1.38 eV. By contrast, Light hole transitions, E_{L1}, E_{L2}, E_{L3}, E_{L4} and E_{L5} are shown as upward peaks at 0.82, 0.89, 1.02, 1.18 and 1.36 eV.

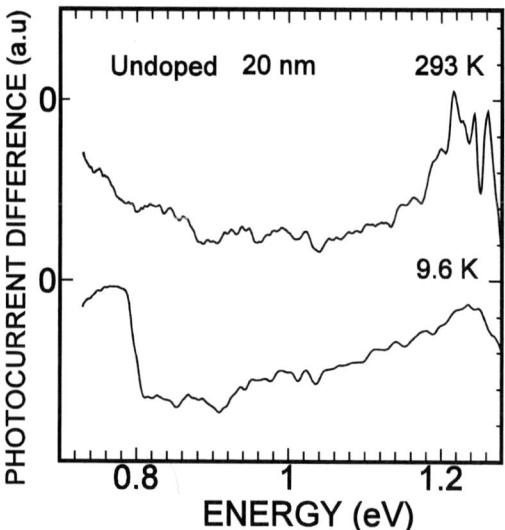

Fig. 3 Photocurrent difference Spectra. The transitions due to the heavy holes are shown as downward peaks, and the transitions due to the light holes are shown as upward peaks

B. Transition Energies

Eigen energies in finite square potential-well model is approximately proportional to square of the quantum number except an eigen energy to close to the top of the quantum well. Consisting of the sum of the eigen enegies of conduction and valence bands and band gap energy of the InGaAs well layer, the transition energies are also proportional to the square of the valence quantum number. Transition energies are shown against the square of the valence quantum number in Fig. 4. Transition energies of the heavy holes that are measured at 293 K are indicated with squares. These transition energies are linearly higher up to square of quantum number, 49. Number of transitions caused by the heavy hole is 7. Light holes transition energies at 293 K are indicated with filled squares. For these light holes transition energies are also linearly higher up to square of quantum number, 25, which Number of light hole transitions is 5. Circles and filled circles indicate respectively transition energies of the heavy hole and the light hole at 9.3 K. Numbers of transitions is respectively 7 and 5. Transition energies at 9.3 K are about 66 eV higher than ones at 293 K by the band gap energy additional. These transition energies that are caused by heavy and light holes are approximately proportional to the square of the quantum number. Transition energies are bent gently downward at higher energy part from linear line. It is reason that the effective mass of electron is nonparabolic and depends on the eigen energy, and the InGaAs layer is finite quantum well.

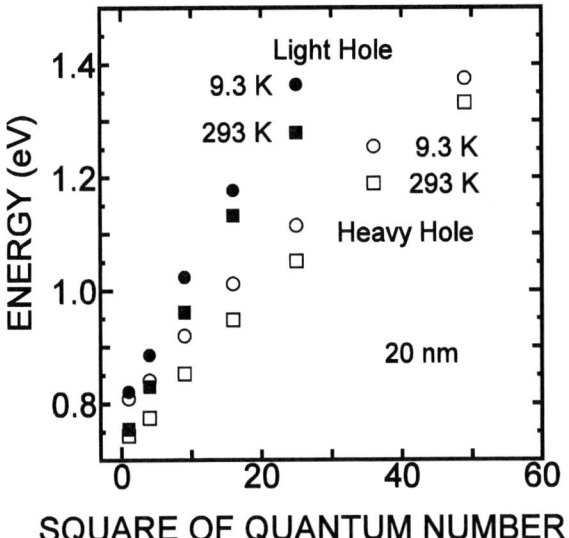

Fig. 4 Transition energies against the square of the valence quantum number. Squares and filled squares indicate respectively transition energies of the heavy hole and the light hole at 293 K. Circles and filled circles indicate respectively transition energies of the heavy hole and the light hole at 9.3 K.

Summary of transition energies are in Table 1. Transition energies of the heavy hole and light hole are arranged in numerical order of quantum number, n, at column 293 K and 9.3 K. Energy differences between higher transition E_n and E_1 are shown in columns of E_n-E_1. Energy per unit quantum number be calculated by dividing the E_n-E_1 of (n^2-1) for each transition. The $(E_n-E_1)/(n^2-1)$ is nearly constant at 0.014 eV for Hn and at 0.025 eV for Ln. Because transition energies is in proportion to square of the quantum number, this assignment of transitions is considered to be reasonable. Further, energy differences between transitions of Hn and Ln are shown in columns of E_{Hn}-E_{Ln}. Energy per unit quantum number be calculated by dividing the E_{Hn}-E_{Ln}. of n^2 for each transition. The $(E_{Hn}-E_{Ln})/(n^2)$ is nearly constant at 0.012 eV. However, in high energy near the edge of the quantum well, which is correspond to large quantum number, the transition energy is slightly lower. In the same way, concerning the spectrum of 9.3 K, the energy per unit quantum number that can be

discussed. The $(E_n-E_1)/(n^2-1)$ is nearly constant at 0.014 eV for Hn and at 0.025 eV for Ln. The $(E_{Hn}-E_{Ln})/(n^2)$ is nearly constant at 0.012 eV. Even if the temperature is changed, the values of $(E_{Hn}-E_{Ln})/(n^2-1)$ and $(E_{Hn}-E_{Ln})/(n^2)$ does not change. This might mean that the quantum well structure does not change even if the change in temperature. Comparing the same transition type for 293 K and 9.3 K, there is energy difference of 66 eV in column $E(9.6K)$-$E(293K)$. This corresponds to the band gap of In$_{0.53}$Ga$_{0.47}$As quantum well.

C. Nonparabolic Effective Mass

It is thought that peak energies of the transitions have nonparabolic nature because the energies don't increase in proportion to square of the quantum number. The effective mass is usually calculated from optical transition energies on using the envelope function approximation [3]. But we used energy differences of the subbands to estimate band offset and effective mass of carriers at the same time. In the result, the band offset was 0.52 eV in conduction band and 0.22 eV in valence band. About the estimated mass, it increased to 50% with energy for electron and was constant for heavy hole and light hole.

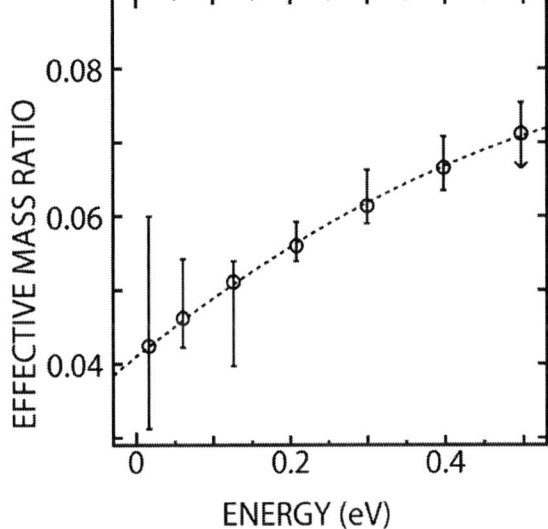

Fig. 3. Nonparabolic electron effective mass

By assuming smooth increase of electron effective mass, the energy bottom is estimated 16 meV below the $E_e(1)$. Eigen energy corresponding to each conduction quantum number, $E_e(n)$, is 0.016, 0.061, 0.126, 0.208, 0.299, 0.398, 0.497 eV, respectively. Effective mass, m_e*, is 0.042, 0.046, 0.051, 0.056, 0.061, 0.067, 0.071 eV and depends on the $E_e(n)$. These electron effective mass is expressed as a formula of energy,

$$m_e*(E \text{ [eV]}) = 0.041\ m_0\ (1+2.2\ E-1.4\ E^2) \qquad (1)$$

978-1-4673-1725-2/12 $31.00 © 2012 IEEE

in the conduction quantum well. If the well width become widely, the effective mass of lowest subband will become almost as much as bulk. Nonparabolic tendency is explicitly observed as the electron effective mass increases toward higher energy. These effective mass were consistent with theoretical electron subbands applying Kane's bulk band theory[4].

IV. SUMMARY

In summary, the nonparabolicity of the electron effective mass was explicitly determined from the experimental transition energies. The electron effective mass smoothly increased from 0.041 m_0 to 0.07 m_0 in the conduction quantum well.

REFERENCES

[1] R. Kohler, A.Tredicucci, F.Beltram, H. Beere, H. Linfield, A.Davies, D. Ritchie, R. Lotti, and F. Rossi,: Nature vol. 417, pp. 156, (2002).
[2] R. Akimoto, B. S. Li, K. Akita, and T. Hasama, Appl. Phys. Lett. vol. 87, pp. 181104, (2005).
[3] N. Kotera and K. Tanaka, Physica E:Low-dimensional Systems and Nanostructures, vol. 32, pp. 199-202 (2006).
[4] K. Tanaka, K. Fujikawa, M. Fujiwara, N. Happo, N. Kotera, Optical Quantum Electronics, vol. 41, pp. 903-912 (2009)

TABLE I
SUMMARY OF TRANSITION ENERGIES

	Heavy Hole Transition Energies (eV)							Light Hole Transition Energies (eV)				
	H1	H2	H3	H4	H5	H6	H7	L1	L2	L3	L4	L5
n	1	2	3	4	5	6	7	1	2	3	4	5
n^2	1	4	9	16	25	36	49	1	4	9	16	25
293 K	0.743	0.775	0.853	0.948	1.052	1.189	1.332	0.755	0.83	0.961	1.132	1.279
E_n-E_1	0.000	0.032	0.110	0.205	0.309	0.446	0.589	0.000	0.075	0.206	0.377	0.449
$(E_{Hn}-E_{Ln})/(n^2-1)$		0.011	0.014	0.014	0.013	0.013	0.012		0.025	0.026	0.025	0.019
E_{Hn}-E_{Ln}								0.012	0.055	0.108	0.184	0.227
$(E_{Hn}-E_{Ln})/(n^2)$								0.012	0.014	0.012	0.012	0.009
9.6 K	0.809	0.841	0.920	1.012	1.115	1.255	1.375	0.821	0.886	1.024	1.177	1.364
E_n-E_1	0.000	0.032	0.111	0.203	0.306	0.446	0.566	0.000	0.065	0.203	0.356	0.478
$(E_{Hn}-E_{Ln})/(n^2-1)$		0.011	0.014	0.014	0.013	0.013	0.012		0.022	0.025	0.024	0.020
E_{Hn}-E_{Ln}								0.012	0.045	0.104	0.165	0.249
$(E_{Hn}-E_{Ln})/(n^2)$								0.012	0.011	0.012	0.010	0.010
$E(9.6K)$-$E(293K)$	0.066	0.066	0.067	0.064	0.063	0.066	0.043	0.066	0.056	0.063	0.045	0.085

Single-Event Transient Sensitivity to Gate Bias in InAlSb/InAs/AlGaSb High Electron Mobility Transistors

V. Ramachandran[1*], R. D. Schrimpf[1,2], R. A. Reed[1,2], E. Zhang[1]
X. Shen[3], S. T. Pantelides[1,3]
D. McMorrow[4], J. Brad Boos[4]

[1]Department of Electrical Engineering and Computer Science, Vanderbilt University, Nashville TN 37235
[2]Institute for Space and Defense Electronics, Vanderbilt University, Nashville TN 37235
[3]Department of Physics and Astronomy, Vanderbilt University, Nashville TN 37235
[4]Naval Research Laboratory, Washington D.C. 20375

Abstract

We have characterized the single-event transient sensitivity to gate bias of InAlSb/InAs/AlGaSb high electron mobility transistors through experiments and simulations. These depletion-mode transistors exhibit increased charge collection as the gate bias moves from depletion toward threshold, similar to the response observed in floating body silicon-on-insulator devices. Maximum charge collection occurs near threshold, decreasing as the gate bias moves toward accumulation. The interplay between the longitudinal electric field in the channel and the vertical electric field underneath the gate affects the net radiation-generated charge in the InAs channel, which is responsible for the observed experimental trends.

I. Introduction

Antimonide-based compound semiconductors (ABCS) have gained increased use over the last decade because of their attractive material and electrical properties. This is primarily because of the excellent transport properties of the ABCS materials among the III-V family that lend themselves to higher frequencies of operation at lower power consumption compared to silicon, germanium, and other III-V materials [1]. In InAlSb/InAs/AlGaSb high electron mobility transistors (HEMTs), the high conduction band offset (ΔE_C = 1.1 eV) ensures strong electron confinement in the InAs quantum well with a very high tunneling barrier. The electron confinement because of the offset translates into both high (25,000 $cm^2/V.s$) room-temperature (RT) low-field mobilities and two-dimensional electron gas (2DEG) densities (8 × $10^{12}/cm^2$) in these devices [2]. These excellent operating characteristics generate interest in the potential use of InAlSb/InAs/AlGaSb HEMTs in space microelectronics, where power is always at a premium.

In this paper, we show the sensitivity of single-particle ionization-induced current transients in InAlSb/InAs/AlGaSb HEMTs to gate bias. We demonstrate through experiments and 2-D technology computer-aided design (TCAD) modeling the influence of both the vertical gate electric field and the horizontal source-to-drain field in modulating the net charge in the InAs channel after a particle strike. As a result, the relatively higher source-to-channel potential barrier lowering in depletion does not lead to a correspondingly higher integrated charge than that at threshold, while a relatively higher net charge in the channel in accumulation also results in less integrated charge than that at threshold.

II. Device description

InAlSb/InAs/AlGaSb HEMT development was first reported by Papanicolau et. al. [3]. The introduction of the InAlSb layer on top reduced problems related to oxidation of the AlSb in AlSb/InAs/AlSb HEMTs and also obviated the need for an extra InAlAs layer deposition over the top AlSb layer. Figure 1 shows the cross-section of an InAlSb HEMT and the corresponding band diagram obtained by using a TCAD simulator.

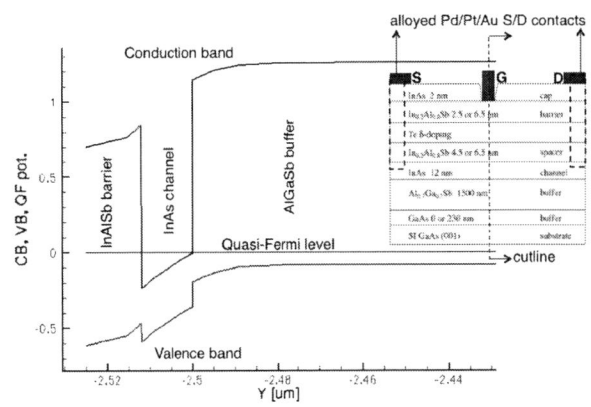

Fig. 1. Band diagram of the InAlSb/InAs/AlGaSb HEMT obtained from a 2-D TCAD simulator at zero bias. The vertical cutline used to obtain the band diagram is shown in the inset.

The 1.5 μm $Al_{0.7}Ga_{0.3}Sb$ buffer layer is grown on top of the starting semi-insulating (SI) GaAs material. The 12 nm InAs channel, the spacer layer, the delta-doping layer, and the

978-1-4673-1725-2/12 $31.00 © 2012 IEEE

In$_{0.2}$Al$_{0.8}$Sb barrier are then grown serially. The delta-doping is carried out using tellurium from a GaTe cell. Next, platinum/palladium/gold (Pd/Pt/Au) source and drain contacts are alloyed in in a 100 Å/200 Å/600 Å ratio by using e-beam evaporation and a 175°C heat treatment [4].

III. Experiments

The InAlSb/InAs/AlGaSb HEMTs were subjected to broadbeam heavy ion experiments at Lawrence Berkeley National Labs. Four HEMTs having threshold voltage values of -0.7 V, -0.6 V, -0.55 V, and -0.55 V were used. All experiments were carried out in air at room temperature and normal incidence. For all experiments, krypton ions from the 16 MeV/nucleon cocktail were used. The total energy and linear energy transfer of each ion were 1 GeV and 25 MeV-cm^2/mg, respectively.

Figure 2 shows the DC characteristics of the four HEMTs prior to irradiation with heavy ions. Before and after every series of ion-induced transient measurements, the HEMTs were tested for their DC characteristics to ensure proper operation. Each HEMT was individually mounted on a customized high-speed transient capture package [5] and the resulting transients

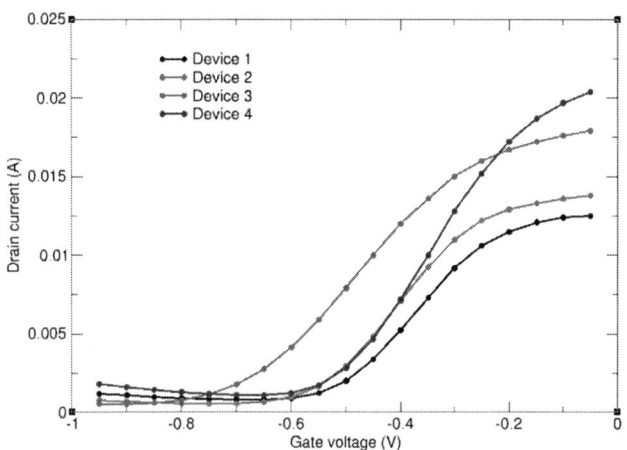

Fig. 2. DC characteristics of the four HEMTs used in the broadbeam experiments.

from the drain contacts of the device under test were captured in real time using a Tektronix TDS6124C high-bandwidth oscilloscope. The simultaneous application of a static DC bias and the corresponding ion-induced drain transient capture on the oscilloscope was facilitated by the use of a bias tee (model 5542-219) from Picosecond Labs [6]. The HEMTs were biased statically at a fixed drain and gate bias and the ion beam turned ON to record transients from the drain. The average flux and total fluence values of the krypton ion beam were 5×10^5 ions/cm^2/s and 1×10^8 ions/cm^2, respectively.

IV. Experimental results

The single-event integrated charge values recorded from the experiments are plotted in Fig. 3 as a function of effective gate bias at a constant drain bias of 0.25 V. Data were analyzed using effective gate bias since it helps in clearly discerning the bias

dependence, independent of threshold voltage variations among individual HEMTs. The integrated charge values were obtained by integrating the drain current transients recorded. The drain current transients were obtained by scaling the heavy ion-induced drain voltage excursion recorded by the oscilloscope and scaling them by the 50 Ω termination resistance.

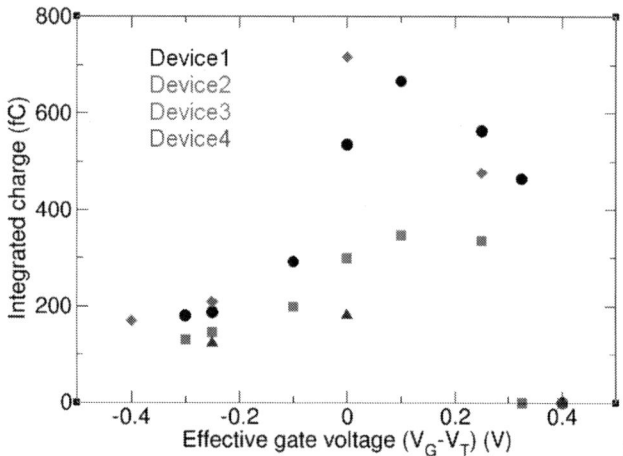

Fig. 3. Integrated charge as a function of effective gate bias, obtained from single-event experiments on four InAlSb/InAs/AlGaSb HEMTs. Drain voltage was 0.25 V for all experiments.

The integrated charge values peak near the threshold voltages of the HEMTs and fall off in both depletion and accumulation. At high accumulation biases, there were no current transients recorded on the drain in some HEMTs, as a result of which the integrated charge drops to zero in Fig. 3. Single-event cross-section values (not shown), defined as the ratio of the number of drain current transients recorded in a given run to the corresponding average fluence, also follow a similar trend as that of the integrated charge. In cases where no transients were recorded, corresponding limiting cross-section values were used.

V. Modeling analysis

A. Modeling approach

In order to investigate the mechanisms responsible for the observed experimental results, the devices were modeled in 2-D using the Sentaurus Device TCAD device simulator from Synopsys [7]. Figure 4 shows the mixed-mode simulation setup used in the modeling analysis. A single-event strike with an ion LET value identical to that used in experiments was simulated, passing through the active region between the gate and the drain. The strike traversed through the entirety of the device including the InAlSb barrier, InAs channel, AlGaSb buffer and the SI GaAs substrate. A 5542-219 Picosecond bias tee circuit model [6] was included to represent the use of the same in the experiments. The 50 Ω resistance used at the output of the bias tee model emulated the termination resistance of the oscilloscope.

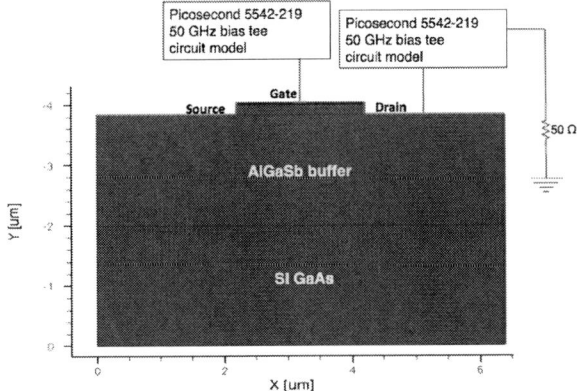

Fig. 4. A 2-D TCAD simulation model of the InAlSb/InAs/AlGaSb HEMT that includes the bias tees and the load resistance emulating the 50 Ω termination on the oscilloscope.

B. Model verification

The 2-D TCAD HEMT model was functionally verified to ensure that it correctly represented the experimental setup both in static (DC) and transient (single-event) modes of operation. First, the DC characteristics of the simulated HEMT were verified to have the same threshold voltage as that of the experimental HEMT, with drain current values being qualitatively comparable to corresponding experimental values. The transient simulation outputs were compared with and without the use of the bias tee circuit model. Figure 5 shows the simulated integrated charge profile with and without the use of the bias tee model, as a function of effective gate bias. The bias

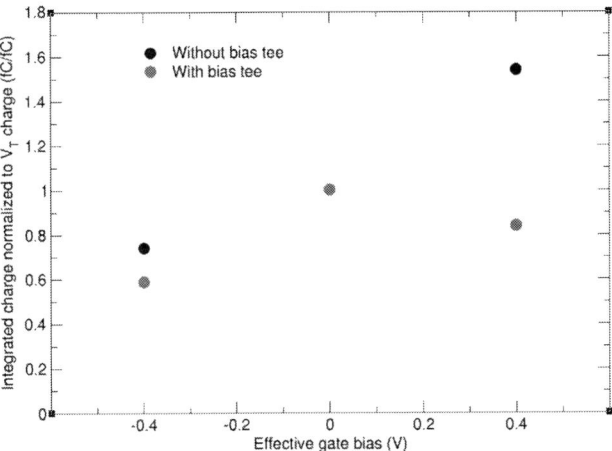

Fig. 5. 2-D TCAD simulation results of an InAlSb/InAs/AlGaSb HEMT. The model without the bias tee shows a monotonic increase of the integrated charge, while that with the bias tee agrees with observed experimental trends.

tee filters out the DC bias in the experiments, so that the integrated charge from the drain current transient can be calculated from a pure transient waveform. If the DC bias is not

removed, the DC level factors into the integrated charge profile, which leads to a monotonic increase from depletion (minimum channel charge) to accumulation (maximum channel charge). This is confirmed in Fig. 5, which shows the results of simulations with and without the bias tee.

C. Mechanisms

In depletion, the channel is depleted of carriers, and the net charge in the channel is less than that at threshold. However, the vertical gate field is stronger than the corresponding field at threshold. It therefore attracts more positive charge underneath the gate. As a result, the potential barrier lowering under the gate after the ion strike is higher in depletion compared to that at threshold, as shown in the horizontal cutline of the channel conduction band of Fig. 6. The "peak" condition referred to in Fig. 6 corresponds to the peak of the simulated single-event drain current transient. The relatively higher depletion gate field keeps the radiation-generated electrons away from the channel thereby resulting in a smaller increase in electron density as compared to that at threshold, as shown in Fig. 7. Therefore, in spite of the relatively larger potential barrier lowering in depletion, the total integrated charge is less in depletion than at threshold.

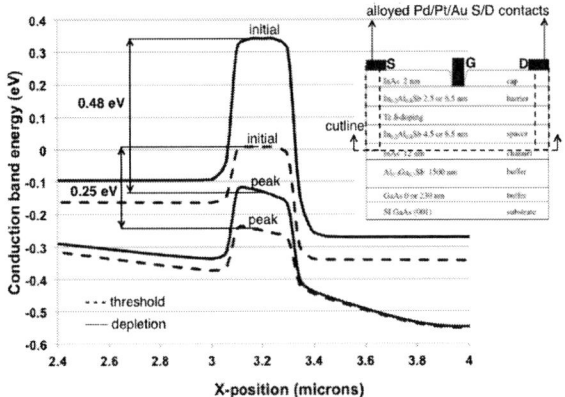

Fig. 6. Source to channel potential barrier lowering comparison between depletion and threshold biases, from 2-D TCAD simulations. The comparison is made between initial condition (reference) and peak of the transient. The entire length of the channel is not shown; a zoomed-in version is displayed.

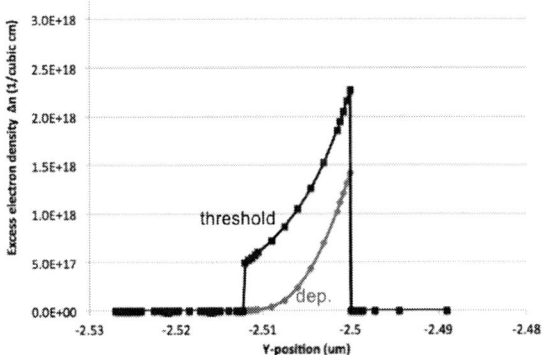

Fig. 7. Excess channel electron density (defined as the difference in electron densities between the peak of strike and initial condition) obtained along a vertical cutline through the center of the channel (as shown in Fig. 1).

In accumulation, the pre-strike net channel charge is higher than that at threshold, as well as at the peak of the simulated strike and 25 ps after the peak of the strike, as seen in Fig. 8. However, the vertical gate field in accumulation is lower than that at threshold at any given time during the strike. It therefore attracts less holes under the gate after the strike. Correspondingly, the potential barrier lowering underneath the gate is lower in accumulation than at threshold, as seen in Fig. 9. The relatively smaller barrier lowering can also be seen near the gate edges as shown in Fig. 10. Additionally, Fig. 9 shows that the change in conduction band slope due to the ion strike is greater at threshold than in accumulation. This implies a larger increase in horizontal channel electric field between the source and drain at threshold than in accumulation, as shown in Fig. 10. The relatively larger potential barrier lowering and the corresponding larger horizontal electric field from source to drain in threshold overwhelm the relatively smaller increase in net channel charge in accumulation. This results in higher injected current from source to drain and therefore more integrated charge in threshold than in accumulation.

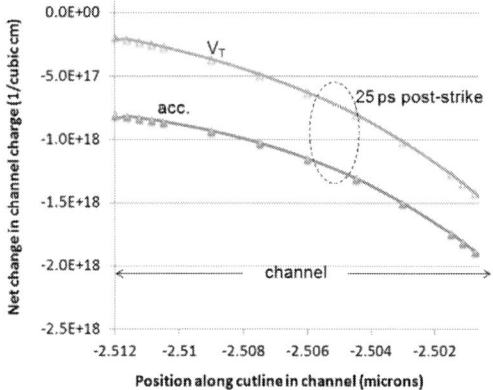

Fig. 8. Change in net channel charge, obtained from a vertical cutline through the center of the channel (as shown in Fig. 1), at 25 ps post peak of strike. The net charge is higher in accumulation.

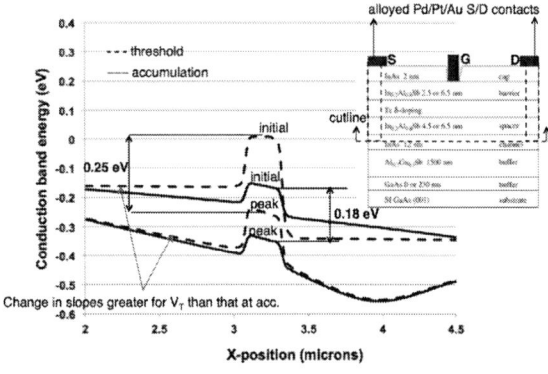

Fig. 9. Source to channel potential barrier lowering comparison between accumulation and threshold biases, from 2-D TCAD simulations. The comparison is made between initial condition (reference) and peak of the transient. The entire length of the channel is not shown; a zoomed-in version is displayed.

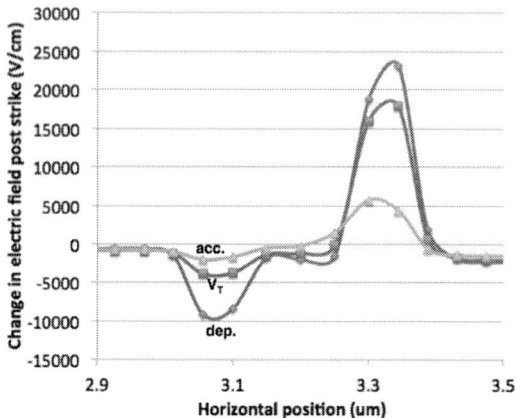

Fig. 10. Change in electric field along a horizontal cutline (identical to that in Fig. 9) through the center of the channel. The barrier lowering at the left gate edge (3.1 μm) is clearly seen for all gate bias conditions, progressively increasing from accumulation to depletion.

VI. Conclusions

The single-particle radiation-induced charge collection in InAlSb/InAs/AlGaSb HEMTs depends strongly on the gate bias. In depletion, a relatively higher vertical electrical field under the gate keeps electrons away from the channel, resulting in less integrated charge than at threshold. In accumulation, the effect of greater net channel charge is countered by a smaller barrier lowering under the gate along with lower source-to-drain electric field in the channel, resulting in less integrated charge than that at threshold.

Acknowledgments

This work was supported in part by the Defense Threat Reduction Agency's 6.1 Basic Materials Research Program. The authors would also like to acknowledge Vanderbilt University's Advanced Computing Center for Research and Education supercomputing cluster on which all TCAD simulations were performed.

References

[1] B. R. Bennett *et. al,* "Antimonide-based compound Semiconductors: A Review", *J. Solid. State Elec.* 2005.

[2] J. Brad Boos *et. al,* "Ohmic Contacts in AlSb/InAs HEMTs for Low-Voltage Operation", *J. Vac. Sci. Technol. B,* 1999.

[3] Papanicolau *et. al,* "Sb-based HEMTs with InAlSb/InAs heterojunction", *Elec. Lett.,* 2005.

[4] J. Brad Boos *et. al.,* "AlSb/InAs HEMTs for Low Voltage, High-Speed Applications", *IEEE Trans. Elec. Dev.,* vol. 49, pp. 1869-1875, 1998.

[5] Wagner *et. al,* "An approach to measure ultrafast-funneling-current transients", *IEEE Trans. Nucl. Sci.,* 1986.

[6] 5542-219 50 GHz bias tee datasheet, *Picosecond Labs Inc.*

[7] Sentaurus Device simulator, *Synopsys Inc.*

Selective Area Heteroepitaxy of InP Nanopyramidal Frusta on Si for Nanophotonics

Wondwosen Metaferia[1], Juha Tommila[2], Himanshu Kataria[1], Carl Junesand[1], Yanting Sun[1],

Mircea Guina[2], Tapio Niemi[2] and Sebastian Lourdudoss[,1*]

[1] Laboratory of Semiconductor Materials, KTH, Electrum 229, 164 40 Kista, Sweden

[2]Optoelectronics Research Centre, Tampere University of Technology, FIN-33101 Tampere, Finland

[*]Corresponding author email address: slo@kth.se

Abstract — **InP nanopyramidal frusta on InP and InP precoated Si substrates were grown selectively from nano-imprinted circular openings in silicon dioxide mask using a low pressure hydride vapor phase epitaxy reactor. The grown InP nanopyramidal frusta, octagonal in shape, were characterized by Atomic Force Microscopy, Scanning Electron Microscopy and Photoluminescence. The growth is extremely selective and uniform over the entire patterned area on both substrates. The measured diagonal of the top surface is 30 nm and 90 nm for the nanopyramidal frusta grown from 120 nm and 300 nm diameter openings, respectively. The size and morphology as well as the optical quality of these pyramidal frusta make them suitable templates for quantum dot structures for nano photonics and silicon photonics.**

I. INTRODUCTION

Semiconductor quantum dot (QD) structures are attractive for nanophotonic application. QD lasers for example, are known to have remarkable reduction in threshold current, temperature-insensitive operation, and large modulation bandwidth [1]-[3]; in addition, luminescent QDs can be used to create single photon source, entangled photon source and quantum bits for quantum computation. In recent years various approaches for formation of QDs have been intensively studied to realize device quality semiconductor QD structures. Among them, the self-assembling technique using the Stranski–Krastanov (SK) growth mode is one of the convenient methods to fabricate high density QDs with good optical quality [4]. However, the uniformity in size and arrangement in location of self-assembled QDs are still far from those required for high-performance optoelectronic devices. Patterned growth of individual QDs provides a scalable approach for the precise positioning of multiple single QDs. This ability of patterned approach can facilitate the fabrication of more complex optoelectronic devices, where multiple dots in certain number are needed to be placed at controlled sites. Selective Area Growth (SAG) can be successfully used to precisely position nano-size InP pyramidal frusta on which single semiconductor QDs are then grown atop, on the flat surface [5]. In addition to providing a high degree of freedom in the control of site and density of the

QDs, this approach also provides other advantages like, avoiding unwanted nearby emitters since part of the sample other than the opening is covered by a dielectric mask, high degree of reproducibility and adequate separation of the QDs from the initial substrates. The latter is advantageous to eliminate the high density of defects at the substrate/QD interface especially in case of heteroepitaxy of QDs. This paper reports on the site controlled growth of InP nanopyramidal frusta (NPF) on Si by SAG. The growth results are compared with the same on InP substrate.

II. EXPERIMENTAL

Pure n-InP (001) misoriented 2° toward <011> and InP precoated Si (001) 4° off oriented toward <111> substrate (called InP(seed)/Si hereafter) were used. Two groups of samples, group A and B, were prepared. Each group contains two samples named as A1 & A2 and B1 & B2. Where A1 & B1 are from pure InP substrate and A2 & B2 from InP(seed)/Si substrate as mentioned above. 150 nm and 80 nm thick SiO_2 layer was deposited on samples of group A and B respectively, by Plasma Enhanced Chemical Vapour Deposition (PECVD). Soft UV nanoimprint lithography (UV-NIL) was utilized to pattern circular openings of 300 nm diameter with center to center spacing of 500 nm on the samples of group A; similarly, openings of 120 nm diameter with center to center spacing of 180 nm were patterned on group B. The NIL process that has been employed is described in [6]. The diameter and spacing of the circular openings were chosen so that the fill factor, ratio of open field to the total area is constant. The summary of sample description is given in Table I. Fig.1 shows the SEM image of nanoimprinted circular openings of diameter 120 nm on on SiO_2 masked sample B1.

SAG of sulphur doped InP with a nominal sulphur concentration (electron concentration) of $5x10^{18}$ cm^{-3} in the openings on these samples was conducted in an Aixtron LP-HVPE reactor [7]. The growth times were 2.5 minutes on samples of group A and 15 minutes on samples of group B. Growth temperature was 590 °C and V/III ratio was 10 (i.e.,

978-1-4673-1725-2/12 $31.00 © 2012 IEEE

PH3/InCl =10) in all cases. InP grown from the openings was characterized by Atomic Force Microscopy (AFM), Scanning Electron Microscopy and micro-Photoluminescence (μ-PL).

Table I
Summary of sample description

Sample group & name	Group A		Group B	
	A1	A2	B1	B2
Substrate	InP	InP(seed)/Si	InP	InP(seed)/Si
Mask thickness(nm)	150		80	
Opening diameter (nm)	300		120	
Spacing between the openings (nm)	500		180	

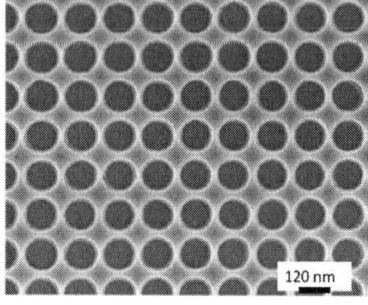

Fig.1 Nanoimprinted circular openings on SiO₂ masked InP for pyramidal QD growth (sample B1).

III. RESULTS AND DISCUSSIONS

Fig. 2 depicts the SEM top view of the patterns after InP growth for 2.5 minutes on samples A1 (Fig.2a left) and A2 (Fig.2a right) and for 15 minutes on samples B1 (Fig.2b left) and B2 (Fig.2b right). These images reveal that the grown InP results in the shape of a pyramidal frustum. Fig. 2(c) shows AFM amplitude image (amplitude voltage of 97.1 mV) of InP grown from a single opening from sample A1. This image clearly confirms the shape of the pyramidal frustum. We call these collectively as Nanopyramidal Frusta (NPF). As seen from the SEM images, the growth from the openings on both InP and InP(seed)/Si substrates is extremely selective with similar growth morphology regardless of the opening diameter. The final shape has an octahedral pyramid frustum (shown in Fig.2(c)) with low index faces surrounding (001) plane at the top. Four of these are identified as {110} and the other four as {111} by measuring the angles of inclinations with respect to the top surface determined by AFM line scans [8]. Such pyramidal morphology with these surrounding

planes has been observed when SAG was conducted in the confined ring openings [9] or circular openings [5]. The formation of such planes has been explained to be due to the crystal symmetry rearrangement within the confined region controlled by the incorporation of rate determining species [5], which in our case is indium. Table II summarizes the dimensions of NPF obtained from AFM measurements. It is apparent from this table that the dimension of the top flat surface depends on the diameter of the circular opening in the oxide mask. By optimizing the diameter of the openings, NPF with flat top surface of dimension suitable to grow a single dot can be grown [10]. It can be noted that no remarkable difference was observed between the pyramids grown on InP substrate and on InP(seed)/Si as shown in Fig. 2 (a) and (b).

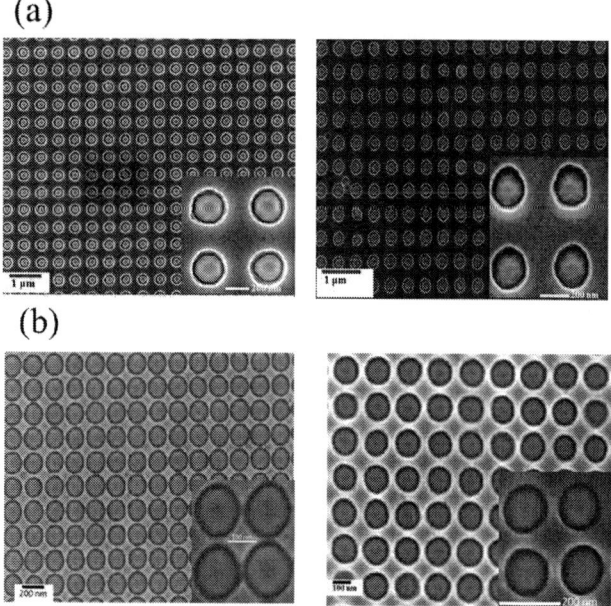

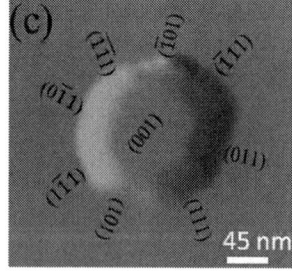

Fig. 2 SEM images of the InP NPF grown for 2.5 minutes on samples A1 (left) and A2 (right) (a), and InP NPF grown for 15 minutes on samples B1 (left) and B2 (right) (b) and AFM amplitude image (amplitude voltage of 97.1 mV) of a single NPF from sample A1 (c).

Since all the samples had the same fill factor, the same growth rate is expected for all [11]. But, in our experiment, the NPF height is nearly the same for the two different growth times for the same fill factor. This suggests that the appearance of slow growing low index planes already at the initial stages of growth leads to little growth even after extended growth.

Table II
Diagonals of the top surface of the NPF

Sample group	Opening diameter (nm)	Average diagonals (nm)	Height of the NPF from the substrate surface (nm)
A	300	90 and 90	80
B	120	30 and 30	90

The grown samples were studied by AFM after removing the oxide mask. In Fig. 3 we choose to present the AFM height image of NPF on A1 (a) and A2 (b). It can be seen from Fig.3 that a good size homogeneity of the NPF is achieved on InP substrate (sample A1); in the case of, those grown on InP (seed)/Si substrate (sample A2), a small size variation is observed. The latter variation is understood to be due to the uneven (rough) surface morphology/topography of the original seed layer, which has a root mean square (RMS) surface roughness (Rq) of ~21 nm and a topographical variation of 50 nm (crest to trough) [12]. More importantly the topography of the seed layer and hence the deposited oxide mask is uneven. The result is that at some regions, it is beyond the flexibility limit of the NIL stamp to get equal size openings on the dielectric mask. Thus the size of the NPF is varying largely due to the uneven size of the imprinted openings. Our ongoing experiments show that this can be further improved through chemical mechanical polishing.

In the case of heteroepitaxy, the top pyramidal surface should contain very low defect density as these are grown from very small areas [13]. In addition the dislocation reduction mechanism involves the termination of threading dislocations at the {111} planes [14] observed on the frustum (Fig.2(c)) so that dislocation density is reduced at the top (001) surface. Hence, this method of fabricating templates for quantum dots on InP/Si is particularly interesting.

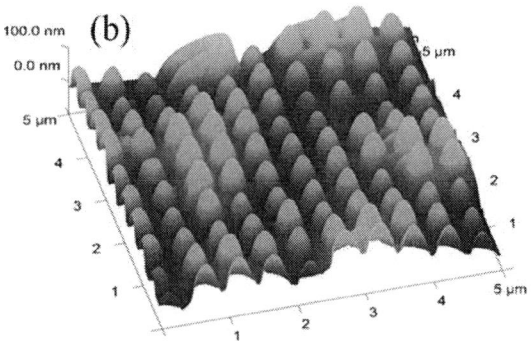

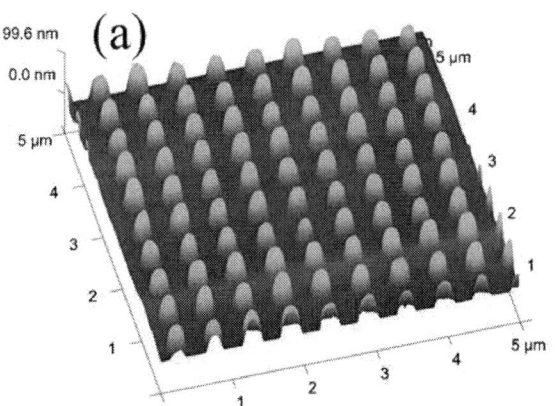

Fig.3 AFM height scans of InP NPF on sample A1 (a) on A2 (b),

PL measurement conducted at room temperature with Ar laser source (514 nm) at 10 mw power confirm that the luminescence intensity is increased after the growth of InP pyramids on InP(seed)/Si . Fig.4. shows PL spectra taken from NPF grown on sample A2. Fig. 4a compares the PL spectra of NPF of sample A2 and of the seed layer; Fig. 4b depicts the normalized PL intensities of NPF and of a reference epitaxial layer of InP on planar InP substrate grown at the same time.

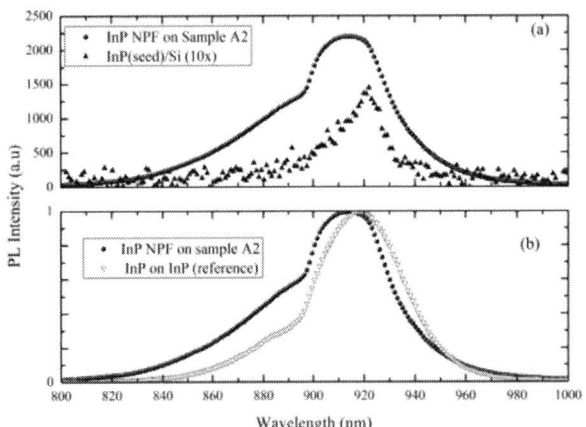

Fig.4 PL spectra from NPF grown on sample A2 (the oxide mask was removed prior to PL measurements) compared with the spectrum from the InP(seed)/Si (a) and normalized PL intensities of the spectra from NPF on sample A2 and reference (InP on planar InP) sample (b).

Clearly the intensity from the NPF is stronger than the InP(seed)/Si layer (the starting material) by more than 15 times. It is however difficult to compare the PL intensity from reference sample (InP on InP) and from the NPF as different effects play a role in enhancing/reducing the PL intensity. On the other hand the spectrum from the NPF follows similar line profile with the reference except that it is to some extent flattened at the peak. The FWHM of the reference sample is 30 nm and that of the NPF is 50 nm. The flattening and the broadening of the peak may be due to the inhomogeneity of amount of dopants incorporated on different facets of the NPF [15]. In our earlier investigation by scanning capacitance microscopy we found that different facet planes can acquire different doping concentrations [16]. Though we did not do similar study here, it has been estimated in that study that the dopant concentration can differ by 3 orders of magnitude between planar surface (i.e., (001) plane) and on certain facets. So the different dopant concentrations in different facets cause different shifts in the PL emission with comparable intensities and hence the flattened peak. The blue shift of the PL emission from the NPF compared to the reference sample is attributed to the Burstein-Moss effect resulting from the enhanced doping concentration in the facets [15].

IV. SUMMARY

In summary, we have demonstrated the growth of InP nanopyramidal templates through SAG on InP and InP on Si. Uniform and site controlled growth of nanopyramidal frusta can be achieved from the smooth surfaces. The PL characterisations show that the frustum has good optical quality. The optical quality and dimension of the grown nanopyramidal frusta are suitable for site controlled growth of quantum dots for future nanophotonic devices.

ACKNOWLEDGEMENT

The authors would like to acknowledge the financial support of the Swedish Foundation for Strategic Research, the Swedish Research Council and Vinnova.

REFERENCES

[1] Y. Arakawa, H. Sasaki, "Multidimensional quantum well laser and temperature dependence of its threshold current", Appl. Phys. Lett. 939 (1982) 40.

[2] Levon V. Asryan, Serge Luryi, "Temperature-insensitive semiconductor quantum dot laser", Solid-State Electronics205–212 (2003) 47.

[3] Bhattacharya P, Klotzkin D, Qasaimeh O, Zhou W, Krishna S, Zhu D, "High-speed modulation and switchingcharacteristics of In(Ga)As–Al(Ga)As self-organized quantum-dot lasers", IEEE JSelect Top Quant Electron 426–38 (2000)6.

[4] Mo, Y. M., Savage , D.E., Swartzenture. B.s., and Lagally, M.G, "Kinetic pathway in Stranski-Krastanov growth of Ge on Si(001l)", Phys. Rev. Lett., 65(1990)102CL1023.

[5] P J Poole, D Dalacu, J Lefebvre and R L Williams, "Selective epitaxy of semiconductor nanopyramids for nanophotonics", Nanotechnology 295302 (2010)21.

[6] A. Tukiainen, J. Viheriälä, T. Niemi, T. Rytkönen, J. Kontio, M. Pessa, "Selective growth experiments on gallium arsenide (1 0 0) surfaces patterned using UV-nanoimprint lithography" Microelectronics Journal 1477–1480 (2006)37.

[7] S. Lourdudoss, O. Kjebon. "Hydride vapor phase epitaxy revisited",IEEE Journal of Selected Topics in Quantum Electronics 49 (1997)3.

[8] W. Metaferia, J. Tommila, C. Junesand, H. Kataria, C. Hu, M. Guina et al. "HeteroepitaxialGrowth of InP on Si through Nanoimprint Lithography", Phys. Stat. solidi C 9, No. 7 (2012).

[9] Y. T. Sun, S. Lourdudoss, M. Avella, and J. Jiménez. « Sulfur-doped indium phosphide on silicon substrate grown by ELOG", Electrochemical and Solid-State Letters, 7 (11) (2004) G269-G271.

[10] J.Yuan, H.Wang, P.J.van Veldhoven and R.Nötzel, "Impact of base size and shape on formation control of multifaceted InP nanopyramids by selective area metal organic vapor phase epitaxy", J. Appl. Phys., 124304 (2009) 106.

[11] F. Olsson, Tiankai Zhu, Gael Mion, S. Lourdudoss, "Large mask area effects in selective area growth", J. Cryst. Growth 24–30 (2006)289.

[12] C.Junesand et al., "Effect of the surface morphology of seed andmask layers on InP grown on Si by epitaxial lateral overgrowth," Journal of Electronic Materials, 2012, DOI: 10.1007/s11664-012-2164-90.

[13] E.A.Fitzgerald et al. "Nucleation mechanisms and the elimination of misfit dislocations at mismatched interfaces by reduction in growth area", J. Appl. Phys., 2220-2237 (1989) 65.

[14] Knall, J., Romano, L. T., Biegelsen, D. K.; Bringans, R. D.; Chui, H. C.; Harris, J. S., Treat, D. W., Bour, D. P., "The use of graded InGaAs layers and patterned substrates to remove threading dislocations from GaAs on Si", J. Appl. Phys 2697–2702(1994)76.

[15] M. Bugajski, W. Lewandowski, "Concentration dependent absorption and photoluminescence of n-type InP", J. Appl. Phys., 57 (1985) 521.

[16] Y.T. Sun, S. Anand, S. Lourdudoss, "Crystallographic orientation dependence of impurity incorporation during epitaxial lateral overgrowth of InP", J. Cryst. Growth 237–239 (2002) 1418.

Characteristics of InP nanoneedles

grown on silicon by low-temperature MOCVD

Kun Li, Fan Ren, Roger Chen, Thai Tran, Kar Wei Ng, and Connie J. Chang-Hasnain[*]

Department of Electrical Engineering and Computer Sciences, University of California at Berkeley, Berkeley, California 94720, United States

[*]*Email: cch@eecs.berkeley.edu*

Abstract — **Recombination dynamics of InP nanoneedles on silicon is studied using time-resolved photoluminescence. Long recombination lifetime ~13 ns and high IQE ~ 10% are obtained for as-grown nanoneedles. Optically pumped nanoneedle laser is demonstrated at room temperature.**

Index Terms — **Indium phosphide, nanostructured materials, optoelectronic devices, photoluminescence, semiconductor lasers, time measurement, ultrafast optics**

.

I.   Introduction

The realization of small, dense laser sources on silicon substrates could open a pathway to equip computer chips with the communication bandwidths of optical data links, but the integration of optical gain materials onto silicon remains difficult due to its lattice mismatch with compound semiconductor materials. One approach to overcome this problem is the hetero-epitaxial growth of low-dimensional materials. However, single crystalline growth is usually limited to films thinner than a critical thickness or to nanostructures whose footprints are smaller than a critical diameter [1]. Recently, we reported on a catalyst-free growth mode on crystalline (111) – silicon that allows scaling the footprint of InGaAs/GaAs nanoneedles and nanopillars to diameters beyond 1 μm while preserving a single-phase crystal structure [2]. Nanopillar lasers grown on (111)-Si were demonstrated by optical pumping at room temperature [3].

InP monolithically grown on silicon can act as a growth template for long wavelength hetero-structure lasers approaching silicon-transparency. With much lower surface recombination rate and therefore higher internal quantum efficiency (IQE) compared with GaAs, InP-based materials also offer greater chance in the field of solar cells. In this paper, we demonstrate InP nanoneedles on silicon grown in a similar core-shell mode with excellent laser and mateiral quality.

II. Laser Characteristics

Grown on surface-roughened silicon substrate with metalorganic chemical vapour deposition (MOCVD) under 425°C in a catalyst-free environment, InP nanoneedle is core-shell structure whose dimensions are linearly scalable with growth time. A very sharp tip and extremely well-faceted hexagonal cross-section can be observed from the SEM images in Fig.1. Each as-grown nanoneedle structure provides a natural optical cavity supporting unique helically propagating resonance modes [3].

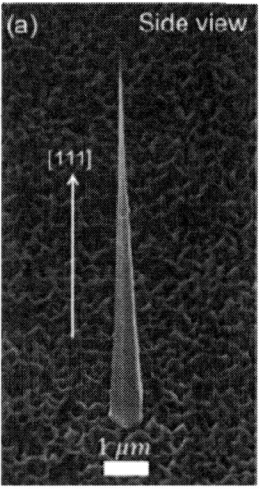 

Fig. 1   (a) Side-view SEM image showing the sharp tip of nanoneedle with tapered angle of ~3°. Nanoneedle was grown on surface-roughened silicon substrate with MOCVD at 425°C for 45 minutes, resulting in a diameter of ~811 nm, and height of ~18.23 μm. Dimensions are scalable with growth time. (b) Top-view SEM image highlighting the extremely well-faceted hexagonal cross-section of the nanoneedle.

Room-temperature laser oscillation is achieved by optically pumping an as-grown nanoneedle from top with a Ti:sapphire mode-locked laser. In Fig.2, camera image shows only a small spot of spontaneous emission when pump power is below

978-1-4673-1725-2/12 $31.00 © 2012 IEEE   85

threshold. A classic signature of laser oscillation can be obtained from the image of strong speckle patterns above threshold. We measured a background suppression ratio of ~22dB from the lasing spectrum.

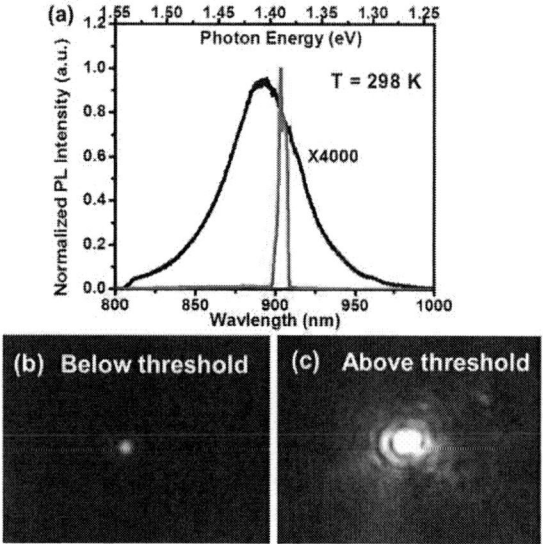

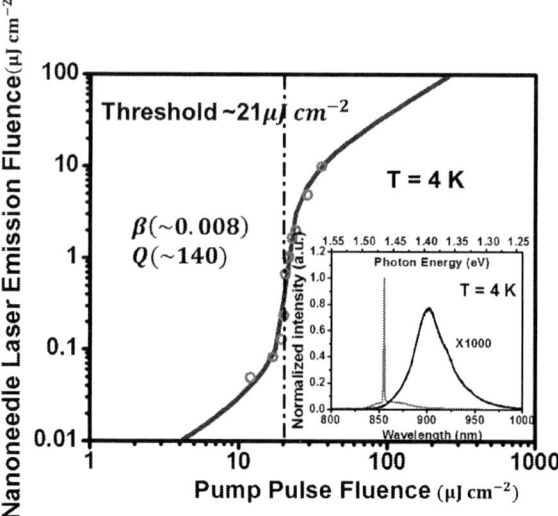

Fig. 3   L-L curve of a nanoneedle laser operating at 4K with a low threshold of ~21 µJ cm⁻² is analyzed with gain model and rate equations to extract estimated values of Q (~140) and β (~0.008). Inset: the corresponding spectra of the nanoneedle laser below (black, magnified X1000) and above (red) threshold.

Fig. 2   (a) Room temperature nanoneedle emission below (black) and above (red) threshold. The spectrum below threshold has been magnified X4000 for visibility. The laser peak clearly dominates above threshold, achieving a ~22dB background suppression ratio. (b) CCD image of nanoneedle emission below threshold shows only a spot of spontaneous emission. (c) Upon lasing, strong speckle patterns appear. Speckle patterns result from the high degree of coherent emission and is a classic signature of laser oscillation.

Additional microphotoluminescence (µ-PL) experiments at 4K combined with rate equation analysis are done so as to extract various laser parameters for further characterization of nanoneedle lasers. We firstly established a correlation between experimental pump levels and carrier density using the spontaneous emission spectrum, and then constructed a gain model with the extracted parameters from our measurements, which gave a threshold gain of ~4800 cm⁻¹ and transparent carrier density of ~1.1×10^{18}cm⁻³. Following a rate equation analysis, the L-L curve from µ-PL measurement can be fitted, revealing an estimation of cavity quality (Q) factor of ~140 and a spontaneous emission coupling factor (β) of ~0.008. From the fitted L-L curve in Fig.3, where both pump power and nanoneedle laser emission are converted into energy per area, we can estimate the threshold of nanoneedle laser as ~21 µJ cm⁻² and energy conversion ratio of ~27% after lasing.

III.   Material Quality

As mentioned above, InP nanoneedle is also an excellent candidate for solar cell applications. Besides the advantage of long recombination lifetime and high IQE resulting from low surface recombination rate, high absorption efficiency can be achieved when nanoneedles are grown into a high-density forest. In order to study carrier recombination lifetime, an avalanche photo-detector (APD) in conjunction with single photon-counting electronics are added into our µ-PL setup. Time-resolved measurements were achieved with time-correlated single photon counting (TCSPC) method at a timing resolution of ~40 ps. Fig.4 shows time-resolved PL decay traces under different pump energies, fitted with bi-exponential decay functions. Under a very low pump energy of ~1.3 µJ cm⁻², carrier recombination lifetime reaches as long as 13.519 ns, with a slow decay component ~37.662 ns and a fast decay component ~8.238 ns. This lifetime is the longest reported for a nanowire/nanoneedle structure, to the best of our knowledge, which clearly indicates superior characteristics obtained with InP nanopillars. Increasing the pump energy to ~13.4 µJ cm⁻², lifetime is reduced to ~2.612 ns due to higher ratio of non-radiative recombination fast decay [4]. A life time as short as 42.1 ps can be detected under ~214.7 µJ cm⁻² pump energy (above laser threshold), confirming the lasing effect.

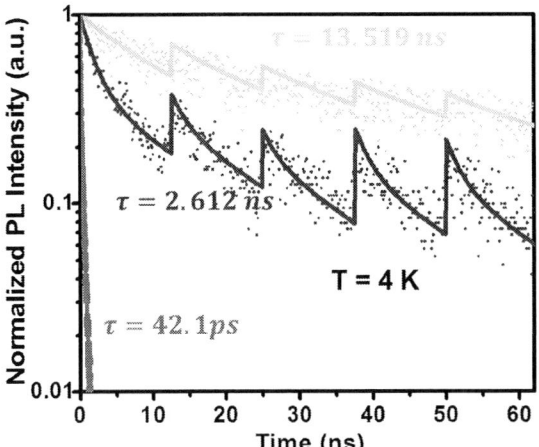

Fig. 4 Time-resolved photoluminescence decay curves under different pump powers, with bi-exponential decay function fitted carrier lifetime of ~13.519 ns at pump energy of 1.3 μJ cm^{-2} (blue), ~2.612 ns at 13.4 μJ cm^{-2} (green), and ~42.1ps at 214.7 μJ cm^{-2} above lasing threshold (red). Pulse picker was used under low pump powers to reduce repetition frequency of pump laser from 80MHz to 8MHz, so that long decay time can be revealed clearly. The smaller peaks after the first peak in green and blue curves were due to insufficient suppression ratio of pulse picker.

Temperature-dependent μ-PL measurement from 4K up to room temperature is shown in Fig. 5, with Arrhenius equation fit [5]. A record value ~10% IQE is obtained, which is twice as high as the best reported 5% IQE value for GaAs nanowires [6].

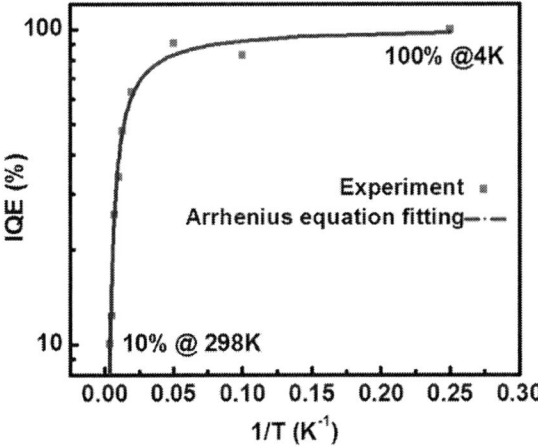

Fig. 5 Internal quantum effienciency (IQE) measurement data extracted from power-dependent photoluminescence measurements under different temperatures, fitted with Arrhenius equation. An IQE of ~10% under room temperature is achieved.

References

[1] L. C. Chuang, et. al. "Critical diamter for III-V nanowires grown on lattice-mismatched substrates," *Appl. Phys. Lett.,* 90, 043115 (2007).

[2] M. Moewe, et. al. "Atomically sharp catalyst-free wurtzite GaAs/AlGaAs nanoneedles grown on silicon," *Appl. Phys. Lett.,* 93, 023116 (2008).

[3] R. Chen, et. al. "All-semiconductor nanolasers on silicon," *IEEE Photonics Society 23rd Annual Meeting,* invited talk WP4, 196109 (2010).

[4] S. Chrankshaw, et. al. "Recombination dynamics in wurtzite InP nanowires," *Phys. Rev.* 77, 235409 (2008).

[5] S. Watanabe, et. al. "Internal quantum efficiency of highly-efficient In$_x$Ga$_{1-x}$N-based near-ultraviolet light-emitting diodes," *Appl.Phys.Lett.* 83, 0003-6951(2003).

[6] C. Patrik, et. al. "Monolithic GaAs/InGaP nanowire light emitting diodes on silicon", *Nanotechnology* 19, 305201 (2008).

Silicon-based long-wavelength III-V quantum-dot lasers

Qi Jiang, Andrew Lee, Mingchu Tang, Alwyn Seeds, and Huiyun Liu*

Department of Electronic and Electrical Engineering, University College London, London WC1E 7JE, United Kingdom; *huiyun.liu@ucl.ac.uk

Abstract — The realization of electrically-pumped lasers on a Si platform will permit the creation of complex optoelectronic circuits, which will enable chip-to-chip and system-to-system optical communication. Direct epitaxial growth of III-V semiconductor materials on Si or Ge substrates is the most promising approach for producing lasers on Si. III-V compound quantum dots – semiconductor nanosized crystals – are very attractive for producing III-V/Si laser diodes with the advantages of lower threshold current density and less sensitivity to defects relative to conventional quantum wells. Here we present studies on the development of InAs/GaAs quantum-dot lasers monolithically grown on Si, Ge, and Ge-on-Si substrates. Room-temperature lasing near the telecommunications wavelength of 1300 nm have been demonstrated at room temperature with low threshold current densities for InAs/GaAs quantum-dot lasers grown on both Si, and Ge substrates.

Index Terms — Quantum dots, Silicon photonics, III-V/Si integration, 1300 nm laser diodes, molecular beam epitaxy.

I. INTRODUCTION

The merger of photonics and electronics into a dual-function platform, the optoelectronic integrated circuit (OEIC) fabricated with existing Si infrastructure can overcome interconnection issues for Si chips while pushing forward silicon microelectronics beyond the classical complementary metal-oxide-semiconductor (CMOS) era. Although great efforts have been devoted to Si-based light generation and modulation technologies over the last 30 years, monolithic growth of electrically pumped lasers on Si substrates remains a 'holy grail' for Si photonics [1]. Because of the indirect bandgap of group IV materials, in which the radiative recombination process for emitters is inefficient, only optically pumped lasers have been demonstrated for devices using Si and Ge as optical gain materials [1]. Monolithic integration of semiconductor III-V compound lasers on a Si platform has long attracted a great deal of attention with the ultimate goal – high-speed and low-power-consumption optoelectronic circuits for Si photonics. III-V quantum dot (QD) lasers have been demonstrated with very low threshold current densities, temperature-insensitive operation above room temperature, and less sensitivity to defects compared with traditional quantum well devices [2,3,4]. Consequently III-V QD technology is one of the vital components for producing high-performance III-V/Si laser diodes. Here, we report the development of long-wavelength InAs/GaAs QD laser diodes on Si platform for silicon photonics with the use of Si, Ge, and Ge-on-Si (Ge/Si) substrates.

II. INAS/GAAS QUANTUM-DOT LASER DIODES GROWN ON SI SUBSTRATES

Phosphorus-doped (100)-orientated Si substrates with 4° offcut towards the [110] plane were used for producing InAs/GaAs QDs directly grown on Si substrates by solid-source molecular beam epitaxy (MBE). Oxide desorption was performed by holding the Si substrate at a temperature of 900 °C for 10 minutes. The Si substrate was then cooled down for the growth of a 30-nm GaAs nucleation layer with a low growth rate of 0.1 monolayers per second (ML/s). An additional 970-nm GaAs layer was grown with a high growth rate of 0.7 ML/s at high temperature. InGaAs/GaAs dislocation filter layers, consisting of two repeats of a five-period (10-nm $In_{0.15}Ga_{0.85}As$/10-nm GaAs) superlattice (SPLs) and 400-nm GaAs, were used. Five InAs/InGaAs dots-in-a-well (DWELL) layers were then grown at optimized conditions as on GaAs substrates, with each layer consisting of 3.0 MLs of InAs grown on 2 nm of $In_{0.15}Ga_{0.85}As$ and capped by 6 nm of $In_{0.15}Ga_{0.85}As$ [3]. Figure 1 shows transmission electron microscopy (TEM) images of 5-layer InAs/InGaAs DWELL structures grown on a Si substrate.

Fig. 1. Dark field (200) TEM cross-sectional images of 5-layer InAs/InGaAs dot-in-a-well structure grown on Si substrate.

We first optimized the growth temperature of the GaAs nucleation layer on Si substrates. Figure 2 shows room-temperature (RT) photoluminescence (PL) spectra of 1300-nm InAs/GaAs QDs grown on Si substrates with different GaAs nucleation temperatures. The GaAs-based InAs QD sample with the same growth conditions is also shown in Fig. 2 as a

978-1-4673-1725-2/12 $31.00 © 2012 IEEE

reference. It should be mentioned that the sample grown on GaAs substrate was fabricated under optimized conditions and represents high optical quality as the 1300-nm InAs/GaAs QD laser diodes based on identical growth parameters give extremely low threshold current density, J_{th} and high output power at RT [3,5]. The RT emission around 1300 nm has been observed for all the samples shown in Fig. 2 with full width at half maximum (FWHM) of ~30 meV. These PL linewidths obtained in Figure 2 are comparable to the values obtained for GaAs-based InAs QDs. Of considerable significance is that the PL intensity of Si-based InAs/GaAs QD ground-state transition is strongly dependent on the nucleation temperature of the initial GaAs layer on the Si substrate, with strongest PL intensity at 400 °C. Further TEM studies indicate that the density of defects generated at the GaAs/Si interface and propagating into GaAs buffer layer is dependent on the growth temperature of the GaAs nucleation layer [6]. The lowest defect density propagating into the GaAs buffer layer is obtained at 400 °C, which is also confirmed by the etch-pit density (EPD) measurements. The defect densities of $1.03 \times 10^7/cm^2$, $6.03 \times 10^6/cm^2$, $8.17 \times 10^6/cm^2$ were obtained for the samples with the initial GaAs layer nucleated at 380, 400, and 420 °C, respectively.

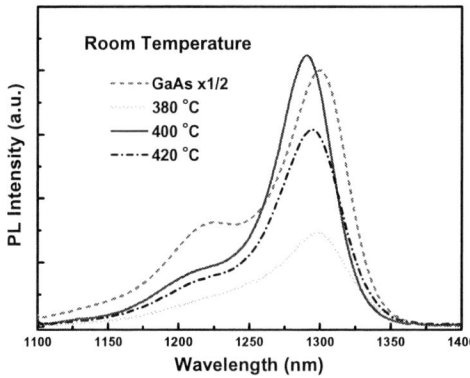

Fig. 2. RT PL spectra of 1300-nm InAs/GaAs QDs directly grown on Si substrates with different growth temperatures for GaAs nucleation layer. The RT PL spectrum of GaAs-based InAs QDs grown witht the same growth conditions is also shown as a reference.

With the use of the optimized GaAs nucleation temperature of 400 °C a standard 1.3-μm DWELL laser structure was then grown on a Si substrate. Our device exhibits ground-state lasing at 1302 nm by electrical carrier injection under pulsed operation, with threshold current density, J_{th} of 725 A/cm² at RT, as shown in Fig. 3. QD laser operation is achieved for heatsink temperatures up to 42 °C, with a characteristic temperature, T_0, of ~44 K between 20 °C and 42 °C.

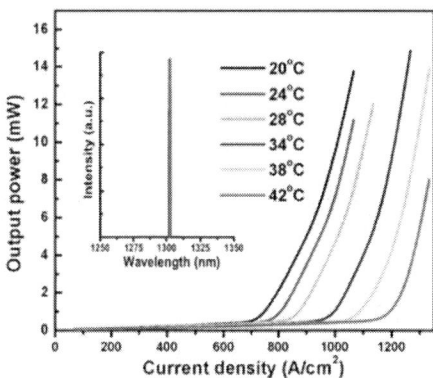

Fig. 3. Light output against current characteristic for InAs/GaAs quantum-dot laser on a Si substrate at various temperatures. The inset shows the laser optical spectrum above threshold at RT.

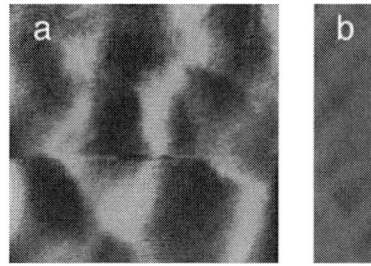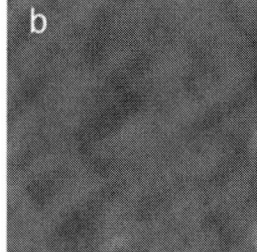

Fig. 4. 5 × 5 m² Atomic Force Microscopy images of the surface morphology for 1.2 m GaAs on Ge substrates with As prelayer (a) and Ga prelayer (b) growth techniques.

III. INAS/GAAS QUANTUM DOTS GROWN ON GE AND GE-ON-SI SUBSTRATES

The main issue for the direct growth of GaAs on Si is the generation of high-density defects due to the lattice mismatch between GaAs and Si. Ge-on-Si can effectively bridge the lattice constant gap between GaAs and Si, because of the relatively small lattice mismatch (0.08%), and closely matched thermal expansion between GaAs and Ge, and the completely miscibility between Ge and Si. Furthermore, Ge-on-Si is becoming a mature technology with low-density threading dislocations. It should also be noticed that, as the scaling of Si microelectronics devices approach the 22 nm node, Ge epilayers may replace Si as p-channel materials in CMOS device on Si substrates, because Ge has a much higher hole mobility than Si. A promising method for incorporating III-V laser diodes on a Si platform is therefore to fabricate III-V materials and devices on Ge-on-Si substrates. However, the growth of GaAs on Ge substrates is not well established so far because of the generation of anti-phase boundaries (APBs) due to the non-polar/polar interface between Ge/GaAs.

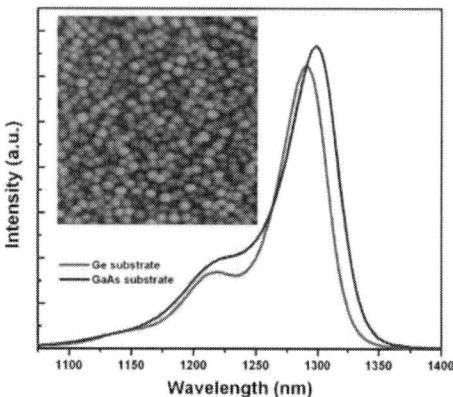

Fig. 5. Photoluminescence spectra comparison for InAs/InGaAs DWELL structures grown on GaAs and Ge substrates, respectively. The inset is $1 \times 1 \ \mu m^2$ AFM image of the surface morphology for uncapped InAs/GaAs QDs grown on a Ge substrate.

The initial step is to investigate solid-source MBE growth of GaAs buffer layers on p-doped (100) Ge (GaAs/Ge) substrates, off-cut 6° towards the [111] plane with As or Ga initial layer. Oxide desorption was performed by holding the Ge substrate at a temperature of 400 $^{\circ}$C. The substrate temperature was then increased to 650 $^{\circ}$C and held at that temperature for 20 minutes. The substrate was then cooled to 380 $^{\circ}$C for the growth of the III-V epitaxial layers [7]. For the Ga prelayer, the base pressure was reduced to below 10^{-10} Torr in the MBE growth chamber before loading the Ge wafer into the growth chamber. To ensure total Ga coverage on the Ge substrate, 1.08 monolayer Ga was first deposited. Alternatively, the Ge surface was terminated with As by opening the valve of the As cracker for 1 minute. After either As or Ga prelayers were deposited, 20 monolayer of GaAs were grown by migration enhanced epitaxy using alternating Ga and As_4 beams, and then addition of the III-V buffer layer at higher temperature. QD laser devices containing five InAs/InGaAs DWELL layers were then grown at optimized conditions as on GaAs substrates, with each layer consisting of 3.0 monolayers of InAs grown on 2 nm of $In_{0.15}Ga_{0.85}As$ and capped by 6 nm of $In_{0.15}Ga_{0.85}As$. Initiation of GaAs with the typical procedure of using self-terminating As layer produces poor GaAs surface morphology due to APBs, see Fig. 4(a), despite the large miscut of the (100) surface. However, initiation of GaAs growth with approximately 1 monolayer of Ga, i.e., Ga prelayer results in a very much smoother surface morphology for the GaAs layer, see Fig. 4(b), which indicates the formation of a single-domain GaAs buffer layer on the Ge substrate [8]. Standard 5-layer InAs/InGaAs DWELL structures were fabricated on the single-domain GaAs buffer layer on Ge substrates. The inset of Fig. 5 shows the $1 \times 1 \ \mu m^2$ Atomic Force Microscopy (AFM) image of the uncapped InAs QDs grown on a Ge

substrate, from which a QD density of about 4.3×10^{10} cm^{-2} is obtained. The InAs QDs randomly distribute on the surface, as on GaAs(100) substrates. This morphology of InAs QDs is significantly different from that of InAs QDs grown on a Ge-on-insulator-on-Si substrate by metal organic chemical vapor deposition, in which APBs were observed and the InAs QDs were lined with a preferential orientation along [1 $\bar{1}$ 0] direction with bimodal size distribution [9]. Figure 5 compares RT PL spectra of the capped InAs QDs grown on Ge and GaAs substrates. The InAs QD PL intensity on the Ge substrate is almost identical to that of QDs grown on a GaAs substrate. The ground-state emission of the QDs grown on the Ge substrate takes place at 1291 nm, and the peak yields a full width at half maximum (FWHM) of 29.8 meV. The FWHM is remarkably narrow, and close to the state-of-the-art value for QDs on GaAs substrates.

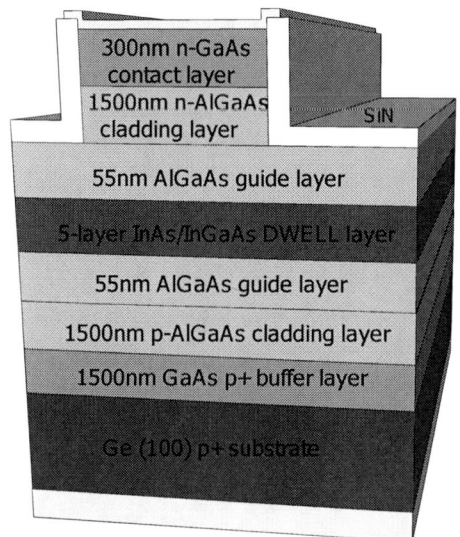

Fig. 6. Schematic showing the layer structure of an InAs/InGaAs dots-in-a-well laser diode on Ge substrate.

Standard 5-layer 1300-nm InAs/InGaAs DWELL lasers were fabricated on the GaAs/Ge substrate. The device structure is shown in Fig. 6. Broad-area devices with cavities of width 50 μm and length 5 mm were fabricated with as-cleaved facets. The laser diode was operated in continuous-wave (CW) condition. Figure 7 shows light output against current characteristic for InAs/GaAs quantum-dot laser grown on a Ge substrate for operating temperatures between 20 and 60 deg. C. The measured RT output power from one facet is close to 28 mW for an injection current of 500 mA, with no evidence of power saturation up to this current. The lasing threshold is 138 mA. J_{th} is 55.2 A/cm^2, which corresponds to about 11 A/cm^2 for each of the five QD laser layers, which is comparable to the best-reported values for GaAs-based QD laser diodes. The inset in Fig. 7 shows the laser optical spectrum above threshold at RT.

978-1-4673-1725-2/12 $31.00 © 2012 IEEE

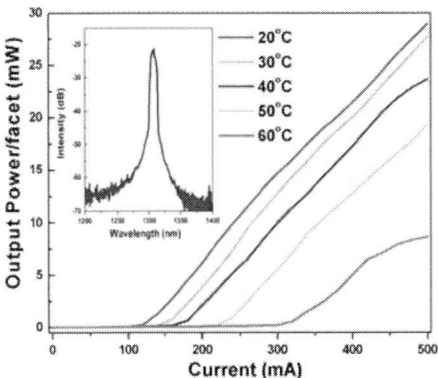

Fig. 7. Light output against current characteristic for InAs/GaAs QD laser on a Ge substrate under CW operation at various temperatures. The inset shows the laser optical above threshold at RT.

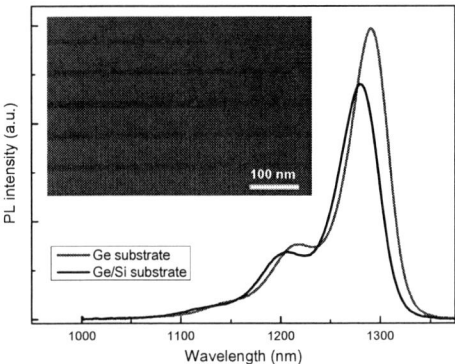

Fig. 8. Photoluminescence spectra are compared for InAs/InGaAs DWELL structures grown on Ge/Si and Ge substrates, respectively. The inset is a cross-sectional TEM image of InAs/GaAs QDs on a Ge/Si substrate.

InAs/GaAs QDs grown on Ge/Si substrate were also studied. To form the Ge/Si virtual substrate, a 2- m thick Ge layer was grown using chemical vapour deposition on phosphorus-doped (100)-oriented Si substrates with a 6° offcut toward [111] planes. The threading dislocation density in Ge epitaxial layer is about $5 \times 10^6/cm^2$. The optimized InAs/GaAs DWELL laser structures were fabricated on Ge/Si. The room-temperature photoluminescence of InAs/GaAs QDs grown Ge/Si substrates is comparable to that grown on Ge substrates, as shown in Fig. 8. InAs/GaAs QD lasers grown on Ge/Si substrates are under investigation.

IV. CONCLUSION

We have presented studies on the development of 1300-nm InAs/GaAs quantum-dot lasers monolithically grown on Si, Ge, and Ge/Si substrates for Si photonics. Room-temperature lasing near 1.3 m has been demonstrated for the devices on both Si and Ge substrates. This study could ultimately form the basis for the monolithic integration of 1300-nm InAs/GaAs QD lasers on Si platform, as well as for the integration of other III-V devices on Si substrates in order to realize the long-dreamed of III-V/Si optoelectronic integrated circuit.

REFERENCES

[1] D. Liang, and J. E. Bowers, "Recent progress in lasers on silicon" *Nature Photon.*, vol. 4, pp. 511-517, August 2010.

[2] T. Badcock, R. Royce, D. Mowbray, M. Skolnick, H. Liu, M. Hopkinson, K. Groom, and Q. Jiang, "Low threshold current density and negative characteristic temperature 1.3 m InAs self-assembled quantum dot lasers," *Appl. Phys. Lett.*, vol. 90, pp. 111102, 2007.

[3] H. Liu, I. Sellers. T. Badcock, D. Mowbray, M. Skolnick, K. Groom, M. Gutierrez, M. Hopkinson, J. Ng, J. David, and R. Beanland, "Improved performance of 1.3 m multilayer InAs quantum-dot lasers using a high-growth-temperature GaAs spacer layer," *Appl. Phys. Lett.*, vol. 85, pp. 704-706, 2004.

[4] R. Beanland, A. Sanchez, D. Childs, K. Groom, H. Liu, D. Mowbray, and M. Hopkinson, "Structural analysis of life tested 1.3 m quantum dot lasers," *J. Appl. Phys.*, vol. 103, pp. 014913, 2008.

[5] H. Liu, D. Childs, T. Badcock, K. Groom, I. Sellers, M. Hopkinson, R. Hogg, D. Robbins, D. Mowbray, and M. Skolnick, "InAs/GaAs quantum-dot lasers with very low continuous-wave room-temperature thresholds currents," *IEEE Photon. Technol. Lett.*, vol. 17, no. 6, pp. 1139-1141, June 2005.

[6] T. Wang, H. Liu, A. Lee, F. Pozzi, and A. Seeds, "1.3- m InAs/GaAs quantum-dot lasers monolithically grown on Si substrates," *Opt. Express.*, vol. 19, no. 12, pp. 11381-11386, June 2011.

[7] T. Wang, A. Lee, F. Tutu, A. Seeds, H. Liu, Q. Jiang, K. Groom, and R. Hogg, "The effect of growth temperature of GaAs nucleation layer on InAs/GaAs quantum dots monolithically grown on Ge substrates," *Appl Phys. Lett.*, vol. 100, pp. 052113, 2012.

[8] H. Liu, T. Wang, Q. Jiang, R. Hogg, F. Tutu, F. Pozzi, and A. Seeds, "Long-wavelength InAs/GaAs quantum-dot laser diode monolithically grown on Ge substrate," *Nature Photon.*, vol. 5, pp. 416-419, July 2011.

[9] D. Bordel, D. Guimard, M. Rajesh, M. Nishioka, E. Augendre, L. Clavelier, and Y. Arakawa, "Growth of InAs/GaAs quantum dots on germanium-on-insulator-on-silicon (GeOI) substrate with high optical quality at room temperature in the 1.3 m band," *Appl. Phys. Lett.*, vol. 96, pp. 043101, 2010.

High Efficiency and Broad-Band Operation of Monolithically Integrated W-Band HBV Frequency Tripler

A. Malko, T. Bryllert, J. Vukusic and J. Stake

Department of Microtechnology and Nanoscience, Chalmers University of Technology,
SE – 412 96 Gothenburg, Sweden
E-mail: malko@chalmers.se

Abstract — We report on a state-of-the-art monolithically integrated heterostructure barrier varactor (HBV) frequency tripler operating in the W-band frequency range. The device utilizes series connection of four HBV diode mesas, with total 12 barriers, and a cross section area of 700 µm². The presented tripler withstands 800 mW input power, while delivering 185 mW of output power at 107 GHz. The corresponding conversion efficiency was measured to be 23% and the circuit exhibited 15% 3-dB bandwidth.

Index Terms — frequency multipliers, heterostructure barrier varactors (HBVs), millimeter wave diodes, III-V semiconductors, varactors, power sources.

I. INTRODUCTION

Higher order harmonic generation by means of nonlinear devices, like Schottky Diodes (SDs) or Heterostructure Barrier Varactors (HBVs), facilitate output power at millimeter and THz frequencies [1]. The generated power is utilized to drive heterodyne mixers used in security imaging, radar, space and radio astronomy applications [2,3]. The SD and HBV based frequency multipliers are also capable to deliver sufficient LO power to the transmit and receive frontend components for the wireless communication [4].

Up to date, most of the HBV frequency multipliers have been fabricated employing discrete diodes that are flip-chip soldered to an external circuit. Using the hybrid technology, a frequency multiplier with 20% efficiency at 113 GHz has been presented [5]. However, the 3-dB bandwidth of the device was limited to 1.5%. The circuit described in [6] is fabricated on an AlN substrate and shows a 17% 3-dB bandwidth centered at 110 GHz, but with a conversion efficiency of only 5%. The very first monolithically integrated HBV frequency multiplier on GaAs, operating at 210 GHz, delivered 1.4 mW output power and had 2.8% efficiency [7]. The use of new material structures and processes has significantly improved the performance of the HBV diodes and circuits. The frequency tripler at 288 GHz reported in [8], is based on InP HBV material transferred to a quartz substrate, and obtained a maximum output power of 6 mW (6% efficiency) and a 3-dB bandwidth of 17%. InP monolithically integrated frequency triplers at 97 GHz [9] and at 282 GHz [10] delivered 85 mW (20% efficiency) and 31 mW (7% efficiency) of output power, respectively.

A pronounced trade-off between the available output power and the bandwidth is a main challenge when designing a HBV

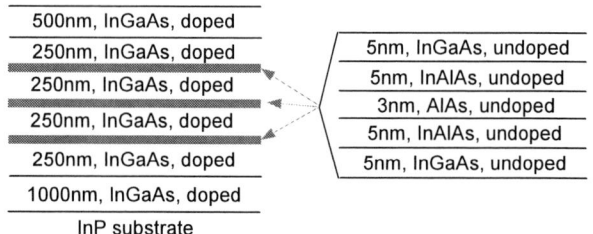

Fig. 1. Schematic of the epitaxial InGaAs/InAlAs HBV material structure on the InP substrate.

frequency multiplier. In this study, we report on a monolithically integrated HBV frequency tripler with 185 mW output power (23% conversion efficiency), and a broad 15% 3-dB bandwidth, constituting the state-of-the-art for W-band HBV frequency triplers.

II. HBV DEVICES AND TRIPLER CIRCUIT

The HBV is a semiconductor device, generating only odd harmonics of the input signal [11]. The nonlinearity is obtained due to a symmetric, voltage-dependent capacitance characteristic, with its maximum at zero bias, which eliminates the need for dc bias network. The capacitance modulation is possible by stacking a semiconductor characterized by high energy bandgap (barrier) in between semiconductors with low bandgap energy. The relatively simple structure of the HBV allows arrangement of several barriers grown by molecular beam epitaxy (MBE). Hence, enhancement of the breakdown voltage and power handling capability of the HBV diode can be achieved [12].

For the presented HBV tripler, the InGaAs/InAlAs/AlAs material structure was epitaxially grown on a lattice-matched 3" InP substrate. A schematic of the HBV material is illustrated in fig.1. The 500 nm and 1000 nm doped $In_{0.53}Ga_{0.47}As$ layers act as contact and buffer layers, respectively. The 250 nm thick, moderately doped $In_{0.53}Ga_{0.47}As$ layers allow for modulation of the capacitance. Depicted in fig. 1, undoped regions of $In_{0.52}Ga_{0.48}As$ and InAlAs/AlAs/InAlAs are used as spacers and barriers, respectively. A pseudomorphic AlAs layer at the center of the barrier increases the barrier height, hence reduces the leakage current of the diode [13].

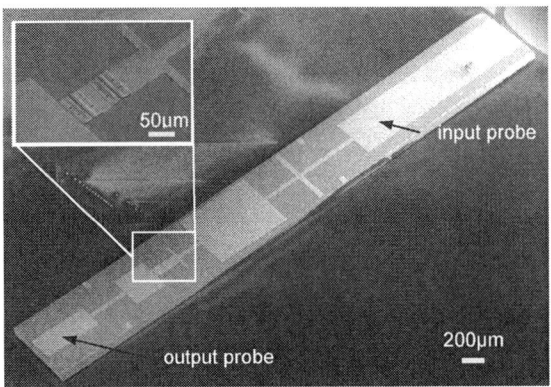

Fig. 2. SEM image of an integrated HBV frequency tripler circuit on InP. The chip dimensions are 3.5mm x 0.8mm x 0.08mm. *Top:* 4 mesa, 12barriers HBV diode, with a cross section area of $7 \times 100 \ \mu m^2$.

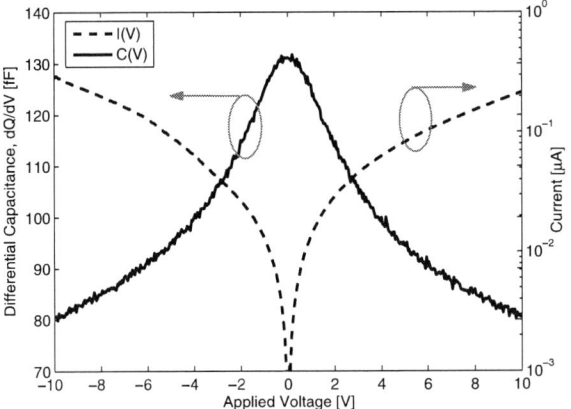

Fig. 3. Capacitance-voltage (solid line) and current-voltage (dashed line) performance of the presented HBV diode. The capacitance-voltage characteristic was measured at 1MHz. The measurements were performed at room temperature and at dark conditions.

The fabrication of the monolithically integrated HBV frequency tripler employs processes of photolithography, ohmic contact e-beam evaporation, dry and wet etching of the diode side-walls. The isolation of the diode mesas was further achieved through wet etching.

The formation of the air bridge connections to the mesas and the microstrip circuits are done by gold electroplating. Subsequently, the chip is diced and lapped down to the thickness of 80 μm. The circuits are mounted in a waveguide block with WR-22 input and WR-10 output with a non-conducting glue. Fig. 2 shows a scanning electron microscopy (SEM) image of a frequency tripler circuit.

The presented HBV device employs a series connection of 4 mesas, with a total of 12 barriers, and a cross section area is $7 \times 100 \ \mu m^2$. The measured differential capacitance for the given device is 130fF at zero bias, see fig. 3.

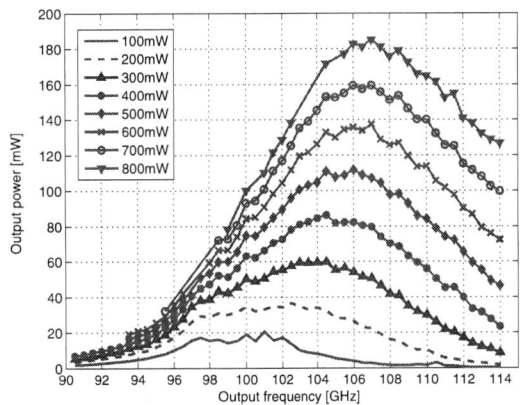

Fig. 4. Measured output power as a function of the output frequency for input signals in the range from 100 mW to 800 mW.

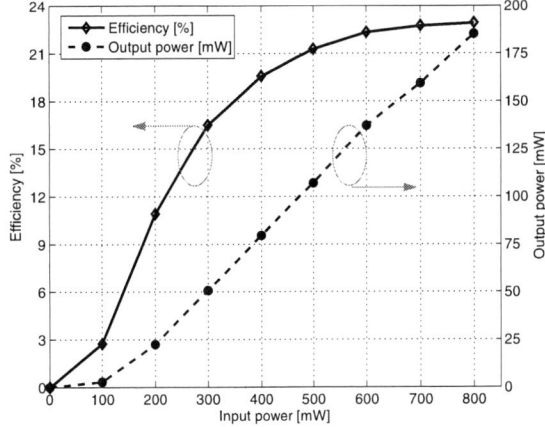

Fig. 5. Measured efficiency (solid lines) and output power (dashed lines) as a function of the input power at 107 GHz.

For the circuit design the quasi-empirical electro-thermal HBV model from Chalmers was implemented [14,15]. Agilent Advanced Design System (ADS) and Ansoft HFSS were employed for the device and circuit simulations. The input and output waveguide signals are coupled via open waveguide probes, marked for clarity in fig. 2. The impedance matching is realized using microstrip elements.

III. RF MEASUREMENTS AND RESULTS

The RF characterization of the frequency tripler was done using the following setup; the input power provided by an Agilent E8257D signal generator was followed by a Spacek Ka-band power amplifier. For an accurate monitoring of the input power a 10dB waveguide coupler was used in between the amplifier and the multiplier. The input and output power was measured by an Agilent E4418B power sensor and an Erickson PM4 power meter, respectively.

The measured characteristics of the output power versus output frequency for the input power in the range 100 mW - 800 mW, are presented in fig. 4. The peak power obtained from this device is 185 mW at 107 GHz for an input signal of 800 mW. The corresponding 3-dB bandwidth at the peak power is 15%. A summary of the maximum available output power and the corresponding conversion efficiency versus the input power at 107 GHz is illustrated in fig 5.

IV. CONCLUSIONS

A W-band monolithically integrated HBV frequency tripler on InP substrate has been demonstrated. The circuit delivers state-of-the-art conversion efficiency of 23%, and 15% 3-dB bandwidth. The demonstrated high power capability of the HBV tripler is believed to greatly enhance the development of solid-state power sources at millimeter and sub-millimeter-wave frequencies. It can be used as the first stage of the multiplier chain to deliver power to heterodyne mixers operating at THz frequencies, or as a broadband, medium power sources for communication applications.

ACKNOWLEDGMENTS

This project was supported by Swedish Research Council (VR).

Authors wish to acknowledge Arne Øistein Olsen from Wasa Millimeter Wave for the assistance during RF measurements.

REFERENCES

[1] P. Siegel, "Terahertz technology," *IEEE Transactions on Microwave Theory and Techniques,* vol. 50, pp. 910-927, 2002.

[2] P. Siegel, "THz instruments for space," *IEEE Transactions on Antenna and Propagation,* vol. 55, pp. 2957-2965, 2007.

[3] H. W. Hübers, "Terahertz Heterodyne Recivers," *IEEE Journal on Selected Topics in Quantum Electronics,* vol. 14, pp. 378-391, 2008.

[4] I. Kallfass, J. Antes, T. Schneider, F. Kurz, D. Lopez-Diaz, S. Diebold, H. Massler, A. Leuther, and A. Tessmann, "All active

MMIC-based wireless communication at 220GHz," *IEEE Transactions on Terahertz Science and Technology,* vol. 1, pp. 477 - 487, 2011.

[5] J. Vukusic, T. Bryllert, T. A. Emadi, M. Sadeghi, and J. Stake, "A 0.2W heterostructure barrier varactor frequency tripler at 113GHz," *IEEE Electron Device Letters,* vol. 28, pp. 340-342, 2007.

[6] T. Bryllert, J. Vukusic, A. Ø. Olsen, and J. Stake, "A broadband heterostructure barrier varactor tripler source," *presented at the IEEE MTT-S International Microwave Symposium,* 2010.

[7] R. Meola and J. Freyer, "210GHz tripler with monolithically integrated single barrier varactors," *Electronic Letters,* vol. 34, pp. 1756-1757, 1998.

[8] T. David, S. Arscott, J. M. Munier, T. Akalin, G. Beaudin, and D. Lippens, "Monolitically integrated circuits incoroporating InP-based heterostrucutre barrier varactors," *IEEE Microwave and Wireless Components Letters,* vol. 12, pp. 281-283, 2002.

[9] J. Vukusic, T. Bryllert, A. Ø. Olsen, and J. Stake, "High-power W-band monolithically integrated tripler," in *34th International Conference on Infrared, Millimeter, and Terahertz Waves,* 2009, pp. 1-2.

[10] J. Vukusic, T. Bryllert, A. Ø. Olsen, J. Hanning, and J. Stake, "Monolithic HBV-based 282GHz tripler with 31mW output power," *IEEE Electron Device Letters,* vol. 33, pp. 800 - 802, 2012.

[11] E. Kollberg and A. Rydberg, "Quantum-barrier-varactor diodes for high-efficiency millimetre-wave multipliers," *Electronic Letters,* vol. 25, pp. 1696-1698, 1989.

[12] J. Stake, S. H. Jones, L. Dillner, S. Hollung, and E. Kollberg, "Heterostrucutre-barrier-varactor design," *IEEE Transactions on Microwave Theory and Techniques,* vol. 48, pp. 677 - 682, 2000.

[13] V. K. Reddy and D. P. Neikirk, "High breakdown voltage AlAs/InGaAs quantum barrier varactor diodes," *Electronics Letters,* vol. 29, pp. 464 - 466, 1993.

[14] L. Dillner, J. Stake, and E. Kollberg, "Modeling of the heterostructure barrier varactor diode," in *International Semiconductor Device Research Symposium,* 1997, pp. 179-182.

[15] M. Ingvarson, J. Vukusic, A. Ø. Olsen, T. A. Emadi, and J. Stake, "An electro-thermal HBV model," *presented at the IEEE MTT-S International Microwave Symposium,* 2005.

Lattice-Matched p$^+$-GaAsSb/i-InAlAs/n-InGaAs Zero-Bias Backward Diodes for Millimeter-wave Detectors and Mixers

Tsuyoshi Takahashi[1,2], Masaru Sato[1,2], Yasuhiro Nakasha[1,2], and Naoki Hara[1,2]

[1]Fujitsu Laboratories Ltd., [2]Fujitsu Limited

10-1 Morinosato-Wakamiya, Atsugi 243-0197, Japan

Abstract — A high voltage sensitivity (β_v) of 2100 V/W at 94 GHz was achieved at zero bias using backward diodes that are based on a type II heterojunction of GaAsSb/InGaAs, which was lattice-matched to an InP substrate. The impedance-matched voltage sensitivity ($\beta_{v,opt}$) of a circular mesa diode with a diameter of 1.6 μm was estimated to be 20,000 V/W. A large curvature coefficient (γ) of −35 V^{-1} was obtained at zero bias. Epitaxial grown lattice-matched backward diodes on the InP substrate have fewer defects and smooth surfaces, indicating that the devices can be stably reproduced.

Index Terms — Backward diodes, GaAsSb, interband tunneling, millimeter wave, sensitivity.

I. INTRODUCTION

Highly sensitive detectors are required in order to rectify RF signals into DC output voltages for the detection millimeter waves. In general, Schottky diodes are used to detect millimeter waves. However, their high impedance property at zero bias may suppress the sensitivity of the detectors. Therefore, several types of zero-bias detectors have been proposed [1]-[3]. In these detectors, GaSb-based backward diodes are attractive because of their high sensitivity in the millimeter-wave region [4], [5]. Using p$^+$-GaSb/i-Al$_{0.1}$Ga$_{0.9}$Sb/i-AlSb/n-InAs backward diodes, a high voltage sensitivity (β_v) of 4600 V/W was reported at 94 GHz [6]. However, the GaSb-based backward diodes fabricated on GaAs substrates are unstable because the lattice constants of the GaSb layer, InAs layer, and GaAs substrate are different.

In this study, we report a β_v of 2100 V/W at 94 GHz by fabricating lattice-matched p$^+$-GaAsSb/i-InAlAs/n-InGaAs backward diodes. A curvature coefficient of −35 V^{-1} was obtained. These properties are applicable for millimeter-wave detectors and mixer ICs [7]. Furthermore, the backward diodes that are lattice-matched to InP are easily integrated with InP-based low-noise amplifiers (LNAs), and can realize superior detection performance.

II. DEVICE FABRICATION

GaAsSb-based backward diodes, which are lattice matched to InP substrates, consist of p$^+$-GaAs$_{0.51}$Sb$_{0.49}$/i-In$_{0.52}$Al$_{0.48}$As/n-In$_{0.53}$Ga$_{0.47}$As heterojunctions. First, a 200-nm-thick i-In$_{0.52}$Al$_{0.48}$As buffer layer was grown on the semi-insulating InP substrate by metal organic chemical vapor deposition (MOCVD). Next, an n-In$_{0.53}$Ga$_{0.47}$As ohmic contact layer, which was doped with Si to the carrier concentration of 2 × 10^{19} cm^{-3}, was grown. The thickness of the low resistance layer was 200 nm. An n-In$_{0.53}$Ga$_{0.47}$As (50 nm) layer, an i-In$_{0.52}$Al$_{0.48}$As (3 nm) layer, and a p$^+$-GaAs$_{0.51}$Sb$_{0.49}$ (30 nm) layer were also grown sequentially, and they act as the backward diode. In these layers, the hole concentration in the p$^+$-GaAs$_{0.51}$Sb$_{0.49}$ layer was maintained at 1 × 10^{19} cm^{-3}, which indicates degeneracy. In contrast, the carrier concentration of the n-In$_{0.53}$Ga$_{0.47}$As was adjusted to 1 × 10^{18} cm^{-3} so that the interband tunneling of electrons could occur from a valence band of the p$^+$-GaAs$_{0.51}$Sb$_{0.49}$ layer to the conduction band of the n-In$_{0.53}$Ga$_{0.47}$As layer though the thin i-In$_{0.52}$Al$_{0.48}$As layer. Fig. 1 shows a band diagram of the GaAsSb-based backward diode in thermal equilibrium. Both the valence band edge of the GaAs$_{0.51}$Sb$_{0.49}$ layer and the conduction band edge of the n-In$_{0.53}$Ga$_{0.47}$As layer were close to a Fermi level (E$_F$), indicating an energy band feature of the backward diode. On the upper side of the p$^+$-GaAs$_{0.51}$Sb$_{0.49}$ layer, contact layers consisting of 8-nm-thick n-In$_{0.8}$Ga$_{0.2}$As (5 × 10^{18} cm^{-3}) followed by 30-nm-thick n$^+$-In$_{0.53}$Ga$_{0.47}$As (2 × 10^{19} cm^{-3}) were formed. This structure enables the ohmic property of the p$^+$-GaAs$_{0.51}$Sb$_{0.49}$ by interband tunneling [2]. The lattice-matched diodes are believed to reduce defects on the surface of the epitaxial wafers.

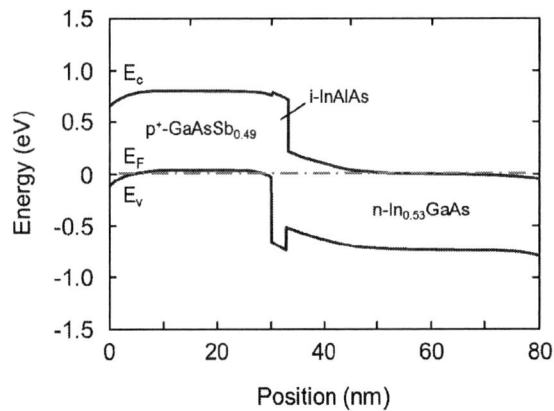

Fig. 1. Calculated band diagrams of GaAsSb-based diodes in thermal equilibrium.

978-1-4673-1725-2/12 $31.00 © 2012 IEEE

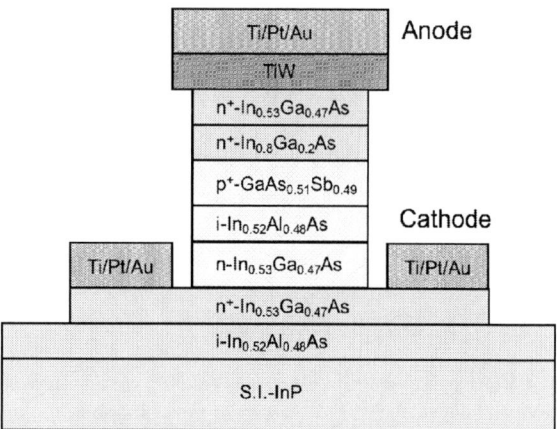

Fig. 2. Cross-sectional diagram of a self-aligned mesa diode. The horizontal distance and circular mesa size were reduced.

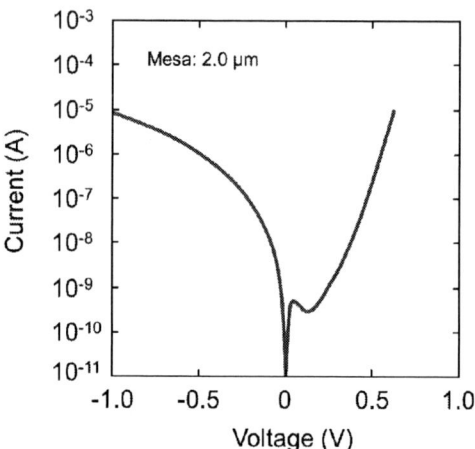

Fig. 3. I–V characteristic of a GaAsSb-based backward diode with a self-aligned structure. The circular mesa diameter is 2.0 μm.

Fig. 2 shows the cross-sectional diagram of the backward diodes, which have a self-aligned electrode structure. First, TiW metal was deposited on the epitaxial wafer by a sputtering method. An anode electrode was formed using the dry etching technique. Next, anisotropic wet chemical etching was performed to form a self-aligned mesa structure using the TiW electrode as the etching mask. After forming the ledge structure at the TiW, a device isolation area was formed by removing an n^+-$In_{0.53}Ga_{0.47}As$ contact layer at the outside of the active area. Ti/Pt/Au (10/15/50 nm) was evaporated to form a self-aligned cathode electrode on the n^+-$In_{0.53}Ga_{0.47}As$ contact layer. The typical distance from the mesa edges to the cathode electrode was 0.3 μm. Finally, the diodes were covered with BCB film, and interconnections were formed by adopting a gold-plating technique.

III. DEVICE CHARACTERISTICS

A. I-V Characteristics

The I–V characteristic of a circular mesa diode with a diameter of 2.0 μm is shown in Fig. 3. A rapid increase in current was observed upon application of a reverse bias. The interband tunneling of electrons occurs from the valence band of p^+-$GaAs_{0.51}Sb_{0.49}$ to the conduction band of n-$In_{0.53}Ga_{0.47}As$. In contrast, a non-linear I–V characteristic was obtained when a forward bias was applied. The curvature coefficient (γ) was obtained from the I–V curvature. γ was defined as

$$\gamma = \frac{d^2I/dV^2}{dI/dV}.\qquad(1)$$

Fig. 4 shows the γ dependence as a function of the applied voltage. The γ of -35 V^{-1} was obtained at zero bias, indicating a superior parabolic characteristic that affects the sensitivity of the diode detectors.

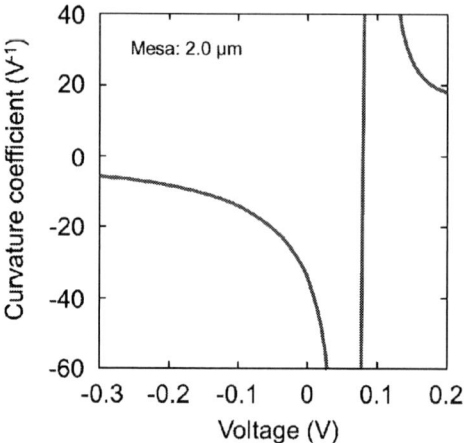

Fig. 4. Curvature coefficient as a function of applied voltage.

B. Sensitivity

The dependence of the detected voltage output (V_{det}) on the input power (P_{in}) was investigated for a diode with a diameter of 2.0 μm at 94 GHz, as shown in Fig. 5. For the measurements, the applied bias was zero.. The diode indicated a linear output characteristic for input powers ranging from −33 to −13 dBm. The polarity of the detected voltage was reversed when P_{in} was smaller than −38 dBm. However, the polarity was unity when the device size was increased up to 3.0 μm. The difference depends on the device size and its I–V curvature. V_{det} was saturated at a P_{in} of about 0 dBm. By using the V_{det} and P_{in} values, the voltage sensitivity β_v should then be calculated. A zero-bias β_v of 2100 V/W was observed at 94 GHz, when the mesa diode having a 2.0 μm diameter was measured. Note that β_v was obtained under unmatched impedance conditions, indicating non-optimized value.

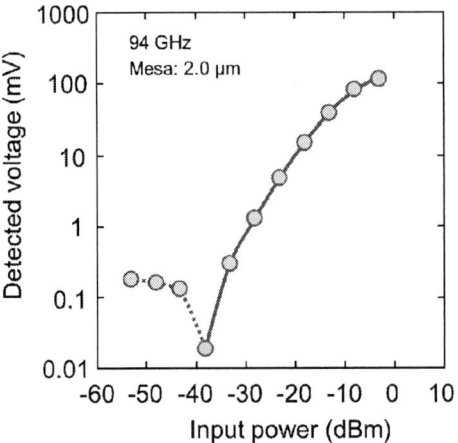

Fig. 5. Dependence of detected voltage on input voltage for a detector diode at 94 GHz. The circular mesa diameter is 2.0 μm. The polarity of the detected voltage was reversed at an input power of −38 dBm.

To obtain the zero-bias voltage sensitivity $\beta_{v,opt}$ impedance matching was assumed. The $\beta_{v,opt}$ was defined as

$$\beta_{v,opt} = \frac{V_{det}}{P_{in}(1-|\Gamma|^2)}, \qquad (2)$$

where Γ is the return loss of the input power caused by an impedance mismatch, that is, S_{11}. Fig. 6 shows the dependence of $\beta_{v,opt}$ on the diameter of the diodes at 94 GHz. A P_{in} of −27 dBm was uniformly applied. Using the diode with a diameter of 2.0 μm, $\beta_{v,opt}$ was calculated to be 12,000 V/W from V_{det}, P_{in} and the return loss (S_{11}) at zero bias. This $\beta_{v,opt}$ indicates a value that is three times higher than that of the previously reported GaAsSb-based diode having the same size [8]. Further improvement of the $\beta_{v,opt}$ up to 20,000 V/W was estimated when the diode diameter was reduced to 1.6 μm. It is believed that further scaling of the mesa size will improve the voltage sensitivity.

The dependence between $\beta_{v,opt}$ and the device parameters was investigated. The $\beta_{v,opt}$ is expressed as

$$\beta_{v,opt} \approx \frac{\gamma}{2/R_j + 8(\pi f C_j)^2 R_s}, \qquad (3)$$

where R_j is the junction resistance, f is frequency, C_j is the junction capacitance, and R_s is the series resistance. The mesa junction area (S_j) is proportional to the C_j. In addition, most of the R_s elements appeared to be due to the contact resistance at the anode electrode, indicating that R_s is inversely proportional to the S_j. Therefore, $C_j^2 \cdot R_s$ in (3) is proportional to the area S_j. Because $2/R_j$ is negligibly small compared to the $8(\pi f C_j)^2 R_s$,

$\beta_{v,opt}$ is proportional to $1/S_j$. Fig. 7 shows $\beta_{v,opt}$ as a function of $1/S_j$. The $\beta_{v,opt}$ for both this work and the previous one indicated a linear relationship. In particular, the $\beta_{v,opt}$ of this work indicated a greater increase with increasing $1/S_j$, indicating that a reduction of the mesa area is effective in improving the sensitivity. A sensitivity exceeding 30,000 V/W at 94 GHz has been estimated for the lattice-matched GaAsSb-based backward diode when the mesa diameter was reduced to less than 1.3 μm. Moreover, $\beta_{v,opt}$ of 52,000 V/W is estimated when the diode diameter is assumed to be 1.0 μm.

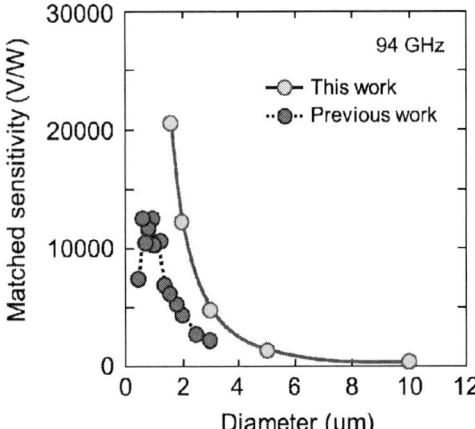

Fig. 6. Dependence of matched voltage sensitivity on the diode mesa diameter, measured at 94 GHz. The return loss S_{11} was considered in obtaining the sensitivity values.

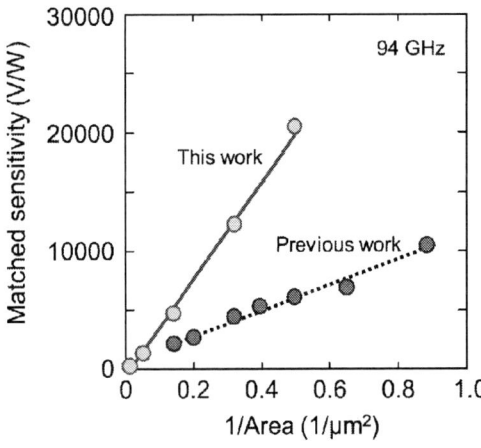

Fig. 7. Dependence of matched voltage sensitivity on the reciprocal of the diode-mesa area.

C. Device Parameters

Using diodes with 1.6 and 2.0 μm diameters, the device parameters were extracted. We used an equivalent circuit model that includes a junction resistance R_j, a junction capacitance C_j, a series resistance R_s, and a parasitic capacitance (C_p) [9]. In addition, a pad capacitance (C_{pad}) and an interconnection inductance (L_p) were considered. R_j was determined to be 8.26 MΩ from dV/dI at zero bias in the DC I–V characteristic. Moreover, C_{pad} and L_p were determined to be 2.92 fF and 2.0 pH, respectively. Table I shows extracted parameters for diodes with 1.6 and 2.0 µm diameter. Although C_j indicated a small value by decreasing the doping concentration in an n-In$_{0.53}$Ga$_{0.47}$As layer, R_s had a considerably large value [10]. An increased contact resistance at the cathode electrode may have caused the increase in R_s, since the R_s value was inversely proportional to the mesa area. A further reduction in the contact resistance is required to improve R_s which can enhance the RF characteristics.

TABLE I
EXTRACTED DEVICE PARAMETERS

Size (µm)	R_j (MΩ)	C_j (fF)	R_s (Ω)	C_p (fF)	C_{pad} (fF)	L_p (pH)
1.6	12.3	2.15	250	0.58	2.92	2.0
2.0	8.26	5.79	116	2.16	2.92	2.0

IV. SUMMARY

In summary, lattice-matched p$^+$-GaAs$_{0.51}$Sb$_{0.49}$/i-n$_{0.52}$Al$_{0.48}$As /n-In$_{0.53}$Ga$_{0.47}$As backward diodes were fabricated and a voltage sensitivity of 2,100 V/W at 94 GHz was achieved at zero bias. A matched voltage sensitivity of 20,000 V/W was estimated using a diode with a 1.6-µm diameter. High sensitivities are required for effective millimeter-wave detection. Furthermore, a curvature coefficient of –35 V^{-1} can be applied to mixers with a low conversion loss.

ACKNOWLEDGMENT

The authors would like to thank T. Ohki for discussion regarding the device process, and T. Ito for his technical assistance in the fabrication of the devices. We would also like to thank Prof. O. Ueda of Kanazawa Institute of Technology for his helpful discussions and encouragement. This work was supported in part by the Ministry of Internal Affairs and Communications, Japan.

REFERENCES

[1] V. T. Vo and Z. Hu, "Optimization and realization of planar isolated GaAs zero-biased planar doped barrier diodes for microwave/millimeter-wave," *IEEE Trans. Microw. Theory Tech.*, vol. 54, pp. 3836–3842, Nov. 2006.

[2] J. N. Schulman and D. H. Chow, "Sb-heterostructure interband backward diode," *IEEE Electron Device Lett.*, vol. 21, pp. 353–355, July 2000.

[3] H. Ito, F. Nakajima, T. Ohno, T. Furuta, T. Nagatsuma, and T. Ishimashi, "InP-based planar-antenna-integrated Schottky-barrier diode for millimeter- and sub-millimeter-wave detection," *Jpn. J. Appl. Phys.*, vol. 47, pp. 6256–6261, 2008.

[4] R. Meyers, P. Fay, J. Schulman, III, S. Thomas, D. Chow, J. Zinck, Y. Boegeman, and P. Deelman, "Bias and temperature dependence of Sb-based heterostructure millimeter-wave detectors with improved sensitivity," *IEEE Electron Device Lett.*, vol. 25, pp. 4–6, Jan. 2004.

[5] N. Su, R. Rajavel, P. Deelman, J. Schulman, and P. Fay, "Sb-heterostructure millimeter-wave detectors with reduced capacitance and noise equivalent power," *IEEE Electron Device Lett.*, vol. 29, pp. 536–539, Jun. 2008.

[6] Z. Zhang, R. Rajavel, P. Deelman, and P. Fay, "Sub-micron area heterojunction backward diode millimeter-wave detectors with 0.18 pW/Hz$^{1/2}$ noise equivalent power," *IEEE Microw. Wireless Components Lett.*, vol. 21, pp. 267–269, May 2011.

[7] M. Morgan and S. Weinreb, "A monolithic HEMT diode balanced mixer for 100-140 GHz," in *Proc. IEEE MTTS Dig.*, 2001, pp. 99–102.

[8] T. Takahashi, N. Hara, M. Sato, T. Hirose, "GaAsSb-based backward diodes for highly sensitive millimeter-wave detectors," *Electron. Lett.*, vol. 45, no. 24, pp. 1269–1270, 2009.

[9] J. N. Schulman, S. Thomas, D. H. Chow, E. T. Croke, H. L. Dunlap, K. S. Holabird, W. M. Clark, "High frequency performance of Sb-heterostructure millimeter-wave diodes," in *Proc. IEEE 14th Int. Conf. IPRM*, 2002, pp. 151–154.

[10] T. Takahashi, M. Sato, T. Hirose, and N. Hara, "Energy band control of GaAsSb-based backward diodes to improve sensitivity of millimeter-wave detection," *Jpn. J. Appl. Phys.*, vol. 49, pp. 104101, 2010.

A *W*-Band InGaAs PIN-MMIC Digital Phase-Shifter Using a Switched Transmission-Line Structure

Jung Gil Yang, Jooseok Lee and Kyounghoon Yang

Department of Electrical Engineering, Korea Advanced Institute of Science and Technology (KAIST)
373-1, Guseong-Dong, Yuseong-Gu, Daejeon, Republic of Korea.
jkyang@kaist.ac.kr

Abstract — This paper describes the design and fabrication of a *W*-band InGaAs PIN-diode MMIC 4-bit phase shifter based on a switched delay line. In order to achieve low insertion loss and good phase shifting characteristics at *W*-band, the topology based on a switched delay-line is employed using a thin-film microstrip line structure. The fabricated phase shifter has demonstrated good performances such as an insertion loss less than 12.7 dB at a frequency range of 81 to 85 GHz with an intrinsic chip size of 1.93 × 0.80 mm². To our knowledge, this is the first InGaAs PIN MMIC digital phase shifter demonstrated up to *W*-band.

Index Terms — *W*-band, InGaAs, PIN-diode, MMIC, phase shifter.

I. INTRODUCTION

Microwave and millimeter-wave phase shifters are essential components, which are used for controlling the phase of signal to change the direction of the radiated beam [1]. Digital phase shifters are usually preferred in phased-array applications because they are more immune to their control voltage noise and temperature variation. There are many different designs for the semiconductor device based digital phase shifters. Some of the more well-known designs are loaded-line phase shifters [2], branch-line phase shifters [3], switched-line phase shifters [4], and high-pass/low-pass filter phase shifters [5]. In particular, the switched-line configuration for millimeter-wave band applications has the advantages of effective phase-shifting characteristics and small chip size due to simple circuit and layout designs. The switched-line phase shifter is a true-time delay (TTD) circuit with the RF performance and bandwidth determined by the high-frequency operation of the switches. From this standpoint, InGaAs PIN diodes with optimized junction areas are considered suitable for millimeter-wave band applications as the active switching devices because of their inherent high cut-off frequency [6].

In this paper, we report on a *W*-band InGaAs PIN MMIC 4-bit phase shifter based on the switched transmission-line. In order to achieve the high-frequency band and low-loss phase shifting performances, a design topology based on the switched delay line has been applied with optimal conditions of the circuit elements. In addition, the PIN-diodes in each bit are designed to be on/off controlled by using only one bias voltage for compact phase shifting operation of the MMIC

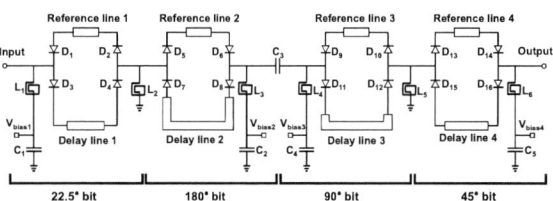

Fig. 1. Circuit schematic of the proposed switched-line type InGaAs PIN MMIC phase shifter.

phase shifter. To the authors' best knowledge, this is the first demonstration of the InGaAs PIN MMIC digital phase shifter operating at *W*-band frequencies.

II. PHASE SHIFTER DESIGN AND MMIC TECHNOLOGY

The circuit schematic of the proposed switched-line type InGaAs PIN 4-bit (180°, 90°, 45°, 22.5°) MMIC phase shifter composed of the transmission lines, series InGaAs PIN diodes, DC blocking capacitors, and spiral inductors is shown in Fig. 1. The InGaAs PIN diodes were used to route signals between the delay line and the reference line in this circuit. In order to realize the compact switching operation, the circuit topology shown in Fig. 1 is proposed based on the configuration of PIN-diodes reversely connected through the integrated bias network. For the single-bit phase shifter sections, when the negative voltage is applied at V_{bias}, the PIN diodes in the reference line are turned on and the PIN diodes in the delay line are turned off. In this case, the RF signals are transmitted through the reference line. When the positive voltage is applied at V_{bias}, the RF signals are propagated through the delay line. The desired phase shift was achieved by changing the length of the delay line in each single-bit section. Its length was designed by the full-wave 3-D EM simulation. Based on this bias-controlled circuit operation, the designated 4-bit phase shifting characteristics can be obtained. In the bias network, the value of inductance for RF-choke inductors ($L_1 \sim L_6$) is set to be 0.31 nH for the lowest RF signal leakage at a center frequency of 83 GHz. At the same time, the DC current is blocked by the MIM capacitor (C_3) placed at the center of the phase shifter. This arrangement of capacitors and

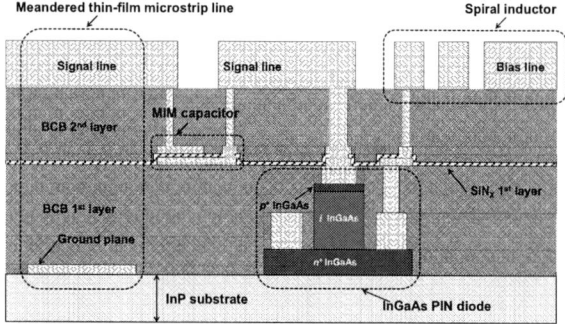

Fig. 2. Schematic cross-section of the BCB-based MMIC multi-layer technology.

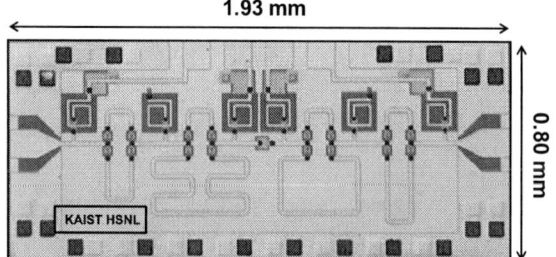

Fig. 3. Microphotograph of the fabricated InGaAs PIN-diode W-band 4-bit MMIC phase shifter.

inductors guarantees the control of each bit individually with the minimized number of devices in the bias network.

The cross-section of the BCB-based multi-layer MMIC technology used to implement the InGaAs PIN-diode based phase shifter is shown schematically in Fig. 2. The device epitaxial layers were grown by MBE on a semi-insulating InP substrate. The InGaAs PIN diodes were fabricated by using optical lithography and wet etching techniques. The MIM capacitors were formed between the two BCB dielectric layers. The meandered thin-film microstrip line was composed of the bottom metal ground plane, 6 μm-thick dielectric layers including the BCB multi-layers, and the 2 μm-thick top signal line. The detailed layer structure and fabrication sequence have been described elsewhere [7]. Fig. 3 shows the microphotograph of the fabricated W-band 4-bit MMIC phase shifter. The chip size of 1.93 × 0.80 mm² was minimized by applying the meandered microstrip-line design using the BCB-based multi-layer fabrication technology. In order to implement a low-loss MMIC phase shifter, the InGaAs PIN-diode with a p-metal size of 6 × 6 μm² was used for the switching element. The achieved on-state resistance of R_{on}=4.5 Ω and off-state capacitance of C_{off}=9 fF for the fabricated device result in a high cut-off frequency over 4 THz. In addition, the thin-film microstrip line (TFMS) with the width of 16 μm was used to obtain the good matching

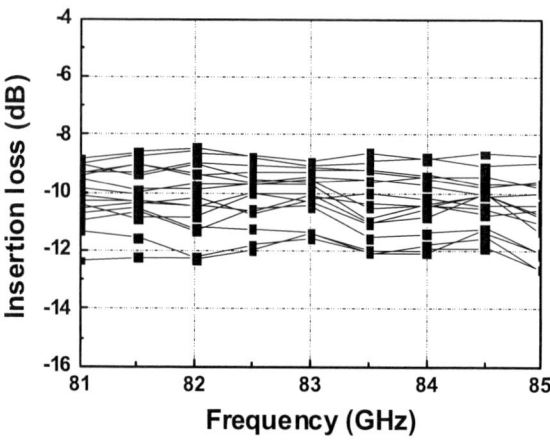

(a)

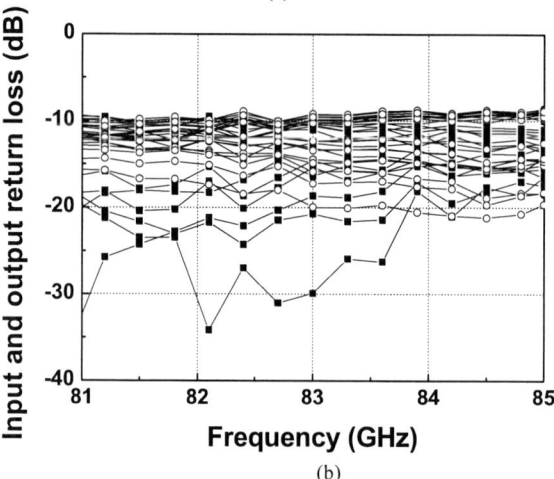

(b)

Fig. 4. (a) Measured 16-state insertion loss and (b) measured 16-state input and output return loss characteristics of the fabricated 4 bit phase shifter.

characteristics. The capacitance value of the integrated MIM capacitor was measured to be 160 pF/mm².

III. MEASUREMENT RESULTS OF THE FABRICATED PHASE SHIFTER

The fabricated W-band 4-bit MMIC phase shifter was measured via on-wafer probing using an Anritsu ME7808A network analyzer and a Cascade Microtech probe station. The switching operation of the InGaAs PIN diode is controlled by the bias voltage of V_{bias} = ±1.28 V with the DC current of 10 mA, which leads to the total DC power consumption of 51 mW. Fig. 4 shows the measured insertion loss and input/output return loss characteristics of the fabricated phase shifter from 81 to 85 GHz. The fabricated 4-bit phase shifter has insertion losses less than 12.7 dB over the frequency

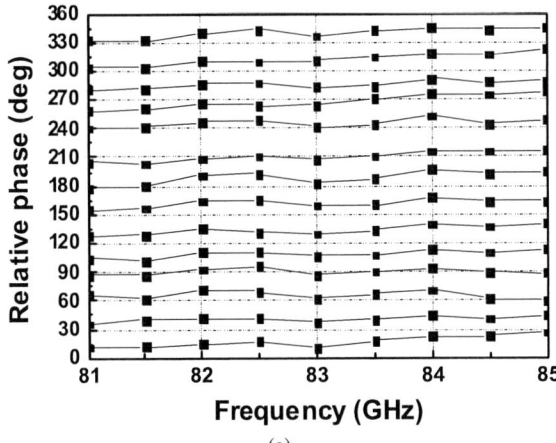

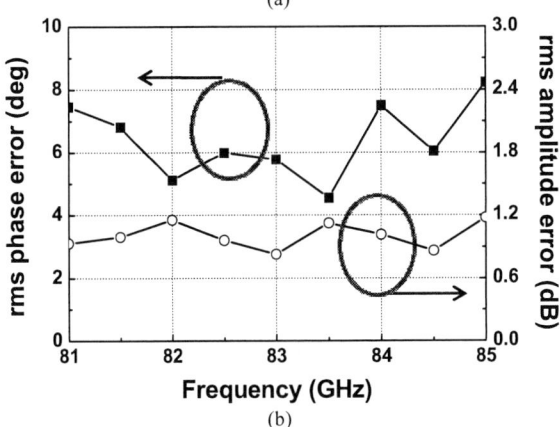

Fig. 5. Measured performances of the 4-bit MMIC phase shifter: (a) relative phase vs. frequency and (b) rms phase & amplitude error vs. frequency.

range with input/output return losses better than 8.5 dB. The measured phase performance and rms deviations of the phase shifter over the frequency band are shown in Fig. 5. As shown in Fig. 5 (a), the accurate phase-shifting characteristics for the 16 states are confirmed. The rms phase error is measured to be less than 8.2° from 81 to 85 GHz. In addition, the measured rms phase error is 5.8° at the center frequency of 83 GHz. For all measured insertion loss characteristics of 16 states, the rms amplitude error is less than 1.18 dB over the frequency range of 81 to 85 GHz. The measured low insertion loss of 12.7 dB and rms phase error of 8.2° are found to be better than those

of the previously reported millimeter-wave band MMIC phase shifters [3, 8].

IV. CONCLUSIONS

A W-band 4-bit MMIC phase shifter composed of the InGaAs PIN diodes was designed and fabricated using the circuit topology of switched transmission line. The fabricated phase shifter showed the overall insertion loss less than 12.7 dB and input/output return losses higher than 8.5 dB from 81 GHz to 85 GHz with a small chip size of 1.93×0.80 mm^2. In addition, the rms phase and amplitude errors were measured less than 8.2° and 1.18 dB over the same frequency range, respectively. The results demonstrate the potential of the InGaAs PIN MMIC technology for implementing high-performance MMIC phase shifters which are essential in millimeter-wave radar system applications.

REFERENCES

[1] S. K. Koul and B. Bhat, Microwave and Millimeter Wave Phase Shifters. Boston, MA: Artech House, vol. 1, pp. 32-59, 1991, ch. 1.

[2] F. Ellinger, H. Jackel, W. Bachtold, "Varactor-loaded transmission-line phase shifter at C-band using lumped elements," *IEEE Trans. Microwave Theory & Tech.*, vol. 51, no. 4, pp. 1135-1140, Apr. 2003.

[3] K. Maruhashi, H. Mizutani, and K. Ohata, "Design and performance of a Ka-band monolithic phase shifter utilizing nonresonant FET switches," *IEEE Trans. Microwave Theory & Tech.*, vol. 48, no. 8, pp. 1313-1317, Aug. 2000.

[4] K. Sun, M. Choi, and D. Weide, "A PIN Diode Controlled Variable Attenuator Using a 0-dB Branch-Line Coupler," *IEEE Microwave & Wireless Components Letters*, vol. 15, no. 6, pp. 440-442, June 2005.

[5] C. F. Campbell and S. A. Brown, "A compact 5-bit phase-shifter MMIC for K-band satellite communication system," *IEEE Trans. Microwave Theory & Tech.*, vol. 48, no. 12, pp. 2652-2656, Dec. 2000.

[6] J. Yang, H. Eom, S. Choi, and K. Yang, "2-38 GHz broadband compact InGaAs PIN switches using a 3-D MMIC technology," *2007 IEEE IPRM*, pp. 542-545, May 2007.

[7] J. Yang, and K. Yang, "Broadband InGaAs PIN traveling-wave switch using a BCB-based thin-film microstrip line structure," *IEEE Microwave & Wireless Comp. Lett.*, vol. 19, no. 10, pp. 647-649, Oct. 2009.

[8] M. Tsai, and A. Natarajan, "60 GHz passive and active RF-path phase shifters in silicon," *2009 IEEE RFIC*, pp. 223-226, June 2009.

Development of a 557 GHz GaAs monolithic membrane-diode mixer

Huan Zhao[1,2], Vladimir Drakinskiy[1,2], Peter Sobis[1-3], Johanna Hanning[1,2], Tomas Bryllert[1,2], Aik-Yean Tang[1,2], Jan Stake[1,2]

[1]GigaHertz Centre, [2]Terahertz and Millimetre Wave Laboratory, Department of Microtechnology and Nanoscience, Chalmers University of Technology, SE-41296 Göteborg, Sweden
[3]Omnisys Instruments AB, SE-43132 V. Frölunda, Sweden

Abstract — We present the development of a monolithically integrated 557 GHz membrane Schottky diode mixer. RF test shows state-of-the-art performance with an optimum receiver noise temperature below 1300 K DSB and an estimated mixer DSB conversion loss of 9 dB and a mixer DSB noise temperature of 1100 K including all losses.

Index Terms — Schottky diodes, membrane circuits, submillimeter wave mixers, receivers

I. INTRODUCTION

There is a growing need for compact, reliable, high performance heterodyne receivers for millimeter and sub millimeter wave applications e.g. earth observation instruments, space science missions, as well as ground-based radiometric applications such as imaging and security scanning [1]. For cost, performance and repeatability reasons advanced terahertz monolithic integrated circuit (TMIC) technology which allows for more functionality, ease of assembly and low loss operation needs development.

Mixers realized with Schottky diodes are the key elements for millimeter and submillimeter wave receiver systems operating at "room temperature". At "low frequencies", up to around 300 GHz, discrete diodes and even HEMT MMIC technology can be used [2]. For high frequency circuits, say above 400 GHz, the designs are typically limited by the thickness of the support substrates, due to parasitics and the onset of higher modes. Membrane monolithic integrated circuit technology has been proposed to overcome these difficulties, and membrane Schottky diode mixers with state-of-the-art performance have already been developed in a wide frequency range [3-5]. For applications in the 520-600 GHz frequency range, which is rich in emission and absorption lines especially the water line at 557 GHz, subharmonic mixers have been investigated based on the membrane technique[5-7], wafer bonding technique [8], and quartz-substrate up-side-down integrated device process [9]. Recently advancement in metamorphic InP-HEMT shows the capability of providing amplifiers well into this frequency range [10]. A packaged InP HEMT amplifier is reported to achieve a noise figure of 13 dB with an associated gain greater than 7 dB at 670 GHz. In this paper, the development of a monolithically integrated 557 GHz subharmonic membrane Schottky diode mixers, utilizing

an anti-parallel Schottky diode, is presented. State-of-the-art mixer performance with a DSB conversion loss of better than 9 dB and an optimum DSB noise temperature of below 1100 K including all losses has been achieved at only 1.5 mW of LO power.

II. MIXER DESIGN

The 557 GHz GaAs membrane integrated subharmonic mixer design employs an anti-parallel Schottky diode. First a diode reference plane is defined and the diode package response is simulated in the 1 GHz - 1.2 THz frequency range to include the circuit response for higher order mixing products. This is done using a 3D-diode model in the HFSS FEM based solver by replacing the two diode junctions with internal 50 Ω ports. When only fundamental propagation modes are assumed, the result is a 4-port S-parameter file that can be used for harmonic balance simulations of the diode in its circuit environment. In this way the "packaged" diode optimum embedding impedances can be investigated using a combination of load-pull type of harmonic balance simulations in ADS, where a standard diode model fitted to existing diodes fabricated at Chalmers is plugged into the S-parameter model of the linear diode simulation. Results from this analysis point out that an anode area of 0.4-0.8 μm² would be optimum and provide feasible diode embedding impedances at both LO and RF frequencies.

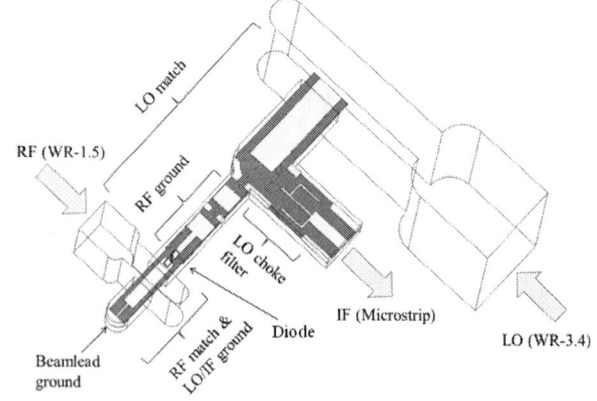

Fig.1: 3D-EM mixer model in HFSS.

The subharmonic mixer was designed to cover a RF band of 530 GHz to 590 GHz with an IF band from DC to +20 GHz. A LO chain consisting of a Heterostructure Barrier Varactor (HBV) diode frequency tripler developed in [11] and a W-band x6 amplifier chain from Fraunhofer IAF was used to pump the 557 GHz mixer. A 'standard' type of mixer topology is chosen consisting of a simple shorted reduced height E-plane waveguide probe for the RF on one side of the diode pair and a RF choke filter together with a LO open reduced height E-plane probe on the other side of the diode pair, at which also the IF signal is extracted, see Fig.1.

The RF probe is designed to provide a good match for the RF signal, entering the RF waveguide, to the diode, simultaneously providing the IF and LO signals with the necessary ground return. The LO ground return is obtained by a $\lambda_{LO}/2$ transformation of the grounded RF probe to a virtual LO ground at the diode reference plane. In a similar fashion the RF choke filter response is transformed to a virtual RF ground at the diode reference plane by a $\lambda_{RF}/2$ long section which also is used for matching of the LO signal to the diode. For a more broadband LO and RF response three open stubs have been added to the $\lambda_{RF}/2$ long section. Moreover the LO probe is also taking part in the LO matching. As a final step a LO choke filter has been designed and co-simulated with the LO probe circuit for optimum IF signal extraction and LO matching.

III. MEMBRANE DIODE FABRICATION

For the 557 membrane Schottky diode mixer, the epitaxy structure grown on GaAs semi-insulating substrate consists of a 3 μm GaAs layer sandwiched between two AlGaAs layers, followed by a thin (~50 nm) n type Schottky layer, and a heavily doped n+ layer for low resistance Ohmic contact on the top. The two AlGaAs layers are used as etch stop layers to form the GaAs membrane structure. A variety of doping concentrations ranging from 3 to 5×10^{17} cm^{-3} were investigated for the n type Schottky layer.

A. Schottky diode fabrication on membrane

The Chalmers diode process is based on electron beam lithography, with a beam spot of less than 5 nm, allowing precise anode and airbridge formation. First the surface is passivated by depositing a thin film of SiO$_2$. Then the Ohmic and Schottky contacts are defined and etched, and Pd/Ge/Au/Pd/Au and Ti/Pt/Au are deposited, respectively. Isolation of diodes is done after fabricating the air-bridges using a selective wet enchant, which stops etching on the top AlGaAs layer. Scanning Electron Microscope (SEM) image of a Schottky diode fabricated on membrane is shown in Fig.2. The anode sizes range from 0.4-0.8 μm^2.

B. Membrane circuit fabrication

After fabricating the Schottky diodes, a post process using either e-beam lithography (EBL) or photolithography (PHL) is used to fabricate the beamleads, waveguide probes, and circuit transmission lines. The membrane substrate is first defined and etched from the front side down to the bottom AlGaAs layer. In the EBL approach, a resist reflow process is used to form a stream-line slope across the mesas and the passive circuits is fabricated by metal deposition followed by a lift-off process, see Fig.3. In the PHL approach, the Au for passive circuits is fabricated by electro-plating followed by a wet etching of the Au seed layer. After finishing either of these two front side processes, the wafer is mounted topside down on a Si wafer using wax. Then the substrate is thinned from the backside down to the buried bottom AlGaAs etch stop layer. Finally the etch stop layer is etched away to release the membrane mixers. Fig.4 shows a SEM image of a released membrane circuit.

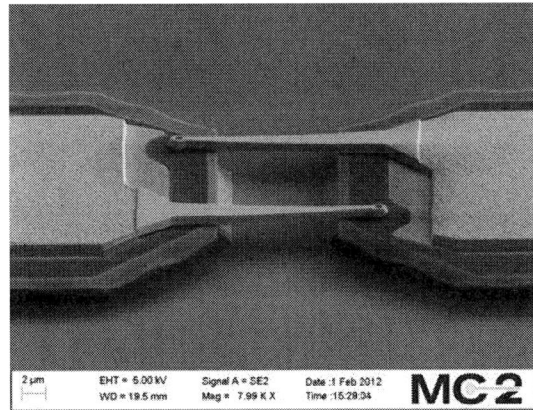

Fig.2: SEM image of an anti-parallel diode with an anode area of 0.8 μm^2 size fabricated at Chalmers.

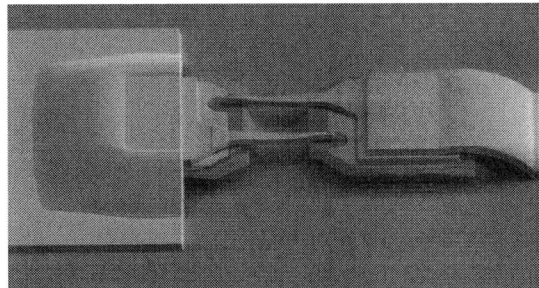

Fig.3. Au connection across the mesa using a resist reflow process.

Fig.4. SEM image of a released membrane circuit.

IV. MEASUREMENTS

On wafer DC characterization is carried out before starting the backside process. The EBL batch has almost 100% yield while the yield for the PHL batch was slightly lower, probably due to damage of anode contacts during the contacting PHL process. For the low doped material the extracted diode series resistance is 9.5 Ω and 18.4 Ω for 0.8 μm^2 and 0.4 μm^2 anode sizes respectively, while the extracted ideality factor is 1.2 for both sizes. Fig.5 shows on-wafer tested I-V characteristic of the anti-parallel Schottky diode. The IV characterization shows good consistency over the entire wafer.

Fig. 6 shows an example of a circuit mounted in the waveguide block. In this case the circuit is fixed in the block by clamping the two block halves and the IF beamlead connection is glued using silverepoxy. The marks on the supporting beamleads in Fig.6 are coming from when assembling the block halves together, showing a good contact when both half blocks are clamped.

Receiver modules with different Schottky layer doping concentrations, sizes and circuit designs have been assembled and tested in a semi-automated hot/cold measurement setup described in [12]. Fig.7 shows a photo of the front end receiver module including the LO multiplier chain and smooth wall waveguide horn as used in the test setup. The local oscillator chain consists of a high efficiency HBV tripler pumped with mHEMT PA and x6 multiplier MMICs from IAF. An external coaxially packaged multioctave (4-16 GHz) custom MMIC hybrid LNA with an external input matching circuit, developed in the Charmant project at Chalmers University of Technology, was used as the IF amplifier. The LNA had a T_{min} of 30 K in the lower IF frequency range. The LO optimum input power measured at the HBV tripler output flange using an Erickson PM4 was about 1.5 mW. The mixer conversion loss is estimated using a 3 dB attenuator between the mixer and external LNA and has been extracted at the minimum IF VSWR range (at around 10 GHz) assuming the 50 Ohm LNA equivalent noise temperature of 50 K.

Initial RF tests show very similar mixer performance for both the EBL process and the PHL process. The measured receiver noise is typically 1500 K DSB covering most of the RF frequency band with a slight shift down in frequency. The optimum receiver noise was measured at 260.6 GHz LO frequency to be 1300 K DSB with an estimated mixer DSB noise and DSB conversion loss of less than 1100 K and 9 dB, respectively, including all losses (optical, horn, waveguide, circuit RF and IF), see Fig.8. There should be further room for improving the receiver noise characteristics and LO power requirement by simply integrating the IF LNA [12] and horn antenna in the mixer module and by reducing the LO waveguide length, that was about 20 mm in the evaluated mixer prototype module. Thereto improved mixer designs, further mixer characterization and new process batches are ongoing.

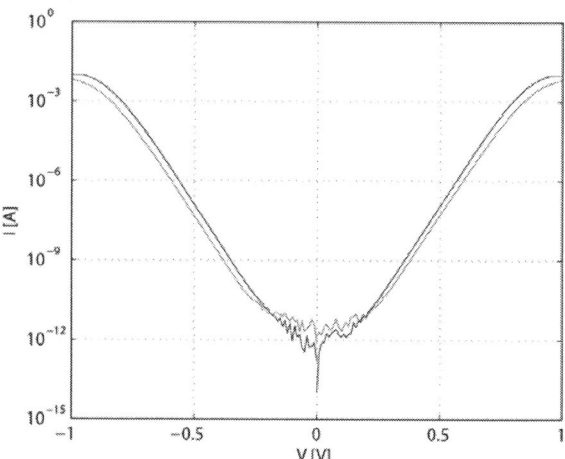

Fig. 5: Typical I-V characteristic of on-wafer tested devices fabricated using EBL, with 0.8 μm^2 anodes in black and 0.4 μm^2 anodes in red.

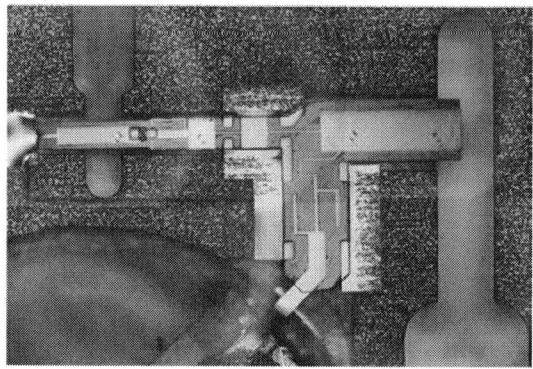

Fig.6: Photograph of an assembled mixer circuit in the waveguide block.

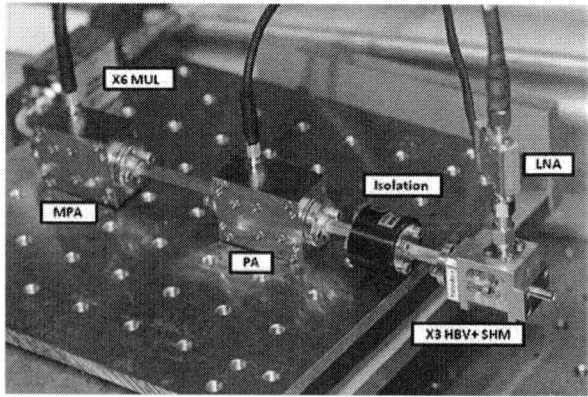

Fig. 7: Photo of the front end receiver module including the LO multiplier chain and external horn as used in the test setup.

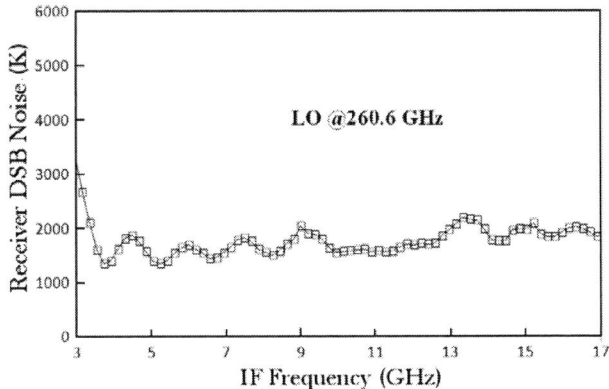

Fig.8: Measured DSB receiver noise temperature over IF frequency showing the IF VSWR between mixer and external IF LNA.

IV. CONCLUSIONS

A THz Schottky diode process line enabling monolithically integrated membrane circuits has been developed. Several batches of monolithically integrated 557 GHz subharmonic membrane Schottky diode mixers utilizing an anti-parallel Schottky diode have been successfully fabricated. The membrane mixer circuits are fabricated by either photolithography or e-beam lithography process allowing for high volume production and scaling to THz frequencies. RF tests show very similar mixer performance for both approaches. State-of-the-art mixer performances of a DSB conversion loss of better than 9 dB and an optimum DSB noise temperature of below 1100 K including all losses has been achieved.

V. ACKNOWLEDGMENT

The work was supported by the EU FP7 project 'Teracomp' under grant no 242424. This work was also carried out in the GigaHertz Centre in a joint project in part financed by Swedish Governmental Agency of Innovation Systems (VINNOVA), Chalmers University of Technology, Wasa Millimeter Wave AB, Omnisys Instruments AB, Low Noise Factory and SP Technical Research Institute of Sweden.

REFERENCES

[1] P. H. Siegel, "Terahertz technology," *IEEE Transactions on Microwave Theory and Techniques,* vol. 50, pp. 910-928, 2002.

[2] P. Sobis, N. Wadefalk, A. Emrich, and J. Stake, 'A Broadband, Low noise, Integrated 340 GHz Schottky Diode Receiver', *IEEE Microwave and Wireless Components Letters,* vol. 22, pp. 366-368, 2012

[3] P. H. Siegel, R. P. Smith, M. C. Graidis, S. C. Martin, '2.5 THz GaAs Monolithic Membrane-Diode Mixer', *IEEE Trans.Microw. Theory Tech.*, vol. 47, pp. 596-604, 1999

[4] B. Tomas, A. Maestrini, J. Gill, C. Lee, R. Lin, I. Mehdi, and P. de Maagt, 'A broadband 835-900 GHz fundamental balanced mixer based on monolithic GaAs membrane Schottky diodes', *IEEE Trans.Microw. Theory Tech,* vol.50, pp. 1917-1924, 2010.

[5] E. T. Schlecht, J. Gill, R. Lin, R. Dengler, and I. Mehdi, 'A 520-590 GHz crossbar balanced fundamental Schottky mixer', *IEEE Microwave and Wireless Components Letters,* vol. 20, pp. 387-389, 2010

[6] B. Tomas, J. Gill, A. Maestrini, C. Lee, R. Lin, S. Sin, A. Peralta, I. Mehdi, 'An integrated 520-600 GHz sub-harmonic mixers and tripler combination based on GaAs MMIC membrane planar Schottky diodes', *35th IRMMW-THz*, 2010

[7] K. Hui, J. L. Hesler, D. S. Kurtz, W. L. Bishop, and T. W. Crowe, 'A micromachined 585 GHz Schottky Mixer', *IEEE Microwave and Guided Wave Letters*. vol. 10, pp.374-376, 2000

[8] S. M. Marazita, W. L. Bishop, J. L. Hesler, K. Hui, W. E. Bowen, T. W. Crowe, 'Integrated GaAs Schottky Mixers by Spin-on-Dielectric Wafer Bonding', *IEEE Trans.Microw. Theory Tech,* vol.47, pp. 1152-1157, 2000.

[9] J. E. Oswald, T. Koch, I. Mehdi, A. Pease, R. J. Dengler, T. H. Lee, D. A. Humphrey, M. Kim, P. H. Siegel, M. A. Frerking, and N. R. Erickson, ' Planar Diode Solid-State Receiver for 557 GHz with State-of-the-Art Performance', *IEEE Microwave and Guided Wave Letters*. vol. 8, pp.232-234, 1998

[10] W. R. Deal, K. Leong, V. Radisic, S. Sarkozy, B. Gorospe, J. Lee, P. H. Liu, W. Yoshida, J. Zhou, M. Lange, R. Lai, and X. B. Mei, 'Low noise amplification at 0.67 THz using 30 nm InP HEMT', *IEEE Microwave and Wireless Components Letters,* vol. 21, pp. 368-370, 2011

[11] J. Vukusic, T. Bryllert, A. Ø, Olsen, J. Hanning, J. Stake, 'Monolithic HBV-based 282-GHz tripler with 31-mW output power', *IEEE Electron Device Letters*, vol. 33, pp. 800-802, 2012

[12] P. Sobis, A. Emrich, and J. Stake, "A low VSWR 2SB Schottky receiver," *IEEE Trans. Terahertz Sci. Technol.*, vol. 1, no. 2, pp. 403–411, Nov. 2011

Fundamental Oscillation up to 1.31 THz in Thin-Well Resonant Tunneling Diodes

H. Kanaya, H. Shibayama, S. Suzuki, and M. Asada

Interdisciplinary Graduate School of Science and Technology, Tokyo Institute of Technology,
2-12-1-S9-3 Ookayama, Meguro-ku, Tokyo 152-8552, Japan

Abstract — Room-temperature fundamental oscillation of up to 1.31 THz was achieved in thin-well resonant tunneling diodes integrated with planar slot antennas. The output powers were ~10 µW at 1.31 THz and around 30 µW in the 0.8–1.1 THz region. This high frequency oscillation with relatively high output power is attributed to a reduction in the intrinsic delay and an increase in the widths of current density and voltage of the negative differential conductance region due to the thin well structure.

Index Terms — oscillators, quantum well devices, resonant tunneling devices, slot antennas, submillimeter integrated circuits.

I. INTRODUCTION

Recently, the terahertz (THz) frequency range has received considerable attention because it can be employed in various applications [1]. In particular, high-capacity short-distance wireless communication is an important application of this range. Demonstrations of THz communication have been intensively conducted [2]–[5]. Previously, we studied wireless communication using resonant tunneling diode (RTD) oscillators in the 500-GHz range and obtained bit rates of up to 3 Gbit/s [5].

Compact, coherent, and high-power solid-state sources are considered as key components for such an application. Because the THz range exists between light waves and millimeter waves, both optical and electronic devices are being investigated as THz sources. For semiconductor single oscillators, THz quantum cascade lasers have been studied from the optical device side [6], [7]. On the electron device side, heterostructure bipolar transistors (HBTs), high electron mobility transistors (HEMTs), and Si CMOS transistors are being studied intensively for the THz sources [8]–[10]. The RTD is also a good candidate; RTDs can oscillate above 1 THz under continuous wave conditions at room temperature and above 400 µW at around 550 GHz [11]–[14].

In this study, we employed thin-well RTDs to enhance these properties; the increases in frequency and power obtained with this structure are attributed to a reduction in the intrinsic delay and large widths of current and voltage of the negative differential conductance (NDC) region. By reducing the well thickness, a fundamental oscillation at 1.31 THz with an output power of ~10 µW was obtained. Oscillation characteristics with different well thicknesses were studied. Theoretical estimation of oscillation frequency and output power were also shown for a thin-well RTD with optimized antenna structure.

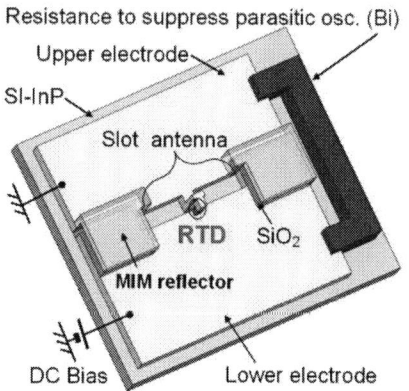

Fig. 1. Schematic structure of the RTD oscillator.

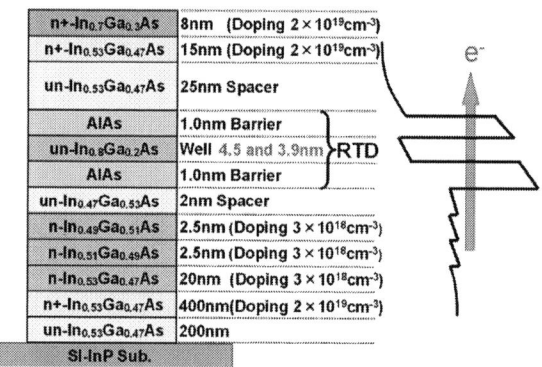

Fig. 2. Layer structure of RTD with thin-well, graded emitter and thick spacer. Band diagram is also shown.

II. DEVICE STRUCTURE

Figure 1 shows the structure of the RTD oscillator. An AlAs/InGaAs double-barrier RTD fabricated on a semi-insulating InP substrate was integrated at the center of a slot antenna. The electromagnetic field of the slot antenna forms a standing wave, and thus the antenna acts as a resonator while at the same time acting as an antenna radiating output power. The electrodes of the RTD are connected to the electrodes on

978-1-4673-1725-2/12 $31.00 © 2012 IEEE

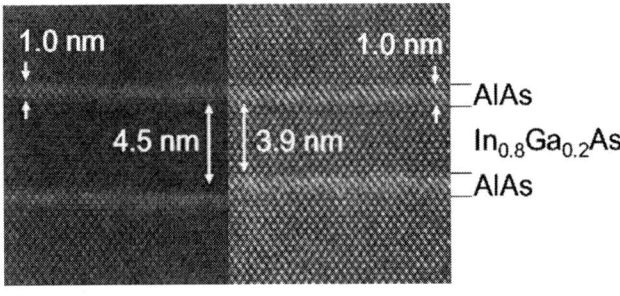

Fig. 3. Cross-sectional lattice images of the 4.5- and 3.9-nm-well RTDs observed by TEM.

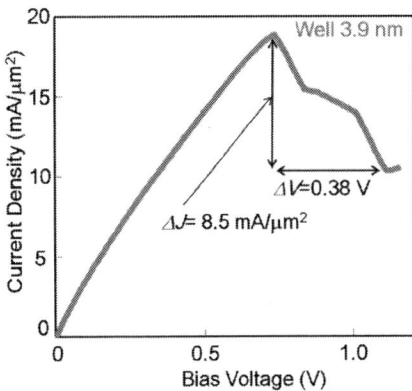

Fig. 4. *I-V* characteristics for the RTD with the 3.9-nm-thick well.

the left- and right-hand sides of the antenna. At both edges of the antenna, the electrodes are overlapped with a SiO_2 layer between them. This structure is used to form reflectors of high-frequency electromagnetic waves and to simultaneously achieve DC bias separation. A parallel resistance element made of bismuth film is connected outside the antenna electrodes to suppress the parasitic oscillation at 2–3 GHz caused by the resonance formed by the external circuits, including the bias supplying lines.

Oscillations can occur if the NDC of the RTD (G_{RTD}) compensates for the radiation loss of the antenna. G_{RTD} degrades with increasing frequency as a result of the intrinsic delay τ, which consists of the resonant tunneling time through the double barriers and the transit time across the collector depletion region. The degradation of G_{RTD} is expressed as $G_{RTD} = G_{RTD0} \times \cos\omega\tau$, where G_{RTD0} denotes the NDC at DC, and ω is the angular frequency. G_{RTD} is also expressed as $(3/2)$ $\cos\omega\tau \times \Delta I/\Delta V$ in approximating the *I-V* curve with a polynomial formula up to the third order, where ΔI and ΔV are the widths of current and voltage of the NDC region. The output power is proportional to $\cos\omega\tau \times \Delta I\Delta V$ or expressed using G_{RTD0} as $(3/2) \cos\omega\tau \times G_{RTD}\Delta V^2$. Large NDC region ($\Delta I$ and ΔV) and small τ are necessary to achieve oscillation with high output power in high frequency region. Details of the analysis of oscillation frequency and output power are described in [13].

The layer structure of the fabricated RTD is shown in Fig. 2. The schematic band diagram is also shown on the right-hand side of Fig. 2. These layers were grown on a semi-insulating InP substrate using molecular beam epitaxy (MBE). The electron direction is from the bottom to the top. A graded emitter structure is employed to suppress the Γ-L valley transition responsible for the transit delay, and a thick spacer reduces the capacitance of the RTDs. Thin barriers are used to reduce the resonant tunneling time through the double barriers. The first resonant level of the well and the peak voltage were reduced by an In-rich strained well structure. A cap layer with an In-rich strained composition and high doping concentration was used to reduce the contact resistance of the electrode.

Fig. 3 shows the cross-sectional lattice images of the RTDs with different well thicknesses observed by a transmission electron microscope (TEM). In this study, we employed RTD oscillators with well thicknesses of 4.5 and 3.9 nm. At a reduced well thickness, a high value of the available current density ΔJ in the NDR region was expected owing to an increase in the full width of the half maximum (FWHM) of the tunneling transmission, associated with an increase in the resonance levels. Additionally, the energy separation between the first and second resonance levels also increases in the thin-well structure. This results in a reduction in the valley current, because the valley current is determined by leakage through the second resonance level. Due to these reasons, large ΔI and ΔV were expected in the thin well structure. The wide FWHM is also effective for reducing the resonant tunneling time, because the resonant tunneling time is inversely proportional to the FWHM.

III. OSCILLATION CHARACTERISTICS

From the *I-V* characteristics shown in Fig. 4, we obtained ΔJ = 8.5 mA/μm² and ΔV = 0.38 V with a 3.9 nm well thickness. ΔJ = 6.1 mA/μm² and ΔV = 0.23 V were obtained with a well thickness of 4.5 nm. By reducing the well thickness, ΔJ and ΔV increased, as we expected. Because the output power is in proportion to the product of ΔI and ΔV, as mentioned above, a higher output power was expected for these thin-well RTDs.

We fabricated the RTD oscillators integrated with a 20-μm-long slot antenna. Fig. 5 shows the dependence of the fundamental oscillation frequency on the mesa area for RTDs with 4.5 and 3.9 nm well thicknesses. The oscillation frequency increased with a decrease in the mesa area owing to a decrease in the capacitance. We obtained fundamental oscillation frequencies of up to 0.77 and 1.31 THz for RTD well thicknesses of 4.5 and 3.9 nm, respectively. The highest oscillation frequency of 1.31 THz with an output power of ~10 μW was obtained with the thin well structure at a mesa area of 0.33 μm². This is currently the highest reported oscillation

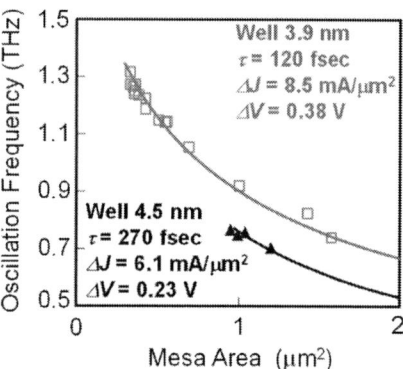

Fig. 5. Dependence of oscillation frequency on RTD mesa area of 4.5- and 3.9-nm-thick wells.

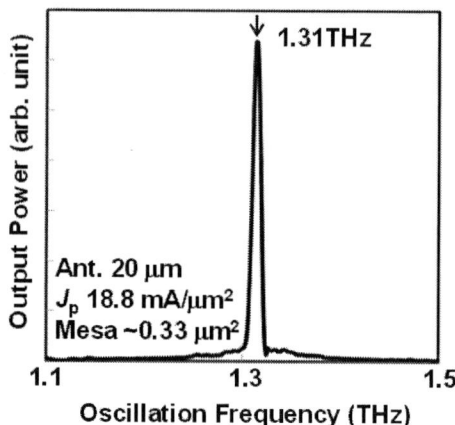

Fig. 6. Measured spectrum of fundamental oscillation of 1.31 THz.

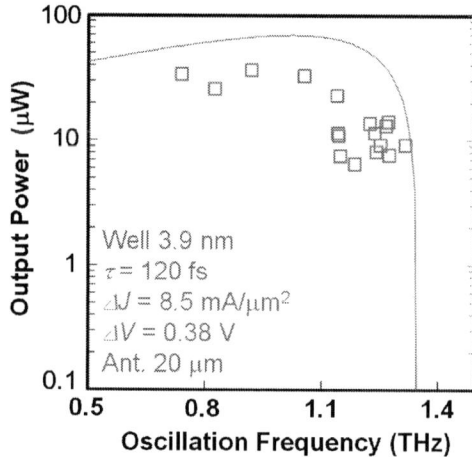

Fig. 7. Experimental and theoretical output power as a function of oscillation frequency for the RTD with 3.9-nm-thick well.

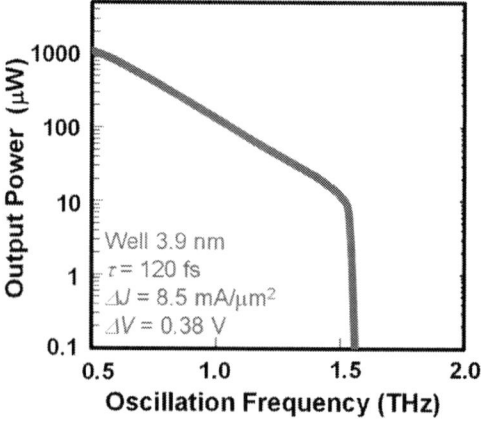

Fig. 8. Calculated oscillation frequency and output power for optimum antenna structure at each frequency using an RTD with 3.9-nm-thick well.

frequency for a room-temperature electronic single oscillator. The oscillation spectrum at 1.31 THz is shown in Fig. 6.

Because the absolute value of G_{RTD} is reduced with an increase in the oscillation frequency owing to intrinsic delay and a decrease in the mesa area, there was a limit to the mesa area below which the oscillation could not be obtained owing to insufficient G_{RTD} to compensate for the radiation loss. This limit was smaller for the thinner well because NDC is less degraded owing to shorter tunneling time. We estimated the intrinsic delay by fitting the theoretical calculation to the experimental results. Details of the estimation method are described in [15]. The fitting curves are also shown in Fig. 5. The estimated intrinsic delays τ were 270 and 120 fs for RTDs with 4.5 and 3.9 nm well thicknesses, respectively. Fig. 7 shows the output power as a function of the oscillation frequency for the RTD with a well thickness of 3.9 nm. The theoretical curve is also shown. An output power of around 30 μW was obtained in the 0.8–1.1 THz region.

We also experimented on an RTD with a well thickness of 3.5 nm. However, heat destruction of the RTD mesa eventually occurred, probably owing to the high current drive. By attaining unbreakable operation with a heat dissipation structure, further increases in the oscillation frequency and output power can be achieved in thinner-well RTDs. Higher frequency oscillation is also expected by optimizing the thickness and material of the collector spacer layer.

In this study, the antenna structure was a fixed 20-μm-long center-fed slot antenna. Higher frequency and output power are expected to be attained by utilizing shorter antennas and offset-fed structures [14]. Fig. 8 shows the theoretically expected output power of an RTD with a 3.9-nm thick well as a function of oscillation frequency, which gives the maximum output power at each frequency by optimizing the antenna length, offset, and mesa area. $\tau = 120$ fs was assumed in the calculation, which was estimated as described above. Higher output powers exceeding 1 mW at ~500 GHz and fundamental oscillation greater than 1.5 THz are expected using RTDs with a 3.9 nm well thickness and optimum antennas.

IV. CONCLUSION

Room-temperature fundamental oscillation of up to 1.31 THz was achieved in thin-well RTDs integrated with planar slot antennas. This is the highest frequency reported for an electronic single oscillator, which bridges the THz gap between semiconductor optical and electronic devices. The output power was ~10 µW at 1.31 THz and around 30 µW in the 0.8–1.1 THz region. This high frequency oscillation with relatively high output power is attributed to a reduction in the intrinsic delay and an increase in the widths of current density and voltage of the negative differential conductance region due to the thin well structure.

Further increases in the oscillation frequency and output power will be achieved using even thinner-well RTDs with optimized structures for effective heat dissipation as well as optimized thickness and material for the collector spacer layer. Additionally, higher output powers exceeding 1 mW at ~500 GHz and fundamental oscillation greater than 1.5 THz are expected using optimum antennas for RTDs with well thicknesses of 3.9 nm.

V. ACKNOWLEDGEMENTS

We thank Emeritus Professors Y. Suematsu and K. Furuya of the Tokyo Institute of Technology for their continuous encouragement. We also thank Professor S. Arai and Associate Professors Y. Miyamoto, M. Watanabe, and N. Nishiyama of the Tokyo Institute of Technology for their fruitful discussions and encouragement. This work was supported by Scientific Grants-in-Aid from the Ministry of Education, Culture, Sports, Science and Technology, Japan; the Industry–Academia Collaborative R&D Program from the Japan Science and Technology Agency, Japan; and the Strategic Information and Communications R&D Promotion Programme from the Ministry of Internal Affairs and Communications.

REFERENCES

[1] M. Tonouchi, "Cutting-edge terahertz technology," *Nat. Photonics*, vol.1, pp. 97-105, February 2007.

[2] T. Kleine-Ostmann and T. Nagatsuma: "A review on terahertz communications research," *J. Infrared Milli. Terahz. Waves*, vol. 32, no. 2, pp. 143-171, January 2011.

[3] H. Song and T. Nagatsuma, "Present and future of terahertz communications," *IEEE Trans. Terahz. Sci. Technol.*, vol. 1, no. 1, pp. 256-263, September 2011.

[4] I. Kallfass, J. Antes, T. Schneider, F. Kurz, D. Lopez-Diaz, S. Diebold, H. Massler, A. Leuther, and A. Tessmann, "All active MMIC-based wireless communication at 220 GHz," *IEEE Trans. Terahertz Sci. Technol.*, vol. 1, no. 2, pp. 477-487, November 2011.

[5] K. Ishigaki, M. Shiraishi, S. Suzuki, M. Asada, N. Nishiyama, and S. Arai, "Direct intensity modulation and wireless data transmission characteristics of terahertz-oscillating resonant tunnelling diodes," *Electron. Lett.*, vol. 48, no.10, pp. 582-583, May 2012.

[6] R. Köhler, A. Tredicucci, F. Beltram, H. E. Beere, E. H. Linfeld, A. G. Davies, D. A. Ritchie, R. C. Iotti, and F. Rossi, "Terahertz semiconductor heterostructure laser," *Nature*, vol. 417, pp. 156-159, May 2002.

[7] B. S. Williams, "Terahertz quantum-cascade laser," *Nat. Photonics*, vol. 1, pp. 517-525, September 2007.

[8] L. A. Samoska, "An overview of solid-state integrated circuit amplifiers in the submillimeter-wave and THz regime," *IEEE Trans. Terahertz Sci. Technol.* vol. 1, pp. 9-24, September 2011.

[9] M. Seo, M. Urteaga, J. Hacker, A. Young, Z. Griffith, V. Jain, R. Pierson, P. Rowell, A. Skalare, A. Peralta, R. Lin, D. Pukala, and M. Rodwell, "InP HBT IC technology for terahertz frequencies: fundamental oscillators up to 0.57 THz," *IEEE J. Solid-State Circuits*, vol. 46, no. 10, pp. 2203-2214, March 2011.

[10] O. Momeni and E. Afshari, "High power terahertz and millimeter-wave oscillator design: a systematic approach," *IEEE J. Solid-State Circuits*, vol. 46, no. 3, pp. 583-597, March 2011.

[11] S. Suzuki, M. Asada, A. Teranishi, H. Sugiyama, and H. Yokoyama, "Fundamental oscillation of resonant tunneling diodes above 1 THz at room temperature," *Appl. Phys. Lett.*, vol. 97, 242102, December 2010.

[12] M. Feiginov, C. Sydlo, O. Cojocari, and P. Meissner, "Resonant-tunneling-diode oscillators operating at frequencies above 1.1 THz," *Appl. Phys. Lett.*, vol. 99, 233506, December 2011.

[13] M. Asada, S. Suzuki, and N. Kishimoto, "Resonant tunneling diodes for sub-terahertz and terahertz oscillators," *Jpn. J. Appl. Phys.*, vol. 47, pp. 4375-4384, June 2008.

[14] M. Shiraishi, H. Shibayama, K. Ishigaki, S. Suzuki, M. Asada, H. Sugiyama, and H. Yokoyama, "High output power (~400 µW) oscillators at around 550 GHz using resonant tunneling diodes with graded emitters and thin barriers," *Appl. Phys. Express*, vol. 4, 064101, May 2011.

[15] A. Teranishi, S. Suzuki, K. Shizuno, M. Asada, H. Sugiyama, and H. Yokoyama, "Estimation of transit time in terahertz oscillating resonant tunneling diodes with graded emitter and thin barriers," *IEICE Trans. Electronics*, vol. E95-C, no. 3, pp. 401-407, March 2012.

Investigation of the Influence of Zn-diffusion profile on the electrical properties of InGaAs/InP photodiodes

A. Djedidi, A. Rouvie, JL.Reverchon, M.Pires, N. Chevalier, D. Mariolle

Affiliations III-V Lab, 1 avenue Augustin Fresnel, 91747 Palaiseau Cedex, France
CEA,LETI, MINATEC Campus, 17 rue des martyrs, 38054 Grenoble Cedex 9, France

Abstract — **Dark current is a drastic specification for any kind of sensor and particularly for imagers dedicated to low light level. We propose here a method for optimizing the InGaAs photodiode properties. An anomalous doping profile feature revealed by scanning capacitance microscopy is counterbalanced by modifying Zn-diffusion process. We analyze the electrical results before further investigation.**

I. INTRODUCTION

Short Wave InfraRed (SWIR) image sensors based on PIN photodiode arrays present a tremendous interest in applications such as enhanced vision systems or low light level imagers. The capability to work at room temperature with dark current (I_{dark}) in the femto Amps range for typical pixel is another motivation for the fast development of this technology.

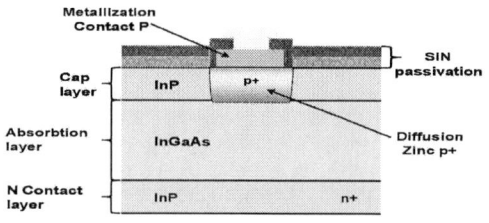

Fig. 1: Schematic cross section of the planar InP/InGaAs photodiode: n$^+$-doped InP substrate, n$^-$-doped InGaAs absorber layer and n-doped InP cap layer. The P-doping is obtained by Zn diffusion.

Figure 1 presents a schematic cross section of our planar InP/InGaAs photodiode. It is a classical PIN structure with high-quality material exhibiting dark current lower than 10 fA for a 15 µm pitch at 20 °C and bias of -0.1V.

Dark current (I_{dark}) and quantum efficiency are key features for the photodiode performance. Diminishing I_{dark} would increase the signal to noise ratio [1]. Its origin is the subject of different studies [2–4]. At ambient temperature and low reverse bias compatible to standard readout circuit, there are different possible localizations and two main contributions to the dark current: radiative and Shockley-Read-Hall (SRH) mechanisms. Other effects like Auger recombination at higher temperature or tunneling at high electric field could be more

important in specific situation, a higher doping level for example.

In order to further improve the photodiode performance, detailed analysis of the material structure with state-of-art characterization techniques have to be conducted. Combined with multidimensional numerical simulation, we can optimize performances while maintaining the relative simplicity of its structure that made this technology renowned.

Each generation-recombination (GR) mechanism has different temperature dependence. So, the activation energy (Ea) is a good indication on its type.

	Temperature dependence	InP	InGaAs
SRH	$\propto T^{1.5} \exp(-Eg/2kT)$	710 meV	445 meV
Radiative	$\propto T^{3} \exp(-Eg/kT)$	1420 meV	810 meV
Tunnel	Weak dependence [2]		

Table.1: GR mechanisms and their temperature dependence. Theoretical value of Ea in InP and InGaAs for an energy gap of (respectively) 1.34 eV and 0.74 eV at 300 K.

Table 1 summarizes some information about the main GR contributions: values for diffusion contribution in InGaAs and SRH generation in InP are close. So it is difficult to distinguish between both contributions.

Furthermore, the dependence of Ea on the geometry parameters (pixel diameter and pitch) allows the identification of dark current origin for both area and perimeter contribution.

Due to the low dark current achieved, activation energy cannot be measured at low temperature. We have studied the current-voltage characteristics over the range of temperature from 283K to 323K and for a large set of photodiode dimensions [5].

The area contribution is diffusion limited; regardless reverse bias, with activation energy equal to the absorber's band gap. The perimeter contribution presents decreasing activation energy when the reverse bias increases. At very low reverse bias, the same diffusion mechanism is also observed for the perimeter contribution. A SRH mechanism is observed when the reverse bias is increasing. It can dominate the diffusion mechanism at 5 V (cf. [5] for detailed investigation). This behavior reveals SRH mechanism in the radial component of the dark current taking place at the interface between the Zn diffusion zone and the InGaAs / InP interlayer. Such generation centers are gradually activated with reverse bias resulting in non academic saturation current in reversed bias.

Following the motivation detailed above, we have investigated the Zn-diffusion profile and the impact of the lateral diffusion on the perimeter contribution to the dark current.

II. ZN-DIFFUSION PROFILE

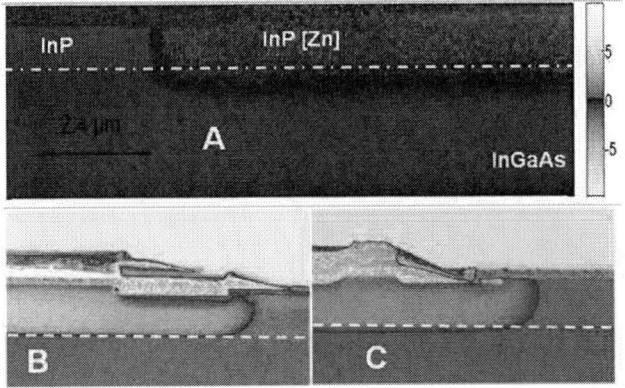

Fig. 2: Zn diffusion profiles for three different samples: (A) standard process with a stronger lateral diffusion in InGaAs; (B) modified process with a limited Zn-diffusion in InGaAs and (C) an interlayer doping at InP/InGaAs interface that suppresses lateral spread. (A) is an SCM measurement (B) and (C) are SEM measurements.

The analysis of the diffusion profile was conducted by Scanning Capacitance Microscopy (SCM) [6] and Scanning Electron Microscopy (SEM) at low voltage (2 keV). The main parameters that influence the diffusion process are diffusion constant and n-doping during epitaxy [7]. A higher n doping is a solution to limit Zn diffusion in specific cases.

Three samples were studied: A reference sample (A) with standard process. A sample (B) with a shorter Zn-diffusion process in order to avoid any lateral diffusion in InGaAs, and a sample (C) with a larger doping introduced in the last 20nm below the InGaAs/InP interface in order to limit the lateral diffusion and to push the depleted region away from the interface prompt to defect localization which can be particularly damaging in photovoltaic mode [5]. We expect a limited lateral Zn diffusion in InGaAs. We want to analyze whether this effect induce different perimeter contribution.

Figure 2 shows the different Zn profile in each sample. The best contrast is achieved with SCM for sample A and with SEM for sample B and C. For the standard sample (A), Zn diffuses laterally with an exponential decay over 1 μm in the InP cap-layer. Vertically, the Zn concentration is constant in InP and decreases exponentially over 200 nm in InGaAs. Another consequence of the diffusion process is a lateral

spread: Zn diffuses laterally 200 nm more in InGaAs than in InP. This effect has already been observed [8]. It could have some consequences on the electric field distribution at InP/InGaAs interface with unexpected local singularities . For sample (B), the diffusion time was optimized in order to limit the Zn diffusion in InGaAs. As a result, we notice that the diffusion stopped at the interface. For sample (C) the lateral spread is limited by the interlayer doping.

In the following, we analyze the electrical properties in order to investigate the influence of the Zn-diffusion profile on the dark current.

III. ELECTRICAL CARACTERISTICS

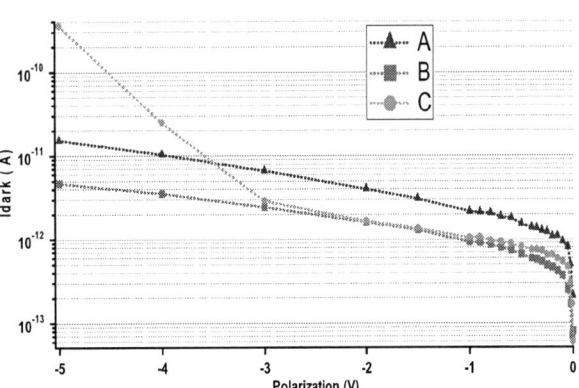

Fig. 3: Dark current (I_{dark}) versus polarization for different 75-μm diameter monopixels.

The Figure 4 shows I-V curve for the three structures measured on a 75μm-monopixel. At -0.2V, I_{dark} is 1.09 pA for the reference structure (A), 0.65 pA for (B) and 0.46 pA for (C).

The slight difference between (B) and (C) is maintained until -3.0V. Then, the current increases dramatically for the structure (C) with n-doped interlayer. Such a phenomenon takes place at reverse bias larger than 10V with classical structure.

This breakdown tunneling current scales with the perimeter. This result is compatible with reference [4] that localize the tunneling in the InGaAs homojunction near the upper InP/InGaAs heterointerface.

We have also studied the variation of dark current with temperature. As mentioned above, the activation energy could give some information on the different contribution to the total current and its variation with polarization.

The Figure 4 shows the activation energy versus polarization for the previous measured samples. There is no difference in activation energy between (B) and (C) except at high reverse bias where the larger tunneling current appears.

It shows that the tunneling current has weaker temperature dependence than SRH or radiative mechanisms.

978-1-4673-1725-2/12 $31.00 © 2012 IEEE   111

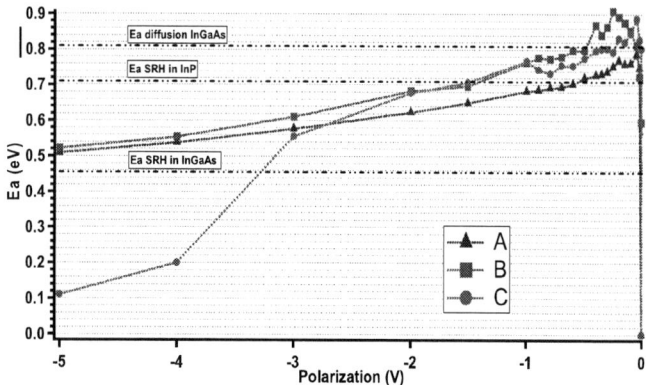

Fig. 4: Activation energy (Ea) versus polarization for the different samples.

At low bias, sample (A) presents a lower activation energy, coherent with a more important SRH contribution to the total current. This trap assisted mechanism is enhanced by the default at InP/InGaAs interface. The modifications in samples (B) and (C) aiming at diminishing these effects are roughly successful.

The activation energy of SRH contribution occurring in InP/Passivation interface being close to the diffusion contribution in InGaAs, we cannot discriminate between them. One possibility is to implement a dual diffusion front or to engage a specific study on dielectric passivation [9].

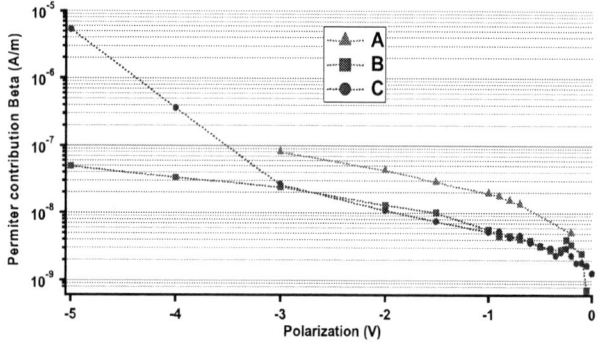

Fig. 5: parameter contribution for the different samples

The activation energy decreases while increasing reverse bias. This is due to the depletion zone increase where SRH generation occurs preferentially. It becomes the main contribution at high reverse bias for samples (A) and (B).

Activation energy from sample (C) drops while increasing reverse bias. This tunneling current has a weak dependence on temperature; its activation energy tends to zero.

In order to analyze the different geometrical contributions, we study the variation of I_{dark} for monopixels with different diameter using a quadratic relation [5]:

$$I_{dark} = \alpha.d^2 + \beta.d$$

We extract the geometrical parameter α and β by fitting I_{dark}/d^2.

The figure 5 shows the different perimeter contribution for each structure at 298K. As expected, he standard structure (A) has a higher contribution than (B) and (C). These results have to be confirmed by further analysis to ensure reproducibility.

The perimeter contribution of sample (C) increases drastically below -3V. This is explained by the tunneling current occurring at the InP/InGaAs interface. The doped interlayer results in a high electric field which enhances the tunneling effect. This contribution is localized and grows linearly with the pixel diameter.

IV. CONCLUSION

An anomalous doping profile is observed by SCM with a standard Zn diffusion process. A first solution is to reduce the diffusion depth by a fine tuning of the diffusion time sensitive to process fluctuation. Another solution is to introduce a n-doped interlayer to modify the anomalous profile and screen the defects present at the InGaAs / InP interface. A result is a slightly lower dark current at moderate bias compatible with readout circuit. The larger current component taking place at larger bias is attributed to breakdown tunneling.

REFERENCES

[1] Rosencher and Vinter, *Optoélectronique : Cours Et Exercices Corrigés*, 2e éd. (Dunod, 2002).
[2] S. Forrest, IEEE Journal of Quantum Electronics **17**, 217 (1981).
[3] J. Boisvert, T. Isshiki, R. Sudharsanan, P. Yuan, and P. McDonald, in *SPIE* (2008), pp. 1–7.
[4] A. Zemel and M. Gallant, Journal of Applied Physics **64**, 6552–6561 (1988).
[5] A. Rouvié, J.-L. Reverchon, O. Huet, A. Djedidi, J.-A. Robo, E. Costard, J.-P. Truffer, T. Bria, J. Decobert, and M. Pires, SPIE (2012).
[6] C. C. Williams, Annual Review of Materials Science **29**, 471 (1999).
[7] K. Kazmierski and B. de Cremoux, Japanese Journal of Applied Physics **25**, 1169 (1986).
[8] H. Yin, Y. Li, W. Wang, H. Xia, X. Li, H. Gong, and W. Lu, in (2010), p. 765820–765820–7.
[9] B. M. Onat, X. Jiang, and M. Itzler, in *LEOS Annual Meeting Conference Proceedings, 2009. LEOS'09. IEEE* (2009), pp. 231–232.

40 Gbit/s identical layer InGaAlAs-MQW electroabsorption-modulated DFB-lasers operating between 1298 nm and 1311 nm

Holger Klein[a], Carsten Bornholdt[a], Georges Przyrembel[a],
Ariane Sigmund[a], Wolf-Dietrich Molzow[a], and Martin Moehrle[a]

[a]Fraunhofer Institute for Telecommunications, Heinrich-Hertz-Institute,
Einsteinufer 37, 10587 Berlin, Germany

Abstract — **We present electroabsorption modulated DFB Lasers using an identical InGaAlAs MQW core for the DFB and the EAM. Excellent 40 Gb/s modulation performance at semi-cooled operation with dynamic extinction ratios exceeding 8.3 dB are demonstrated.**

I. INTRODUCTION

1300 nm electro-absorption modulated DFB lasers (EML) are key components in transmission systems for long reach (up to 10 km) and extended reach (up to 40 km) applications as defined in the IEEE 802.3ba Standard. Commonly used EMLs are regarded as high-cost components due to their separate epitaxial butt-coupling growth process [1] to allow for separate optimization of the DFB laser and the electro-absorption modulator (EAM).

In contrast to that we have developed EMLs that use a single InGaAlAs MQW waveguide for both the DFB laser and the electro-absorption modulator enabling an easy temperature controlled operation and a low-cost fabrication similar to a conventional DFB laser diode.

II. DEVICE STRUCTURE

To avoid separate epitaxial growth steps for the DFB and the EAM section a double-stack MQW layer structure for 1300 nm EMLs was proposed and demonstrated in [2]. As an alternative we employ an identical Al-containing MQW layer stack for the DFB and the EAM section [3] (cf. Fig. 1). Benefits of the identical layer structure are low internal optical reflection and no optical coupling losses at the DFB/EAM interface. The EML waveguide core consists of an compressively strained InGaAlAs MQW grown in a MOVPE reactor. The InGaAlAs material system provides high temperature lasing and electro-absorption properties that are superior to those of InGaAsP. The DFB-lasers are realized using a $\lambda/4$ phase shifted index coupled grating. The DFB- and EAM sections are electrically isolated by an isolation groove with a separation resistance exceeding 50 kΩ. To achieve a large modulation bandwidth a low parasitic capacitance of the EAM electrodes is required. The EAM is therefore realized in a ground-signal-ground (GSG) contact configuration (cf. Fig. 2) using benzocyclobutene (BCB) as an isolation layer underneath the signal electrode. The DFB section is 300 µm long whereas for the EAM electrode lengths

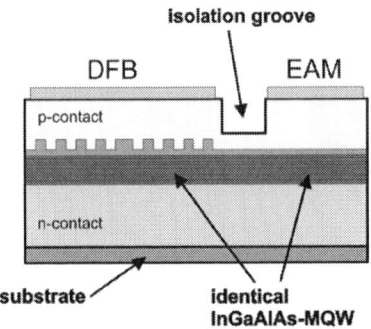

Fig. 1. Schematic structure of the EML chip with identical InGaAlAs-MQW waveguide core in the DFB and the EAM section.

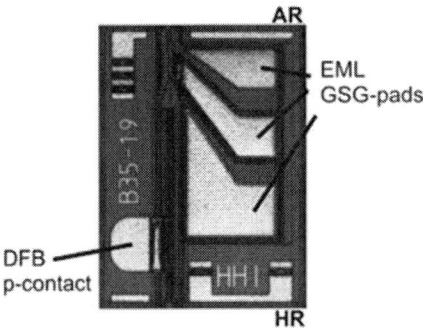

Fig. 2. Microscopic image of a processed EML chip showing the DFB p-contact and the Ground-Signal-Ground-RF-contact of the EAM section.

of 80 µm and 100 µm have been realized. The spectral detuning between the four DFB-wavelengths and the InGaAlAs MQW band gap is optimized for semi-cooled operation conditions between 35...55°C. To avoid optical back reflections into the EAM and DFB section the output waveguide is tilted by 5°.

III. DC CHARACTERISTICS

To perform DC and RF characterization the EML devices have been mounted on gold plated copper heat sinks for good heat dissipation during operation. Fig. 3 shows the optical output powers vs. injection currents of four EMLs emitting at 1297.8 nm, 1302.4 nm, 1306.3 nm and 1310.6 nm. The PI-

978-1-4673-1725-2/12 $31.00 © 2012 IEEE

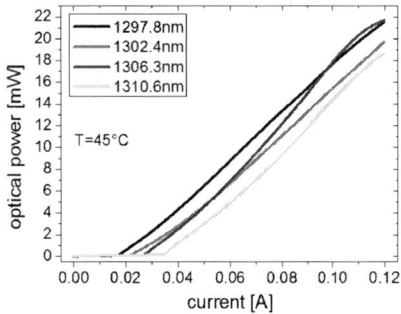

Fig. 3. Power vs. injection current characteristics of four EML devices emitting at four different wavelengths. T=45°C.

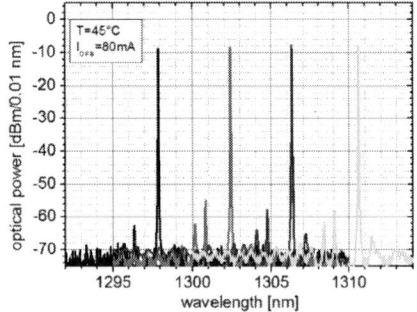

Fig. 4. Wavelength spectra of four EML devices with side mode suppression ratios (SMSR) > 40 dB. T=45°C, I_{DFB}=80 mA.

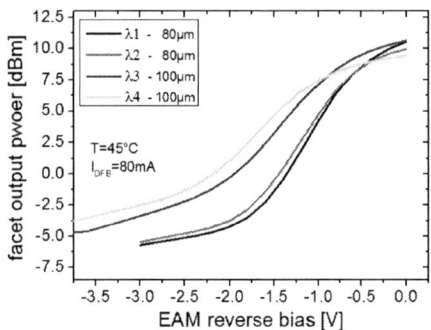

Fig. 5. Facet output power vs. reverse bias of EML devices at the four respective wavelength λ1, λ2, λ3 and λ4. T=45°C, IDFB=80mA.

curves were measured at 45°C operation temperature using an integrating sphere. The laser threshold currents are 18 mA, 22 mA, 28 mA and 35 mA, respectively. The optical output powers at 100 mA drive current are 14 mW, 15 mW, 17.5 mW and 18 mW with an applied EAM bias voltage of 0 V. Fig. 4 shows the optical spectra of four EML devices measured at 45°C applying drive currents of 80 mA to the DFB laser section. The single-mode operation of the four EMLs is shown by side mode suppression ratios (SMSR) of 53 dB, 47 dB, 49 dB and 50 dB, respectively. The static extinction ratios and the required drive voltage of the EMLs can be determined

from the measurements shown in Fig. 5. The facet output power is plotted against the EAM reverse bias voltage where the measurements have been performed with an integrating sphere using devices with an EAM electrode length of 80μm for the wavelengths λ_1 and λ_2 and with EAM electrode length of 100μm for the emission wavelengths λ_3 and λ_4. The laser current was 80mA and the temperature set to 45°C in all four cases. As expected from the nature of the Quantum-Confined Stark effect (QCSE) which is responsible for the electro-absorption in the MQW the highest absorption efficiency is observed for wavelength λ_1 with the lowest detuning to the photoluminescence wavelength of the MQW. To achieve static extinction ratio of 10dB a reverse bias voltage of -1.3V is required. The EML emitting at wavelength λ_2 has a slightly lower efficiency and requires a reverse voltage of -1.46V to reduce the optical output power by 10dB. The absorption efficiency of the QCSE drops quickly going to larger wavelengths. To maintain a good static extinction and a low drive voltage EML devices with an electrode length of 100μm have been chosen for λ_3 and λ_4. The required reverse bias voltages for 10dB extinction are -1.86V for wavelength λ_3 and -2.32V for wavelength λ_4.

IV. RF CHARACTERISTICS

The small signal electro-optic response of the EMLs has been measured using an Agilent Lightwave Component Analyzer (LCA) N4373B from 100 MHz up to a frequency of 50 GHz. An especially built RF probe head with an integrated 50Ω impedance matching resistor was used to contact the G-S-G pads of the EAM. To evaluate the influence of the emission wavelength on the high frequency behavior only devices with an EAM electrode length of 80 μm are compared. Fig. 6 shows the electro-optic frequency response measurements of EML devices for the four emission wavelengths shown before. The measured 3-dB bandwidths are 29 GHz, 30 GHz, 32 GHz and 33 GHz for 1297.8 nm, 1302.4 nm, 1306.3 nm and 1310.6 nm emission wavelength, respectively. The frequency dip around 34 GHz present in all four measurements is associated to the RF probe head. The shown 3dB-bandwidths are sufficient for 40 Gbit/s data transmission. The DFB current was set to 80 mA and the EAM was biased at -1.5 V for every measurement. The temperature was set to 45°C.

Large signal eye diagram measurements were performed using an Anritsu bit pattern generator MP1775A BPG and a MP1801A multiplexer at 39.8 Gbit/s using a PRBS 2^{31}-1 bit sequence and a DFB current of 80 mA. The electrical pulse patterns were amplified using an SHF 804EA broadband amplifier. A Praseodym-doped fiber amplifier (PDFA) at the output of the modulator ensured a constant optical power level of 8 dBm. The eye diagrams were recorded using an Agilent 86100D Infiniium DCA with an 86116C-040 module. The eye diagrams obtained from four EML devices emitting at all four wavelengths λ1, λ2, λ3 and λ4 are shown in Fig. 7. For the

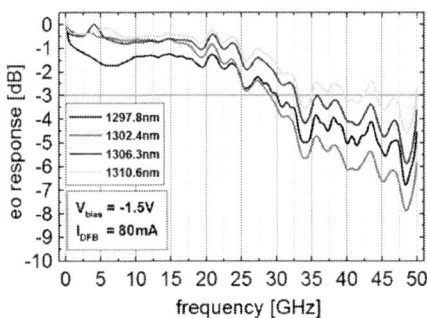

Fig. 6. Electro-optic small signal responses of EML devices with 80 μm electrode length. V_{bias}=-1.5 V, I_{DFB}=80 mA, T=45°C

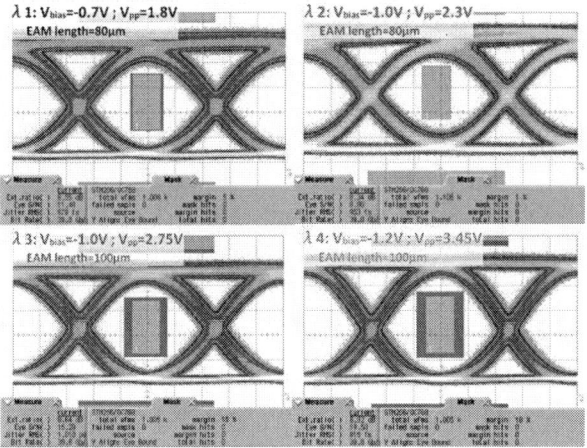

Fig. 7 Large signal eye diagrams of EML devices at four wavelengths. PRBS 2^{31}-1. I_{DFB}=80 mA, T=45°C.

two lower wavelengths EAM electrode lengths of 80 μm have been used while for the two larger wavelengths devices with 100 μm electrode length have been chosen to increase the dynamic extinction ER_{dyn} and reduce the drive voltage V_{pp}.

The measured eye diagrams of four EML devices show dynamic extinction ratios of 8.5 dB, 8.3 dB, 8.6 dB, 8.9 dB for 1297.8 nm, 1302.4 nm, 1306.3 nm and 1310.6 nm emission wavelength, respectively. The required drive voltages where 1.8 V, 2.3 V, 2.75 V and 3.45 V for emission wavelengths λ1, λ2, λ3 and λ4, respectively. Error-free operation at all four wavelengths was demonstrated by STM-256 mask margin tests.

V. CONCLUSION

We have successfully developed and fabricated identical layer InGaAlAs-MQW electro-absorption modulated DFB lasers for 40 Gbit/s. Our EML devices from a single wafer show 3dB-bandwidths over 29 GHz and dynamic extinction ratios ER_{dyn}>8.3 dB between 1298 nm and 1311 nm emission wavelength at an operating temperature of 45°C. The design is promising for a monolithically integrated EML-array for 10x40 Gbit/s = 400 Gbit/s transmission systems using 10 WDM channels with 200 GHz channel spacing.

Acknowledgements:
This work has been funded by the Future Fund of the Land Berlin co-sponsored by the European Fund for Regional Development (EFRE) in the project *100x100 Optics*.

REFERENCES

[1] T. Saito et. al., "Clear eye opening 1.3μm-25 / 43Gbps EML with novel tensile-strained asymmetric QW absorption layer", European Conference on Optical Communication ECOC, Vienna, Austria, paper 8.1.3 (2009)

[2] B.K. Saravanan et. al., "Integrated DFB laser electro-absorption modulator based on identical MQW-double stack active layer for high-speed modulation beyond 10 Gbit/s", 16th International Conference on Indium Phosphide and Related Materials, Kagoshima, Japan, p236f (2004)

[3] M. Möhrle, G. Przyrembel, C. Bornholdt, A. Sigmund, W.-D. Molzow, H. Klein, "Low-Cost 25Gb/s 1300nm Electroabsorption-Modulated InGaAlAs RW-DFB-Laser", 23th International Conference on Indium Phosphide, Berlin, paper Mo-1.12 (2011)

978-1-4673-1725-2/12 $31.00 © 2012 IEEE

Reflective Amplified Modulator Operating at 40 Gbps up to 85°C as Colorless Transceiver for Optical Access Networks

K. Ławniczuk[1,2], O. Patard[3], R. Guillamet[3], N. Chimot[3], A. Garreau[3], C. Kazmierski[3], G. Aubin[4] and K. Merghem[4]

1. *Eindhoven University of Technology, P.O. Box 513 5600 MB Eindhoven, the Netherlands*
2. *Warsaw University of Technology, Pl. Politechniki 1, 00-661 Warsaw, Poland*
3. *III-V lab, Common laboratory of Alcatel-Lucent Bell Labs France', 'Thales Research and Technology' and 'CEA Leti', Route de Nozay, 91460 Marcoussis,France*
4. *CNRS-Laboratory for Photonics and Nanostructures, Route de Nozay, 91460 Marcoussis, France*

Abstract — **In this paper we report a 40 Gb/s operation of Remote Amplified Modulator at the temperature up to 85°C within the C- and L-band spectral ranges. The presented device was fabricated using an indium phosphide (InP) monolithic integration platform which relies on AlGaInAs quantum well active material, gap engineering by Selective Area Growth and low-parasitic RC semi-insulating buried heterostructures. We investigated the high temperature operation capabilities of the device as well as chirp and Rayleigh scattering effects in a bi-directional transmission. This 40 Gb/s remote amplified modulator could operate at fastest short sections of next-generation wavelength division multiplexing (WDM) optical access networks or in WDM routers as a part of a colorless transceiver.**

Index Terms — **Remote amplified modulator, electro-absorption modulator, indium phosphide, optical access networks, photonic integrated circuit.**

I. INTRODUCTION

Realization of the next-generation optical access networks requires offering multiple services, flexibility, continuous increase of bandwidth demands with concurrent integration of a large number of optical components within single electro-optical circuits and providing relatively low costs of maintenance. Additionally, novel optical devices should support the modulation speed from 10 Gb/s to 25 Gb/s and later also to 40Gb/s. Therefore, remotely controlled wavelength-agnostic reflective modulators such as RSOAs (Reflective Semiconductor Optical Amplifier) or REAM-SOAs (Reflective Electro-Absorption Modulator integrated with Semiconductor Optical Amplifier) seem to be very attractive to the industry as a reliable low-cost/consumption solution for wavelength division multiplexing (WDM) passive optical networks (PON), WDM Fiber-to-the-Home (FTTH), WDM Radio-over-Fibre (RoF) or other WDM flexible short-reach network transceivers.

We have previously proposed to separate the function of modulation and amplification in a two-electrode REAM-SOA monolithically integrated photonic circuit [1]. Due to a high bandwidth operation of an electro-absorption

modulator, 10 Gb/s colorless amplified remote operation was demonstrated without any additional electronics [2, 3]. More recently, we presented a component working at 10 Gb/s over 80 nm at 25°C and over 30 nm in the 20°-70°C temperature range [4].

In this paper we demonstrate speed scalability up to 40 Gb/s of REAM-SOA compatible with a very large working spectral range, up to 60 nm within the telecommunication C- and L-bands. Additionally, separated amplification and modulation functions allow indoor uncooled operation, as verified during experiments up to the temperature of 85°C. In fabrication process, we took the advantage of a single active layer growth, Selective Area Growth gap engineering and a single-regrowth semi-insulating buried heterostructure technologies [5], available on our new photonic indium phosphide-based platform. In our measurement experiments at 40 Gb/s we investigated the high temperature influence on the REAM-SOA operation, as well as Rayleigh scattering effect in a bi-directional transmission for determining the level of insertion penalties.

II. FABRICATION

The photograph of the fabricated Reflective Amplfied Modulator is presented in Fig. 1. The device monolithically integrates a 70 μm long reflective electro-absorption modulator (REAM) with a 400 μm long Semiconductor Optical Amplified (SOA) and a tapered output waveguide tilted at 7° to provide relaxed tolerances for antireflection coatings as well as a better coupling of the light to an optical

Fig. 1. Photograph of a Reflective Amplified Modulator integrating REAM, SOA and tilted output tapered waveguide.

fiber. Its multiple-quantum well (MQW) active structure is based on AlGaInAs/InP materials which provide a better electron confinement compared to InGaAsP-based structures. The MQW is composed of ten tensile-strained AlGaInAs-wells and eleven compressive-strained barriers and is located between two InGaAsP separate confinement heterostructure layers. In order to obtain amplification in the EAM working spectral range, the SOA gain maximum was positively detuned. We realized this gap wavelength engineering by Selective Area Growth in the metal-organic Vapor Phase Epitaxy (MOVPE). The measured material gap wavelength are λ_{EAM} = 1453 nm and λ_{SOA}= 1523 nm, giving a detuning value of 70nm. Larger detuning of more than 100 nm required for a transparent passive waveguide was realized also during the same SAG step. After the first epitaxial growth, a 2 μm height ridge was etched by the Inductively Coupled Plasma Reactive Ion Etching. After removing the mask in the passive section, the ridge was selectively buried by a semi-insulating (SI) material to obtain a reduced capacitance for the EAM. The SI MOVPE regrowth was chloride-assisted to limit an overgrowth on the top of the ridge. The SI regrowth consisted of a combination of Fe doped InP layer and Ru doped InP layer as an interdiffusion barrier. Both of the materials have a high resistivity superior to 10^8 Ω.cm.

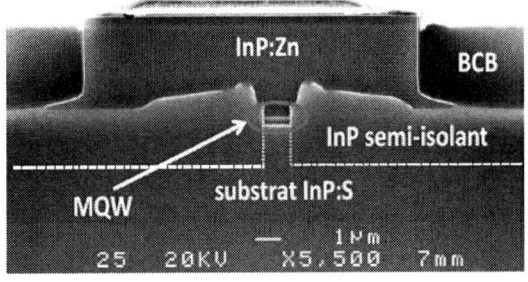

Fig. 2. The SEM photograph of the device after cleaving through the EAM section following the chemical revelation.

The p-type contact layer was re-grown in a third MOVPE epitaxial step. The contact separation between the EAM and the SOA is realized by proton implantation resulting in an inter-section resistance of $6.10^5\Omega$. The EAM facet was coated with a high reflection coating ant the output guide with a anti-reflection coating. The SEM photograph of the EAM cross-section after the chemical revelation is presented in Fig. 2.

III. MEASUREMENTS

The device was mounted on a high frequency submount and the EAM was bonded to the RF transmission line providing 50Ω termination and thereby impedance matching. The chip was soldered with 23° tilt in order to make the fiber alignment and coupling more efficient. The photograph of the bonded REAM-SOA is presented in Fig. 3.

The experimental setup used for the bit-error-rate (BER) measurements and eye-diagrams observation at 40 Gb/s

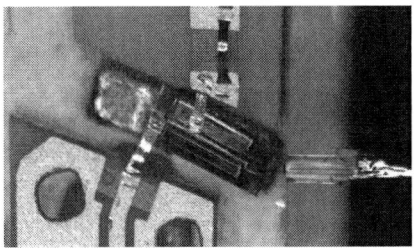

Fig. 3. Photograph of the bonded REAMSOA. The length of the device is 610μm (SOA 400 μm and EAM 70 μm). Anti-reflective coating was introduced at the output facet of the chip.

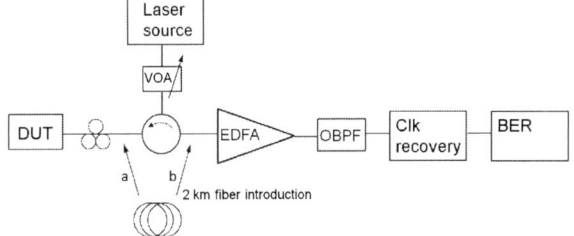

Fig. 4. The experimental setup used to measure BER at 40 Gbps.

is represented in Fig. 4. The injected wavelength and the power were controlled with a tunable laser source and a variable optical attenuator (VOA). We first performed measurements without optical fiber in back-to-back (B2B) configuration. Then, by introducing 2 km long optical SM fiber in the position *b* (see Fig 4) we examined the chromatic dispersion penalty induced by the modulator chirp alone in order to dissociate this figure from that introduced by Rayleigh backscattering. Both chirp and Rayleigh effects are present with the fiber in the position *a* (see Fig. 4) affecting bidirectional transmission. We also investigated the temperature and the wavelength dependence on BER and eye diagrams versus the received optical power. The modulated and amplified light from the REAM-SOA was detected with an avalanche photodiode (APD) and directed either to the error detector or to an oscilloscope for the eye diagram observation. Depending on the wavelength and the temperature, the SOA current and the reverse bias voltage of the EAM were adjusted and optimized for the transmission. Due to a remaining polarization sensitivity of this prototype device, we inserted a polarization controller within the experimental setup. The light was coupled into and from the device using a lensed fiber.

First static experiments of the EAM showed a low capacity of the modulator of 0.14 pF while biasing the modulator at -2.0V. This measurement allowed estimating the bandwidth of the device at the level of 30 GHz. Additionally we performed measurements of a static extinction ratio of EAM (Fig. 5). The high extinction ratio of more than 20 dB over a 40 nm spectral range was obtained. The insertion gain up to 10 dB at 1550 nm and lossless operation over 50 nm covering the C- and L-band part of the spectrum while driving the SOA with 120mA are presented in Fig. 6.

978-1-4673-1725-2/12 $31.00 © 2012 IEEE

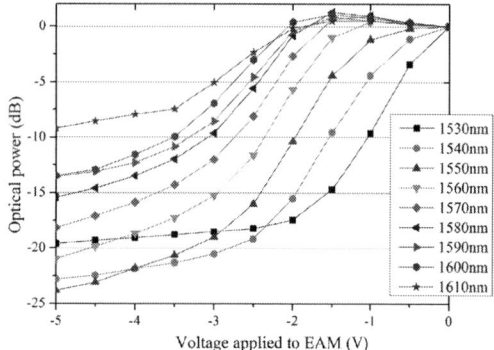

Fig 5. Measured static extinction ratio of EAM for a different wavelength injected to the device.

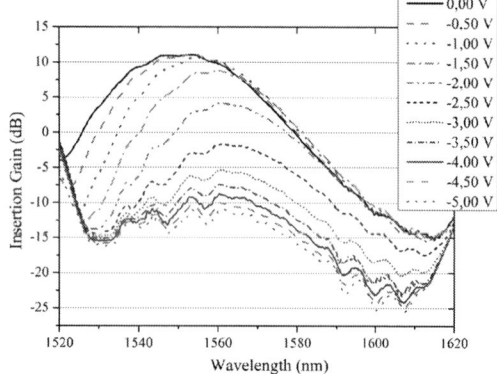

Fig 6. Obtained spectra of the insertion gain while driving the SOA with 120mA and changing the bias voltage of EAM.

A. 10 Gb/s experiments

As a part of the device characterization experiments we performed measurements at 10 Gb/s data rate. We investigated the device colorless behavior by measuring the error-free transmission power level while changing the operational temperature. The obtained spectral characteristics are presented in Fig. 7 and show that the device can operate at the room temperature in the wide spectral range of more than 40 nm with the input power level below -25 dBm in B2B configuration.

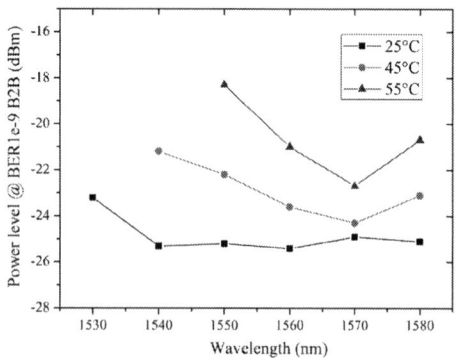

Fig. 7. Detected power level at the BER of 1e-9 in back-to-back configuration at different temperatures.

B. 40 Gb/s experiments

At the data rate of 40 Gb/s we performed measurements of BER and obtained eye-diagrams in B2B configuration, as well as by introducing 2 km long optical SM fiber in the position *a* and *b* (see Fig. 4.). The eye-diagrams obtained for various temperatures in B2B are presented in Fig. 8. We achieved colorless operation up to 85°C at the modulation data rate of 40 Gbps. The voltage applied to the EAM varied for the 1570 nm between -1.5V (at 25°C) and -0.7V (at 85°C). The applied peak-to-peak modulation voltage was set to 1.8V and the current injected to the SOA was between 50 mA and 80 mA (depending on the temperature). The observed eye shift to the longer operational wavelengths was caused mainly by a local heating of the structure. Such red-shift is expected and understood from the thermal variations of the material bandgap. In spite of this shift, it was possible to cover a spectral range of 25 nm up to 85°C while modulating the EAM at 40 Gbps. Open eyes attest the high temperature potential of the device.

Basing on the experimental results of the eye-diagrams we measured the BER at 25°C and at 65°C for the following

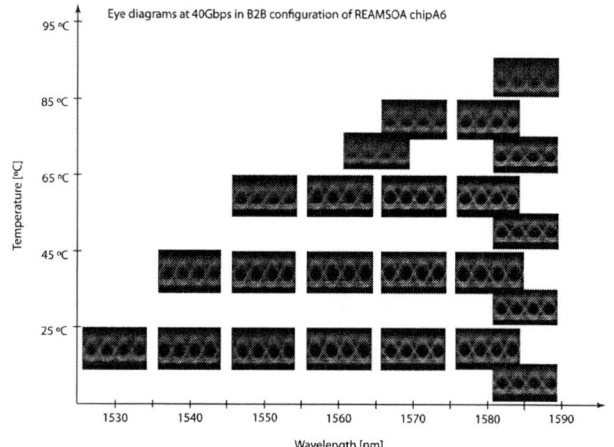

Fig. 8. Chart of the eye-diagrams at 40 Gb/s obtained in B2B configuration.

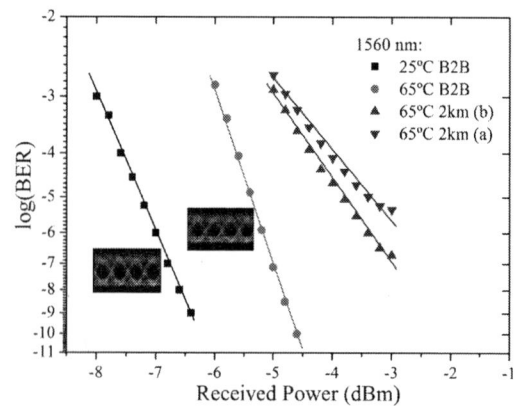

Fig. 9. BER and eye-diagrams at 40 Gb/s measured at two temperatures 25 °C and 65 °C for the wavelength of 1560 nm.

wavelengths 1550 nm, 1560 nm and 1570 nm. In Fig 9 the 40 Gb/s transmission behavior at 1560 nm is represented by BER measurements. Typical thermal penalty of 1.6 dB between 25°C and 65°C is observed in B2B configuration. After 2 km the transmission limit is rather dominated by the EAM chirp than by the Rayleigh backscattering which adds less than 0.5 dB to the overall penalty. We assume that the detected penalties after transmission through the 2 km SMF in the position a are introduced by both the Rayleigh backscattering and the chromatic dispersion effects.

We also investigated the device colorless behavior by measuring the ratio of the 40 Gb/s modulated output power to the fed of continuous power as a function of temperature. Fig. 10 shows the insertion gain obtained at 25°C and 45°C, as well as lossless operation at 65°C and up to 9 dB loss at 85°C over large C/L spectral range.

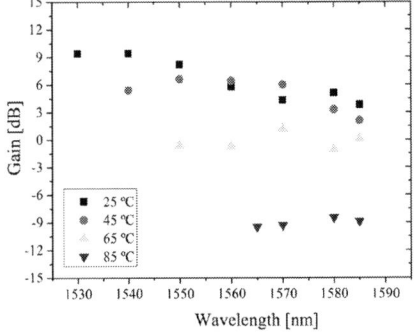

Fig 10. Insertion gain at 40 Gb/s as a function of wavelength and temperature. We observed lossless operation up to 65°C.

IV. CONCLUSIONS

We presented a monolithically integrated fast reflective electro-absorption modulator with SOA and a passive waveguide taper. The device was fabricated on the flexible photonic integrated circuit InP platform by using AlGaInAs QW active-passive structure, SAG bandgap engineering and semi-insulating buried heterostructure technologies. This InP-based platform provides the low RC EAM and fast optical amplification SOA suitable for data rates operation from 10 Gb/s to 40Gb/s.

The performed measurements show the 40 Gb/s open eye operation of the device after transmission through 2 km long fiber in the spectral range of more than 50 nm at the temperature of 25°C and at the range of 25 nm up to 85°C. The colorless and uncooled operation was achieved and showed a potential of RAM required for low cost/consumption

WDM transceivers. Additionally the BER measurements showed up to 3 dB cumulated penalties induced by the EAM chirp and bidirectional transmission. These results confirm the effectiveness of the monolithic integration platform on InP for high temperature operation and fast modulation type components.

The obtained results show a very promising performance of the device to be used in the next generation wavelength division multiplexing passive optical networks as colorless and uncooled transceivers working at the data rate up to 40 Gb/s. However, it is clear that next improvements are required for the actual network implementations. We believe that the component gain can be increased and therefore the backscattering impact could be reduced. Also, the 40 Gb/s transmission performance would have to be improved by electronics FEC, dispersion compensation and equalization in order to enter the PON standards.

ACKNOWLEDGEMENT

The authors are grateful to Nadine Lagay, Florence Martin, Daniele Carpentier, Natasha Keffen and Francis Poingt for their technical support.

REFERENCES

[1] A. Garreau, J. Decobert, C. Kazmierski, M-C. Cuisin, J-G. Provost, H. Sillard, F. Blache, D. Carpentier, J. Landreau, P.Chanclou, "10Gbit/s Amplified Reflective Electroabsorption Modulator for Colourless Access Networks," *Proc. IPRM 2006*, pp. 168-170, 7-11 May 2006.

[2] D. Smith, I. Lealman, X. Chen, D. Moodie, P. Cannard, J. Dosanjh, L. Rivers, C. Ford, R. Cronin, T. Kerr, L. Johnston, R. Waller, R. Firth, A. Borghesani, R. Wyatt, A. Poustie, "Colourless 10Gb/s reflective SOA-EAM with low polarization sensitivity for long-DWDM-PON networks," *Proc. ECOC 2009*, pp 8.6.3, 20-24 Sept. 2009.

[3] D.C. Kim, H-S. Kim, K.S. Kim, B-S. Choi, J-S. Jeong, O-K. Kwon, "10Gbps SOA-REAM using Monolithic Integration of Planar Buried Heterostructure SOA with Deep-Ridge Waveguide EA Modulator for Colourless Optical Source in WDM-PON," *Proc. ECOC 2011*, 18-22 Sept 2011.

[4] N. Dupuis, J. Decobert, C. Jany, F. Alexandre, A. Garreau, N. Lagay, F. Martin, D. Carpentier, J. Landreau and C. Kazmierski, "10-Gb/s AlGaInAs Colorless Remote Amplified Modulator by Selective Area Growth for Wavelength Agnostic Networks," *Photonics Technology Letters*, vol.20, N°21, pp. 1808-1810, 2008.

[5] O. Patard, F. Alexandre, F. Martin, J. Decobert, N. Lagay, R. Guillamet and C. Kazmierski "REAM-SOA integrating Ru doped InP current blocking layer," *Proc. IPRM 2011*, 22-26th May 2011, Post deadline paper.

InP Photonic Integrated Circuit with an AWG-like design for Optical Beam Steering

Weihua Guo[1], Pietro R. A. Binetti[1], Chad Althouse[1], Huub P.M.M. Ambrosius[2], Leif A. Johansson[1], member, IEEE, and Larry A. Coldren[1], Fellow, IEEE

[1]Department of Electrical and Computer Engineering, University of California Santa Barbara, CA93106, USA

[2]Electrical Engineering Department, Eindhoven University of Technology, Eindhoven, NL

Abstract — **Optical beam steering through an InP PIC with an AWG-like design has been demonstrated. Good far-field pattern has been kept without resetting the phase shifter currents when changing the input wavelength to steer the beam.**

Index Terms — **Photonic integrated circuits, optical beam steering, LIDAR.**

I. INTRODUCTION

Similar to electronically scanned phased array Radar, electronically controlled optical beam steering is useful for light detecting and ranging (LIDAR). It also finds applications in 3D imaging, precision targeting, guidance and navigation, etc. Different methods have been tried to achieve electronically controlled optical beam steering [1-3]. In [4-5] we demonstrated electronically controlled 2D optical beam steering through an InP photonic integrated circuit (PIC). The critical part of the PIC is a waveguide array with embedded 2nd-order gratings for out-of-plane emission. An array of phase shifters is used to add a phase slope across the waveguide array so as to steer the beam perpendicularly to the waveguide in the array (lateral direction) [4]. The input wavelength is changed to steer the beam along the waveguide in the array (longitudinal direction) because the emission angle of the grating depends on wavelength [4]. Because of phase errors caused by imperfect fabrication, we normally need to optimize the phase shifter currents to generate a good beam—narrow beam width and high side-lobe suppression for the far-field pattern. Because it is a 2D scan it would be preferable not to reset the phase shifter currents when changing wavelength. In this work, we show that through an AWG-like design, i.e. all the channels use bends with the same radius and the same length, the far-field pattern of the beam can keep a good shape and high side-lobe suppression when changing the input wavelength even when we do not reset the phased shifter currents.

II. PIC LAYOUT

The PIC layout is shown in Fig. 1. After amplification by a preamplifier—the leftmost semiconductor optical amplifier (SOA), the input signal is split into eight channels by a 1×8

splitter consisting of cascaded 1×2 MMIs. Each channel then has its own SOA and phase shifter to boost the power and control the phase. After passing through some additional waveguides consisting mostly of bends with the same radius and the same total bending angles (1.5π), the signal enters into the waveguide array region as shown in Fig. 1.

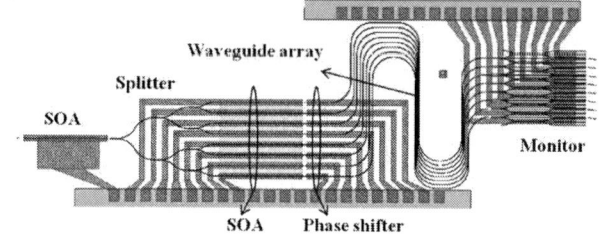

Fig. 1. Layout of the PIC.

Each waveguide in the waveguide array is passive and has embedded 2nd-order gratings etched into the upper optical confinement layer of the waveguide core and buried by the regrown P-doped cladding layers. These gratings scatter the signal from the waveguide out of the plane, upward and downward. There is an aperture in the N-contact metal which is on the backside of the thinned-down substrate. This aperture is aligned to the gratings so that the downward emission can transmit through and be detected in our measurement setup. The additional bending waveguides are added to make all the channels have the same length from the end of the phase shifters to the beginning of the 2nd-order gratings. They are needed because the spacing between the SOA array and the waveguide array for emission is different: the former is 100 μm but the latter is only 5.5 μm. After passing through the waveguide array for emission, some signal which is still left in the waveguide passes through additional bending waveguides and then enters into the monitor array. These additional bending waveguides are needed due to the same reason as those added before the grating array. The monitors integrated on chip are used to monitor the phase difference between adjacent channels through interferometer structures formed by adjacent channels. All waveguides used in the PIC including both active and passive waveguides are deep ridge waveguides which allow small bend radius—200 μm—to be used. Because of the relatively long bends and the tight bend radius, it is better to use the same bend radius and the same total bend length in each channel so that their influence on phase can cancel each other. This is a rule quite generally followed by AWG designs especially when waveguides with strong optical confinement are used [6].

III. PIC FABRICATION

The PIC as being introduced in the above section includes both passive and active waveguides. The active-passive integration has been realized by using the quantum well intermixing technique [7-8]. Starting from a base wafer structure as shown in Fig. 2 (a), phosphorous ions are implanted into the top sacrificial InP layer to produce defects in the area intended to be passive. These defects are then thermally driven down through the quantum well layers to cause component mixture between the barrier and well materials. This is the so-called quantum well intermixing process which increases the bandgap of the quantum wells in the implanted area. After intermixing, the top InP sacrificial layer is removed by wet etching then the 2nd-order gratings are patterned by using electron beam lithography and etched into the upper optical confinement layer of the waveguide core. After that a P-doped InP cladding layer is re-grown above followed by the highly P-doped InGaAs contact layer. The full wafer structure after regrowth is schematically shown in Fig. 2 (b). The waveguides are patterned by using the I-line wafer stepper and are etched by using ICP with the $Cl_2/H_2/Ar$ plasma [9]. The etch depth is about 5 μm so all the waveguides including both active and passive waveguides are deeply etched ridge waveguides. After waveguide etching is the SiN_x isolation layer deposition followed by via opening for the P-metal and P-metal deposition. Then the wafer is thinned down and N-metal is put on the backside of the wafer. An aperture has been opened in the N-metal which is aligned to the grating area to allow the downward emission from the grating to transmit through.

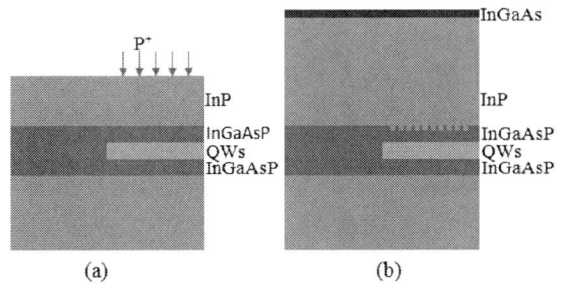

Fig. 2. Schematic wafer structure before regrowth (a) and after regrowth (b).

III. MEASUREMENT

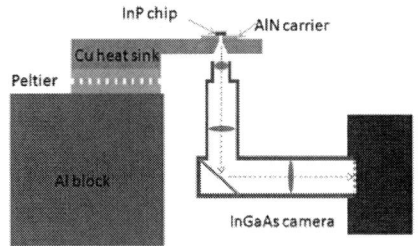

Fig. 3. Schematic of the measurement setup.

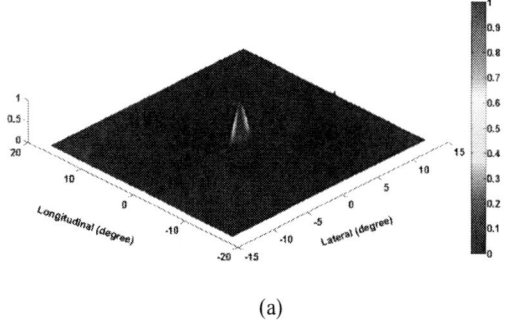

(a)

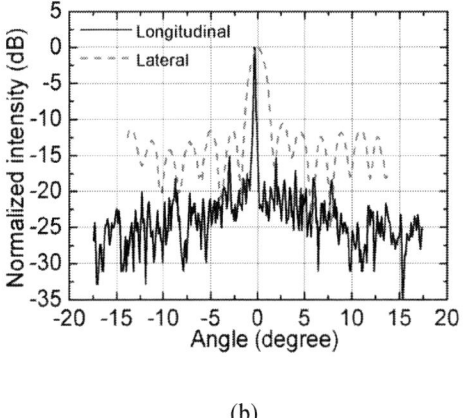

(b)

Fig. 4. (a) Far-field pattern when the beam pointing at zero degree in the lateral direction and the input wavelength is 1540 nm; (b) Intensity distribution across the peak in the lateral and longitudinal direction.

The measurement setup is schematically shown in Fig. 3. The PIC is mounted facing-up onto an AlN carrier. There are holes in the carrier and the Cu heat sink which are aligned to the aperture of the N-contact metal of the PIC so that the downward emission can be captured by the imaging system as shown in Fig. 3. The imaging system consists of three lenses with folded optical path and an InGaAs infrared camera. The first lens projects the far-field pattern of the emission onto its lower focal plane. The far-field pattern is then magnified and

978-1-4673-1725-2/12 $31.00 © 2012 IEEE   121

projected onto the detector plane of the infrared camera by the other two lenses. The imaging system is able to capture the far-field from -17 to 17 degrees in the longitudinal direction and from -14 to 14 degrees in the lateral direction with an angle resolution of ~0.1 degree. For the following measurement the on-chip SOAs are all biased at 100 mA. The eight channel SOAs are biased by a single current source. To account for the series resistance difference of the SOAs, a variable resistor network has been used as demonstrated in [5]. The eight phase shifters are controlled by eight current analog outputs from a DAC card. The input signal is from an external cavity tunable laser fiber coupled into the chip. The tuning range is set from 1530 to 1560 nm.

First, the input wavelength is set to 1540 nm. The phase shifter currents are optimized by the Particle Swarm Optimization (PSO) algorithm to make the beam point at zero degree in the lateral direction and to maximize the side-lobe suppression ratio in the angle range from -14 to 14 degrees [4]. The resulted 2D far-field pattern is plotted in Fig. 4 (a) which shows that a nice clean beam has been obtained. The intensity distribution across the peak in the lateral and longitudinal direction is shown in Fig. 4 (b). A side-lobe suppression of 10 dB has been achieved. The 3-dB beam width in the lateral direction is 1.7 degrees which is in close agreement of the theoretical value of 1.8 degrees, assuming the emission amplitude is uniform across the eight channels. In the longitudinal direction the beam is much narrower because a very long (500 μm) and slowly attenuating grating has been used. Then, we steer the beam to different angles in the lateral direction by repeating the above optimization process around the lateral angles from -6 to 6 degrees with a step of 2 degrees. The results are shown in Fig. 5. The array spacing is 5.5 μm which determines that the angle spacing between two adjacent diffraction orders is 16 degrees, as it can be clearly seen from Fig. 5.

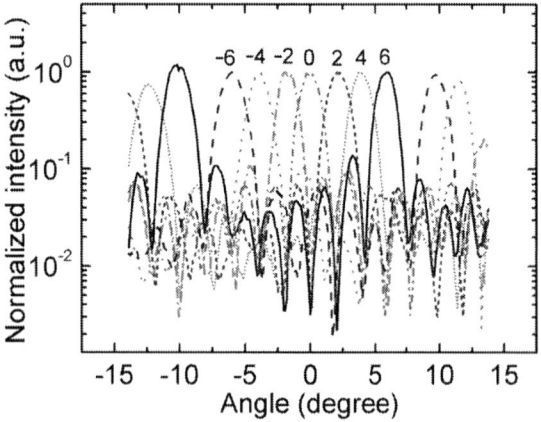

Fig. 5. Intensity distribution in the lateral direction when the beam steered to lateral angles from -6 to 6 degrees with a step of 2 degrees.

After this, we steered the beam back to the lateral zero-degree angle and then changed the wavelength from 1530 to 1560 nm. Without changing the phase shifter currents, the far-field pattern in the lateral direction is recorded and plotted in Fig. 6. When changing the wavelength, the gain of the SOAs on-chip varies due to the limited gain bandwidth and the fixed current injections to the SOAs. This influences the output power as seen from Fig. 6. But the 3-dB beam width and the side-lobe suppression keep nearly the same, ~2.0 degrees and 10 dB, respectively. For the wavelength of 1550 nm which is the gain peak, the emission is a little bit too strong to saturate the infrared camera which causes the peak of the recorded field distribution to be cut-off. The design not only keeps the length of all the channels the same but also uses the same bends: the same bend radius and the same total bending angles. Furthermore all the SOAs and phase shifters are arranged to have the same positions relative to the splitter and to the grating array. All of these help to achieve the effect shown in Fig. 6. But there is also one penalty from these additional waveguides—the added loss. Our passive waveguides typically have a loss of about 2 dB/mm. So these additional waveguides contribute to a loss of about 4 dB due to their ~2 mm length.

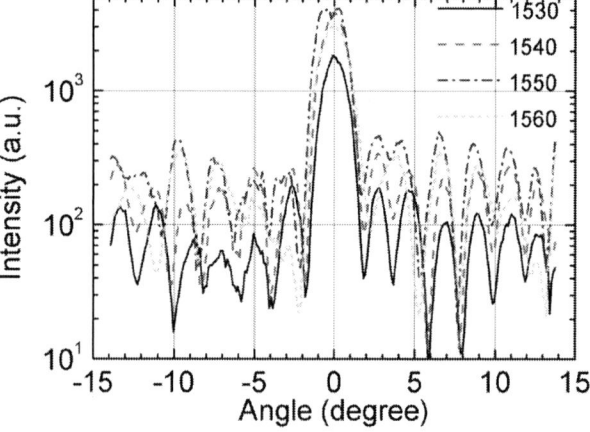

Fig. 6. Far-field pattern in the lateral direction for the input wavelength of 1530, 1540, 1550, and 1560 nm.

IV. SUMMARY

In summary we have demonstrated an InP PIC for optical beam steering: when we change the input wavelength to steer the beam the array of phase shifters does not need to reset the bias currents in order to preserve the beam shape and the side-lobe suppression. The AWG-like design employed in the PIC uses bends with the same radius and the same total length in each channel. This ensures that the influences from these bends cancel with each other even when a very tight bend radius—200 μm—has been used.

ACKNOWLEDGEMENT

This work is supported by DARPA SWEEPER project.

REFERENCES

[1] D. P. Resler, D. S. Hobbs, R. C. Sharp, L. J. Friedman, and T. A. Dorschner, "High-efficiency liquid-crystal optical phased-array beam steering," *Opt. Lett.*, vol. 21, no. 9, pp. 689-691, May 1996.

[2] A. Tuantranont, V. M. Bright, J. Zhang, W. Zhang, J. A. Neff, Y. C. Lee, "Optical beam steering using MEMS-controllable microlens array," *Sens. Actuators A: Phys.*, vol. 91, no. 3, pp. 363-372, Jul. 2001.

[3] K. Van Acoleyen, W. Bogaerts, R. Baets, "Two-dimensional dispersive off-chip beam scanner fabricated on silicon-on-insulator," *IEEE Photon. Technol. Lett.*, vol. 23, no. 17, pp. 1270-1272, Sep. 2011.

[4] W. H. Guo, P. R. A. Binetti, C. Althouse, A. Bhardwaj, J. K. Doylend, H. P. M. M. Ambrosius, L. A. Johansson, and L. A. Coldren, "InP photonic integrated circuit for 2D optical beam steering," Post-deadline paper, IEEE Photonics 2011 (IPC11), Arlington, Virginia, USA, 2011.

[5] W. H. Guo, P. R. A. Binetti, C. Althouse, H. P. M. M. Ambrosius, L. A. Johansson, and L. A. Coldren, "Improved performance of optical beam steering through an InP photonic integrated circuit," CW1K, CLEO, San Jose, CA, USA, 2012.

[6] H. Takahashi, I. Nishi, and Y. Hibino, "10 GHz spacing optical frequency division multiplexer based on arrayed-waveguide grating," *Electron. Lett.*, vol 28, no. 4, pp. 380-382, Feb. 1992.

[7] E. J. Skogen, J. S. Barton, S. P. Denbaars, and L. A. Coldren, "A quantum-well-intermixing process for wavelength-agile photonic integrated circuits," *IEEE J. Select. Topics Quantum Electron.*, vol. 8, no. 4, pp. 863-869, Jul./Aug. 2002.

[8] P. R. A. Binetti, M. Z. Lu, E. J. Norberg, R. S. Guzzon, J. S. Parker, A. Sivananthan, A. Bhardwaj, L. A. Johansson, M. J. Rodwell, and L. A. Coldren, "Indium phosphide photonic integrated circuits for coherent optical links," *IEEE J. Select. Topics Quantum Electron.*, vol. 48, no. 2, pp. 279-291, Feb. 2012.

[9] J. S. Parker, E. J. Norberg, R. S. Guzzon, S. C. Nicholes, and L. A. Coldren, "High verticality InP/InGaAsP etching in Cl₂/H₂/Ar inductively coupled plasma for photonic integrated circuits," *J. Vac. Sci. Technol. B*, vol. 29, no. 1, Jan./Feb. 2011.

Design, fabrication, and preliminary test results of a new InGaAsP/InP high-Q ring resonator for gyro applications

Francesco Dell'Olio, Caterina Ciminelli, Mario N. Armenise,
Francisco M. Soares*, Wolfgang Rehbein*

Optoelectronics Laboratory, Politecnico di Bari, Via Orabona 4, 70125 Bari, Italy
(*) Fraunhofer Institute for Telecommunications, Heinrich-Hertz-Institut, Einsteinufer 37, 10587 Berlin,
Germany

Abstract — **Design, fabrication and initial characterization of large-size InGaAsP/InP ring resonators for gyro applications are reported in this paper. The devices configuration includes a ring with a radius of 13 mm and a straight bus waveguide with tapered ends. Four cavities with the same radius and different values of the bus/ring gap have been fabricated by metal-organic vapour-phase-epitaxy, standard photolithography and reactive ion etching. Characterization results show that the resonator with nominal gap = 1.444 µm has a quality factor exceeding 7×10^5 and resonance depth close to 10 dB.**

Index Terms — **Gyroscopes, Integrated optoelectronics, Optoelectronic and photonic sensors, Optoelectronic devices.**

I. INTRODUCTION

In the last few years, complexity of InP photonic integrated circuits (PICs) is quickly growing with a rate similar to that one of microelectronic integrated circuits and the development process of some InP PICs has already achieved the commercialization stage [1]. For example, InP PICs technology allowed the fabrication of compact (footprint < 10 cm^2) transmitters/receivers for coherent optical fiber communications [2], WDM (wavelength division multiplexing) routers [3], and optical arbitrary waveform generators [4]. Application domain of InP PICs is now limited to high-capacity telecom networks but benefits associated to monolithic integration of tens or hundreds of optoelectronic devices suggest the PICs technology extension to high resolution sensing.

Miniaturized rate-grade MEMS gyroscopes are low-cost devices with a wide application spectrum including robotics, medical instrumentation, automotive, and customer electronics [5]. Although a few tactical-grade MEMS gyros with bias drift of some °/h and resolution > 50 °/h have been recently launched on the market, miniaturization of both tactical-grade and inertial grade angular velocity sensors with applications in defense & aerospace industry is still a technological challenge.

Scaling of conventional photonic gyroscopes, i.e. the FOG (Fiber Optic Gyroscope) and the RLG (Ring Laser Gyroscope), through integrated optical technologies is one of the most promising approach for developing compact, low-power, and reliable medium/high performance gyros

(resolution ≤ 10 °/h). For this reason R&D activity on this topic is attracting an increasing effort [6].

Semiconductor ring lasers (SRLs) [7] are under investigation since four decades and their application in the field of angular velocity sensing has been theoretically and experimentally investigated [8-9]. Backscattering and mode competition within the active laser cavity have been identified as the main physical effects limiting the performance of those sensors which never demonstrated their functionality at reasonable values of angular velocity [10].

In the last decade, several phase-sensitive integrated optical gyroscopes including multi-rings CROW structures have been proposed and theoretically investigated [11]. Periodic modulation of coupling coefficients and chirping of the rings radii have been recently proposed for the CROW gyros performance enhancement [12].

Resonant micro optical gyros (RMOGs) [13], which are based on passive resonators serving as sensing element, are considered ideal candidates for optoelectronic gyroscopes miniaturization. However, RMOGs prototypes reported in literature, having a resolution of some hundreds of °/h [14], are complex modules including several packaged optoelectronics components in different technologies which are connected by optical fibers.

The idea of a monolithic integrated gyroscope to be realized on a single InP chip by using the PICs technology has been proposed in [15] and an ESA-founded project aiming at the experimental demonstration of this concept is ongoing [16].

The key element of an InP PIC for angular velocity sensing is the InP ring resonator, whose size and performance strongly influence the gyroscope resolution. In fact, the sensor resolution δΩ is given by:

$$\delta\Omega = \frac{\sqrt{2}\,c}{Q\,d}\,\Pi \qquad (1)$$

where

$$\Pi = \sqrt{Bh\,v_0/\eta_{pd}P_{pd}} \qquad (2)$$

d is the ring diameter, Q is the resonator quality factor, c is the speed of light in vacuum, B is the sensor bandwidth, h is the Planck's constant, v_0 is the gyro operating frequency (= 193

THz, corresponding to the operating wavelength of 1.55 μm), η_{pd} is the quantum efficiency of the photodiodes included in the readout optoelectronic system, and P_{pd} is the average optical power at the photodiodes input.

Assuming B = 1 Hz, η_{pd} = 0.9, and P_{pd} = 1 mW, we need a ring diameter > 10 mm and a resonator quality factor > 10^6 to achieve $\delta\Omega \leq 10$ °/h. Since fabrication of high-Q InGaAsP/InP integrated cavities is very challenging because InGaAsP/InP waveguides usually exhibit propagation loss values > 1 dB/cm, the fulfillment of those requirements needs a notably technological effort.

The best performing InP ring resonator so far reported was realized by a buried waveguide with propagation loss = 1.7 dB/cm and exhibits a quality factor of 113,000 [17]. The resonator radius is 0.2 mm. InP integrated resonators including a semiconductor optical amplifier which partially compensates the propagation loss were reported in [18]. These cavities have Q = 220,000.

In this paper we report on InP ring resonators (see Fig. 1) properly designed to serve as sensing element in a PIC for angular velocity sensing with target resolution = 10 °/h.

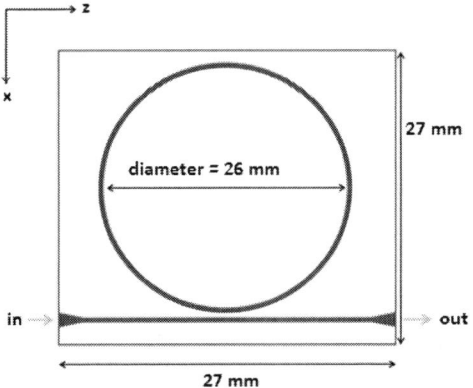

Fig. 1.   Configuration of the fabricated chips.

II. CHIPS DESIGN

The fabricated chips include a ring with a radius of 13 mm evanescently coupled to one straight bus waveguide. The configuration including one bus was preferred to the one with two buses because the former exhibits a higher quality factor. To assure an appropriate uniformity of the technological process, the radius value has been chosen so that the cavity fits in ¼ of the 3-inch InP wafer.

An InGaAsP/InP rib waveguide with etch depth = 300 nm and width = 2000 nm has been selected for the device. The 1 μm thick InGaAsP guiding layer has the band-gap wavelength = 1.06 μm. The quasi-TE mode supported by the guiding structure (see Fig. 2) exhibits a confinement factor exceeding 80 % and a propagation loss, estimated by the 3D model proposed in [6], in the range 0.5 ÷ 0.8 dB/cm.

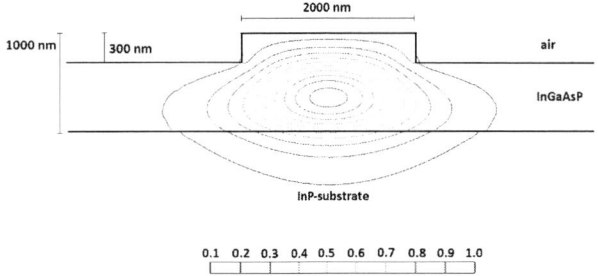

Fig. 2.   Quasi-TE mode supported by the selected InGaAsP/InP rib waveguide. Level curves relevant to the normalized squared module of the electric field are shown.

To enhance the fiber/waveguide coupling, a taper has been designed and realized at the two ends of the straight waveguide. The waveguide width has been tapered up to 8000 nm (core diameter of a standard single mode fiber) and the taper length is 0.2 mm. Optical propagation within the taper has been simulated by the 3D BPM (Beam Propagation Method). Distribution of the quasi-TE optical mode at taper input and output is shown in Fig. 3. Loss due to the propagation within the taper is less than 1 % and the overlap integral between the taper output field and the field of a standard single mode fiber is equal to 22 %.

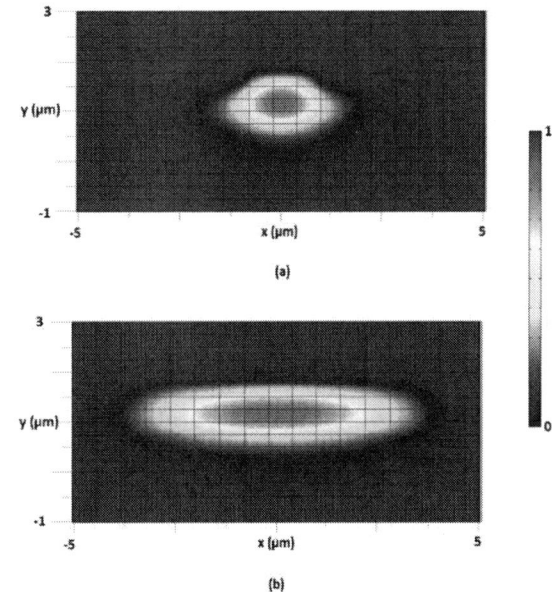

Fig. 3.   Optical field distribution at taper input (a) and output (b).

The gap g between the ring resonator and the straight bus waveguide, which determines the efficiency η of the bus/cavity coupler, was designed by imposing a resonance depth of 8 dB.

The relationship between the efficiency η and the gap g was evaluated by using the Coupled Mode Theory (CMT).

Since the coupler is formed by a the straight bus waveguide and a bent waveguide (see Fig. 4), the distance d_w between the waveguides is given by:

$$d_w(z) = \frac{d}{2} + \left(g + \frac{w}{2}\right) - \sqrt{\left(\frac{d}{2}\right)^2 - z^2} \qquad (3)$$

where w is the waveguides width.

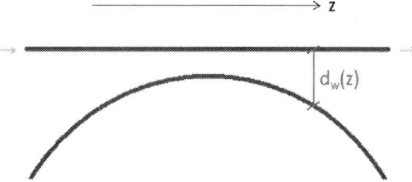

Fig. 4. Coupler included in the fabricated resonators.

The coupling coefficient κ depends exponentially on d_w and can be written as:

$$\kappa(z) = A \exp\left[B \, d_w(z)\right] =$$
$$= A \exp\left\{B\left[\frac{d}{2} + \left(g + \frac{w}{2}\right) - \sqrt{\left(\frac{d}{2}\right)^2 - z^2}\right]\right\} \qquad (4)$$

where A and B are coefficients which depend on the geometrical and optical properties of the WGs forming the coupler.

The coupler was modeled through the following coupled equations:

$$\begin{cases} \dfrac{da_1(z)}{dz} = -i \, \kappa(z) \, a_2(z) \\[2mm] \dfrac{da_2(z)}{dz} = -i \, \kappa(z) \, a_1(z) \end{cases} \qquad (5)$$

where a_1 and a_2 are the amplitudes of the two modes propagating within the two waveguides and i is the imaginary unit.

The coupling efficiency dependence on g was derived by solving Eq. (3) with the appropriate boundary conditions. Then we calculated both the coupler efficiency value (η_{opt}) allowing a resonance depth = 8 dB and the relevant value of the gap between the ring and the straight bus waveguide (g_{opt}).

The dependence of η_{opt} and g_{opt} on the waveguide propagation loss α (we varied α in the range 0.5÷1 dB/cm) is shown in Fig. 5. As α increases, the optimum coupler efficiency increases, reaching values > 50 % when α > 0.8 dB/cm. The optimum value of the gap decreases when α increases and it is always > 0.8 μm, which is the minimum gap allowed by the selected technological process.

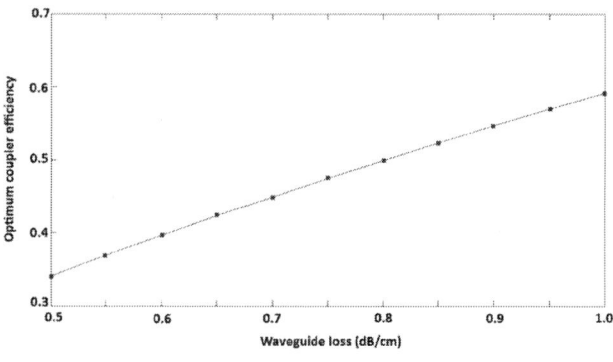

(a)

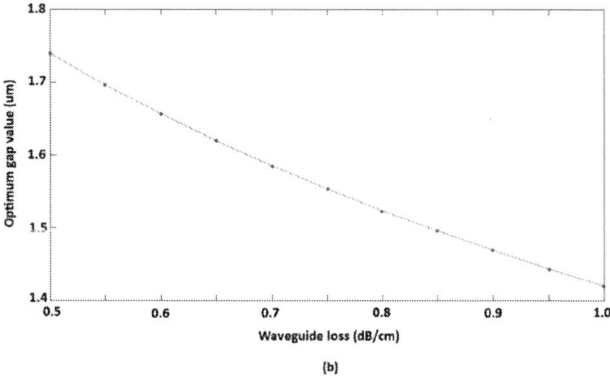

(b)

Fig. 5. Optimum coupler efficiency (a) and optimum gap value (b) vs. waveguide loss.

Since the numerically predicted propagation loss is in the range 0.5÷0.8 dB/cm range, to optimize the design of the sensor we have considered four different values of the gap between the bus and the ring: $g_1 = 1.739$ μm (relevant to α = 0.5 dB/cm), $g_2 = 1.620$ μm (relevant to α = 0.65 dB/cm), $g_3 = 1.524$ μm (relevant to α = 0.8 dB/cm), and $g_4 = 1.444$ μm (relevant to α = 0.95 dB/cm).

III. Chips Fabrication

The InGaAsP layer has been grown on top of an InP substrate in a metal-organic vapour-phase-epitaxy (MOVPE) reactor. The ring resonator has been fabricated using a single standard-lithography process. The waveguides were anisotropically etched in a reactive-ion etcher using a thin silicon-nitride layer as mask. After processing, the wafer was cleaved and an anti-reflection coating was deposited at the input- and output waveguide facet of the ring resonator.

Initial characterization of the ring resonator was done by measuring the transmission spectrum of the ring resonator. The wavelength of the optical input was swept around 1550 nm with a 1pm resolution, and the intensity at the output was measured with a photodetector. Fig. 3 below shows the measured transmission spectrum of the ring resonator, with a

978-1-4673-1725-2/12 $31.00 © 2012 IEEE

gap of 1.444 μm, for a TE polarized input signal. The resonance dips caused by the coupling of the light from the bus to the ring can be clearly seen occurring at a period of around 8 pm, which well agrees with the expected interference from the ~8.17-cm long ring waveguide. The slight curve roughness can be attributed to the scanning of the tunable laser which was set to a minimum resolution of 1 pm. As Fig. 3 shows, the measured resonance depth is better than 9 dB. The resonances full width at the half maximum is 2.1 ± 0.4 pm and the cavity quality factor is 740,000 ± 20%. These experimental results show a very good agreement with our numerical predictions.

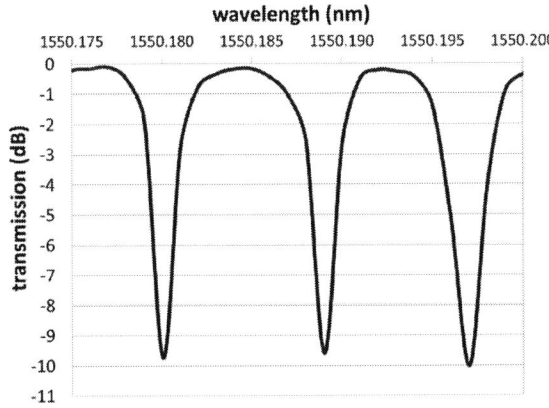

Fig. 5. Measured TE transmission spectrum of the ring resonator with a gap of 1.444 μm between the ring and the bus..

IV. CONCLUSIONS

We have designed, fabricated, and preliminarily characterized an InP-based ring resonator with a ring radius of 13 mm. The device size is 27mm x 27mm, and we were able to achieve resonance peaks with a depth larger than 9 dB and a width less than 2.5 pm, for TE polarized input light. The cavity Q-factor (=740,000 ± 20%) is at least six times larger than the state-of-the-art.

IV. ACKNOWLEDGEMENTS

The work has been partially funded by European SpaceAgency (ESA) under MiOS project n. 4000102311/10/NL/PA.

REFERENCES

[1] M. Smit, J. van der Tol, M. Hill, "Moore's law in photonics," *Laser Photonics Rev.*, vol. 6 , pp. 1-13, 2012.

[2] C. R. Doerr, "High Performance Photonic Integrated Circuits for Coherent Fiber Communication," Optical Fiber Communication Conference, paper OWU5, 2010.

[3] S. C. Nicholes, M. L. Mašanović, B. Jevremović, E. Lively, L. A. Coldren, D. J. Blumenthal, "The World's First InP 8x8 Monolithic Tunable Optical Router (MOTOR) Operating at 40 Gbps Line Rate per Port," Optical Fiber Communication Conference, paper PDPB1, 2009.

[4] F. M. Soares, et al., "Monolithically Integrated InP Wafer-Scale 100-Channel × 10-GHz AWG and Michelson Interferometers for 1-THz-Bandwidth Optical Arbitrary Waveform Generation," Optical Fiber Communication Conference, paper OThS1, 2010.

[5] C. Acar, A. Shkel, *MEMS Vibratory Gyroscopes*. New York: Springer-Verlag, 2009.

[6] C. Ciminelli, F. Dell'Olio, C. E. Campanella, and M. N. Armenise, "Photonic technologies for angular velocity sensing," *Advances in Optics and Photonics*, vol. 2, pp. 370-404, 2010.

[7] S. Liao, S. Wang, "Semiconductor injection lasers with a circular resonator," *Appl. Phys. Lett.*, vol. 36, pp. 801–803, 1980.

[8] M.N. Armenise, V.M.N. Passaro, F. De Leonardis, M. Armenise, "Modeling and Design of a Novel Miniaturized Integrated Optical Sensor for Gyroscope Applications," *J. Lightwave Technol.*, vo. 19, pp. 1476-1494, 2001.

[9] M. Armenise, P.J.R. Laybourn, "Design, simulation of a ring laser for miniaturized gyroscopes," *Proc. SPIE*, vol. 3464, pp. 81–90, 1998.

[10] M.N. Armenise, C. Ciminelli, F. De Leonardis, V.M.N. Passaro, "Quantum effects in new integrated optical angular velocity sensors," Proc. Int. Conference on Space Optics, pp. 595-597, 2004.

[11] J. Scheuer, A. Yariv, "Sagnac Effect in Coupled-Resonator Slow-Light Waveguide Structures," *Phys. Rev. Lett.*, vol. 96, 053901, 2006.

[12] C. Sorrentino, J. R. E. Toland, C. P. Search, "Ultra-sensitive chip scale Sagnac gyroscope based on periodically modulated coupling of a coupled resonator optical waveguide," *Opt. Express*, vol 20, pp. 354-363, 2011.

[13] C. Ciminelli, F. Peluso, M. N. Armenise, "A new integrated optical angular velocity sensor," *Proc SPIE.*, vol. 5728, pp. 93–100, 2005.

[14] H. Mao, H. Ma, Z. Jin, "Polarization maintaining silica waveguide resonator optic gyro using double phase modulation technique," *Opt. Express*, vol. 19, pp. 4632- 4643, 2011.

[15] F. Dell'Olio, C. Ciminelli, V. M. N. Passaro, and M. N. Armenise, "Optical angular velocity sensors and related read-out systems for new generation gyroscopes," 1st Networking/Partnering Day 2010, Noordwijk, Nederland, January 28, 2010.

[16] ESA-founded MiOS project, n. 4000102311/10/NL/PA.

[17] S. J. Choi, K. Djordjev, Z. Peng, Q. Yang, S. J. Choi, P. D. Dapkus, "Laterally Coupled Buried Heterostructure High-Q Ring Resonators," *IEEE Photon. Tech. Lett.*, vol. 16, pp. 2266-2268, 2004.

[18] S. J. Choi, Z. Peng, Q. Yang, E. H. Hwang, and P. D. Dapkus, "A High-Q Wavelength Filter Based on Buried Heterostructure Ring Resonators Integrated With a Semiconductor Optical Amplifier", *IEEE Photon. Tech. Lett.*, vol. 17, pp. 2101-2103, 2005.

978-1-4673-1725-2/12 $31.00 © 2012 IEEE

Double-Layer Stepped Si(100) for III-V-on-Silicon Integration

Henning Döscher[1,2], Peter Kleinschmidt,[1,3] Sebastian Brückner,[1,2] Oliver Supplie,[1] Anja Dobrich,[1] and Thomas Hannappel[1,2,3].

[1]*Helmholtz-Zentrum Berlin für Materialien und Energie, D-14109 Berlin, Germany*
[2]*Technische Universität Ilmenau, Institut für Physik, D-98693 Ilmenau, Germany*
[3]*CiS Forschungsinstitut für Mikrosensorik und Photovoltaik, D-99099 Erfurt, Germany*

Abstract — We demonstrate the formation of anomalous atomic double-layer surface steps on 2° misoriented Si(100) in hydrogen process ambient. Employing a contamination-free sample transfer, low energy electron diffraction and atomic resolution scanning tunneling microscopy reveal dimer rows running parallel to the step edges, i.e. D_A type steps, thought to be energetically unfavorable. Based on the interaction of the Si(100) surface with the hydrogen ambient, we propose a model whereby step formation results from interplay of surface vacancy generation, diffusion, and annihilation at the step edges. Reflection anisotropy spectroscopy enables in situ observation of Si(100) surface formation and confirms our model.

I. INTRODUCTION

Integration of III-V semiconductor with silicon substrate has been a subject of intense investigations. Difficulties arising from III-V on IV integration include lattice mismatch, different thermal expansion coefficients and the transition from a non-polar substrate to a polar material [1]. Assuming a uniform interface, III-III bonds or V-V bonds, corresponding to the formation of anti-phase domains, result in the III-V material if the silicon substrate exhibits atomic single-layer steps [2]. In this model, anti-phase domains can be avoided if double layer steps are achieved on the substrate surface prior to III-V epitaxy [3,4]. In ultrahigh vacuum (UHV), the problem of double-layer step formation is well investigated and established processes exist to produce suitable silicon substrates. On the one hand, type B double layer steps, where dimer rows of the reconstructed Si(100) surface above the step edge are perpendicular to the edge, are energetically favorable over single layer steps [5]. On the other hand, this regime is easily accessible by annealing the Si(100) surface in UHV if the step density is sufficient, i.e. the misorientation exceeds 2° [6]. Other techniques exist which allow double-layer step preparation in UHV, including electro-migration [7], application of surface stress [8] or ion-bombardment [9]. However, conditions in the metalorganic vapor phase epitaxy (MOVPE) environment differ, and similar Si(100) substrate preparation methods are not readily available there.

Preparation of clean Si(100) surfaces in an MOVPE reactor has been demonstrated [10]. This was achieved by annealing a Si(100) wafer, which had previously been wet-chemically treated [11], in hydrogen ambient at temperatures around 1000°C. The resulting surfaces were atomically flat and free of any contamination. However, we found that the applied process conditions resulted in a Si(100) surface which was completely covered by Si-H monohydride where each of the two dangling bonds of a surface Si dimer was saturated with a single hydrogen atom [12]. It is known from investigations in UHV, that the monohydride surface step structure is governed by different energetics than the hydrogen-free surface [13],[14]. In these investigations, conditions where double-layer steps are energetically favorable were not achieved, and instead a mixture of different step types prevailed even at large misorientations [15]. The other preparation methods for double layer steps in UHV mentioned above are not easily incorporated in standard MOVPE equipment. Still, generation of double layer steps has been reported in the MOVPE environment [4]. This was achieved in Si(100) substrates which were slightly misoriented in [011] direction, but due to the limited resolution of the applied AFM techniques the step type could not be specified.

II. EXPERIMENTAL

We investigated the preparation of Si(100) substrates with 0.1°, 2° and 6° misorientation in [011] in a commercial, horizontal MOVPE reactor (Aixtron AIX-200). A contamination-free transfer system to UHV [16] was applied in order to access surface sensitive UHV techniques including scanning tunnelling microscopy (STM), low energy electron diffraction (LEED) and photo-electron spectroscopy (XPS, UPS). Optical in-situ control of the MOVPE process was available via a commercial reflection anisotropy spectroscopy (RAS) setup (Laytec EpiRAS 200), which was benchmarked to the surface science instruments.

III. RESULTS AND DISCUSSION

We investigated Si(100) preparation of substrates with 0.1°, 2° and 6° misorientation in [011] and applied various surface science methods via a contamination free sample transfer [16]. Fig. 1 (a) shows the surface of a sample misoriented 6° in [011], prepared by deoxidation under H_2, growth with SiH_4

978-1-4673-1725-2/12 $31.00 © 2012 IEEE

and annealing. The image shows a regularly stepped surface where step edges run from top to bottom, perpendicular to the direction of the misorientation. The terraces consist of stripes which run perpendicular to the step edges and, based on their size, can be identified as dimer rows. They cover the majority of the surface, but in a few places on the surface dimer rows are missing. However, in most places the dimer rows cover the width of the terraces so that steps separating these terraces are double-layer steps of the D_B type (where dimer rows run perpendicular to the step edge).

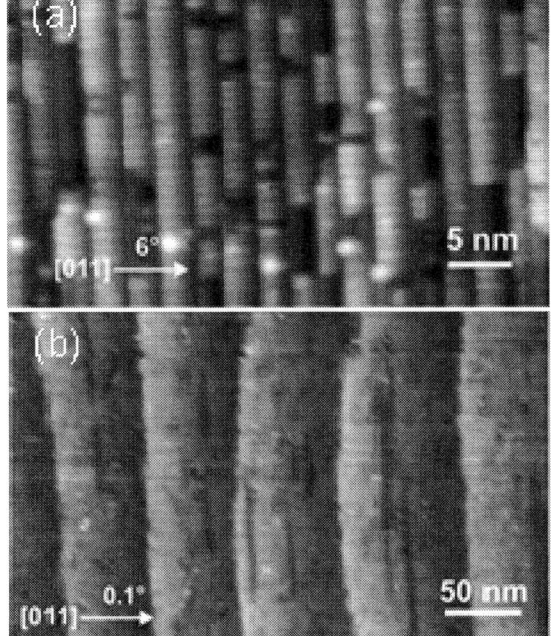

Fig. 1. STM images of MOVPE prepared Si(100) surfaces with 6° (a) and 0.1° (b) misorientation in [011] direction.

Samples misoriented 0.1° in [011] were prepared by an identical process to that used for the 6° samples. Figure 1(b) shows an STM image of a region of this surface where the terrace structure approximately corresponds to the expected gradient based on the specified misorientation. The vertical distance between the large terraces equals two atomic layers. Dimer rows on the terraces run parallel to the step edge, suggesting that the observed steps correspond to type D_A steps, in contradiction to results previously reported in literature [6,15]. Closer inspection of the step edge, however, reveals that in most places an intermediate ledge exists consisting of dimer rows extending from the A type edge, only a few dimers long. The step edge therefore is not truly a double layer step, but instead the image reflects a drastically reduced type B terrace.

Using the same process we prepared a surface with 2° misorientation and found that, in contrast to 0.1° and 6°

misorientation, a surface characterized by single-layer steps resulted. This was confirmed by LEED measurement (not shown here) where a (2×1)/(1×2) surface reconstruction was found, with approximately equal intensities of the two domains, indicating equal proportions of the two possible surface dimer orientations. Aiming to achieve a higher domain ratio and, correspondingly, a surface with a higher content of double layer steps, we varied our process conditions and studied the effect on surface quality and step structure. We found that a modified process with different parameters during cooling at the end of the process led to a substantial improvement in the case of the 2° misoriented surface. In fact, atomic resolution STM (not shown here) evidenced true D_A type steps in many locations, based on step height and orientation of the dimer rows and on absence of an intermediate ledge.

IV. CONCLUSION

We have developed processes to prepare double layer steps on Si(100) substrates with 0.1°, 2° and 6° misorientation in [011]. Different preparation procedures were established and process conditions were found resulting in D_B or D_A type steps (depending on preparation parameters and substrate orientation) with a strongly suppressed minority domain, which should be ideally suited for subsequent APD-free III-V on Si(100) heteroepitaxy.

REFERENCES

[1] H. Kroemer, *J. Cryst. Growth 81*, 193-204 (1987).
[2] S. F. Fang, K. Adomi, S. Iyer, H. Morkoc, H. Zabel, C. Choi, and N. Otsuka, *J. Appl. Phys. 68*, R31-R58 (1990).
[3] H. Döscher, T. Hannappel, B. Kunert, A. Beyer, K. Volz, and W. Stolz, *Appl. Phys. Lett. 93*, 172110 (2008).
[4] B. Kunert, I. Németh, S. Reinhard, K. Volz, and W. Stolz, *Thin Solid Films 517*, 140 (2008).
[5] D.J. Chadi, *Phys. Rev. Lett. 59*, 1691-1694 (1987).
[6] B. S. Swartzentruber, Y.-W. Mo, M. B. Webb, and M. G. Lagally, *J. Vac. Sci. Technol. A 7*, 2901-2905 (1989).
[7] T. Doi, M. Ichikawa, S. Hosoki, and K. Ninomiya, *Phys. Rev. B 53*, 16609 (1996).
[8] D Fujita, M Kitahara, K Onishi, and K Sagisaka, *Nanotechnology 19*, 025705 (2008).
[9] P. Bedrossian and T. Klitsner, *Phys. Rev. Lett. 68*, 646 (1992).
[10] H. Döscher, S. Brückner, A. Dobrich, C. Höhn, P. Kleinschmidt, and T. Hannappel, *J. Cryst. Growth 315*, 10 (2011).
[11] A. Ishizaka, *J. Electrochem. Soc. 133*, 666 (1986).
[12] H. Döscher, A. Dobrich, S. Brückner, P. Kleinschmidt, and T. Hannappel, *Appl. Phys. Lett. 97*, 151905 (2010).
[13] Sukmin Jeong and Atsushi Oshiyama, *Phys. Rev. Lett. 81*, 5366 (1998).
[14] F. A. Reboredo, S. B. Zhang, and Alex Zunger, *Phys. Rev. B 63*, 125316 (2001).
[15] A. R. Laracuente and L. J. Whitman, *Surface Science 545*, 70 (2003).
[16] T. Hannappel, S. Visbeck, L. Töben, and F. Willig, *Rev. Sci. Instrum. 75*, 1297 (2004).

Atomic-plane-thick reconstruction across the interface during heteroepitaxial bonding of InP-clad quantum wells to Si

A.Talneau, C.Roblin, A.Itawi, O. Mauguin, L.Largeau, G.Beaudouin, I.Sagnes and G. Patriarche

Laboratoire de Photonique et de Nanostructures, route de Nozay, F-91460 Marcoussis

corresponding author : anne.talneau@lpn.cnrs.fr

Abstract — Monolithic integration of InP-based materials on Si will allow lasers as well as optical amplifiers operating at 1.55µm to be efficiently included in photonic integrated circuits. We demonstrate here oxide-free heteroepitaxial bonding of InP-clad GaInAs quantum wells to Si, with an atomic-plane-thick reconstruction across the InP-Si interface. The wells photoluminescence emitted-wavelength demonstrates no shift after bonding. Several InP surface activation procedures have been investigated. This bonding approach is compatible with guiding designs including a nanostructuration, thus enabling specific designs associated to desirable optical functions.

Index terms — III-V integration on Si, photonic integrated circuits, heteroepitaxial bonding, interface TEM analysis, surface reconstruction

I. INTRODUCTION

The future of integrated optics will be on Silicon, due to the very efficient performances of silicon photonics, and the well established Si processing technology. But emission and isolation are still optical functions that cannot be performed efficiently by Si-based materials. Bonding III-V InP-based materials is thus still mandatory for emission in the 1.55µm wavelength domain. Including an intermediate amorphous bonding layer [1-3] or creating in-situ an oxide layer [4] relaxes bonding conditions (temperature, surface preparation, contact conditions) but an oxide-free interface has been demonstrated to produce the best electrical interface [5]. When investigating advanced optical designs, it is preferable to avoid any low-optical-index layer in the core of the guiding layers formed by the two bonded materials. Additionally, any structuration included for a dedicated optical function will be preserved, thus enabling more complex optical functions to be included, thus improving the photonic integrated circuit functionality.

We investigate here, at die size (~1cm²), the heteroepitaxial bonding of InP layers and InP-clad GaInAs quantum wells to Si under oxide-free surface preparation and oxide-free bonding

conditions. We demonstrate an InP-Si reconstructed interface across a single atomic layer without amorphous layer and the preservation of the photoluminescence wavelength of bonded wells close to the surface (80nm). In-depth characterization of the bonded interfaces through nano-indentation is under investigation, and further characterization of bonded wells will be performed.

II. EXPERIMENTS

InP and Si materials have a large lattice difference of 8.1%. When there is no intermediate layer at the interface of InP and Si materials in order to compensate for this large lattice mismatch, bonding has to be performed at a temperature at least above 500°C, in order to allow In and P atoms to move on their lattice at the atomic scale so as to re-arrange their structural lattice for matching the Si lattice. We only consider In and P atoms since such a temperature is too low to allow any Si atom displacement. In order to avoid any oxidation at the interface during bonding, both oxide-free surfaces preparation as well as oxide-free bonding conditions are mandatory.

Both experimental procedures are described below.

We have followed the well documented oxide-free Si surface preparation [6]. First a solvent ultrasonic cleaning is performed in order to eliminate any dust that could be produced after cleaving, then the so-called RCA cleaning eliminates any possible contamination and finally HF (4%) dipping during 30s, with a short 5s rinsing in DI water. The low HF concentration is chosen to prevent any surface roughness formation [6]. The short rinsing duration is preferred since a long stay in water could re-produce –OH links, thus embedding O atoms at the interface [7]. After this desoxidation procedure, the Si surface is perfectly hydrophobic; such a deoxidized surface can be preserved during at least 30mn without any re-oxidation, thus the Si surface is prepared first [6].

Several InP-surface preparations have been investigated. None of the hydrophobic InP surfaces described in [8] could be obtained. In order to make sure that the surface was oxide-free, XPS characterization were performed. The kinetic of InP surface re-oxidation has also be investigated, demonstrating that a concentrated HF solution and a short waiting time should be preferred (to be published). We thus perform HF desoxidation with a concentrated HF (40%) dipping, during 1mn, a short rinsing time, and nitrogen drying. The InP wafer is put immediately in contact to Si, and the bonding procedure is started.

Bonding is performed at 550°C during 90mn, under a N2 flux, or under vacuum, in order to prevent any oxidation. A 0.2MPa pressure is applied, in order to keep the two wafers in physical contact, the applied pressure has to be low since there is no intermediate layer. The pressure uniformity has been checked to ensure a homogeneous contact all along the two atomic surfaces.

A stack of four InGaAs wells has been bonded on an Si substrate. The wafer is MOVPE grown, the layer stack has been dedicated to bonding investigation, thus including a 300nm thick InGaAs sacrificial layer after the buffer layer in order to allow the chemical InP substrate removing by selective etching after bonding. The membrane is composed of four GaInAs wells with GaInAsP barriers, with an InP bonding layer 71nm thick keeping the wells close to the bonded interface. All the layers are lattice matched on the InP substrate.

After bonding, the InP substrate is chemically removed with an HCl solution, and the GaInAs stop etch layer is removed by selective wet etching.

III. Bonded wafer characterization

A. TEM analysis of the bonded interface

The bonded wafer stack is detailed in Fig.1-a. An in depth Scanning Transmission Electron Microscope –STEM– analysis of the bonded interface (Fig.1-b, Fig.2-a) has been performed on a JEOL 2200FS equipment, including an X ray analysis. A thin cut has been prepared by Focused Ion Beam (FIB) and then observed at 200keV. On Fig.1-b, the four wells are visible, as well as an In inclusion at the bonded interface.

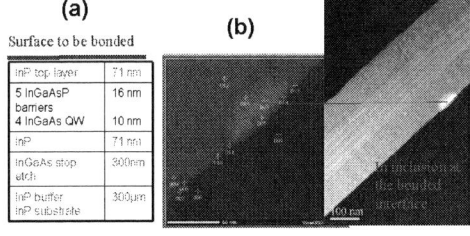

Fig.1 (a) : bonded wafer stack   (b) Bright Field STEM micrograph of the bonded stack and interface with an In inclusion visible by X ray

On Fig.2-a , both crystalline lattices of Si (bottom right) and InP (top left) are visible, without any amorphous layer at the interface. A sharp 8% lattice variation corresponding to the lattice difference has been measured by geometrical phase analysis (GPA) cartography (Fig.2-b). Scarce Oxygen inclusion could be found, the most often found were In inclusions, as shown in Fig.1-b. All interface reconstruction sequels are thus contained in a single atomic layer across the interface.

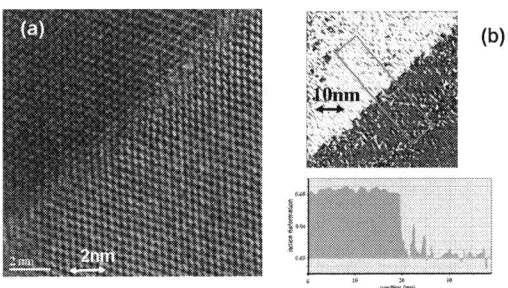

Fig. 2. (a) Bright Field STEM micrograph of the bonded interface showing both Si lattice (right) and InP lattice (left). (b) GPA cartography showing the abrupt interface transition between the two crystal lattices.

Nano-identation measurement is under progress in order to characterize the residual strain at the interface.

B. Photoluminescence of the bonded quantum wells

The bonding dedicated wafer includes four quantum wells close to the bonding interface (InP bonding layer 71nm thick) in order to measure the impact of the bonding procedure on the quantum well luminescence. Fig.3 shows the photoluminescence (PL) spectra at 300K under CW conditions for both the reference (solid line), grown without the InGaAs stop etch layer during a dedicated growth under identical conditions, and the bonded stack (dashed line). Both spectra are obtained under the same pumping condition (30μW) in the same setup; the detector sensitivity is changed, being 50 in the case of the reference wafer, and 10 in the case of the bonded wafer.

The maximum luminescence energy exhibits a very small shift, which evidences that strain is limited at the interface and does not spread to the wells through the 71nm thick InP layer. This is of interest since such an InP thick bonding layer will be included in an electrically driven integrated device for current injection. The lower PL intensity measured for the bonded layer could firstly originate from the final InP-air interface liberated after etching out the sacrificial InGaAs layer. It could however also stem from inter-diffusion between wells and barriers or residual strain in the wells. Low-temperature PL and lifetime measurements are in progress in order to identify the mechanisms responsible for the measured reduced PL intensity. These results, as well as the nano-identation characterization, will give a full insight on the reconstructed interface.

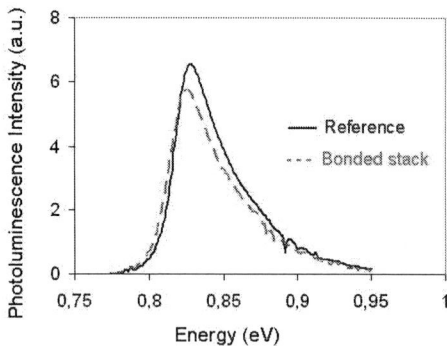

Fig. 3. Photoluminescence spectra at 300K : (solid line) reference wafer with the same QW stack, grown separately without the etch stop, the detector sensitivity is set to 50;(dashed-line) bonded wafer, the detector sensitivity is set to 10.

A first successful bonding of an InP membrane on a patterned Si substrate has been recently obtained, opening the route to advanced designs for integrated devices including an embedded nanostructuration.

V. CONCLUSION

We have demonstrated heteroepitaxial bonding without any intermediate layer of InP-clad GaInAs quantum wells on Si, with an atomic-plane-thick reconstruction across the InP-Si interface. Bonding is performed at high temperature (550°C) under oxide-free conditions. Oxide-free surface preparation of the InP bonding layer has been deeply investigated. GaInAs bonded wells are also shown to preserve their luminescence energy and a decent efficiency.

Bonding on patterned Si surface has to be validated for an active stack including quantum wells, in order to demonstrate that this bonding technique is promising for integrated devices including an internal nanostructuration for dedicated optical functions.

REFERENCES

[1] G. Roelkens et al., Materials Today, 10, **36** (2007).
[2] A.W. Fang et al., Materials Today, 10, **28** (2007).
[3] Van Campenhout, et al., Optics Express, **15**,6744 (2007).
[4] K. Tanabe et al., Scientif. Reports, **2**:349 (2012) DOI10.1038.
[5] D. Pasquariello et al., Mat. Sci and Eng **B80**, 134 (2001) .
[6] Q.-Y.Tong and U. Gösele, Semiconductor Wafer Bonding, Wiley, New York, 1998.
[7] G. W.Trucks et al., Phys.Rev.Lett., **65**,504 (1990).
[8] D. Pasquarielo, et al., Jpn.J.Appl.Phys.,**40**,4837 (2001).

Electrical conduction property at InAs/Si(111) interface by selective-area MOVPE

S. Watanabe[1*], K. Watanabe[2], A. Higo[2], M. Sugiyama[1], Y. Nakano[1,2]

School of Eng[1] & RCAST[2], The University of Tokyo.

Abstract — Ohmic I-V characteristics at interface between n-type InAs and n-type Si have been obtained. Single-domain InAs islands (1μm in diameter and 0.5 μm in height) have been grown epitaxially on Si(111) by selective-area MOVPE. After annealing, linear I-V characteristic was observed with excellent reproducibility and the current through the electrode depends on the number of InAs islands beneath the surface contact metal. The resistivity of InAs was estimated to be $0.0662~\Omega \cdot cm$. The value corresponds to a carrier concentration of approximately $2 \times 10^{16}~cm^{-3}$. This is reasonable as the intrinsic carrier concentration for an un-doped InAs layer.

Index Terms — Indium compounds, infrared detectors, silicon, epitaxial layers, CCD image sensors, tunneling.

I. INTRODUCTION

III-V semiconductors have several merits which cannot be obtained with Si, such as direct bandgaps, which are variable by composition and high electron mobility. In order to include III-V semiconductors physicality in Si devices, a lot of methods for III-V layer growth on Si have been vigorously investigated.

One attractive application is an integrated middle-infrared photodetectors for solar-blind object detection. Currently, such detectors are fabricated by separately assembling both absorption layers, for instance pn-junction of HgCdTe, and Si read out integrated circuit (Si ROIC). When the layer is attached to Si ROIC by In bump, short or open of the circuit at the interface tends to occur due to non-uniformity of In bump. Eventually, these problems directly lead to defects of pixels and reliability of a device degrades. Meanwhile, we aim at an InAs/Si monolithic multichannel photoconductor without In bump. Here, we intend to use InAs as an infrared absorber with integrated circuits made of Si. Device fabrication with InAs substrate is not straightforward, while Si has established sophisticated circuit technology for image capture such as charge coupled devices (CCDs). An InAs/Si monolithic photodetector can take advantage of the both merits of InAs and Si and will allow us the implementation of a mid-infrared CCD.

Fig. 1 shows InAs islands obtained by selective-area heteroepitaxial growth on n-type Si(111) with circular open areas of 1 μm in diameter. A single InAs island is attained as an absorber for one pixel and a Si charge-transfer circuit can be implemented beneath the InAs island. For the implementation of InAs/Si monolithic photodetectors, it is vital to investigate the vertical current conduction at InAs/Si interface obtained by selective-area heteroepitaxial growth, but little work has been presented on that interfacial conduction issue.

Most of successful report focused on the current conduction parallel to a III-V/Si interface for the purpose of implementing electron channels with III-V semiconductors for MOSFETs. We here investigated the vertical current conduction at InAs/Si(111) interface obtained by selective-area heteroepitaxial growth by metal-organic vapor phase epitaxy (MOVPE), which we have already reported [1][2].

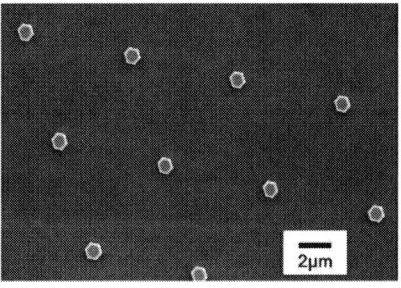

Fig. 1. SEM image of selective-area heteroepitaxial growth of InAs on Si

II. EXPERIMENTAL PROCEDURE

InAs islands were obtained by selective-area MOVPE on n-type Si(111), as shown in Fig. 1. The substrate used for selective-area growth of InAs by MOVPE was covered with mask pattern made of SiO_2, as shown in Fig. 2. The mask pattern consisted of 1 μm-diameter circular growth areas arranged at 7 μm intervals among the centers of neighboring circles. The height of the InAs pillars on Si was approximately 500 nm from SiO_2 surface.

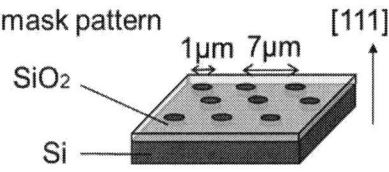

Fig. 2. A structure of the Si(111) substrate with circular open areas of 1 μm in diameter to be used for selective-area growth by MOVPE

The electron concentration of Si was $2.4 \times 10^{19}~cm^{-3}$. We needed as high doped n-type Si substrate as possible in order to realize tunneling electron transport across the interface between InAs and Si substrate, as shown in Fig. 3.

978-1-4673-1725-2/12 $31.00 © 2012 IEEE

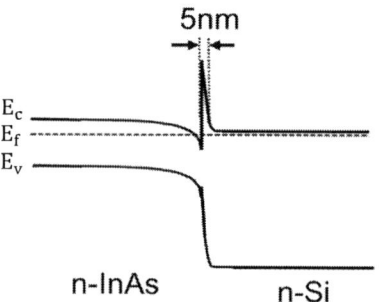

Fig. 3. Energy band diagram at n-type InAs (carrier concentration 2×10^{16} cm^{-3}) and n-type Si (2.4×10^{19}cm^{-3})

MOVPE of InAs was carried out at a total pressure of 10 kPa and a temperature of 610 °C. During the growth of InAs, the partial pressure of trimethylindium (TMIn) and tertiary-butylarsine (TBAs) were 0.0443 and 5.40 Pa, respectively. This condition is suitable for covering the Si area with InAs perfectively. Before the growth of InAs, we applied pre-flow with tertiary-butylphosphorous (TBP) in order to promote the nucleation of InAs on the Si area [1]. In addition to the pre-flow, the partial pressure of TMIn and TBAs were 0.0681 and 2.76 Pa, respectively for the initial 20 s because we needed a single InAs island nucleated on each Si circular area. These 2 steps growth were significant to attain uniformity of InAs islands.

As for I-V measurement, we firstly examined the vertical current between a single InAs island and Si by conductive atomic force microscopy (AFM), which could apply a voltage on a single InAs island by putting a cantilever on the top of the island. Besides we made a measurement of I-V at an InAs island through electrode. The ohmic metals of Ti/Au (110 nm/1000 nm) were deposited directly on the substrate surface, as shown in Fig. 4, by electron-beam evaporation. A single InAs island was isolated by focused ion beam (FIB) and I-V characteristics was obtained by conductive AFM. For this measurement, however, the contact resistance between the cantilever and the metal on the InAs island seemed to be larger than the resistance of the InAs/Si structure itself.

This is why we then tried to characterize the conductive property for a number of InAs islands by making an electrode covering them. We deposited metals in a similar way through a shadow mask, which had two sizes of circles: 0.3 mm (small) and 0.5 mm (large) in diameter, respectively. An area of the large electrode was 2.78 times as large as the small one. One electrode contained 1665 InAs islands for the small electrode and 4627 for the large electrode, respectively. After that, we annealed these specimens for 5 min under nitrogen. It has been reported that the contact between Ti and n-type InAs is improved due to annealing because In atoms diffuses into the Ti layer [3]. However, deterioration of the ohmic contact resistance occurred when annealing temperature was more than 350 °C because of the formation of TiAs and TiInAs sublayers. Therefore, we carried out annealing at 250 and 300 °C.

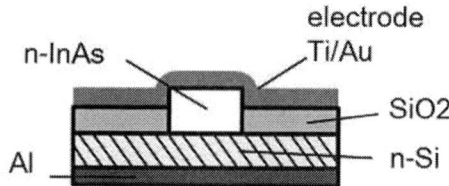

Fig. 4. Cross-sectional diagram of an InAs island on Si with Ti/Au (110 nm/1000 nm) and Al (300 nm) electrodes on the surface and the back side, respectively

III. RESULTS AND DISCUSSION

We confirmed current conduction between a single InAs island and Si substrate by the conductive AFM with its cantilever on the top of the InAs surface. However the I-V characteristics were not ohmic, as shown in Fig. 5, which characteristic was obtained only after the application of voltage above 3.0 V several times. Therefore, the problem was probably due to native oxide layer on the surface of InAs and on the back side of Si.

Then, in order to resolve this problem, a single InAs island on Si was isolated by FIB after evaporation of the electrodes and annealing at 300 °C. We obtained the ohmic characteristic, as shown in Fig. 6. This result indicates the I-V characteristic at interface between InAs and Si was clearly ohmic. The resistance value was approximately 7.4×10^6 Ω as estimated from the slope.

The resistance value itself cannot be regarded as the genuine resistance for the InAs/Si structure due to contact resistance of the cantilever and the object.

We needed to measure the resistance of the InAs/Si structure with a negligible parasitic resistance than the conductive AFM measurement. We therefore characterized the conductive property for a number of InAs islands under a large electrode. For a circular electrode (0.3 mm in diameter) after annealing at 300 °C, current conduction

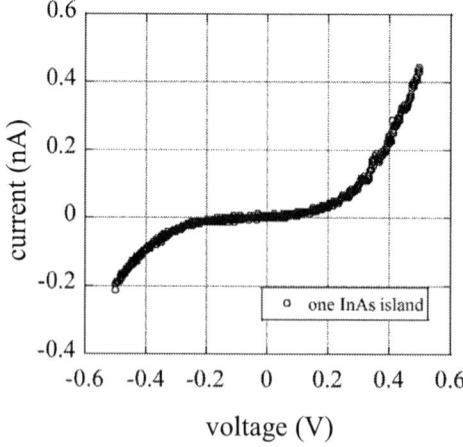

Fig. 5. The I-V profile obtained with applying voltage from -0.5 V to 0.5 V by putting cantilever on the InAs surface after applying approximately 3.0 V several times

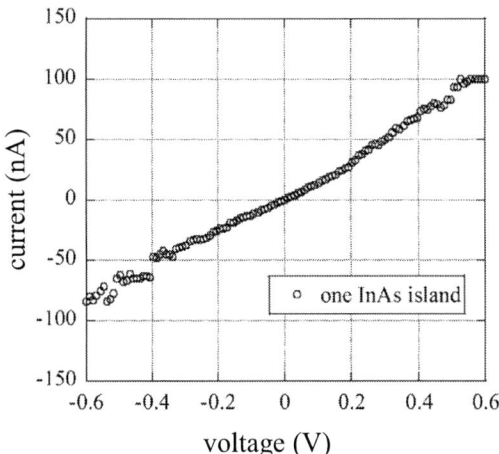

Fig. 6. The I-V ohmic profile obtained through the electrode with applying voltage from -0.5 V to 0.5 V by putting cantilever on the InAs surface after annealing at 300 °C

through the electrode for plural InAs islands exhibited, a clear ohmic I-V characteristic as shown in Fig. 7. We restricted the current value within ±500 mA in order to avoid destruction of the structure. That is why the plot is limited to voltage range from –0.3 V to 0.27 V. The resistance value was 0.635 Ω as estimated from the slope. We consider that the ohmic I-V characteristic was obtained because native oxide layer on the surface of InAs and on the back side of Si was removed after annealing the annealing. When this resistance value is multiplied by the number of InAs islands under an electrode (1665), we would obtain the resistance for a single InAs island, 1.06×10^3 Ω. This value is much smaller compared with the value obtained with the conductive AFM measurement (approximately 7.4×10^6 Ω) for a single InAs island on Si. This is because the contact resistance between the cantilever and metal (Au) was so large due to a small contact area and low pressure of the cantilever.

We needed to confirm whether the resistance value of the I-V profile in Fig. 7 is reliable. As for the annealing temperature, 300 °C annealing led to larger current than 250 °C, as shown in Fig. 8, indicating that 300 °C annealing is necessary for obtaining enough ohmic contact at the electrodes. We also measured the I-V characteristics for the large (0.5 mm) electrode. The current through the large electrode was 2.7 times larger than through the small (0.3 mm) electrode at a given voltage, as expected from the area ratio of the electrodes.

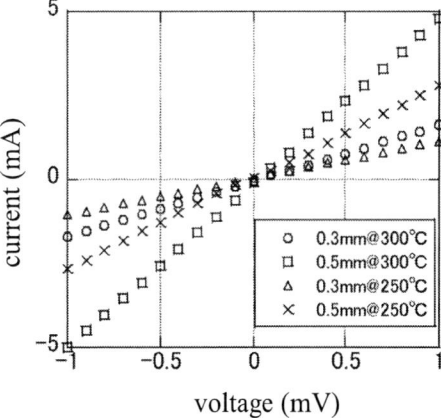

Fig. 8. The I-V characteristics through the large (0.5 mm) and the small (0.3 mm) electrode after annealing at 300 °C and 250 °C

The value with 300 °C annealing was compared with the value for the patterned Si substrate without the growth of InAs, as shown in Fig. 9. As for InAs on Si and pattern Si, the resistance value was 0.59 Ω and 0.337 Ω, respectively. As the result, the resistance for the 1665 InAs islands beneath one electrode was estimated to be 0.253 Ω. The resistivity of InAs was deduced to be 0.0662 Ω·cm, which corresponds to a carrier concentration of approximately 2×10^{16} cm^{-3}. The value is reasonable as the intrinsic carrier concentration of un-doped InAs.

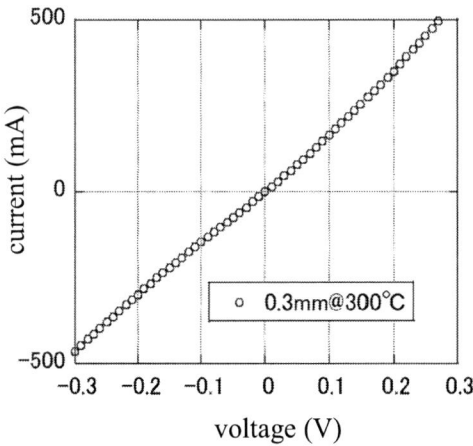

Fig. 7. The I-V ohmic characteristic through the small (0.3 mm) electrode with applying voltage from -0.3 V to 0.27 V after annealing at 300 °C

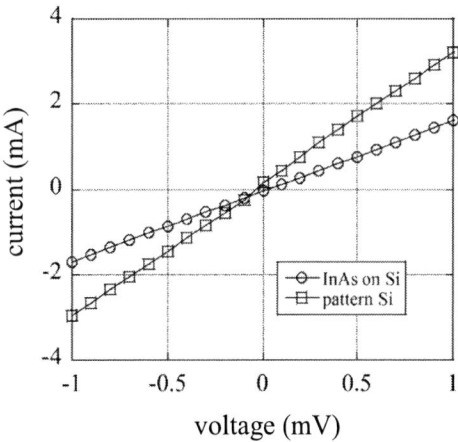

Fig. 9. The I-V characteristics of InAs on Si and pattern Si without growth of InAs through the small (0.3 mm) electrode after annealing at 300 °C

IV. CONCLUSIONS

The vertical current conduction at the InAs/Si interface, which was obtained by the selective-area heteroepitaxial growth of InAs on n-type Si substrate, was investigated. The electrical conduction property at n-InAs/n-Si interface was ohmic, and the contact between Ti and InAs needs nitrogen annealing to be ohmic. The ohmic profile would probably be due to tunneling electron transport across the interface between InAs and Si substrate. The value of current at a given voltage was almost proportional to the area of the electrode, indicating that all the InAs islands under the electrode were equally conductive. Moreover the resistivity of InAs by heteroepitaxial growth was estimated to be $0.0662\ \Omega \cdot cm$, which is an adequate value for an un-doped InAs layer.

ACKNOWLEDGEMENT

The authors would like to thank Dr. Takuji Takahashi of Institute of Industrial Science, The University of Tokyo for assistance with use of conductive AFM.

REFERENCES

[1] Y. Kondo, M. Deura, Y. Terada, T. Hoshii, M. Takenaka, S. Takagi, Y. Nakano, and M. Sugiyama, "Initial growth of InAs on P-terminated Si(111) surfaces to promote uniform lateral growth of InGaAs micro-discs on patterned Si," *J. Cryst. Growth.* **312** (2010) 1348-1352

[2] T. Hoshii, M. Deura, M. Sugiyama, R. Nakase, S. Sugahara, M. Takenaka, Y. Nakano, and S. Takagi, "Epitaxial lateral overgrowth of InGaAs on SiO2 from (111) Si micro channel areas," *Phys. Status. Solidi (C)*, **5** (2008) 2733-2735

[3] C. Lee, K. Jaw, and C. Tsai, "Thermal stability of Ti/Pt/Au ohmic contacts on InAs/graded InGaAs layers," *Solid State Electron*, **42** (1998) 871-875

Interface and Surface Dielectric Anisotropies of GaP/Si(100)

O. Supplie[1], T. Hannappel[1,2], M. Pristovsek[3], and H. Döscher[1,2]

[1] *Helmholtz-Zentrum Berlin für Materialien und Energie, Institut für Solare Brennstoffe und*
Energiespeichermaterialien, Hahn-Meitner-Platz 1, 14109 Berlin, Germany
[2] *Technische Universität Ilmenau, Institut für Physik, Postfach 10 05 65, 98684 Ilmenau, Germany*
[3] *Technische Universität Berlin, Institut für Festkörperphysik EW6-1, Hardenbergstraße 36, 10623 Berlin, Germany*

Abstract — **In situ control of the formation of the III-V/Si(100) heterointerface is crucial for high-efficiency opto-electronic devices. We analyzed in situ reflectance anisotropy spectra of pseudomorphical GaP/Si(100) films which deviate from the RA spectra of well-known GaP/Si(100). We derived an equation [2] to separate different contributions, in particular surface and interface contributions. The resulting GaP/Si(100) surface dielectric anisotropy allows for a precise numerical in situ quantification of the APD content and agrees with the GaP(100) surface anisotropy. Deviations between the homo- and heteroepitaxial RA spectra can be attributed to an anisotropic GaP/Si(100) interface which we addressed in situ using RAS.**

I. INTRODUCTION

The small lattice mismatch between GaP and silicon allows pseudomorphic growth of significantly thick GaP films on silicon substrates. GaP/Si(100) is both a promising model system in order to study the polar on non-polar heteroepitaxy in situ during metalorganic vapor phase epitaxy (MOVPE). Reflection anisotropy spectroscopy (RAS) probes the normalized difference $\Delta r/r$ of light reflected along two perpendicular surface directions under normal incidence [3]. For cubic symmetry, optical anisotropy induced by surface reconstructions often determines the features of the RA spectra, e.g. for the P-rich, (2x2)/c(4x2) reconstructed GaP(100) surface characterized by H-stabilized, buckled P-dimers [4,5].

During the nucleation phase, anti-phase boundaries (APBs), defined by undesired homopolar bonds, emerge at the edges of single-layer steps at the Si(100) surface [6]. Typically APBs propagate in growth direction separating incongruent anti-phase domains (APDs) in subsequently grown layers. Due to the tetrahedral coordination, mutually perpendicular orientated P-dimers reflect APDs at the GaP/Si(100) surface. Lateral integration over the spot size therefore leads to a reduced RAS signal. Consequently, linear scaling the RA spectra to that of a single domain reference allows for in situ APD quantification [8]. In addition, pseudomorphically grown GaP films are sufficiently transparent to allow internal reflection at the buried heterointerface. Convolution with a so-called relative reflectance spectrum corrects interference effects via the normalization semi-empirically [1], but residual deviations suggest an additional anisotropy.

In the following, we apply an optical model to calculate dielectric surface and interface anisotropies from measured complex RAS signals of six heteroepitaxial samples differing in the GaP film thickness [2].

II. EXPERIMENTAL

GaP/Si(100) samples with varying GaP films have been analyzed. To ensure comparability, all samples were subsequently grown with equal process conditions in a commercial AIX-200 MOVPE reactor. Exactly oriented GaP(100) substrates were used as homoepitaxial reference. For the heteroepitaxy, 0.1° towards [011] off-oriented Si(100) substrates were wet-chemically cleaned, thermally deoxidized and overgrown with a Si buffer using silane as precursor. After a pulsed GaP nucleation, the heteroepitaxial GaP films were grown with tertiarybutylphosphine (TBP) and triethylgallium (TEGa) in H_2 ambient. Finally, the P-rich surface was prepared by cooling under TBP supply and annealing without TBP to desorb excess phosphorous. A LayTec EpiRAS spectrometer was used to monitor the whole MOVPE process. The spectra used for calculation were taken at 50°C. Both the imaginary and the real part have been measured but the Kramers-Kronig imaginary part was used for the calculations.

III. RESULTS AND DISCUSSION

Heteroepitaxial RAS signals deviate both in lineshape and amplitude from the GaP(100) reference and the deviations depend on the thickness of the GaP film, as shown in Fig. 1 for RA spectra of the P-rich surface of both a 14nm and a 26nm thick GaP film on Si(100) in comparison to GaP(100).

In general, RA spectra $\Delta r/r$ measured at a thin film structure contain signal contributions from different sources of optical anisotropy, including those from the surface, the interface and the bulk of the film. Based on earlier work of Yasuda [8] and Hunderi et al. [9], we recently derived an equation to decompose the measured RA spectra. If bulk effects like e.g. ordering, uniaxial strain or doping can be neglected, then the RAS signal may be expressed as superposition of weighted dielectric surface and interface anisotropies (SDA / IDA).

978-1-4673-1725-2/12 $31.00 © 2012 IEEE

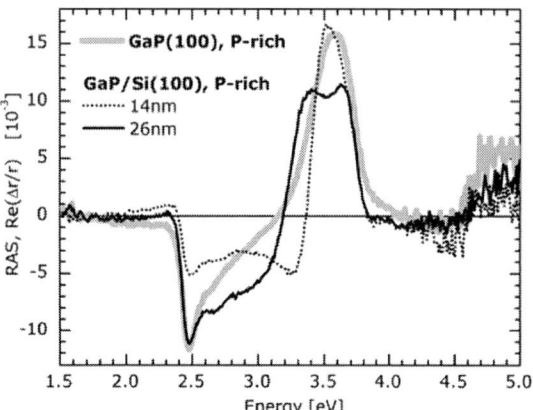

Fig. 1. RA spectra of GaP/Si(100) compared to GaP(100).

Except for both SDA and IDA being independent of GaP film thickness, no atomistic assumptions have to be made. The weighting factors may be calculated from bulk optical properties if the film thickness is known. In consequence, both SDA and IDA can be calculated from two measured complex RA spectra solving a system of equations without any additional fitting. Introducing a scaling factor in the SDA, we obtain the APD content at the GaP/Si(100) surface numerically.

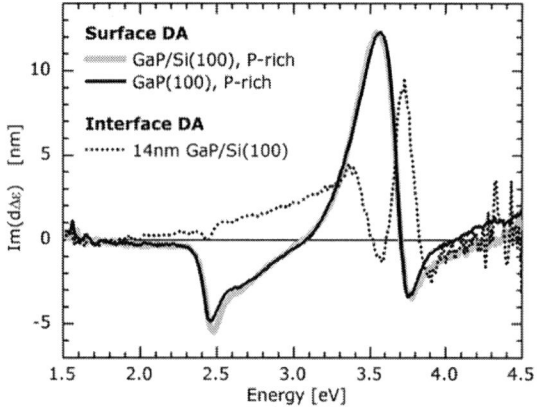

Fig. 2. SDA spectra of GaP/Si(100) (arithmetic mean) compared to the GaP(100) reference and the IDA spectrum of 14nm GaP/Si(100).

Fig. 2 shows the calculated SDA from a set of six GaP/Si(100) samples (arithmetic mean of all possible 15 combinations, black line). The plotted imaginary part of the DA relates to the real part of the RA spectrum [15]. Despite a small redshift of the GaP/Si(100) SDA, both lineshape and amplitude agree well with the homoepitaxial GaP(100)

reference (grey line). This good agreement in the SDA motivates the use of the homoepitaxial reference SDA as additional input parameter for the calculation of the IDA. With scaling factors used for the SDA, the IDA is calculated for each GaP/Si(100) sample individually and the result for the 14nm thick GaP film on Si(100) is shown in Fig. 2 (dotted line). The IDA spectra of the other GaP films differ slightly but are a qualitatively similar.

IV. CONCLUSION

RAS monitoring during GaP/Si(100) heteroepitaxy enables in situ access to the dielectric anisotropy of the heterointerface. Considering anti-phase disorder within optical model calculations, a heteroepitaxial SDA can be extracted from the measured data and agrees well with that of the GaP(100) reference. Doing so, the APD content at the GaP/Si(100) surface can be obtained in situ. We calculated a GaP/Si(100) IDA that can explain deviations remaining after semi-empirical correction. Future work is directed to understand the physical origin of the IDA. Precise RAS measurements, in particular of the imaginary part, and complementary results from other interface characterization methods are required.

REFERENCES

[1] H. Döscher and T. Hannappel, "In situ reflection anisotropy spectroscopy analysis of heteroepitaxial GaP films grown on Si(100)," *Journal of Applied Physics*, vol. 107, p.123523, 2010.

[2] O.Supplie, T. Hannappel, M. Pristovsek, and H. Döscher, "In situ access to the dielectric anisotropy of buried III-V/Si(100) heterointerfaces," *Physical Review B*, vol. 86, p. 035308, 2012.

[3] P. Weightman, D.S. Martin, R.J. Cole, and T. Farrell, "Reflection anisotropy spectroscopy," *Reports of Progress in Physics*, vol. 68, p.1251, 2005.

[4] P.H. Hahn, W.G. Schmidt, F. Bechstedt, O. Pulic, and R. Del Sole, "P-rich GaP(001) (2x1)/(2x2) surface: A hydrogen-adsorbate structure determined from first-principle calculations," *Physical Review B*, vol. 68, p.033311, 2003.

[5] L. Töben, T. Hannappel, K. Möller, H. Crawack, C. Pettenkofer, F. Willig,, "RDS, LEED and STM of the P-rich and Ga-rich surfaces of GaP(100)," *Surface Science*, vol. 494, p.L755, 2001.

[6] H. Kroemer, "Polar-on-nonpolar epitaxy," *Journal of Crystal Growth*, vol. 81, p.193, 1987.

[7] H. Döscher, T. Hannappel, B. Kunert, A. Beyer, K. Volz, and W. Stolz, "In situ verification of single-domain III-V on Si(100) growth via metal-organic vapor phase epitaxy," *Applied Physics Letters*, vol. 93, p.172110, 2008.

[8] T. Yasuda, "Interface, surface and bulk anisotropies of heterostructures," *Thin Solid Films*, vol. 313, p.544, 1998.

[9] O. Hunderi, J.-T. Zettler, and K. Haberland, "On the AlAs/GaAs (001) interface dielectric anisotropy," *Thin Solid Films*, vol. 472, p.261, 2005.

[10] J.D. McIntyre and D.E. Aspnes, "Differential reflection spectroscopy of very thin surface films," *Surface Science*, vol. 24, p.417, 1971.

Metal-Clad Photonic Crystal Membrane Nanolasers

Joshua D. Sulkin[1] and Kent D. Choquette*

Department of Electrical and Computer Engineering
University of Illinois, Urbana, IL 61820
*choquett@illinois.edu
[1]now with Space Exploration Technologies, Hawthorne, CA 90250

Abstract — **We report laterally injected photonic crystal membrane light emitters that are clad on one surface by metal. Light emission from an electrically injected photonic crystal defect cavity is reported. Carriers are injected via a lateral diode that is formed in an InGaAsP membrane using ion implantation. The membrane is removed from its native InP substrate and placed on a thin dielectric layer above Au. The Au-coated substrate provides a large index contrast and good thermal conduction. A cavity resonance in the DC electroluminescence and photopumped laser operation are both observed at room temperature.**

Index Terms — **photonic crystal, semiconductor membrane emitter, nanolasers.**

I. INTRODUCTION

Photonic crystal membrane lasers offer many features that make them good candidates for future on-chip optical communication applications. These include small mode volume [1], high quality (Q) factor [2], low threshold power, and a lithographically tunable wavelength [3]. Recent results have shown that room temperature, continuous wave electrical injection can be achieved in photonic crystal membrane lasers using an ion-implanted lateral diode [4]. The main advantage of this lateral current injection (LCI) design [5] is that it eliminates the need for a central post under the cavity, which would otherwise decrease the Q-factor [6]. Moreover, it has also been demonstrated that metal confinement can significantly reduce the modal volume of a nanolaser [7].

A suspended membrane design is used in [6], which offers the highest Q-factor. However, a suspended membrane also suffers from poor heat dissipation and may be difficult to integrate onto a foreign substrate. Bonding a photonic crystal membrane to a high thermal conductivity substrate has been shown to greatly improve heat dissipation, though previous results have all been for optically pumped lasers [8],[9].

Herein we present results of LCI photonic crystal devices in a semiconductor membrane clad by Au bonded to a foreign substrate. We consider both photonic crystal defect cavities as well as photonic crystal line defect heterostructures which have been shown to exhibit significant in-plane emission [10], important for coupling laser emission into an on-chip waveguide. Our metal-clad devices lase under photopumped operation and exhibit a cavity mode under DC electrical injection at room temperature. The room temperature photopumped performance is encouraging for nanolaser diodes.

II. DESIGN AND FABRICATION

An illustration of a bonded LCI photonic crystal device is shown in Fig. 1. The active material is a quantum well InGaAsP membrane with a peak photoluminescence wavelength around 1350 nm. Fabrication begins by defining doped *n*-type and *p*-type regions of the lateral diode using ion implantation. Si is used for the *n*-type region, while a co-implantation of Be and P are used for the *p*-type region. More details on the implantation and metallization processes can be found in [5],[11].

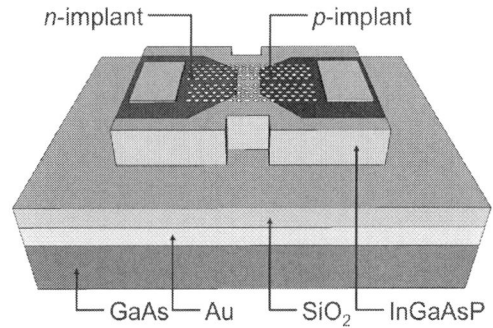

Fig. 1: Illustration of an ion-implanted LCI photonic crystal membrane bonded to SiO_2 with a Au layer separating the oxide from the GaAs substrate.

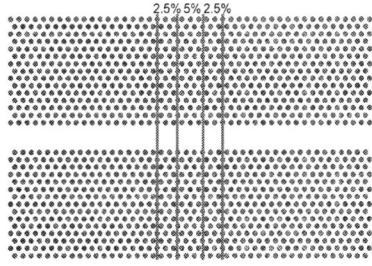

Fig. 2: Sketch of a LDH photonic crystal cavity with three hole rows removed. The percentages indicate the amount the pitch is displaced in order to form a cavity.

978-1-4673-1725-2/12 $31.00 © 2012 IEEE

The photonic crystal pattern is defined by electron beam lithography and an isolation mesa is defined by photolithography. Both are transferred into the membrane via an inductively-coupled plasma reactive-ion etch. Several different photonic crystal designs are used, including the line defect heterostructure [10] and D4 cavities [3],[6]. For the line defect heterostructure, three rows of holes are removed, as depicted in Fig. 2. Removing several rows of holes increases the amount of semiconductor area, which reduces series resistance and improves heat dissipation, though at the cost of a lower Q-factor. A cavity is formed in the direction along the missing rows by varying the hole pitch in the middle of the device [10].

Trenches are etched into the device mesa around the junction to eliminate leakage current paths parallel to the photonic crystal cavity. Previous work on LCI membrane lasers [5] has demonstrated that current confinement trenches are important to reduce leakage current via parallel current paths around the cavity and increase current density in the photonic crystal cavity. An obvious solution to this problem would be to etch a trench that extends all the way from the edge of the photonic crystal to the edge of the membrane, thus isolating the *n* and *p* regions, except through the photonic crystal. Unfortunately, such a trench makes the membrane much more likely to crack or cleave during the bonding process.

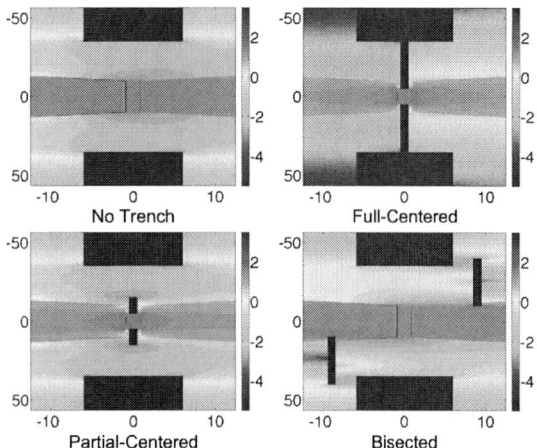

Fig. 3: Simulated current density magnitude (kA/cm² on a log scale) in membranes with various trench designs. Values on the horizontal and vertical axes are distances in μm.

We have developed a two-dimensional finite-element-model Ohm's law simulation to model resistance and current flow in the membrane using various trench designs. Fig. 3 shows the calculated magnitude of the current density for a membrane with no trenches, trenches extended from the mesa edge to cavity (full centered), partial trenches near the cavity (partial centered) and trenches offset laterally

from the junction (bisected). The conductivities used in the model were determined from experimental transmission line measurements. The membrane with no trench and the bisected trench have the lowest center current density, while the full centered trenches creates the highest current density through the photonic crystal cavity. The partial-centered design exhibits the best tradeoff between high current density in the cavity with a relatively robust membrane mechanical structure.

Once diode and photonic crystal fabrication steps are complete, the sample is coated in Apiezon W wax [12] and the InP substrate is removed using a HCl:DI etch. The wax-embedded mesas are pressed against the foreign substrate. Finally the wax is removed using methylene chloride, leaving the mesas bonded to the new substrate via Van der Waals forces [12]. More details of the bonding procedure can be found in [11].

The foreign substrate used here is GaAs coated in Au and SiO_2. Many prior examples of bonded photonic crystal membranes use a sapphire substrate because of its high thermal conductivity [8]. However, sapphire also has a high refractive index, which reduces the Q-factor. SiO_2 has both a lower refractive index and thermal conductivity [9]. For our metal-clad membranes, Van der Waals bonding between two Au layers is used. The InGaAsP membrane is sometimes separated from the Au layer by a combination of 475 nm of Al_2O_3 and SiO_2 layer or the Au is directly in contact with the membrane. The Au layer is used to prevent coupling of the optical mode into the higher index GaAs substrate, which allows for a thinner SiO_2 layer, thus improving the thermal properties while maintaining a large index contrast.

III. CHARACTERIZATION AND RESULTS

Characterization of the devices is performed at room temperature on the stage of a micro-photoluminescence system. Electrical probes are brought in contact to provide a DC current, and the emitted light is collected with an objective lens positioned over the sample. All of the photonic crystal heterostructure and defect cavities are found to lase under pulsed optical pumping, and many of the metal-clad membranes lase under continuous wave optical pumping.

Electrically injected lasing has not been observed; however, simultaneous electrical and optical pumping demonstrates that electrical injection does lead to radiative recombination in the photonic crystal cavity and optical amplification is obtained [13]. Electroluminescence showing the cavity resonance is observed for line defect devices bonded to metal-backed SiO_2 and is shown in Fig. 4. At 1.0 mA and higher, a small peak from a cavity mode (circled) can be seen above the quantum well spontaneous emission. Similar devices bonded to sapphire substrates do

978-1-4673-1725-2/12 $31.00 © 2012 IEEE   140

not show evidence of a cavity mode. Devices on both substrate types lase continuous wave optically pumped, though no evidence of electrically injected lasing is observed.

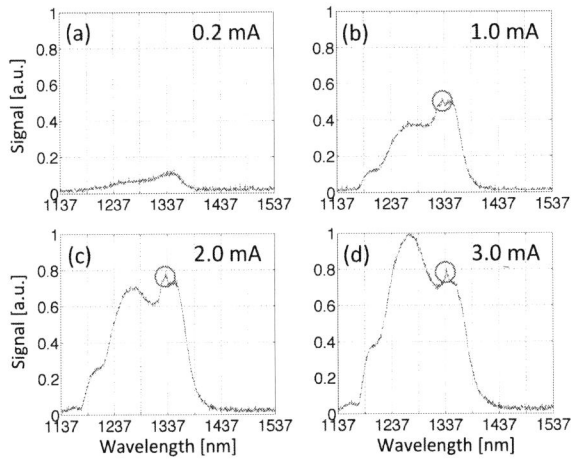

Fig. 4: Spectra from line defect heterostructure with DC electrical injection at room temperature, with currents of (a) 0.2 mA, (b) 1.0 mA, (c) 2.0 mA, and (d) 3.0 mA. The cavity optical resonance peak is circled.

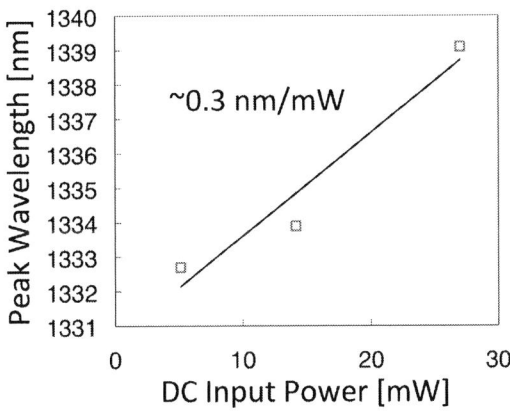

Fig. 5: Peak cavity mode emission wavelength versus DC input power.

Fig. 5 shows the peak wavelength of the line defect cavity spectra at various levels of input power (current x voltage). The slope of this line can be used to calculate the thermal impedance of the laser using [8]:

$$Z_{th} = \frac{\Delta T}{\Delta P} = \frac{\Delta \lambda}{\Delta P} \cdot \frac{\Delta T}{\Delta \lambda} \quad (1)$$

where ΔT is the change in temperature, $\Delta \lambda$ is the change in wavelength, and ΔP is the change in input power. Assuming $\Delta \lambda / \Delta T$ is equal to 0.05 nm/K [8], the thermal impedance is

6.0 K/mW. This is lower than the thermal impedance of 11.0 K/mW reported for an optically pumped photonic crystal laser bonded to a thick SiO_2 layer on Si [9]. In that case, a thick SiO_2 layer was used to isolate the photonic crystal mode from the high index substrate, rather than a Au layer.

Though electrically injected lasing was not achieved, a cavity mode was observed under DC electrical injection. The results in Fig. 4 show that a significant fraction of the injected carriers recombine in the doped regions, as indicated by the large peak that is blue-shifted from the cavity resonance. The peak is blue shifted because quantum well intermixing causes the bandgap to increase in the doped regions [14]. Photopumped experiments have been conducted to test whether a photonic crystal membrane device can still lase when in direct contact with a metal layer. Fig. 6 shows the continuous wave optically pumped lasing spectrum of a D4 cavity directly bonded to a Au-coated substrate. This device has a higher threshold power compared to devices bonded to SiO_2 or sapphire [11], as is expected due to the increased optical loss. Electrically injected lasing may be achieved if carrier confinement is improved, perhaps by increasing the amount of intermixing or by using a buried heterostructure [14].

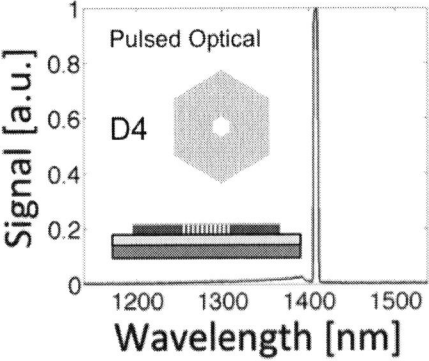

Fig. 6: Metal-clad D4 photonic crystal defect lasing continuous wave via photo-pumping at room temperature.

IV. CONCLUSIONS

In order for photonic crystal membrane lasers to be used in future on-chip optical interconnect applications, it will be necessary to integrate them onto a foreign substrate such as Si. We have reported an electrically injected photonic crystal light-emitting device bonded to a foreign substrate. This was made possible by a LCI diode defined by ion implantation and a substrate composed of a thin SiO_2 layer on top of a Au isolation layer, which allowed for a large index contrast and low thermal impedance. The metal-clad membrane nanolaser leads to both a reduced modal volume and an efficient means for current injection for photonic crystal diode laser diodes.

Acknowledgments: This work was supported by the National Science Foundation (award no. ECCS 07-25515).

REFERENCES

[1] K. Nozaki, T. Ide, J. Hashimoto, T. Baba, T., and W. H. Zheng, "Photonic crystal point-shift nanolaser with ultimate small modal volume," *Electron. Lett.* **41**, pp. 843-845 (2005).

[2] M. H. Shih, W. Kuang, A. Mock, M. Bagheri, E. H. Hwang, J. D. O'Brien, and P. D. Dapkus, "High-quality-factor photonic crystal heterostructure laser," *Appl. Phys. Lett.* **89**, pp. 101-104 (2006).

[3] M. Loncar, T. Yoshie, A. Scherer, P. Gogna, and Y. Qiu, "Low-threshold photonic crystal laser," *Appl. Phys. Lett.* **81**, pp. 2680-2682 (2002).

[4] S. Matsuo, K. Takeda, T. Sato, M. Notomi, A. Shinya, K. Nozaki, H. Taniyama, K. Hasebe, and T. Kakitsuka, "Room-temperature continuous-wave operation of lateral current injection wavelength-scale embedded active-region photonic-crystal laser," *Opt. Exp.* **20**, pp. 3773-3780 (2012).

[5] C. M. Long, A. V. Giannopoulos, and K. D. Choquette, "Lateral current injection photonic crystal membrane light emitting diodes," *J. Vac. Sci. Tech. B* **28**, pp. 359-364 (2010).

[6] H. G. Park, S. H. Kim, M. K. Seo, Y. G. Ju, S. B. Kim, and Y. H. Lee, "Characteristics of electrically driven two-dimensional photonic crystal lasers," *IEEE J. Quantum Elect.* **41**, pp. 1131-1141 (2005).

[7] M. T. Hill, Y. S. Oei, B. Smalbrugge, Y. Zhu, T. J. De Vries, P. J. Van Veldhoven, F. W. M. Van Otten, T. J. Eijkemans, J. P. Turkiewicz, H. De Waardt, E. J. Geluk, S. H. Kwon, Y. H. Lee, R. Totzel, and M. K. Smit, "Lasing in Metallic-Coated Nanocavities," *Nature Photonics* **1**, pp. 589-594 (2007).

[8] M. H. Shih, Y. C. Yang, Y. C. Liu, Z. C. Chang, K. S. Hsu, and M. C. Wu, "Room temperature continuous wave operation and characterization of photonic crystal nanolaser on a sapphire substrate," *J. Phys. D* **42**, pp. 1-5 (2009).

[9] M. H. Shih, A. Mock, M. Bagheri, N. K. Suh, S. Farrell, S. J. Choi, J. D. O'Brien, and P. D. Dapkus, "Photonic crystal lasers in InGaAsP on SiO_2/Si substrates and its thermal impedance," *Opt. Exp.* **15**, pp. 227-232 (2007).

[10] A. Mock and J. D. O'Brien, "Strategies for reducing the out-of-plane radiation in photonic crystal heterostructure microcavities for continuous wave laser applications," *J. Lightwave Tech.* **28**, pp. 1042-1050 (2010).

[11] J. D. Sulkin and K. D. Choquette, "Hybrid Integration of Photonic Crystal Membrane Lasers via Post-Process Bonding," *IEEE Photonics Journal* **3**, pp. 375-380 (2011).

[12] E. Yablonovitch, D. M. Hwang, T. J. Gimitter, L. T. Florez, and J. P. Harbison, "Van der Waals bonding of GaAs epitaxial liftoff films onto arbitrary substrates," *Appl. Phys. Lett.* **56**, pp. 2419-2421 (1990).

[13] J. D. Sulkin and K. D. Choquette, "Lateral Current Injection Photonic Crystal Membrane Emitters Bonded to Sapphire," IEEE Photonics Conference, Arlington, VA (Oct. 2011).

[14] S. Charbonneau, E. S. Koteles, P. J. Poole, J. J. He, G. C. Aers, J. Haysom, M. Buchanan, Y. Feng, A. Delage, F. Yang, M. Davies, R. D. Goldberg, P. G. Piva , and I. V. Mitchell, "Photonic integrated circuits fabricated using ion implantation," *IEEE J. Sel. Topics in Quantum Electron.* **4**, pp. 772-793 (1998).

10-Gbit/s Direct Modulation of Optically Pumped InGaAlAs Multiple-Quantum-Well Photonic-Crystal Nanocavity Laser up to 100°C

Tomonari Sato,[1,3]* Koji Takeda,[1,3] Hiromitsu Imai,[1] Akihiko Shinya,[2,3] Kengo Nozaki,[2,3] Hideaki Taniyama,[2,3] Koichi Hasebe,[1,3] Wataru Kobayashi,[1] Takaaki Kakitsuka,[1,3] Masaya Notomi,[2,3] and Shinji Matsuo[1,3]

[1]NTT Photonics Labs., [2]NTT Basic Research Labs., [3]Nanophotonics Center, NTT Corporation
3-1 Morinosato-Wakamiya, Atsugi, Kanagawa 243-0198, Japan
*sato.tomonari@lab.ntt.co.jp

Abstract — **We demonstrate the 10-Gbit/s direct modulation of optically pumped photonic-crystal (PhC) nanocavity lasers at up to 100°C by using an InGaAlAs multiple-quantum-well (MQW) structure as an active region. The device exhibits an output power exceeding 20 μW and a maximum 3-dB bandwidth of 8.3 GHz at an operating temperature of 100°C. In addition, when the laser is modulated at 10 Gbit/s, low energy costs of 7.2 fJ/bit at 25°C and 23.4 fJ/bit at 100°C are successfully achieved. Thus an InGaAlAs MQW PhC nanocavity laser is a promising candidate for the light source of a photonic network on a CMOS chip.**

Index Terms — **Semiconductor lasers, Photonic crystals, Photonic integrated circuits.**

I. INTRODUCTION

A photonic network on a CMOS chip has attracted a lot of attention with a view to increasing transmission capacity, which is limited by the power consumption of electrical interconnects. The optical devices used for the photonic network must operate at an energy cost of less than 10 fJ/bit [1]. In this context, we have developed lambda-scale embedded active-region PhC (LEAP) lasers with an InGaAsP-based MQW active region in an InP-PhC slab [2]. This enables us to reduce the thermal resistance and to increase the carrier confinement of the laser. The LEAP laser has exhibited 20-Gbit/s modulation with an energy cost of 8.76 fJ/bit at room temperature (RT) with optical pumping [3]. Recently, we have also demonstrated an electrically driven LEAP laser that operates under an RT continuous-wave (CW) condition [4]. However, important issues remain including the need for high-speed direct modulation at high temperature, which is required because the temperature of CMOS chips reach 80°C [5]. Although we have already demonstrated CW operation at temperatures up to 90°C for an optically pumped device, the 3-dB bandwidth was limited to 7.0 GHz at 66°C [6]. Furthermore, the active region of an electrically driven device requires higher temperature characteristics than that of an optically pumped device because the active region temperature increases owing to the resistance of p-type and n-type doping layers. To overcome this, here we use an InGaAlAs-based MQW as the active region of an optically pumped LEAP laser. An InGaAlAs MQW can suppress the electron overflow from

quantum wells at high temperature, and it has a large differential gain for high-speed modulation, because its conduction-band offset is larger than that of an InGaAsP-based MQW [7]. However, it is difficult to employ an InGaAlAs MQW as an embedded active region owing to the easy oxidation of the material containing Al. Furthermore, since the width of the active region of the LEAP laser (~0.3 μm) is much smaller than that of a conventional laser, the condition of the interface between an InGaAlAs-based active region and an InP regrown layer will affect the laser characteristics. In this paper, we report the characteristics of optically pumped LEAP lasers with an InGaAlAs MQW active region and demonstrate 10-Gbit/s direct modulation up to 100°C.

II. DEVICE STRUCTURE AND FABRICATION

Figure 1 shows scanning electron microscopy (SEM) images of a fabricated LEAP laser. We use an active region that consists of three 6-nm-thick compressively strained InGaAlAs quantum wells and four 10-nm-thick tensile-strained InGaAlAs barriers with a band-gap wavelength of 1.1 μm. The photoluminescence peak wavelength of the MQW was 1.55 μm. The active region was defined with an SiN_x mask using

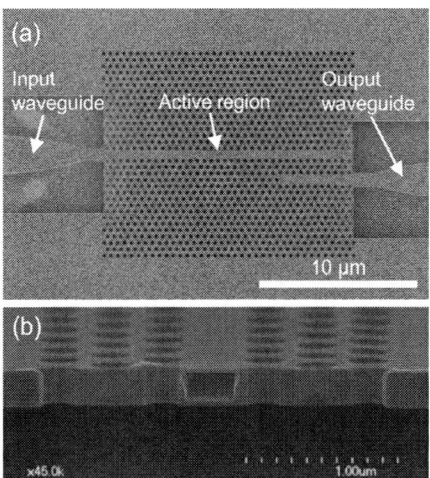

Fig. 1.   SEM image of a fabricated LEAP laser: (a) top view (b) cross-sectional view.

978-1-4673-1725-2/12 $31.00 © 2012 IEEE

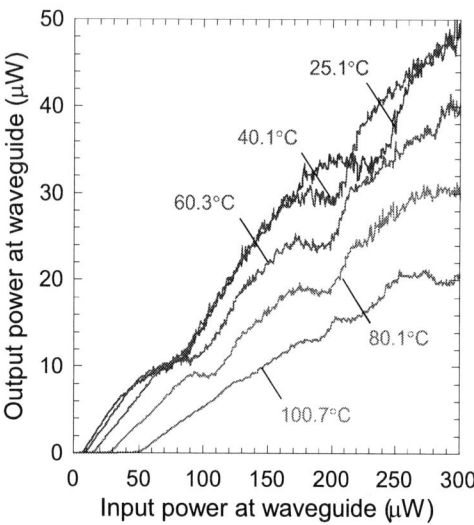

Fig. 2. LL characteristics of the LEAP laser at various temperatures.

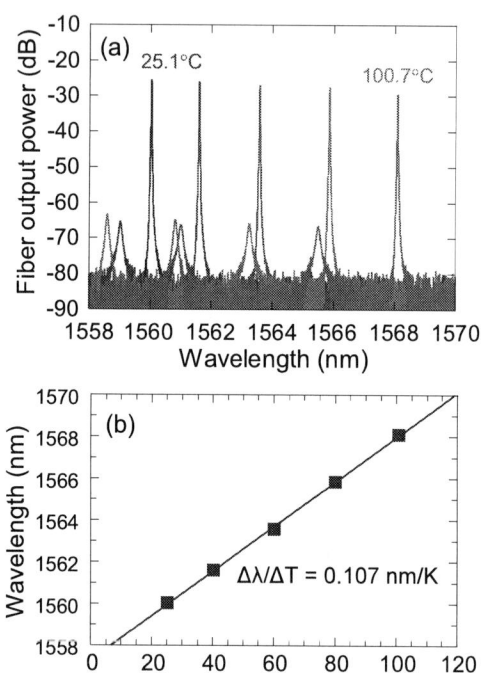

Fig. 3. (a) Lasing spectra of the LEAP laser. (b) Temperature dependence of the lasing wavelength.

electron-beam lithography and then etched with inductively coupled plasma reactive ion etching (ICP-RIE) and selective wet chemical etching techniques. An InP layer was then selectively regrown by using metalorganic vapor phase epitaxy. After removing the SiN_x mask, we grew an InP layer to obtain a flat surface. Air holes were etched with ICP-RIE and an air-bridge structure was fabricated by the selective wet chemical etching of an InGaAs sacrificial layer through the air holes. By using this process, an active region with a volume of $5.2 \times 0.3 \times 0.15$ μm^3 was embedded within a line defect waveguide in an InP-PhC slab. The diameter of the air holes and the lattice constant, which means the distance between the air holes of the PhC, were 200 and 438 nm, respectively.

III. DEVICE CHARACTERISTICS

A. Static Characteristics

First, we measured the light-in/light-out (LL) characteristics and the lasing spectra of the LEAP laser at various temperatures ranging from 25.1 to 100.7°C under CW operation as shown in Figs. 2 and 3, respectively. A 1.06-μm DFB laser was employed for optical pumping. The pump light was injected into the input waveguide from the fiber by using a collimator lens, a 3-μm wide waveguide and a tapered waveguide, as shown in Fig. 1 (a). The total coupling loss from the fiber to the line-defect input waveguide was 10 dB. The facet of the output waveguide was coated with an anti-reflection film whereas that of the input waveguide was as cleaved. Thus, the laser output light from the output line-defect waveguide was collected into a single-mode fiber with an 8.5-dB coupling loss. The sample was clamped on a stage with a Peltier temperature controller. All the powers reported in this paper are the optical powers at the input or output waveguide. As shown in Fig. 2, the threshold input power of

the LEAP laser at 25.1°C was 7.3 μW, and that at 100.7°C was 54.0 μW. We obtained an output power of over 20 μW even at a high temperature of 100°C. These results indicate that we could obtain good characteristics for the LEAP laser with the InGaAlAs-based MQW active region, which is difficult to employ as an embedded active region. Figure 3 (a) shows the lasing spectra of the LEAP laser operating at an input power of -7.8 dBm. The lasing wavelengths at 25.1 and 100.7°C were 1560.036 and 1568.088 nm, respectively. Figure 3 (b) shows the temperature dependence of the lasing wavelength for the LEAP laser. The wavelength was exactly proportional to the operating temperature at a slope of 0.107 nm/K, which is almost the same as that of the InGaAsP MQW LEAP laser [6]. Although the maximum operating temperature

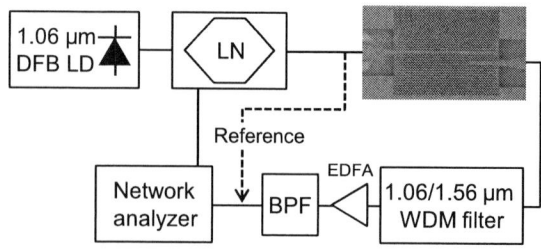

Fig. 4. The measurement setup for the frequency responses of the LEAP laser.

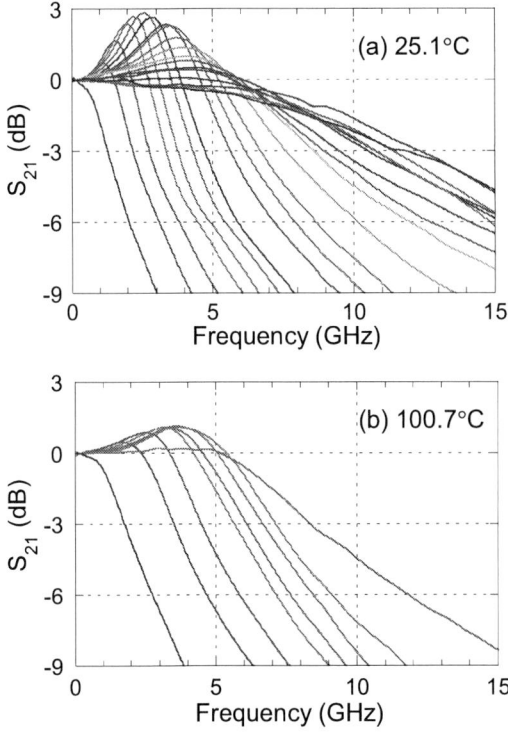

Fig. 5. Small signal responses of the LEAP laser at (a) 25.1°C and (b) 100.7°C. The maximum 3-dB bandwidths are 12.3 and 8.3 GHz at 25.1°C and 100.7°C, respectively.

was limited to 100°C because of our measurement system, we believe that the fabricated LEAP laser can be operated above 100°C. To the best of our knowledge, this is the first demonstration of a PhC nanocavity laser that can operate above 100°C in a CW condition.

B. Dynamic Characteristics

Next, we measured the small-signal response of the LEAP laser at 25.1 and 100.7°C. The measurement setup is shown in Fig. 4. To measure the responses, the 1.06-μm pump light was modulated with a LiNbO₃ (LN) modulator using a network analyzer. The pump light was filtered from the output light using a 1.06/1.56-μm WDM filter. The lasing light was amplified through an erbium doped fiber amplifier (EDFA) and detected by the network analyzer. We subtracted the frequency response of the LN modulator from the measured data, which contained the responses from both the modulator and the device, to extract the responses of the LEAP laser. Figure 5 (a) and (b) show the results at 25.1 and 100.7°C, respectively. The input power was changed from -22.5 to -6.6 dBm at 25.1°C, and -12.1 to -7.2 dBm at 100.7°C. The 3-dB bandwidth increased with increasing input power, and we obtained maximum 3-dB bandwidths of 12.3 and 8.3 GHz at

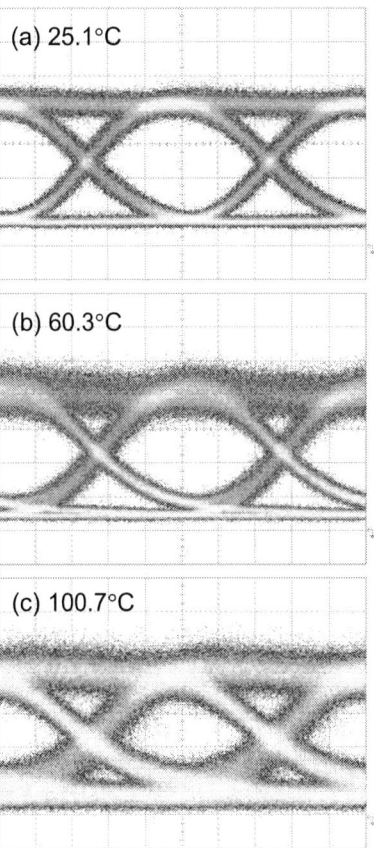

Fig. 6. Eye diagrams for 10 Gbit/s NRZ signals at operating temperatures of (a) 25.1°C, (b) 60.3°C, and (c) 100.7°C.

25 and 100.7°C, respectively. For the InGaAsP-based MQW LEAP laser shown in [6], the maximum 3-dB bandwidth at an operating temperature of 66°C was 7.0 GHz, which was lower than that of the InGaAlAs MQW LEAP laser at 100°C. The increase in the 3-dB bandwidth of the InGaAlAs MQW LEAP laser at a high operating temperature is achieved because of the high output power and large differential gain of the InGaAlAs-based MQW active region.

Finally, we measured the direct modulation response of the fabricated laser. Figure 6 shows eye diagrams of a 1.56-μm output light modulated by 10-Gbit/s NRZ signals with a pseudo-random bit sequence of 231-1 at operating temperatures of (a) 25.1°C, (b) 60.3°C and (b) 100.7°C. As shown in these figures, clear eye openings were observed. This indicates that the output power at a high temperature of 100°C is sufficient to detect a 10-Gbit/s signal. We believe that this is the first demonstration of a 10-Gbit/s modulated nanocavity laser operating above 100°C. In addition, the minimum input powers at which clear eye opening was observed were -11.4 dBm at 25.1°C, -10.9 dBm at 60.3°C and -6.3 dBm at 100.7°C.

978-1-4673-1725-2/12 $31.00 © 2012 IEEE   145

Thus, the energy costs for transferring a single bit at 25.1, 60.3 and 100.7°C were estimated to be 7.2, 8.1 and 23.4 fJ, respectively. We also achieved low power consumption operation for a LEAP laser with an InGaAlAs MQW active region.

IV. CONCLUSION

We have demonstrated the 10-Gbit/s direct modulation of an optically pumped LEAP laser at an operating temperature of 100°C using an InGaAlAs-based MQW as an active region. The LEAP laser exhibits an output power of more than 20 µW at a high operating temperature of 100°C. In addition, we obtained the dynamic responses of the LEAP laser at 25 and 100°C with maximum 3-dB bandwidths of 12.3 and 8.3 GHz, respectively. Furthermore, when the LEAP laser was modulated at 10 Gbit/s, we achieved low energy costs of 7.2 fJ/bit at 25.1°C and 23.4 fJ/bit at 100.7°C. Hence, an InGaAlAs MQW LEAP laser is a promising candidate for the light source of a photonic network on a CMOS chip.

ACKNOWLEDGMENTS

We thank Y. Shouji, and K. Ishibashi for fabricating the device. This work was supported by the New Energy and Industrial Technology Development Organization (NEDO).

REFERENCES

[1] D. A. B. Miller, "Device requirements for optical interconnects to silicon chips," *Proc. IEEE,* vol. 97, pp. 1166-1185, July 2009.

[2] S. Matsuo, A. Shinya, T. Kakitsuka, K. Nozaki, T. Segawa, T. Sato, Y. Kawaguchi, and M. Notomi, "High-speed ultracompact buried heterostructure photonic-crystal laser with 13 fJ of energy consumed per bit transmitted," *Nat. Photonics,* vol. 4, no. 9, pp. 648-654, September 2010.

[3] S. Matsuo, A. Shinya, C.-H. Chen, K. Nozaki, T. Sato, Y. Kawaguchi, H. Taniyama, and M. Notomi, "20-Gbit/s directly modulated photonic crystal nanocavity laser with ultra-low power consumption," *Opt. Express,* vol. 19, no. 3, pp. 2242-2250, January 2011.

[4] S. Matsuo, K. Takeda, T. Sato, M. Notomi, A. Shinya, K. Nozaki, H. Taniyama, K. Hasebe, and T. Kakitsuka, "Room-temperature continuous-wave operation of lateral current injection wavelength-scale embedded active-region photonic-crystal laser," *Opt. Express,* vol. 20, no. 4, pp. 3773-3780, February 2012.

[5] H. F. Hamann, A. Weger, J. A. Lacey, Z. Hu, P. Bose, E. Cohen, and J. Wakil, "Hotspot-limited microprocessors: direct temperature and power distribution measurements," *IEEE J. Sol.-State Circuits,* vol. 42, no. 1, pp. 56-65, January 2007.

[6] K. Takeda, T. Sato, T. Kakitsuka, A. Shinya, K. Nozaki, C.-H. Chen, H. Taniyama, M. Notomi, and S. Matsuo, "High-temperature operation of photonic-crystal lasers for on-chip optical inter connection," *IEICE Trans. Electron.,* vol. E95-C, no. 7, pp. 1244-1251, July 2012.

[7] T. Tadokoro, T. Yamanaka, F. Kano, H. Oohashi, Y. Kondo, and K. Kishi, "Operation of a 25-Gb/s direct modulation ridge waveguide MQW-DFB laser up to 85°C," *IEEE Photon. Technol. Lett,* vol. 21, no. 16, pp. 1154-1156, August 2009.

Multi-stack quantum cascade lasers

Romain Blanchard[1], Christian Pfluegl[2], Laurent Diehl[2], Russell D. Dupuis[3], Federico Capasso[1]

[1]Harvard University, Cambridge, MA 02138, U.S.A.
[2]Eos Photonics, Cambridge, MA 02138, U.S.A.
[3]Georgia Institute of Technology, Atlanta, GA 30332, U.S.A.

Abstract — We demonstrate a double-waveguide quantum cascade laser (QCL) consisting of two full broadband QCLs vertically integrated into a single monolithic device. Up to 1.1W peak power at room temperature is obtained for this record thick QCL. Coupling between the two laser waveguides is minimized to reduce gain competition. Simultaneous lasing on Fabry-Perot modes separated by as much as 360 cm^{-1} is obtained. This design opens the route to high-power ultra-broadband mid-infrared sources.

Index Terms — Quantum cascade lasers, Semiconductor waveguides, Infrared spectra.

I. INTRODUCTION

Quantum cascade lasers (QCLs) are very efficient sources of mid-infrared light now reaching continuous-wave Watt-level output powers [1]. Their range of applications covers free space communication to chemical sensing [2]. Here we are interested in the development of QCLs lasing simultaneously over a broad spectral range. Such devices could be used as a bright source in Fourier-transform infrared spectrometers and enable new applications, such as solvent-based process monitoring.

We propose here a novel waveguide structure to obtain broad and bright emission from a single monolithic device by growing two full QCLs, each with its own optical waveguide, on top of each other. A thick InP spacer separating the active regions of the two QCLs is used to prevent coupling between the two optical waveguides. We demonstrate excellent room temperature performance from this record-thick device, grown by metal-organic chemical vapor deposition [3] (MOCVD). The ability to grow by MOCVD such thick QCL structure is particularly interesting for industrial production prospects. The device demonstrated here offers the advantages of a small footprint in addition to a small and reliable spacing between active regions, enabling to obtain good beam quality, for example using a microlens array.

II. BROADBAND QUANTUM CASCADE LASERS

One of the main strategies used to obtain broadband lasing in QCLs consists of stacking in series active regions operating at different wavelengths, in order to form a heterocascade [4]. This strategy provides a broad gain spectrum over which lasing mode proliferation is enabled by a low threshold for the spatial hole burning instability in QCLs. Obtaining simultaneous lasing over a broad spectrum requires having a flat net gain in order to ensure that all lasing wavelengths have the same threshold. Furthermore, cross-absorption and gain competition between the different cascades needs to be mitigated. In particular, gain competition in heterocascade QCLs has been shown to lead to lasing on higher order lateral modes [5], thereby reducing the laser brightness. With carefully designed active regions, simultaneous lasing over a range as wide as 360 cm^{-1} with a small gap of 36 cm^{-1} has been reported [6].

III. WAVEGUIDE DESIGN

The structure is composed of four different cascade designs distributed over the two gain media (see Fig. 1). All cascades are based on a bound-to-continuum design [7], characterized by an inherently broad gain spectrum, respectively centered at 135, 145, 165 and 175 meV.

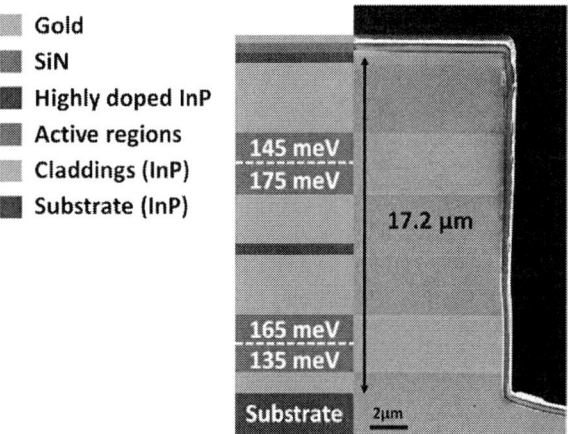

Fig. 1.   Schematic of the waveguide structure and scanning electron microscope image (SEM).

In order to reduce gain competition, which increases with increasing overlap of the gain curves, spectrally adjacent active regions were separated from each other in two different waveguides separated by a thick indium phosphide (InP) spacer. Contrary to previous works using indium phosphide (InP) inter-stacks in a QCL structure [8], the thick InP spacer used in this work to separate the active regions of the two QCLs is intended to create two uncoupled optical waveguides.

978-1-4673-1725-2/12 $31.00 © 2012 IEEE

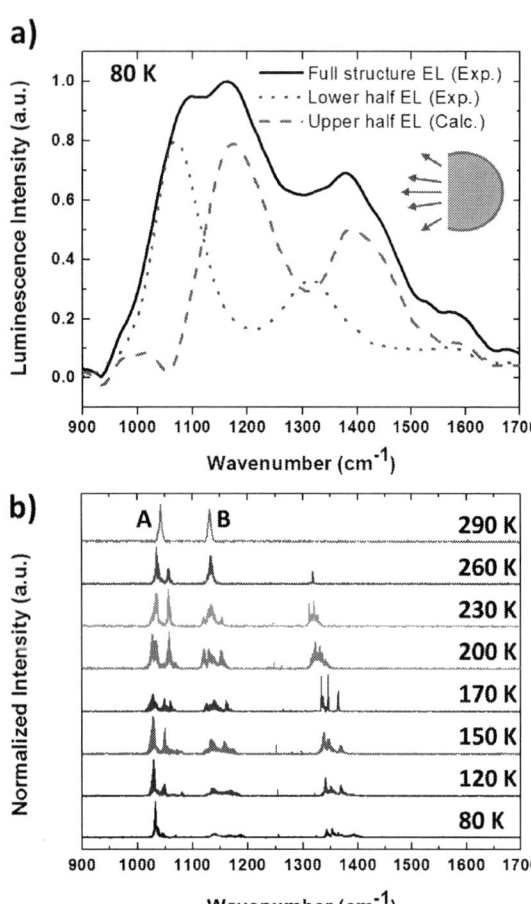

RIE on top of the laser ridges. Top and bottom metallization consisted of Ti(15 nm) and Au(450 nm) layers. The fabricated devices were cleaved to 3 mm length, high-reflection coated (Al₂O₃/Au) on the back facet and indium mounted epi-side up on copper heat sinks. The devices were characterized in pulsed operation with 100 ns pulse width and 20 kHz repetition rate.

IV. RESULTS

A broad electroluminescence (EL) spectrum spanning 1025 to 1475 cm^{-1} is obtained for the full double-waveguide structure (see Fig. 2(a)). Two EL peaks are observed for each waveguide, with the long-wavelength peak about twice as strong as the short-wavelength peak in both cases. We believe that this can be explained by resonant cross-absorption of the short wavelengths in the long-wavelength stages, resulting in a significantly lower net gain at short wavelengths. While gathering spectrally adjacent stages in the same waveguide could reduce cross-absorption, it would increase gain competition. For this reason, cross-absorption would be better addressed with a careful design of the stages. In future work, we will focus on improving the designs of the individual active regions, in particular by implementing innovative solutions for broadband lasing. The combination of proven broadband designs [6] with our double-waveguide QCL approach is a straightforward path towards monolithic QCL sources with unprecedented lasing bandwidth.

Fig. 2. (a) Measured EL spectra at 80 K from the full double-waveguide structure (black curve) and from the lower waveguide only (red dotted curve), with upper waveguide etched away. Calculated difference of the two spectra as an approximation of the EL from the upper waveguide only (blue dashed curve). The measured spectra are taken for two cleaved mesas of similar sizes with similar current flowing through them. The corresponding bias for the full structure is approx. 20 V, comparable to the 19 V bias used for the lasing spectra in (b). EL is shown at 80 K for direct comparison with the lasing spectrum at 80 K. (b) Lasing spectra at different temperatures for a 23 μm wide and 3 mm long laser, operated in pulsed mode (100 ns, 20 kHz). The laser is driven at a constant bias of 19 V (current approx. 2 A). A high-reflectivity coating was deposited on the back facet of the laser. The letters A and B allow identification of the two groups of modes.

Fabrication of the lasers started with the etching of 17 μm-deep trenches by reactive ion etching (RIE) using a photolithography-defined SU8 mask. The side-walls remain straight and smooth over the whole depth, as seen in Fig. 1. The trenches defined waveguides 20, 23, and 26 μm-wide. A 300 nm-thick SiN insulation layer was then deposited by plasma-enhanced chemical vapor deposition and opened by

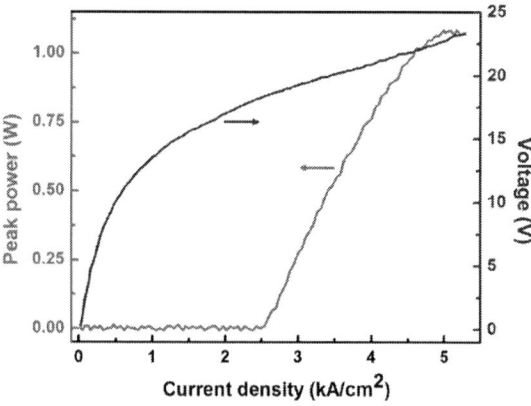

Fig. 3. (a) Light intensity (peak power) (red) and voltage (blue) versus current density for a representative device (with high-reflectivity coating) operated in pulsed mode (100 ns, 20 kHz) at room temperature.

Ridge lasers were fabricated and characterized in pulsed operation with 100 ns pulse width and 20 kHz repetition rate. Fig. 2 (b) shows lasing spectra obtained at different operating temperatures. Up to 260 K, a temperature that can be reached with thermoelectric coolers, we can distinguish three main islands of Fabry-Perot modes, with lasing modes separated by up to 360 cm−1. For temperatures above 260 K, only the two long-wavelength islands survive. Experimental evidence suggests that the group of modes B is lasing mainly in the upper waveguide where it experiences the most gain, whereas the group of modes A is lasing mainly in the lower waveguide.

The light-current-voltage characteristics shown in Fig. 3 demonstrate good lasing performance from both gain media, at room temperature and in pulsed operation (100 ns/20 kHz). Using a short-pass filter at 9.3 μm (between the group of modes A and B), we observed that the threshold is the same for both groups of modes and that the power is approximately equally divided between them. This indicates that both waveguides have comparable optical losses.

Remarkably, the good performance of the device, in particular of the group of modes B for which gain is provided by the last stages grown, indicates that high quality growth can be sustained for such thick structure and that our waveguide design maintains low optical losses.

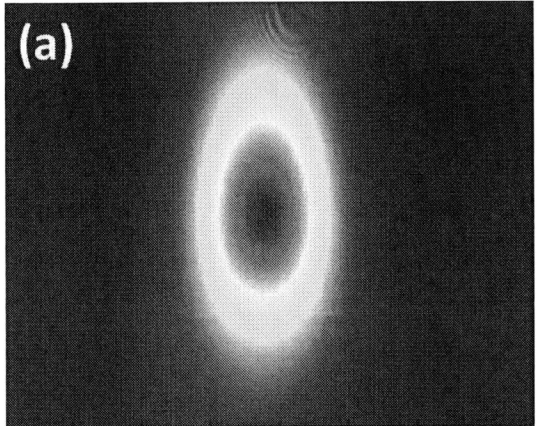

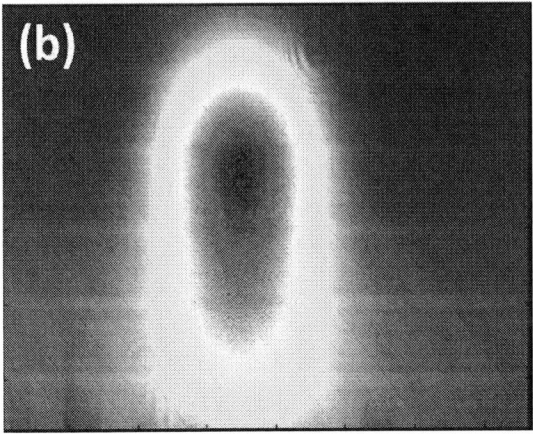

Fig. 4. Far-field of a representative device as measured with a mid-infrared camera placed approx. 1cm away from the laser facet. The laser is operated at room-temperature with a repetition rate of 20kHz and a duty cycle of 2% (a) or 1% (b). Steering is observed in the vertical direction (perpendicular to the grown layers).

The far-field of a representative laser operated at room-temperature was measured with a mid-infrared camera placed in front of the laser facet (see Fig. 4). Lasing on the fundamental lateral mode is observed. However, steering of the beam in the direction perpendicular to the epitaxial layers is observed. This indicates that at least some amount of coupling exists between the two waveguides. We noted different steering angles for different duty cycles, suggesting an effect of temperature on the coupling behavior.

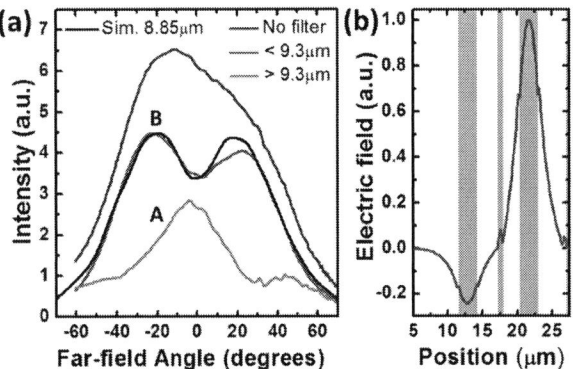

Fig. 5. (a) Far-field of the representative device of Fig. 3, operated in the same conditions. In blue, without filter; in red, with a short-pass filter at 9.3 μm. In green, the calculated difference of the two previous curves (blue minus red), illustrating the far-field of modes A. In black, simulated far-field of the linear combination of the two lowest-order orthogonal TM waveguide modes fitting best to the experimental data. The simulations are done at the central wavelength of the group of modes B (8.85 μm). The far-fields are measured by scanning a mercury cadmium-telluride detector located 10 cm away from the laser, along a direction perpendicular to the growth planes. (b) Instantaneous electric field (at a time maximizing field amplitude) across the structure, along the direction perpendicular to the layer planes. The two active regions are in light blue, the central plasmon layer in green and the top gold contact in yellow.

In order to quantify the coupling, we measured far-field line-scans along the direction perpendicular to the layer planes, for a representative laser operated at room-temperature (see Fig. 5). We observe that the emission of the group of modes B features two peaks separated by a shallow dip. Using finite-difference simulations (Lumerical FDTD), we solved for the two lowest-order orthogonal TM modes supported by the double-waveguide structure. We then computed the far-fields (obtained by near-field to far-field transformations) of linear combinations of these two modes in order to identify the superposition corresponding best to the experimental far-field. The best fit is shown in black in Fig. 3(b), and the corresponding mode profile is shown in Fig. 3(c). The intensity is mostly confined to the upper waveguide, with only about 6% of the power in the lower waveguide. The two lobes of the electric field in the two waveguides are π-shifted one with respect to the other. Lasing on such mode is favored as it corresponds to a minimum of field at the center of the spacer region where we introduced a highly doped lossy layer. A thicker highly-doped layer with a larger doping level between the two active regions could easily reduce the coupling between the two waveguides. In contrast, the far-field calculated for the group of modes A indicates that light is emitted mainly from one waveguide, the lower waveguide in our case.

V. CONCLUSION

In summary, we have presented a new double-waveguide QCL design. This small-footprint design provides flexibility to distribute multiple QCL stages over two weakly-coupled optical waveguides, in order to obtain broadband emission while dealing with common issues of heterocascade QCLs, such as cross-absorption and gain competition. We demonstrated that good room temperature performance is preserved with this record thick device grown by MOCVD. Full decoupling of the waveguides was not achieved at all wavelengths but this can easily be accomplished by increasing the thickness and/or doping of the cladding layers.

REFERENCES

[1] A. Lyakh, R. Maulini, A. Tsekoun, R. Go, S. Von der Porten, C. Pfluegl, L.Diehl, F. Capasso, and K. N. Patel, *Proc. Natl. Acad. Sci. U.S.A.*, vol. 107, p. 18799, 2010.

[2] R. F. Curl, F. Capasso, C. Gmachl, A. A. Kosterev, B. McManus, R. Lewicki, M. Pusharsky, G. Wysocki, and F. K. Tittel, *Chem. Phys. Lett.*, vol. 487, p. 1, 2010.

[3] Y. Huang, J.-H. Ryou, R. D. Dupuis, C. Pfluegl, F. Capasso, K. Sun, A. M. Fischer, and F. A. Ponce, *J. Cryst. Growth*, vol. 316, p. 75, 2010.

[4] C. Gmachl, D. L. Sivco, J. N. Baillargeon, A. L. Hutchinson, F. Capasso, and A. Y. Cho, *Appl. Phys. Lett*, vol. 79, p. 572, 2001.

[5] M. Geiser, C. Pfluegl, A. Belyanin, Q. J. Wang, N. Yu, T. Edamura, M. Yamanishi, H. Kan, M. Fischer, A. Wittmann et al., *Opt. Express*, vol. 18, p. 9900, 2010.

[6] A. Hugi, R. Terazzi, Y. Bonetti, A. Wittmann, M. Fischer, M. Beck, J. Faist, and E. Gini, *Appl. Phys. Lett.*, vol. 95, p. 061103, 2009.

[7] J. Faist, M. Beck, T. Aellen, and E. Gini, *Appl. Phys. Lett.*, vol. 78, p. 147, 2001.

[8] A. Bismuto, T. Gresch, A. Bachle, and J. Faist, *Appl. Phys. Lett.*, vol. 93, p. 231104, 2008.

Design of high-current L-valley GaAs/AlAs$_{0.56}$Sb$_{0.44}$/InP (111) ultra-thin-body nMOSFETs

Saumitra Mehrotra*, Michael Povolotskyi*, Jeremy Law[†], Tillmann Kubis*, Gerhard Klimeck*, and Mark Rodwell[†]

* Network for Computational Nanotechnology, Purdue University, West Lafayette, IN 47907, USA

[†] Department of Electrical and Computer Engineering, University of California, Santa Barbara, CA 93106, USA

Abstract— **We propose and analyze a high-current III-V transistor design using electron transport in the Γ- and L-valleys of (111) GaAs. Using $sp^3d^5s^*$ empirical tight-binding model for band-structure calculations and the top-of-the-barrier transport model, improved drive current is demonstrated using L-valley transport in a strained GaAs channel grown on an (111) InP substrate. At a body thickness of 2 nm the (111)GaAs/InP MOSFET design outperforms both (100) Si and (100) GaAs/InP for all EOTs larger than 0.3nm.**

Index Terms— **MOSFET, L-valley, tight-binding, GaAs, InP, top-of-the-barrier**

I. INTRODUCTION

III-V nMOSFETs have small transport effective mass that provides high electron velocities and high on-state currents. However, small effective mass also leads to a small semiconductor density of states, and consequently III-V channels provide no benefit over Si for EOT < 0.6 nm [1]. This loss of state density can be compensated by using the highly anisotropic L-valley for electron transport [2]. Confining the channel along the (111) direction leads the L-valley to have a large confinement mass and much smaller in-plane transport mass. At some channel thickness, the Γ- and L-valleys are aligned in energy, increasing the state density and on-current [3]. Simulations in [3] ignored interactions of the channel wavefunction with gate dielectric and the well bottom barrier: here we report practical L-valley GaAs channel designs incorporating AlAs$_{0.56}$Sb$_{0.44}$ barriers to set the boundary conditions for Γ-L alignment.

II. DEVICE STRUCTURE

Interaction of the channel wavefunction with the amorphous gate dielectric is difficult to compute, hence ideal hydrogen-terminated semiconductor interfaces are often assumed in simulations. To prevent this interaction from changing the Γ-L energy alignment and dispersion, the designs here reported use thin AlAs$_{0.56}$Sb$_{0.44}$ cladding layers to strongly attenuate the channel wavefunction at the dielectric-semiconductor interface. Fig. 1 shows the device geometries under study in this paper. A single-gate (SG) MOSFET consists of a biaxially strained (3.67% mismatch with InP) GaAs channel grown on a 5nm AlAs$_{0.56}$Sb$_{0.44}$ barrier layer, lattice matched to InP. Two monolayers of AlAs$_{0.56}$Sb$_{0.44}$ serve as a cap-layer. Similarly, for a double-gate (DG) MOSFET a biaxial strained GaAs with AlAs$_{0.56}$Sb$_{0.44}$ cap layer on top and bottom are assumed.

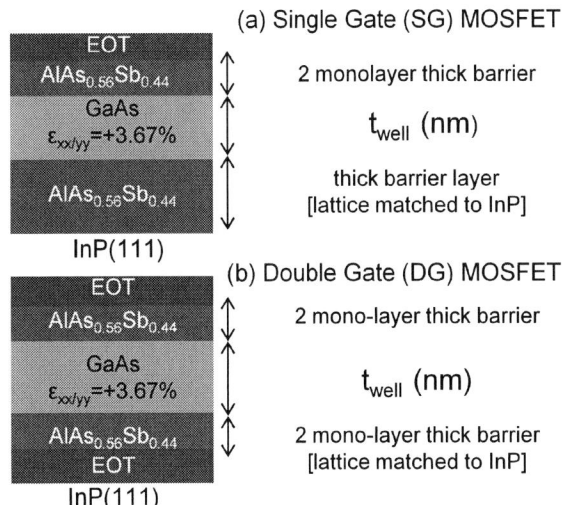

Fig. 1. GaAs/AlAs$_{0.56}$Sb$_{0.44}$/InP (111) MOSFETs based on (a) a single gate with strained GaAs channel and AlAs$_{0.56}$Sb$_{0.44}$ as a capping and barrier layer (b) a double gate structure with strained GaAs channel AlAs$_{0.56}$Sb$_{0.44}$ capping layer.

Similar designs are used for GaAs/InP (100) and Si (100) (no cap layer) for comparison of device performance.

III. SIMULATION METHODOLOGY

Band structure calculations of the SG/DG structures (Fig. 1) are performed using an $sp^3d^5s^*$ empirical tight-binding model including spin-orbit coupling. GaAs has a smaller lattice constant than InP, leading to an in-plane tensile strain ≃3.67% in GaAs. The biaxial strain is modeled as a homogenous strain tensor that affects the original atomic positions [4]. Strain effects are taken into account according to the Boykin model [5]. Both the channel material, GaAs and the capping layer AlAs$_{0.56}$Sb$_{0.44}$ are included in the simulation domain. Fig. 2 explicitly shows the effect of including a barrier layer on the band structure calculations. For a 2 nm thick, GaAs thin body structure grown on InP(111) with idealized hydrogen-terminated interfaces, the conduction band minima is formed by the L-valley states. The inclusion of thick AlAs$_{0.56}$Sb$_{0.44}$ layers on top and bottom reduces effective confinement and the Γ valley becomes the lowest-energy band. It should be noted that, electronic effects due to strain are included only

978-1-4673-1725-2/12 $31.00 © 2012 IEEE

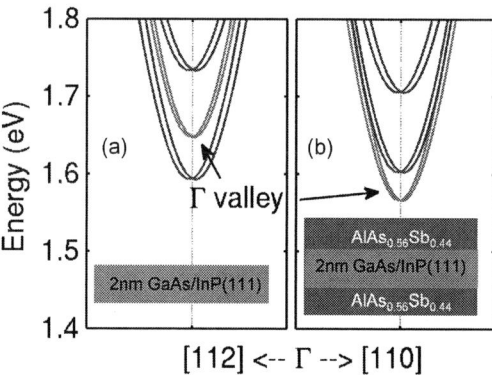

Fig. 2. Bandstructure calculations for 2 nm thick and biaxialy strained (a) GaAs/InP(111) and (b)GaAs/InP(111) with 5 nm thick AlAs$_{0.56}$Sb$_{0.44}$ layers.

for the GaAs material [5]. The AlAs$_{0.56}$Sb$_{0.44}$ layer, that is lattice matched to InP is not affected by strain.

To compare the device performance, 2D E-k relations are calculated for different device structures. The energy dependent density of states (DOS) and carrier velocity are then extracted from the band structure information. For GaAs/InP(111) case, $< 110 >$ orientation is considered to be the transport orientation while for GaAs/InP(100) and Si(100) cases $< 100 >$ is taken as the transport orientation. The DOS(E) and velocity(E) information are then used to calculate the on-state current (I_{on}) using the ballistic top-of-barrier transport model implemented in NEMO5 simulation package [6], [7].

IV. RESULTS

Fig. 3(a) shows the bandstructure calculations for GaAs UTB teminated by 2 - monolayer AlAs$_{0.56}$Sb$_{0.44}$ and lattice matched to InP. From the GaAs body thickness dependent band structure calculations it is revealed that L-valley minima transistor can be reached by confining GaAs to a 2 nm body thickness (marked in Fig. 3(a)). This particular structure

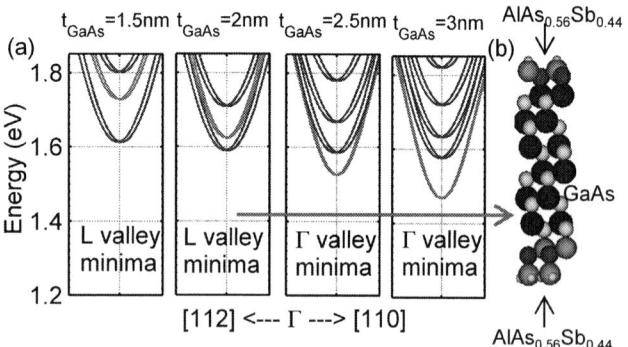

Fig. 3. Tight-binding bandstructure ($sp^3d^5s^*$) calculations for (111) oriented GaAs terminated with 1-2 monolayer Al$_{0.56}$As$_{0.44}$Sb lattice matched to InP. The blue lines correspond the L valley states, while the red lines correspond to the Γ valley states.

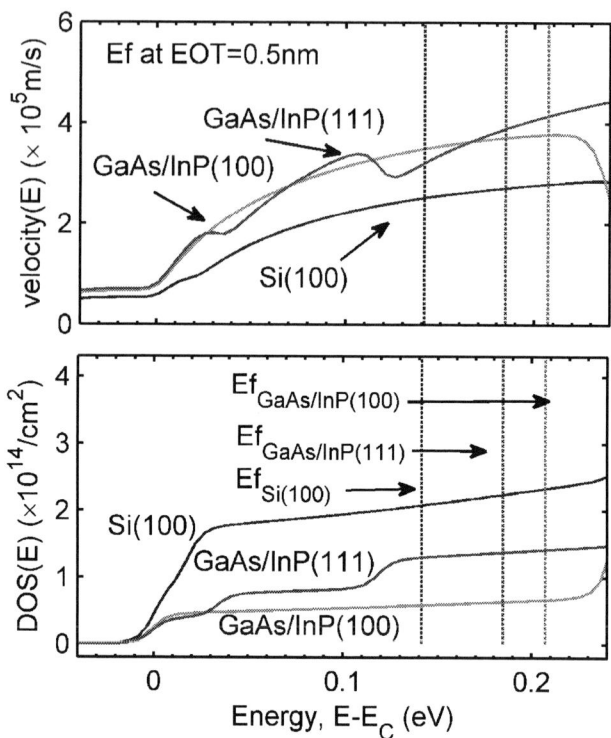

Fig. 4. Density of states and carrier velocity for 2 nm thick GaAs/InP(111), GaAs/InP(100) and Si(100) structures. Fermi level E_f position in the on state for DG MOSFET and EOT=0.5 nm is also shown. Energy axis has been adjusted to the conduction band edge, E_C

is used for GaAs/InP(111) device performance comparison throughout this paper.

Fig. 4 compares state density and carrier velocity for the different channel designs. It is readily observed that Si(100) exhibits higher DOS and lower velocity when compared to GaAs/InP(100) which exhibits a high velocity but suffers from lower DOS. The small state density leads to the condition popularly known as 'DOS bottleneck' associated with low carrier effective mass III-V materials [1]. The GaAs/InP(111) transistor design aims to offest both the issues as will be shown later.

As a next step, ON state currents are calculated for 2 nm body structures with the channel material as GaAs/InP(111), which is the L-valley minima case, GaAs/InP (100) and Si (100). The on-state current is defined as the current at $V_{ds}=V_{gs}=0.5$V, with threshold voltage V_{th} set so that $I_{off} = 0.1\mu A/\mu m$. The on-state currents are calculated for different EOT values ranging from 0.3nm to 1.1nm. The degrading effect of the AlAs$_{0.56}$Sb$_{0.44}$ cap layer on the capacitive coupling of the GaAs channel to the gate is considered by increasing the EOT per gate by 0.1nm.

Fig 5 shows the gate capacitance (C_G) normalized to oxide capacitance (C_{OX}) for the DG MOSFET structures. It can

978-1-4673-1725-2/12 $31.00 © 2012 IEEE

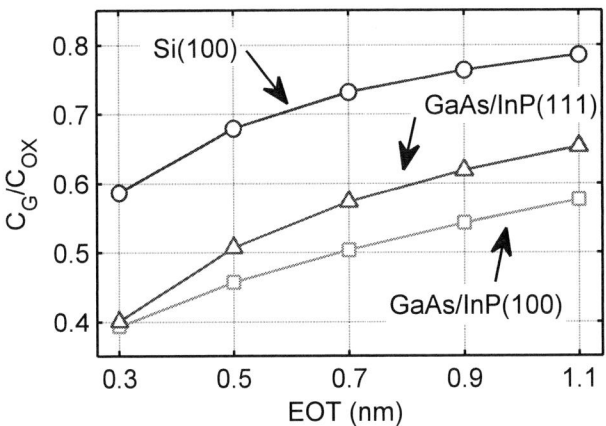

Fig. 5. Gate capacitance normalized with oxide capacitance is shown for DG MOSFET. Gate capacitance calculated V_{gs}=0.5V and V_{ds}=0.05V

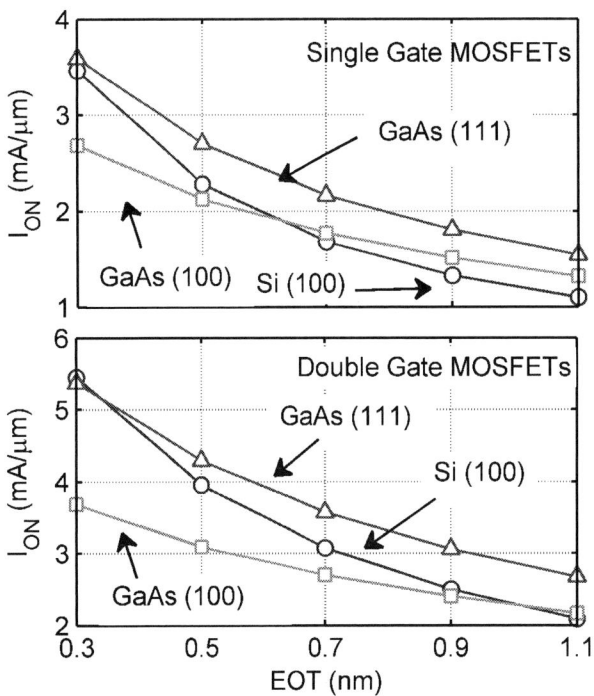

Fig. 6. ON state currents calculated for different EOT values for (a) single gate and (b) double gate structures.

be deduced from Eq (2) a high C_{DOS} (directly related to DOS(E) ,Eq (1)) leads to C_G being closer to C_{OX}, or lower a DOS will lead to degraded C_G and ultimately on-state current (Eq 3). In Eq (2) D_{mean} is the mean wavefunction depth in the semiconductor and ϵ_s is the semiconductor permittivity. As expected GaAs/InP(100) which has a small confinement mass or lowest DOS has the smallest C_G. This is the 'DOS bottleneck' effect, which practically nullifies the expected gains from increasing the gate dielectric capacitance. This drawback can be minimized by utilizing L-valley minima transistor designs that allows multiple subbands close to band edge. The L-valley minima design of GaAs/InP(111) exhibits an increased C_G when compared to GaAs/InP(100). However, Si(100) still exhibits the highest gate capacitance owing to its large state density (Fig (4)). It should be noted that the GaAs/InP cases have further degraded capacitance due to the presence of AlAs$_{0.56}$Sb$_{0.44}$ cap layer.

$$C_{DOS} = q^2 \frac{dn_s}{dE_f} \qquad (1)$$

and,

$$C_G = (C_{DOS}^{-1} + C_{OX}^{-1} + D_{mean}/\epsilon_s)^{-1} \qquad (2)$$

$$I_{ON} = qC_G(V_{gs} - V_{th}). \times v_{avg} \qquad (3)$$

The final computed I_{on} values are shown in Fig. 6. GaAs/InP(111) exhibits an improved gate capacitance over GaAs/InP(100) and at the same time shows a higher carrier velocity over Si(100) MOSFET. This fact leads GaAs/InP(111) MOSFET to perform better than both GaAs/InP(100) and

Si(100) MOSFET structures at higher EOT values. For SG and DG MOSFET designs, Si(100) surpasses GaAs/InP(100) in $I_{on} \simeq 1.0$ nm and $\simeq 0.6$ nm EOT, respectively. GaAs/InP(111) has a smaller density of states than Si (100), hence loses its advantage over Si at ultra thin EOT values. GaAs/InP(111) exhibits the highest on-current; only at $\simeq 0.3$ nm EOT is Si comparable. At EOT=0.5 nm, considered feasible for MOS-FETs at L_g=5nm, the GaAs/InP(111) design delivers 8.5% higher I_{on} than Si(100) DG MOSFETs and 18% higher than Si(100) SG MOSFETs confirming the efficacy of the Γ-L channel designs (see [8]).

V. CONCLUSIONS

GaAs/AlAs$_{0.56}$Sb$_{0.44}$/InP (111) channel designs are presented for high current MOSFETs. Using $sp^3d^5s^*$-SO bandstructure calculations it is showed that at $\simeq 2$ nm body thickness L valley minima can be acieved in GaAs/AlAs$_{0.56}$Sb$_{0.44}$/InP (111) quantum wells. Later, using the top-of-the-barrier transport model it is shown that GaAs/AlAs$_{0.56}$Sb$_{0.44}$/InP (111) orientation outperforms both Si(100) and GaAs/AlAs$_{0.56}$Sb$_{0.44}$/InP(100) at similar channel thicknesses for EOT >0.5nm. These results could be useful in design of future nanoscale device applications.

VI. ACKNOWLEDGMENT

The work is supported by National Science Foundation (award number ECCS-1125017)

REFERENCES

[1] M. Fischetti, L. Wangt, B. Yut, C. Sachs, P. Asbeckt, Y. Taurt, and M. Rodwell, "Simulation of electron transport in high-mobility mosfets: Density of states bottleneck and source starvation," in *Electron Devices Meeting, 2007. IEDM 2007. IEEE International*, dec. 2007, pp. 109 –112.

[2] M. Rodwell, W. Frensley, S. Steiger, E. Chagarov, S. Lee, H. Ryu, Y. Tan, G. Hegde, L. Wang, J. Law, T. Boykin, G. Klimek, P. Asbeck, A. Kummel, and J. Schulman, "III-V FET channel designs for high current densities and thin inversion layers," in *Device Research Conference (DRC), 2010*, june 2010, pp. 149 –152.

[3] R. Kim, T. Rakshit, R. Kotlyar, S. Hasan, and C. Weber, "Effects of surface orientation on the performance of idealized III-V thin-body ballistic n-mosfets," *Electron Device Letters, IEEE*, vol. 32, no. 6, pp. 746 –748, june 2011.

[4] R. I. Cottam and G. A. Saunders, "The elastic constants of gaas from 2 k to 320 k," *Journal of Physics C: Solid State Physics*, vol. 6, no. 13, p. 2105, 1973. [Online]. Available: http://stacks.iop.org/0022-3719/6/i=13/a=011

[5] T. B. Boykin, G. Klimeck, R. C. Bowen, and F. Oyafuso, "Diagonal parameter shifts due to nearest-neighbor displacements in empirical tight-binding theory," *Phys. Rev. B*, vol. 66, p. 125207, Sep 2002.

[6] A. Rahman, J. Guo, S. Datta, and M. Lundstrom, "Theory of ballistic nanotransistors," *Electron Devices, IEEE Transactions on*, vol. 50, no. 9, pp. 1853 – 1864, sept. 2003.

[7] S. Steiger, M. Povolotskyi, H.-H. Park, T. Kubis, and G. Klimeck, "Nemo5: A parallel multiscale nanoelectronics modeling tool," *Nanotechnology, IEEE Transactions on*, vol. 10, no. 6, nov. 2011.

[8] "International technology roadmap for semiconductors," http://www.itrs.net/Links/2011ITRS/Home2011.htm/.

High Performance Substitutional-Gate MOSFETs Using MBE Source-Drain Regrowth and Scaled Gate Oxides

Sanghoon Lee[1], A. D. Carter[1], J. J. M. Law[1], D. C. Elias[1], V. Chobpattana[2], Hong Lu[2], B. J. Thibeault[1], W. Mitchell[1], S. Stemmer[2], A. C. Gossard[2], and M. J. W. Rodwell[1]

[1]Departments of Electrical and Computer Engineering and of [2]Materials University of California, Santa Barbara, CA 93106 USA

Abstract — **We report high transconductance MOSFETs with scaled dielectric (1.2 nm EOT) fabricated using a substitutional-gate process with MBE source/drain regrowth. A 50 nm-L_g and 1.2 nm EOT device shows 0.8 mA/μm on-current at V_{gs}-V_{th} = 0.8 V and V_{ds} = 0.5 V and 1.0 mS/μm peak transconductance at V_{ds} = 0.5 V which are 25% and 40% higher than those of a 1.65 nm EOT control device, respectively. Transmission line method (TLM) measurements indicate 0.8 Ohm-μm² metal-semiconductor contact resistivity and 18 Ohm sheet resistance of the regrown N+ source-drain contact layer.**

I. INTRODUCTION

Due to the low transport mass in $In_{1-x}Ga_xAs$-based materials, InGaAs MOSFETs can provide higher on-state current and transconductance than reference Si channel devices [1]-[3], given low dielectric interface trap density (D_{it}) [4] and low source/drain access resistivity [5]. According to FET scaling laws [5], high on-current and integration density in scaled devices requires the following: (1) low source/drain contact and access resistivities, (2) heavily-doped source and drain regions for adequate carrier supply (3) low D_{it}, and (4) small equivalent oxide thickness (EOT) and thin channel layers for high channel charge density

Thin gate dielectrics are critical not only for proper FET electrostatics at a short gate length but also for high on-current and transconductance. Considering gate-channel capacitances (Fig. 1), given a 7.5 nm thick channel, and estimating D_{it} as $5\cdot10^{12}$/cm², we calculate from ballistic transport theory that reducing the EOT from 1.65 nm (3.3/1.5 nm Al_2O_3/HfO_2) to 1.16 nm (1.1/4.0 nm Al_2O_3/HfO_2) should increase the 2D electron density (N_s) by ~15%. In the absence of S/D resistance, the thinner dielectric should increase current density and transconductance ~25%. Here we report a high-transconductance (1.0 mS/μm peak transconductance at V_{ds} = 0.5 V) device with a scaled 1.16 nm EOT dielectric, fabricated using a substitutional-gate scheme and using MBE source-drain regrowth.

II. DEVICE FABRICATION

The epitaxial layers, grown by MBE, has a InP (100) semi-insulating (S. I.) substrate, a 400 nm unintentionally doped (UID) $In_{0.52}Al_{0.48}As$ buffer/barrier layer, a 3 nm Si-doped $(6.0\cdot10^{12}$/cm²) $In_{0.52}Al_{0.48}As$ pulse doping layer, a 1 nm UID

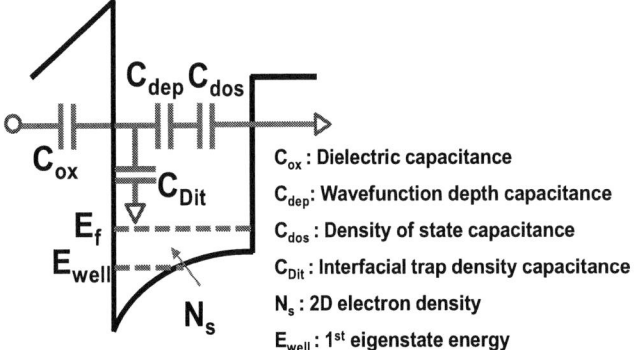

Fig. 1.   Circuit model of QW-MOSFET electrostatics.

C_{ox} : Dielectric capacitance
C_{dep} : Wavefunction depth capacitance
C_{dos} : Density of state capacitance
C_{Dit} : Interfacial trap density capacitance
N_s : 2D electron density
E_{well} : 1st eigenstate energy

$In_{0.52}Al_{0.48}As$ setback layer, and a 2 nm $In_{0.53}Ga_{0.47}As$ / 2.5 nm InAs / 3 nm $In_{0.53}Ga_{0.47}As$ composite channel. Fig. 2 summarizes the process flow. 300 nm of SiO_2 as a dummy gate and 20 nm of Cr as a hard mask were deposited by PECVD and e-beam evaporation, respectively. A Cr hard mask was pattern by e-beam lithography and inductively coupled plasma (ICP) dry etching, and subsequently SiO_2 was dry-etched by ICP using the patterned Cr hard mask. The Cr hard mask was then removed by photo resist planarization [7]. Prior to loading into the MBE chamber, the semiconductor surface was oxidized by exposure to UV ozone and subsequently etched in 10:1 DI H_2O:HCl. Approximately 60 nm of Si doped (5-$10\cdot10^{19}$ /cm³) InAs was non-selectively grown on the sample. Amorphous InAs on top of the dummy gate was then removed by photo resist planarization [7]. The sample was wet-etched to form device mesas. The SiO_2 dummy gate was then removed using a buffered oxide etch and a few drops of Tergitol as a surfactant. Prior to gate dielectric deposition, native oxide on the channel was removed in dilute HCl. Using atomic layer deposition (ALD), 3.3 nm Al_2O_3 and 1.5 nm HfO_2 (1.65 nm EOT) were deposited on the control sample and 1.1 nm Al_2O_3 and 4.0 nm HfO_2 (1.16 nm EOT) deposited on the experimental sample. The samples were then annealed at 400°C for 1 hour in forming gas. Approximately 80 nm of Ni was thermally deposited as the gate electrode. Finally, 20 nm Ti / 20 nm Pd / 130 nm Au was lifted off for source/drain metallization by e-beam evaporation. The schematic cross-section of the device is shown in Fig. 3.

978-1-4673-1725-2/12 $31.00 © 2012 IEEE

- **Pattern Dummy Gate (PECVD SiO$_2$ + Dry-etch)**
- **Regrow Source/Drain (MBE)**
- **Isolate Mesa**
- **Remove Dummy Gate**
- **Deposit Gate Dielectric (ALD)**
- **Lift-off Gate Metal (Thermal Evaporation)**
- **Lift-off S/D Metal (e-beam Evaporation)**

Fig. 2. Process flow.

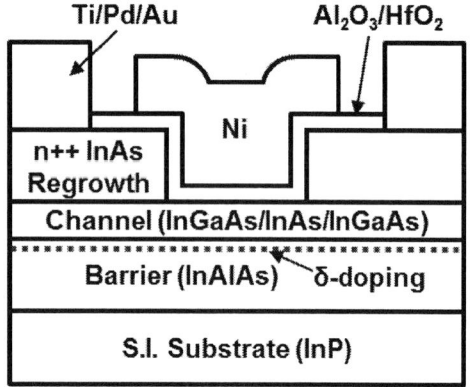

Fig. 3. Schematic cross-section of the substitutional-gate MOSFET.

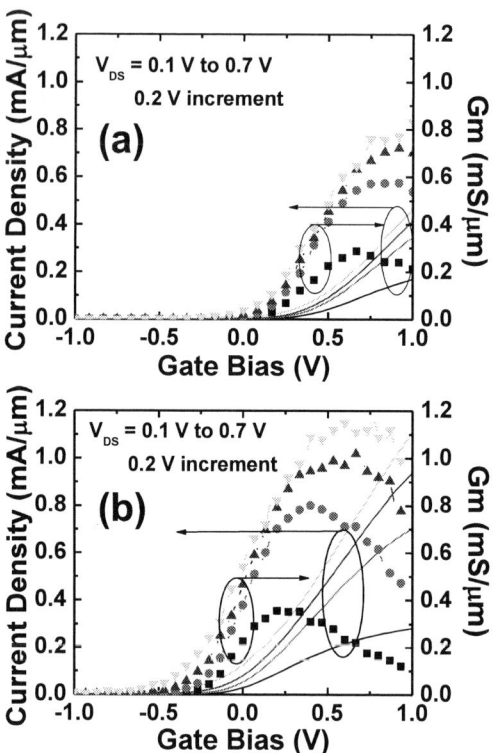

Fig. 4. Transfer (I_d-V_{gs} and g_m-V_{gs}) characteristics for a 50 nm-L_g device with 3.3/1.5 nm Al$_2$O$_3$/HfO$_2$ (a) and a 50 nm-L_g device with 1.1/4.0 nm Al$_2$O$_3$/HfO$_2$ (b)

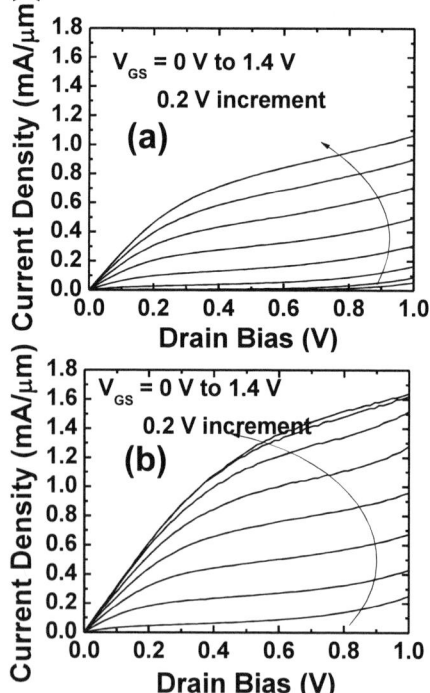

Fig. 5. Output (I_d-V_{ds}) characteristics for a 50 nm-L_g device with 3.3/1.5 nm Al$_2$O$_3$/HfO$_2$ (a) and a 50 nm-L_g device with 1.1/4.0 nm Al$_2$O$_3$/HfO$_2$ (b)

III. RESULTS AND DISCUSSION

Fig. 3 compares transfer characteristics (I_d-V_{gs} and g_m-V_{gs}) for a 50 nm-L_g (as drawn) device with 1.65 nm EOT (control sample) and a 50 nm-L_g (as drawn) device with 1.2 nm EOT (experimental sample). Threshold voltages extracted from linear extrapolation of the I_d-V_{gs} characteristics are ~0.4 V for the 1.65 nm EOT sample and ~0 V for the 1.16 nm EOT sample. The peak transconductance of the 1.16 nm EOT device is approximately 1.0 mS/µm at V_{ds} = 0.5 V which is approximately 40 % larger than that of the 1.65 nm EOT control. As shown in Fig. 4, the 1.16 nm EOT device shows approximately 0.8 mA/µm on-current at V_{gs}-V_{th} = 0.8 V and V_{ds} = 0.5 V, a bias at which impact ionization does significantly increase I_{on}, outperforming the 1.65 nm EOT control device by 25%. Fig. 5 shows log(I_d)-V_d plots of the 50 nm-L_g devices. The 1.65 nm EOT control and 1.16 nm EOT experimental samples both show ~200 mV/dec sub-threshold swing (SS) at V_{ds} = 0.1 V, which results both from short channel effects, back barrier leakage, and D_{it}. Approximately 60 mV hysteresis is observed near the V_{th}. Subthreshold characteristics of 200 nm-L_g devices are represented in Fig. 7.

978-1-4673-1725-2/12 $31.00 © 2012 IEEE   156

Devices show a minimum SS of ~170 mV/dec for the 1.65 nm EOT control device and ~130 mV/dec for the 1.16 nm EOT experimental device, respectively. Fig. 8 shows the gate leakage current for both cases. At gate biases larger than 0.3 Volts, the 1.16 nm EOT device shows smaller gate leakage current than the 1.65 nm EOT device. This is a consequence of the thinner dielectric. However, for both samples the gate leakage current is negligible, <10 nA/μm at all gate biases.

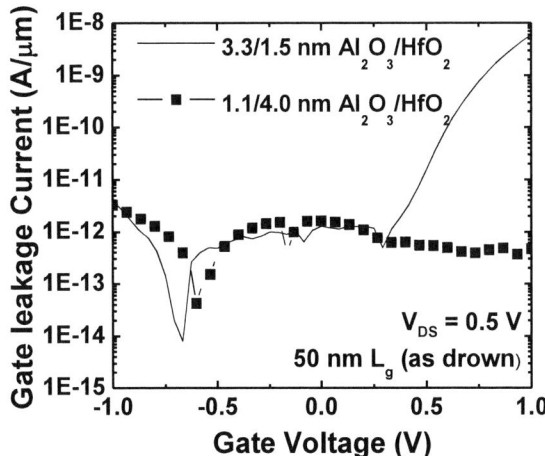

Fig. 8. Gate leakage current for both a 50 nm-Lg device with 3.3/1.5 nm Al_2O_3/HfO_2 (Solid line) and a 50 nm-Lg device with 1.1/4.0 nm Al_2O_3/HfO_2 (Square)

Fig. 9 shows transmission line method (TLM) measurement data for n++ InAs regrown source-drain, from which 0.8 Ohm-μm² of metal-semiconductor contact resistivity and 18 Ohm of sheet resistance are determined, respectively.

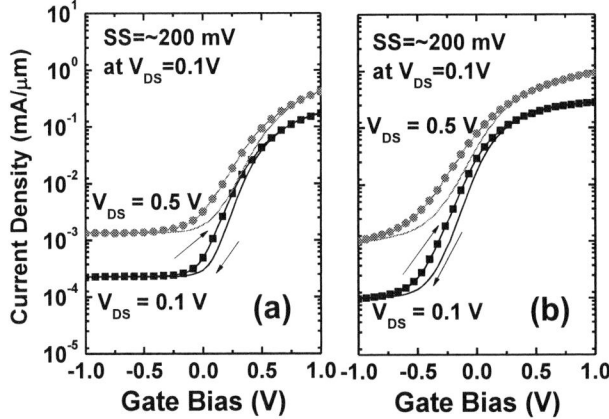

Fig. 6. Sub-threshold (log(I_d)-V_{gs}) characteristics for a 50 nm-L_g device with 3.3/1.5 nm Al_2O_3/HfO_2 (a) and a 50 nm-L_g device with 1.1/4.0 nm Al_2O_3/HfO_2 (b). Solid lines and lines with symbols represent forward and reverse sweeps, respectively.

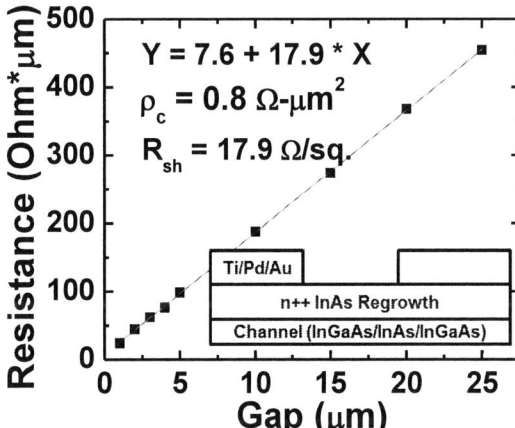

Fig. 9. Transmission line method (TLM) measurement of the n++ regrown source-drain layer. The metal-to-semiconductor contact resistivity and sheet resistance are 0.8 Ohm-μm² and 17.9 Ohm/square, respectively.

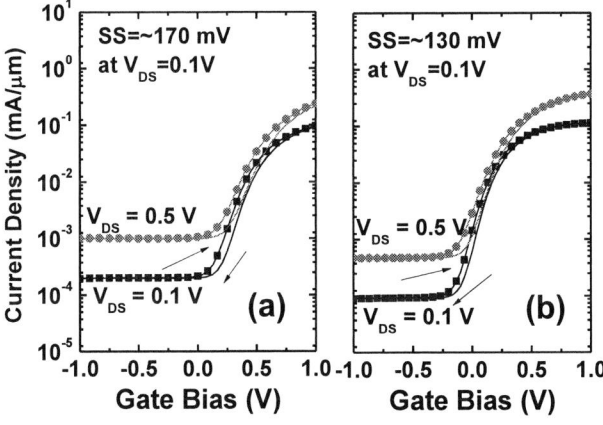

Fig. 7. Sub-threshold (log(I_d)-V_{gs}) characteristics for a 200 nm-L_g device with 3.3/1.5 nm Al_2O_3/HfO_2 (a) and a 200 nm-L_g device with 1.1/4.0 nm Al_2O_3/HfO_2 (b). Solid lines and lines with symbols represent forward and reverse sweeps, respectively.

IV. CONCLUSION

We have demonstrated high transconductance III-V MOSFETs with scaled gate dielectric using substitutional–gate scheme and MBE source drain regrowth technique. An experimental sample with 1.1/4.0 nm Al_2O_3/HfO_2 shows 1.0 mS/μm peak transcondance at V_{ds} = 0.5 V and 0.8 mA/μm on-current at V_{gs}-V_{th} = 0.8 V and V_{ds} = 0.5 V.

978-1-4673-1725-2/12 $31.00 © 2012 IEEE   157

ACKNOWLEDGMENTS

This work was supported by the SRC Non-classical CMOS Research Center (Task 1437.006). A portion of this work was done in the UCSB Nanofabrication facility, part of the NSF funded NNIN network and MRL Central Facilities supported by the MRSEC Program of the NSF under award No. MR05-20415

REFERENCES

[1] M. Radosavljevic, et al., "Advanced high-K gate dielectric for high-performance shortchannel In0.7Ga0.3As quantum well field effect transistors on silicon substrate for low power logic applications," in *IEDM Tech. Dig.*, Dec. 2009, pp. 319–322.

[2] M. Egard, L. Ohlsson,M. B. M. Borg, F. Lenrick, R. Wallenberg L.-E. Wernersson, and E. Lind, "High transconductance self-aligned gate-last surface channel In0.53Ga0.47As MOSFET," in *IEDM Tech.Dig.*, pp.13.2.1-13.2.4, Dec. 2011

[3] Yonai, Y.; Kanazawa, T.; Ikeda, S.; Miyamoto, Y.; , "High drain current (>2A/mm) InGaAs channel MOSFET at V_D=0.5V with shrinkage of channel length by InP anisotropic etching," Electron Devices Meeting (IEDM) vol., no., pp.13.3.1-13.3.4, 5-7 Dec. 2011

[4] A. D. Carter, W. Mitchell, B. Thibeault, J. Law, and M. J.W. Rodwell, "Al2O3 Growth on (100) In0.53Ga0.47As Initiated by Cyclic Trimethylaluminum and Hydrogen Plasma Exposures, " *Applied Physics Express,* Vol. 4, Issue 9, pp. 091102-091102-3 2011.

[5] Mark Rodwell, et al. "III-V FET Channel Designs for High Current Densities and Thin Inversion Layers" IEEE Device Research Conference, June 21-23, 2010

[6] U. Singisetti, et al., "In$_{0.53}$Ga$_{0.47}$As Channel MOSFETs with Self-Aligned InAs Source/Drain Formed by MEE Regrowth," *IEEE Electron Device Lett.,* vol.30, no.11, pp.1128-1130, Nov. 2009

[7] G. Burek, Y. Hwang, A. Carter, V. Chobpattana, J. Law, W. Mitchell, S. Stemmer, M. J. W. Rodwell "Influence of gate metallization processes on the electrical characteristics of high k/In0.53Ga0.47As interfaces," *Journal of Vaccum Science and Technology B*. Vol. 29, No. 4, Jul 2011

Epitaxy of III-V based channels on Si and transistor integration for 12-10nm node CMOS

Matty Caymax, Clement Merckling, Gang Wang[1], Tommaso Orzali[2], Weiming Guo, Wilfried Vandervorst[3], Johan Dekoster, Niamh Waldron and Aaron Thean.

Imec Belgium, Kapeldreef 75, B-3001 Leuven, Belgium; [1]Now with MEMC Electronic Materials, Inc; [2]Now with Aixtron AG, Aachen, Germany; [3]Also with KULeuven, Dpt of Physics, Leuven, Belgium

Abstract — **Moore's Law describes the scaling of Si-based CMOS technology in terms of performance, power consumption, area and cost. As we have reached the physical limits of scaling Si channels, alternative materials with higher carrier mobility such as Ge and IIIV compound semiconductors are in order. This paper reviews some of imec's work on introducing $In_{0.53}Ga_{0.47}As$ in a manufacturable and integratable way into mainstream Si-based CMOS technology. Several major issues are known: dielectric/IIIV interface passivation, mismatch of lattice and crystal structure between IIIV and Si, small bandgap leading to enhanced leakage,... We will discuss mainly the epitaxial growth aspects and the integration of IIIV materials in Si MOSFET devices, and point out some more unexpected materials and device issues.**

Index Terms — **CMOS, compound semiconductors.**

I. INTRODUCTION

Over the past five decades, exponential gains in compute power have fuelled unprecedented progress in innovation and economical growth which has provided considerable benefits to society and has also enabled entirely new businesses such as e-commerce, social networking and mobile devices. This expected progress however now is at stake: we have reached the limit of what seems possible with traditional and equivalent scaling of silicon based MOS devices. The main rationale behind scaling is that with every subsequent node, the circuits' performance increases by about 25%, while power consumption is reduced by 20%, area by 50%, and cost by 15%. Slowing down any of these drivers will slow down scaling. Dimensional scaling was sufficient down to the 130nm node, from where equivalent scaling took over: straining the Si channel boosts mobility and hence drive currents (90-65nm node), high-k oxides help reducing gate leakage (65-45nm node), Trigate or FinFET transistors [1] allow better electrostatic control (22-16nm nodes). Below 16nm however, further progress will need even higher carrier mobilities than what can be delivered by strained Si, and alternative channel materials have been proposed over the last couple of years – Ge for pMOS, IIIV and especially InGaAs for nMOS [2].

Whereas the road to integrating Ge channels on Si for enhanced pMOS is getting clear [3], several hurdles still have to be taken before IIIV materials will make their way to integration on Si. An important requirement is that in all cases, these materials must be deposited epitaxially on Si. Because of their low cost, high mechanical strength, extremely high quality and all this on large diameters (currently 300mm, 450mm in near future), Si substrates are the unrivalled workhorse for high volume CMOS manufacturing. The lattice mismatch with Si (4% for Ge, 8% for InP, 12 % for InAs) and the different crystal structure (non-polar covalent diamond lattice for Si and Ge, polar zincblende for IIIV) are an important source of crystalline defects (dislocations, stacking faults, anti-phase boundaries). Another problem area is the passivation of the MOS gate stack interface. The interface quality (D_{it}) of Ge and IIIV MOS structures is no real show-stopper anymore but should still be improved for higher mobility and lower sub-threshold swing (SS), especially when targeting the combination of low D_{it} and low EOT (<1nm). More importantly, more work is needed to understand and control border trap (or slow trap or oxide trap) issues. Also, low contact resistance for n-type contacts on IIIVs needs further improvements.

II. GROWTH OF IIIV ON SI FOR INTEGRATION IN CMOS

It is anticipated that these highly performing devices will be used only in those circuits of a CPU or analog/RF front end chips where high speed is required, while their more conservative Si counterparts are available for all non-core peripheral functions such as I/O, see Fig 1. Therefore, a scheme allowing co-integration of Ge and IIIV together with Si channels is highly desired. Selective epitaxial growth (SEG,

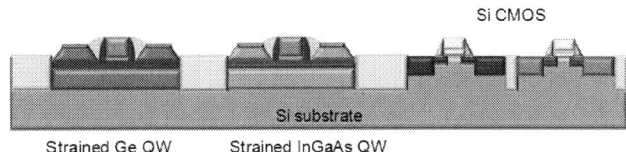

Figure 1Ge, IIIV and Si FET co-integrated on Si substrate

978-1-4673-1725-2/12 $31.00 © 2012 IEEE

or Selective Area Growth, SAG) of the IIIV material in narrow trenches between SiO$_2$ isolation structures allows defects to be confined and trapped at the vertical interface between the growing IIIV and the oxide [4], coined as "Aspect Ratio Trapping" [5]. Since the standard vertical device isolation in state-of-the-art Si CMOS circuits uses Shallow Trench Isolation (STI), removal of the Si in between the STI oxide is an easy and cost-effective way of creating trenches. Moreover, in this way one follows exactly the standard design rules for that specific node, ensuring a seamless and elegant integration of these IIIV devices in Si CMOS circuits. Fig. 2 outlines the main steps towards full channel growth comprising formation of the trenches, Ge seed deposition, InP filling of the trench, InP CMP and channel stack growth.

A. Trench etching and Ge seed deposition, substrate orientation

Starting from a standard STI substrate, trenches are created by etching out Si by means of a thermal HCl etch in a production type group IV epitaxial tool (ASM Epsilon single wafer epi reactor). Typically, epitaxial growth of the polar IIIV layers on non-polar group IV substrates (Si, Ge) results in the formation of anti-phase boundaries [6] in {111} or {110} planes. This can be avoided by the use of

(100) oriented substrates which have been cut under a small angle (4-6°) towards the [111] or [110] axis, resulting in numerous double-atomic steps on the surface after the standard pre-epitaxial bake [7]. As Ge is more prone to double-step formation than Si, we deposit a thin epitaxial seed layer of Ge first in the group IV epi reactor before transfering the substrate to a IIIV MOCVD reactor (Aixtron Crius). This off-orientation substrate approach however has as drawback that different defect structures and surface morphology will result depending on the orientation of the trench (parallel or perpendicular to the surface step orientation). Therefore, we developed an alternative technique of step creation on on-axis cut (001) wafers [8] based on a curved Ge surface which follows the contours of a cup-shaped Si trench bottom. This cup shape is created by

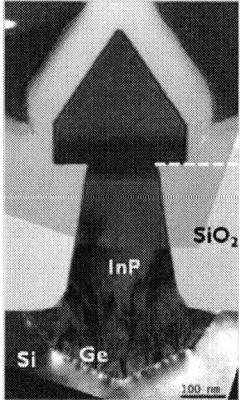

Figure 2 InP grown selectively on top of Ge seed layer in trench. The trench underetches the SiO2. InP overgrows the STI oxide, indicated by dotted line

performing the HCl etch reaction in the kinetic etch control regime, which results in the formation of {111} and {113} facets at the trench bottom, see Fig.3. Next, a thin Ge layer (~50nm) is grown epitaxially on the faceted Si bottom. After transfering the wafer to the IIIV MOCVD reactor and pre-epi bake at 720C, the Ge suface shows an elliptical shape due to surface tension. The Ge layer is found to be fully relaxed (Fig.4) after this bake showing a dense netwerk of interfacial misfit dislocations and threading dislocations.

B. InP SEG inside STI trench

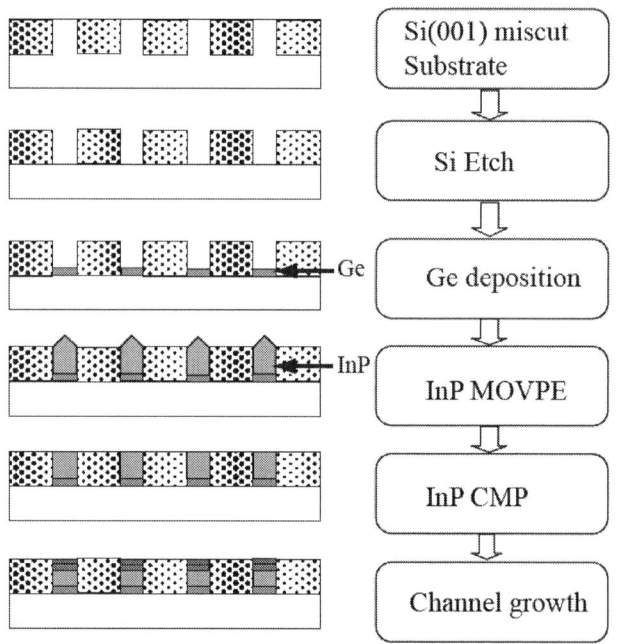

Figure 3 Process flow for IIIV SEG and ART in narrow STI trenches

Figure 4 Longitudinal composite x-TEM picture of InP/Ge in 10x0.2 μm² trench, showing the highly defective region in the lower InP layer. The top part still shows some defects in {111} planes and an APB resulting in a surface step

The Ge surface is annealed in tertiarybutylarsine TBA to create double-atomic steps between the terraces resulting from the intersection of the curved Ge surface with the (001) surface. The InP deposition then starts with a low temperature seed layer (420°C, 30nm thick) to suppress In mobility and so create a smooth InP starting layer surface. The main InP buffer layer is then grown at 640°C for better crystalline quality. Fig.3 shows clearly the very high density of crystalline defects in the lower part of the InP layer, partially seeded from the threading defects in the Ge layer and partially accompanying the Ge-InP lattice mismatch relaxation. The top part of the layer however is virtually defect free due to the defect trapping effect at the oxide side walls. The longitudinal cross-section in Fig. 4 however shows there are still defects in the InP layer, with an estimated density of ~5x10⁸/cm². Typical defect densities are ~ 1-2x10⁹/cm². The 'averaged' X-ray diffractogram of the InP/Ge/Si structure in a series of narrow trenches (500-20nm) in Fig. 5 shows well-defined fully relaxed InP and Ge diffraction peaks next to the Si substrate peak.

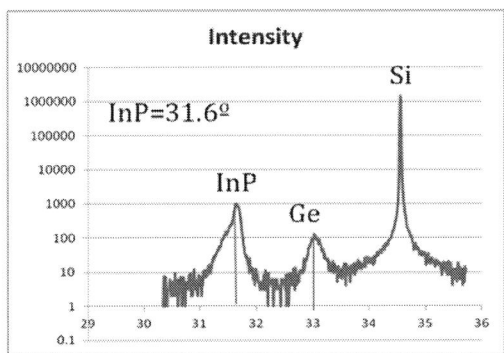

Figure 5 (004) XRD diffractogram of InP/Ge/Si filled array of trenches

C. InP CMP, InGaAs channel layer growth

Since facets are very difficult to avoid (Fig. 3), we prefer to grow the InP out of the trench, chemical-mechanical polish (CMP) the surface, and then finalize the

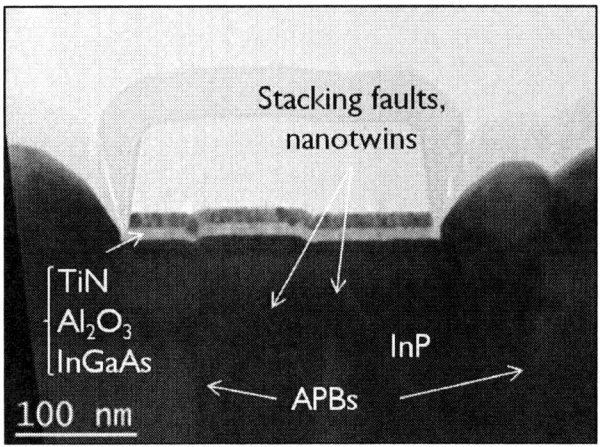

Figure 6 x-TEM of completed gate stack (TiN, Al2O3) on In$_{0.53}$Ga$_{0.47}$As channel on InP buffer in STI trench, also showing extended defects (twins, SFs, APBs) in the epi layer stack. Left and right from gate stack n⁺ InGaAs source.drain regions are visible

layer structure with the InGaAs channel layer. As shown in Fig.6, the remaining extended defects can cause steps in the channel surface which are duplicated in the gate stack.

III. INGAAS MOSFET INTEGRATION

The final device architecture for IIIV (and Ge) devices in 12-10nm node CMOS will be FinFET devices. For this first integration exercise which should be completely compatible with a regular manufacturing process flow, we chose however for a simpler planar implant-free quantum well device, see Fig.7. After InP CMP and regrowth of a 15nm InGaAs channel layer, a 10nm ALD Al$_2$O$_3$/10 TiN gate metal/80nm hard mask SiO$_2$ stack is deposited. After gate stack patterning and Al$_2$O$_3$/TiN dry etch, the remaining Al$_2$O$_3$ is removed to allow epitaxial growth of n⁺ in-situ doped InGaAs source/drain regions after formation of Si$_3$N$_4$ spacers. Removal of the SiO$_2$ hardmask on top of the gate stack is followed by pre-metal dielectric, Ti/TiN liner, W plugs to

978-1-4673-1725-2/12 $31.00 © 2012 IEEE

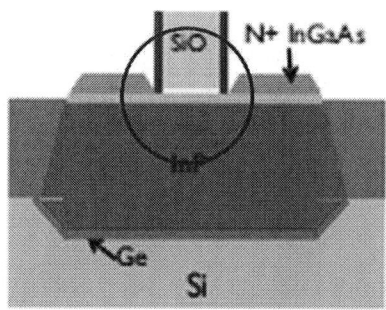

Figure 7 Integration of InGaAs channel in MOS device before back-end metal (cross-section perpendicular to Figs.2&3); an x-TEM view of the part in the circle is shown in Fig.6. The layer between the n+ InGaAs S/D and the InP buffer is the undoped InGaAs channel

form metal contacts to gate and source/drain and completed with a standard level one Cu metal damascene back-end [9]. Fig. 7 shows a cross-section of the device before the metal back-end.

Although simplified gate stack and contact modules have been used preventing fully optimized device operation, a more serious issue has been encountered which is heavy leakage between source and drain, hiding almost entirely drain current modulation by the gate bias. The root cause for this leakage has been found to be inter-diffusion at the InP/Ge interface, resulting in an all n-type doped leakage path into the Si substrate. PL measurements [9] indicate an active n-type doping level in InP on the order of $10^{17}/cm^3$, probably due to diffusion of Ge which is an n-type dopant into the InP layer. Scanning Spreading Resistance Measurements (SSRM) [11, 12] on the cross section of this device (Fig. 8) show a netto n-type doping in the Si-substrate, apparently resulting from P-diffusion through the Ge into the Si. This is confirmed by SIMS measurements on a device sample after stripping the metal, InP and field dielectric showing clearly the P diffusion profile in the Si, see Fig. 8. The physical mechanism of this P

diffusion is not yet well understood in view of the very limited thermal budget the wafer has seen during processing. Recent studies indicate a defect (and stress?) mediated interdiffusion operative at very low temperatures [13].

IV. CONCLUSIONS

Extending the ITRS roadmap and CMOS scaling beyond the 16nm node will come to an end unless the industry considers introducing Ge and IIIV high mobility channel materials in their process flows. We show how SEG of IIIVs (similar to Ge) on standard STI starting substrates is a very powerful approach to integrating these materials in a manufacturing compatible way into VLSI process flows. Major issues which require further work are the defectivity in the IIIV layers as well as the unexpected important interdiffusion phenomena leading to unacceptable leakage levels.

ACKNOWLEDGEMENTS

The authors thank the imec Core Partners with the Logic-DRAM Program.

REFERENCES

[1] http://eda360insider.wordpress.com/2011/05/19/3d-thursday-intel-and-finfets-tri-gate-transistors

[2] M. Heyns and W.Tsai, *Mrs Bulletin 34*, 485-492, 2009.

[3] M.Caymax, G.Eneman, F.Bellenger, C.Merckling, A.Delabie, G.Wang, R.Loo, E.Simoen, J.Mitard, B.De Jaeger, G.Hellings, K.De Meyer, M.Meuris, M.Heyns, "Germanium for advanced CMOS anno 2009: a SWOT analysis", *IEDM Technical Digest*, pp.461-464, 2009.

[4] E.A. Fitzgerald and N.Chand, *Electron. Mater*, 20, 839-53, 1991.

[5] J.Z. Li, J. Bai, C. Major, M. Carroll, A. Lochtefeld, Z. Shallenberger, *Journal of Applied Physics*, 103 (10) 106102-106102-3, 2008.

[6] H. Kroemer, *J. Cryst. Growth* 81, (1987) 193

[7] T. E. Crumbaker, M. J. Hafich, G. Y. Robinson, K. M. Jones, M. M. Aljassim, A. Davis, and J. P. Lorenzo, *Appl. Phys. Lett.* 59, 1090, 1991

[8] G.Wang, M.Leys, D.Nguyen, R.Loo, G.Brammertz, O.Richard, H.Bender, J.Dekoster, M.Meuris, M.Heyns, M.Caymax, *Journal of the Electrochemical Society*, 157, pp.H1023-H1028, 2010.

[9] N. Waldron, G. Wang1, N.D. Nguyen, T. Orzali, C.Merckling, G. Brammertz, P. Ong, G. Winderickx, G. Hellings, G.Eneman, M. Caymax, M. Meuris, N. Horiguchi and A. Thean, *ECS Transactions* 45, 115, 2012

[10] D.Lin, N.Waldron, G.Brammertz, K.Martens, W.Wang, S.Sioncke, A.Delabie, H.Bender, T.Conard, W.Tseng, S.Lin, K.Temst, A.Vantomme, J.Mitard, M.Caymax, M.Meuris, M.Heyns, T.Hoffmann, *ECS Transactions* 28, 173, 2010

[11] D.Alvarez, J.Hartwich, M.Fouchier, P.Eyben, W.Vandervorst, *Applied Physics Letters* 82(11) 1724-1726 (2003)

[12] P.De Wolf, M.Geva, T.Hantschel, W.Vandervorst, R.B.Bylsma, *Applied Physics letters* 73(15) 2155-2157 (1998)

[13] W.Vandervorst, P.Eyben, B.Douhard, K.Arstila, C.Merckling, T.Orzali, to be published

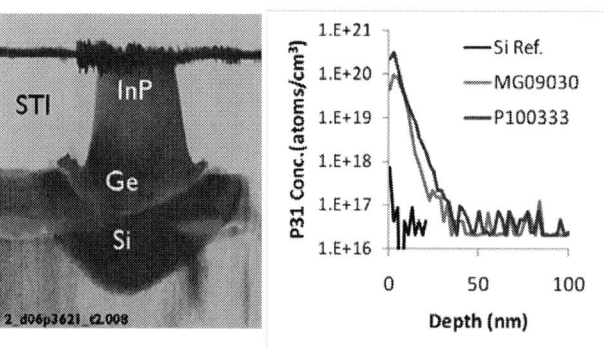

Figure 8 Left pane: SSRM on InP/Ge/Si in trench showing highly conductive Ge and Si regions; Right pane: SIMS measurements of P diffusion into the Ge layer after stripping metal, field oxide and InP from two different device samples

Novel atomic layer deposited thin film beryllium oxide for InGaAs MOS Devices

D. Koh[1], J. H. Yum[1,2], T. Akyol[1], D. A. Ferrer[1], M. Lei[3], Todd. W. Hudnall[5], M. C. Downer[3], C. W. Bielawski[4], R.Hill[2], G. Bersuker[2] and S. K. Banerjee[1]

[1]Microelectronics Research Center, Department of Electrical and Computer Engineering, The University of Texas at Austin, Austin, Texas 78758, USA

[2]Sematech Inc., 2706 Montopolis Drive, Austin, Texas 78741, USA

[3]Department of Physics, The University of Texas, Austin, Texas 78712, USA

[4]Department of Chemistry, The University of Texas, Austin, Texas 78712, USA

[5]Department of Chemistry and Biochemistry, Texas State University, San Marcos, TX 78666 USA

Abstract — Metal-oxide-semiconductor (MOS) capacitors and field-effect-transistors (FET) with atomic layer deposited (ALD) BeO and Al_2O_3 dielectric layers on InGaAs substrate were fabricated for electrical characterization and device performance comparison. Physical characterization of BeO grown film on Si and InGaAs substrates were done using TEM, AFM, and XPS. BeO devices show decent C-V results, lower dielectric leakage current and interface defect density. High thermal stability, large energy band-gap and strong diffusion barrier properties of BeO along with ALD self-cleaning effects on the interface makes it an excellent candidate for a dielectric or interface passivation layer for InGaAs MOS devices.

Index Terms — ALD, Beryllium Oxide, high- κ, MOS Capacitance, III-V MOSFETs

I. INTRODUCTION

Superior electron transport properties of InGaAs over Si make it attractive for CMOS technology beyond 22 nm-node. The bane of III-V MOSFET technology is the lack of a suitable gate insulator with minimal defects on the semiconductor/oxide interface [8-10]. Introduction of Atomic Layer Deposition (ALD) and high-κ materials mitigated this chemical barrier and made the III-V channels a component of the MOSFET menu. A proper high-κ material should satisfy several requirements such as chemical and thermal stability, high energy band-gap, low interface defect density and roughness, high mobility of charge carriers and low leakage current density.[1,2] Currently, the most promising high-κ gate oxide dielectrics are Al_2O_3, HfO_2 and ZrO_2. However, these dielectric films still contain the limitation of the performance and reliability of MOS devices by suffering from high density of electron traps, significant changes in interface and bulk properties after high temperature annealing and charge carrier mobility degradation due to remote Coulomb scattering.

Recently, by employing ALD beryllium oxide (BeO) as a gate dielectric, excellent results such as smaller capacitance-voltage (C-V) hysteresis, lower frequency dispersion, interface defect density (D_{it}) and gate leakage current have been achieved compared to Al_2O_3 with same effective oxide thickness (EOT) [3,4]. In addition, BeO has excellent thermal stability on Si and III-V substrates and large energy band-gap (10.6 eV) [5]. It is an excellent oxygen diffusion barrier because of short bond length and strongly covalent bonding of Be-O and dense structure [6]. Dielectric constant of BeO is about 6.8, comparable with that of ALD grown Al_2O_3 (7.1 nm). Low phonon scattering and high thermal conductivity attributes of BeO allows it to be an excellent choice for high performance and low power dissipation devices.

In this work, we report physical analysis of epitaxial BeO films grown on the InGaAs substrate, electrical characterization and comparison of surface channel InGaAs MOSFETs with ALD BeO and Al_2O_3 as a high- κ gate dielectric.

II. EXPERIMENTAL

An ALD BeO precursor is not commercially available. By a novel process, $Be(CH_3)_2$ (dimethylberyllium) was synthesized from $BeCl_2$ using Grignard metathesis. The purity of this precursor, as synthesized, is very low and it contains significant amount of Et_2O and carbon impurities which required sublimation of this precursor in the ALD system before use.

Semi-insulating (100) InP wafers with 200 nm undoped $In_{53}Ga_{47}As$ epi-layer grown on them were used for our gate-last MOSFET fabrication process. Surface of the InGaAs layer was cleaned of native oxides in 1% HF solution for 1 minute. A capping layer of 100Å ALD-Al_2O_3 dielectric was deposited to prevent out diffusion and surface degradation during the activation annealing. On lithography layer 1, alignment marks were formed by wet-etching Al_2O_3 and InGaAs in BOE and $HCl+HNO_3+H_2O$ (1:1:1) solutions, respectively. On layer 2, source/drain (S/D) regions were defined and doped with Si ion implantation at 35 keV, 5 x $10^{14}/cm^2$. After removal of photoresist, S/D activation anneal was done at 700 °C for 10-15s and the capping Al_2O_3 layer was removed using buffered oxide etchant (BOE). Surface was cleaned and passivated by 1 minute dip in 1% HF solution and 10 minute dip in 20% $(NH_4)_2S$ solution at room temperature. Subsequently, 50-90 Å BeO or Al_2O_3

gate dielectrics were grown in the ALD system at 250°C.

For ALD BeO growth, Be(CH₃)₂ precursor was heated up to about 150°C and pulsed into the ALD chamber for 0.4s alternating with 0.025s water pulses. Next, a post-deposition (rapid thermal) annealing (PDA) step was performed at 500°C for 2 minutes in N_2 ambient, followed by a 2000Å thick TaN metal-gate deposition using a dc magnetron sputtering system. On layer 3, TaN gate was patterned and etched by CF_4 plasma RIE process. On layer 4, S/D contacts patterned and BeO/Al_2O_3 dielectric layer is etched in BOE. For S/D metallization 40/10/50nm AuGe/Ni/Au was deposited using e-beam evaporation. Process was completed with lift-off in acetone and annealing at 425 °C for 30s.

A schematic of the resulting device cross-section and the ring-type MOSFET device layout are illustrated in Figure 1. Cross-sectional Transmission Electron Microscopy (TEM) images of BeO on InGaAs and Si are shown in Figure 2, demonstrating the layer-by-layer crystalline growth of ALD BeO and a sharp interface with InGaAs surface. Surface Atomic Force Microscope (AFM) images of ALD grown BeO and Al_2O_3 films on InGaAs substrate in Figure 2 compare surface roughness values. BeO deposition indicates a very low RMS surface roughness of 0.19 nm in 3X3 μm scan area.

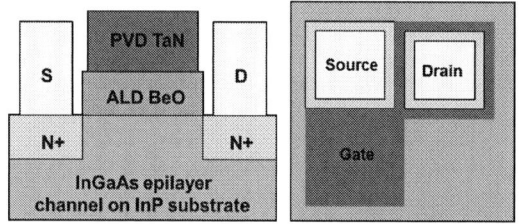

Fig. 1. Schematic of device cross-section and the ring-type MOSFET device layout.

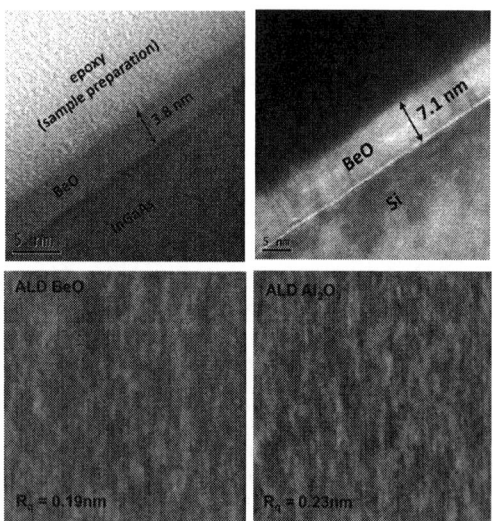

Fig. 2. High resolution cross-sectional TEM analysis of ALD BeO grown on InGaAs and Si substrates. AFM and surface roughness analysis of ALD BeO and Al_2O_3 grown on InGaAs substrate.

III. RESULTS & DISCUSSION

X-ray photoelectron spectroscopy (XPS) results in Figure 3 demonstrate the interface quality comparison between Al_2O_3 and BeO samples with and without PDA. First, we see that BeO is indeed successfully deposited by ALD using our dimethyl beryllium precursor. Second, it has more efficient self-cleaning effect on InGaAs substrate by removing its native oxides (In_xO_y, Ga_xO_y and As_xO_y) [7]. Third, BeO has better thermal stability compared to ALD Al_2O_3 by showing less oxide diffusion after PDA.

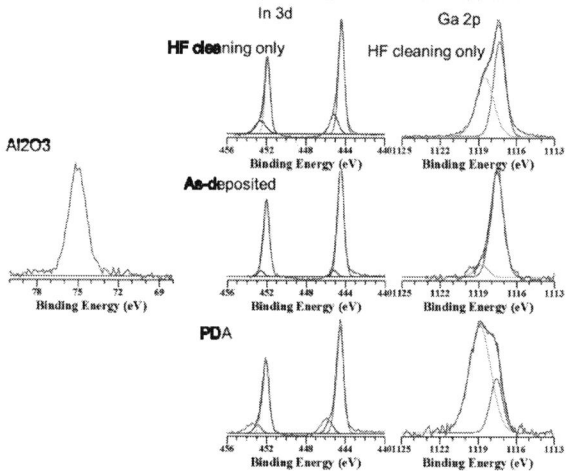

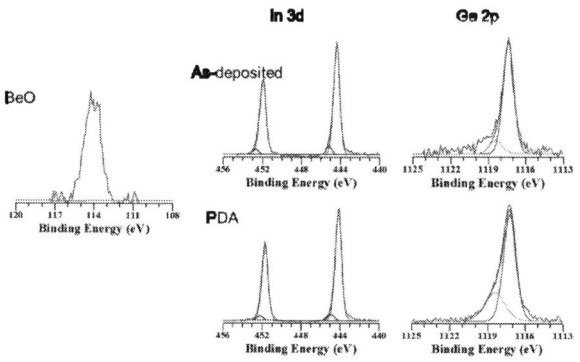

Fig. 3. XPS analysis of ALD grown Al_2O_3 and BeO interfaces with InGaAs substrate.

C-V and I-V characteristics of MOS Capacitors with ALD BeO dielectric are superior to its Al_2O_3 counterpart in terms of lower frequency dispersion, interface state density (D_{it}) and leakage current (Figure 4). D_{it} at midgap calculated by conductance method for BeO and Al_2O_3 are 7.44×10^{11} cm²/eV and 1.05×10^{12} cm²/eV, respectively. This indicates a cleaner InGaAs-dielectric interface for ALD BeO compared to Al_2O_3. Lower frequency dispersion after high temperature annealing is a result of BeO being a better oxygen diffusion barrier than Al_2O_3.

978-1-4673-1725-2/12 $31.00 © 2012 IEEE

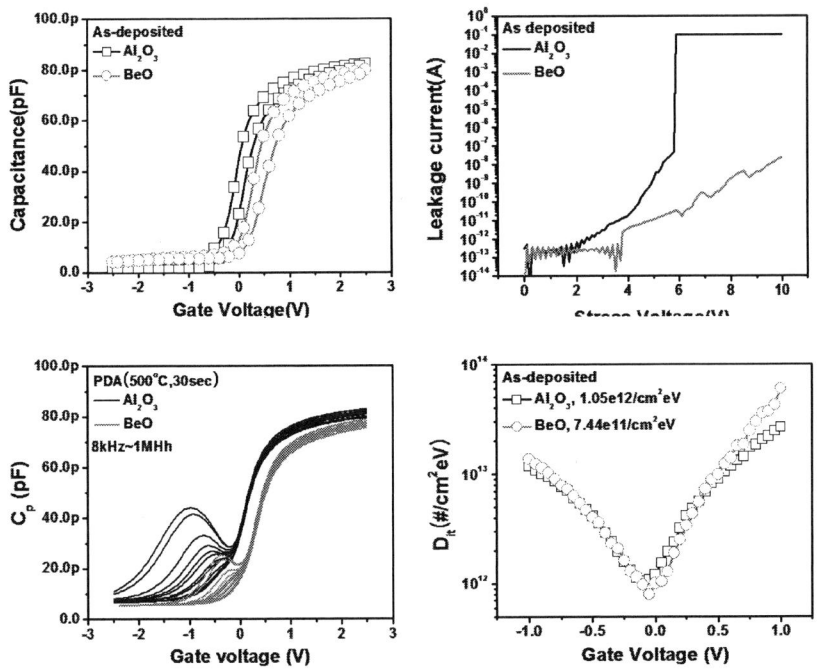

Fig. 4.   C-V and I-V characteristics of n-InGaAs MOS Capacitors with BeO and Al$_2$O$_3$

Figure 5 summarizes the dc output characteristics of n-channel MOSFETs fabricated with ALD Al$_2$O$_3$ and BeO gate dielectrics. Channel carrier mobility is generally regarded as one main figure of merit benchmarking the MOSFET performance. Calculated effective mobility of InGaAs n-MOSFETs with BeO dielectric (using split C-V method) exhibit maximum electron mobility of 1477 cm^2/V.s, which is about 26% higher than that of Al$_2$O$_3$ with 1172 cm^2/V.s

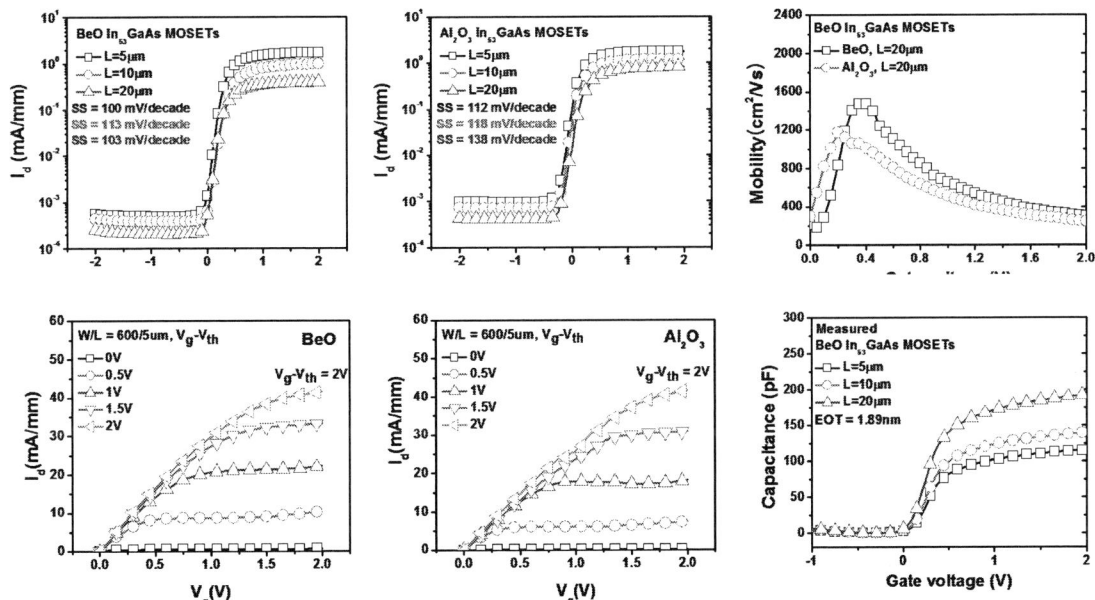

Fig. 4.   DC output characteristics of n-channel InGaAs MOSFETs with BeO and Al2O3 dielectrics.

978-1-4673-1725-2/12 $31.00 © 2012 IEEE

The minimum subthreshold swing (SS) of InGaAs MOSFETs (L=5um) with BeO dielectric is 100 mV/decade, compared to that with Al_2O_3 (112 mV/decade). BeO MOSFETs also show better performance characteristics in terms of lower threshold voltage, higher maximum transconductance, drive current and current on/off ratio. We have also investigated stress-induced flat-band shift and leakage current and time dependent dielectric breakdown, and confirmed that the BeO/InGaAs interface is much more stable than Al_2O_3/InGaAs. This is a result of less charge trapping and less oxygen vacancies within the BeO film. Even though our ALD BeO quality and fabrication process were not fully optimized, it can be concluded from these MOSFET characteristics that ALD BeO has great potential for application in III-V technology.

IV. CONCLUSION

In summary, for the first time, we have fabricated and characterized surface channel n-MOSFETs with InGaAs channel and ALD BeO dielectric. These MOSFETs show improved device characteristics compared to that for Al_2O_3. Introducing ALD BeO as a gate dielectric or interface passivation layer (IPL) may be the key solution to overcome the performance limit of current III-V MOS devices with high-κ dielectrics

ACKNOWLEDGEMENTS

This work was supported in part by DARPA, Micron Foundation, Robert Welch Foundation grant F-1038, and NSF grant DMR-0706227

REFERENCE

[1] M. Houssa et al., Materials Science and Engineering: R: Reports 51, 37-85 (2006).
[2] E. Gusev et al., "Ultrathin high-K metal oxides on silicon: processing, characterization and integration issues" Microelectronic Engineering, 59, 341-349 (2001).
[3] J. H. Yum and T. Akyol et al., "Atomic layer deposited beryllium oxide: Effective passivation layer for III-V metal/oxide/semiconductor devices" J. App. Phys., 109, 064101 (2011).
[4] T. Akyol and J. H. Yum et al., "Introduction of ALD Beryllium oxide gate dielectric for III–V MOS devices" 69th Annual Device Research Conference (2011).
[5] S. Oktyabrsky and P. Ye, Fundamentals of III-V Semiconductor MOSFETs (Springer, 2010), p 94.
[6] J. Robertson et al., "High dielectric constant oxides" The European Physical Journal Applied Physics 28, 265-291 (2004)
[7] W. Melitz and J. Shen et al., J. App. Phys., 110, 013713 (2011)
[8] K. Kalna, and J. A. Wilson, et al., "Monte Carlo simulations of high-performance implant free In0.3Ga0.7As nano-MOSFETs for low-power CMOS applications," IEEE Trans. Nanotechnol., vol. 6,no. 1, pp. 106–112, Jan. 2007.
[9] M. Passlack, and R. Droopad, et al., "High mobility NMOSFET structure with high-k dielectric," IEEE Electron Device Lett., vol. 26, no. 10, pp. 713–715, Oct. 2005.
[10] P. Zimmerman, "High performance Ge pMOS devices using a Sicompatible process flow," IEDM Tech. Dig., 2006, pp. 1–4

Sulfur cleaning for (100), (111)A, and (111)B InGaAs surfaces with In content of 0.53 and 0.70 and their Al₂O₃/InGaAs MOS interface properties

*Masafumi Yokoyama [1, *], Noriyuki Taoka[1], Rena Suzuki[1],*
Osamu Ichikawa[2], Hisashi Yamada[2], Noboru Fukuhara[2], Masahiko Hata[2],
Masakazu Sugiyama[1], Yoshiaki Nakano[1], Mitsuru Takenaka[1], and Shinichi Takagi[1]

[1]The University of Tokyo, 7-3-1 Hongo, Bunkyo-ku, Tokyo 113-8656, Japan,
Phone: +81-3-5841-6733, Fax: +81-3-5841-8564
[2]Sumitomo Chemical Co. Ltd., 6 Kitahara, Tsukuba, Ibaraki 300-3294, Japan

[*]yokoyama@mosfet.t.u-tokyo.ac.jp

Abstract — **We have studied the impact of sulfur cleaning for (100), (111)A, and (111)B InGaAs surfaces with In content of 0.53 and 0.70 and their Al₂O₃/InGaAs MOS interfaces using an (NH₄)₂Sₓ solution. We found that the sulfur cleaning can effectively remove native oxides from InGaAs surfaces and passivate the InGaAs surfaces by sulfur atoms. Combining with trivalent oxide passivation using Al₂O₃, the Al₂O₃/InGaAs MOS interfaces with sulfur cleaning can realize good interface properties with the small frequency dispersion of the capacitance *versus* voltage ($C - V$) curves in accumulation region with independent of In content and the surface orientations.**

Index Terms — **III-V semiconductor materials, Semiconductor-insulator interfaces, MOS devices, CMOS technology, Capacitors, surface engineering .**

I. INTRODUCTION

III-V compound semiconductor channel MOSFETs are one of the promising key technologies to enhance the performance of CMOS devices in post scaling generation [1]. InGaAs is a promising channel material because of the high electron mobility, light electron mass and their good Al₂O₃/InGaAs MOS interfaces formed by atomic-layer-deposition (ALD) [2]-[6]. We have demonstrated a 1-nm-capacitance-equivalent-thick HfO₂/Al₂O₃/In₀.₅₃Ga₀.₄₇As MOS structure with low interface trap density (D_{it}) and low gate leakage current density [7]. On the other hand, it is also interesting to use the surface orientations other than (100). It has been reported that (111) InGaAs MOSFETs exhibit higher electron mobility even under high effective field than the Si universal mobility and the one in (100) InGaAs MOSFETs [8]-[10]. It was reported for (111)A GaAs that this surface orientation reduces the D_{it} values [11]. In contrast, it has also been reported that a (111) InGaAs surface has exhibited D_{it} values of as high as the (100) one [10]. From the viewpoint of the D_{it} reduction, the surface treatment prior to ALD-Al₂O₃ is quite important. Here, sulfur cleaning is one of the most common pre-treatment for InGaAs gate stacks. However, the impact of the sulfur cleaning for

(100), (111)A, and (111)B InGaAs surfaces and their MOS interface properties have not been well studied yet.

In this paper, we have examined the impact of sulfur cleaning on (100), (111)A, and (111)B InGaAs surfaces with In content of 0.53 and 0.70 by (NH₄)₂Sₓ solution and their Al₂O₃/InGaAs MOS interface properties. As a result, we have found that the sulfur cleaning realizes the small frequency dispersion of $C - V$ curves in accumulation region, indicating good MOS interface properties.

II. SULFUR CLEANING

The experiments were carried out for *n*-type 500-nm-thick In₀.₅₃Ga₀.₄₇As and In₀.₇₀Ga₀.₃₀As layers with doping concentration (N_D) of 3×10^{16} cm⁻³ grown on (100), (111)A, and (111)B InP wafers by metal organic chemical vapor deposition. After wafer surfaces were pre-cleaning, Al₂O₃ gate dielectrics were deposited by 10 cycles for x-ray photoelectron spectroscopy (XPS) measurements and by 90 cycles for MOS capacitors by using ALD at 200 ºC. The deposition rate of Al₂O₃ on (100), (111)A, and (111)B InGaAs surfaces was ~ 0.11 nm/cycle, irrespective of the InGaAs surfaces. Here, the following three pre-cleaning conditions were used; (1) only NH₄OH cleaning: 29% NH₄OH for 1 min and deionized water (DIW) rinse for 1 min, (2) only (NH₄)₂Sₓ cleaning: 0.6 – 1.0% (NH₄)₂Sₓ for 10 min and DIW rinse for 1 min, and (3) S cleaning: 29% NH₄OH for 1 min and DIW rinse for 1 min followed by 0.6 – 1.0% (NH₄)₂Sₓ for 10 min and DIW rinse for 1 min. We also prepared the wafers with native oxide layers. InGaAs MOS capacitors were fabricated with Al gate electrodes and followed by rapid thermal annealing (RTA) at 350 ºC for 10 sec.

We have investigated the impact of sulfur cleaning for the InGaAs surfaces. Figures 1(a) and 1(b) show As 2$p_{3/2}$ and S 2$p_{3/2}$ photoelectron spectra for native oxides, only NH₄OH, only (NH₄)₂Sₓ, and S cleaning on (100) InGaAs surfaces and S cleaning on (111)A, and (111)B InGaAs surfaces, respectively.

We found that the $(NH_4)_2S_x$ solution effectively removes III-V oxides form the InGaAs surfaces regardless of the surface orientation. On the other hand, the sulfide peaks are found after $(NH_4)_2S_x$ cleaning, which is independent of the NH_4OH treatment and the surface orientation. The sulfide peaks are confirmed after ALD-Al_2O_3 deposition and after RTA process regardless of the surface orientation. Figure 2(a) shows the As $2p_{3/2}$ photoelectron spectra for only NH_4OH, only NH_4OH with ALD-Al_2O_3, S cleaning, and S cleaning with ALD-Al_2O_3 on (100) InGaAs. Figures 2(b) and 2(c) show the As $2p_{3/2}$ photoelectron spectra for only NH_4OH, S cleaning, and S cleaning with ALD-Al_2O_3 on (111)A and (111)B InGaAs surfaces, respectively. The arsenide oxide peaks from As_2O_5 were eliminated by $(NH_4)_2S_x$ cleaning. On the other hand, the arsenide oxide peaks from As_2O_3 are still observed for NH_4OH cleaning and ALD. In addition, the arsenide oxide peaks from As_2O_3 slightly increase after ALD, indicating that the InGaAs surfaces were passivated by As_2O_3, which is a trivalent oxide as Al_2O_3 [4]-[7]. The sulfide peaks are found after $(NH_4)_2S_x$ cleaning, which is independent of the surface orientation. The sulfide peaks were confirmed after ALD-Al_2O_3 deposition independent of surface orientation. These characteristics are basically the same among (100), (111)A, and (111)B. As a result, we can conclude that $(NH_4)_2S_x$ solution even with low concentration of 1% can remove the III-V oxides effectively.

Next, we also studied the impact of S cleaning for (100), (111)A, and (111)B InGaAs MOS interfaces with In content of 0.70. We also prepared the samples with native oxides, S cleaning, S cleaning with ALD-Al_2O_3 as the same manner as In content of 0.53. Figure 3 shows As $2p_{3/2}$ and S $2p$ photoelectron spectra of (100), (111)A, and (111)B InGaAs with native oxides, S cleaning, and S cleaning with Al_2O_3, respectively. Figures 3(a), 3(c), and 3(e) are for As $2p_{3/2}$ and Figs. 3(b), 3(d), and 3(f) are for S $2p$ photoelectron spectra, respectively. We also found that S cleaning effectively removes III-V oxides form the InGaAs surface independent of the surface orientations as InGaAs with In content of 0.53.

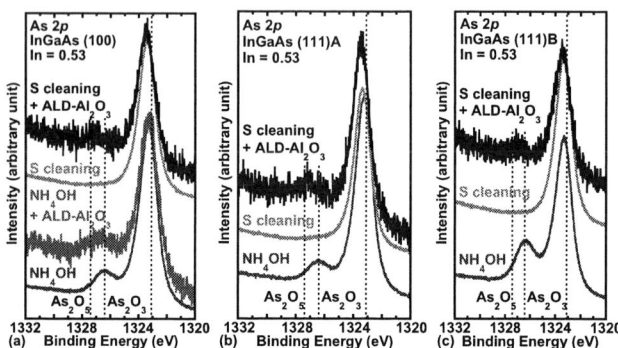

Fig. 2 As $2p_{3/2}$ photoelectron spectra for (a) (100) InGaAs surfaces with only NH_4OH, NH_4OH with ALD-Al_2O_3, S cleaning, and S cleaning with ALD-Al_2O_3 and (b) (111)A and (c) (111)B InGaAs surfaces with only NH_4OH, S cleaning, and S cleaning with ALD-Al_2O_3, respectively

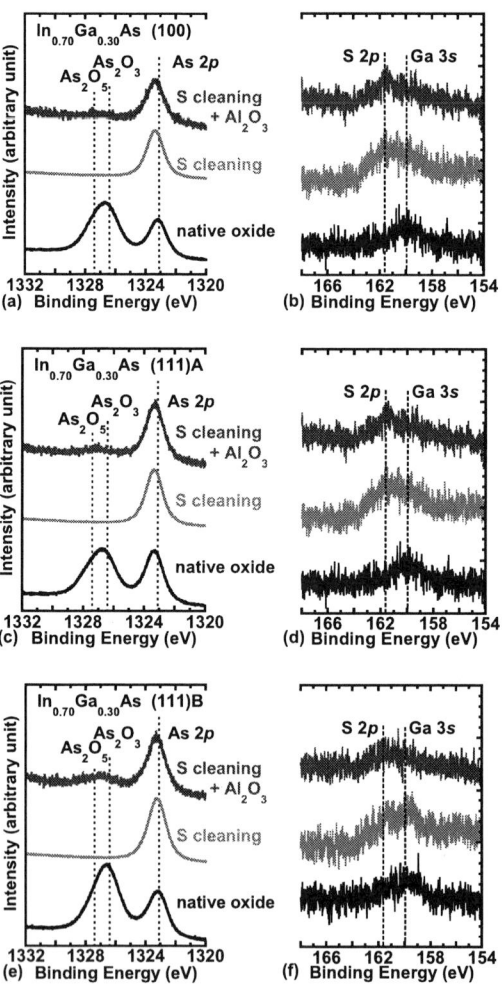

Fig. 3 XPS spectra of InGaAs surfaces with and without S cleaning. As $2p_{3/2}$ and S $2p$ photoelectron spectra for (a) and (b) (100), (c) and (d) (111)A, and (e) and (f) (111)B InGaAs surfaces with native oxide, S cleaning, and S cleaning with Al_2O_3, respectively.

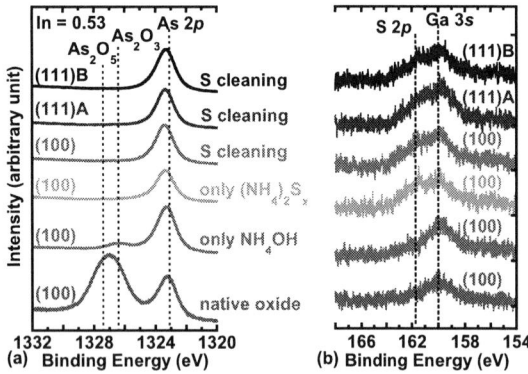

Fig. 1 XPS spectra of InGaAs surfaces with and without S cleaning. (a) As $2p_{3/2}$ and (b) S $2p_{3/2}$ photoelectron spectra for native oxide, only NH_4OH, and only $(NH_4)_2S_x$ for (100) and S cleaning for (100), (111)A, and (111)B InGaAs surfaces, respectively.

III. MOS INTERFACE PROPERTIES

We have studied the $Al/Al_2O_3/InGaAs$ MOS interface properties with In content of 0.53. Figures 4(a), 4(b), and 4(c) show the frequency dispersion with frequency of 1 kHz, 10 kHz, 100 kHz, and 1 MHz of $C - V$ curves of $Al/Al_2O_3/InGaAs$ MOS capacitors with In content of 0.53 using (100), (111)A, and (111)B orientations, respectively, with S cleaning. Here, the solid and broken curves indicate the $C - V$ curves measured by sweeping the voltage from accumulation voltage (V_{acc}) to inversion voltage (V_{inv}) and from V_{inv} to V_{acc}, respectively. The steepness of the $C - V$ curves is identical among the three surface orientations. The obtained $C - V$ characteristics, such as good stretch-out and small frequency dispersion in the accumulation and the depletion region, are basically the same among (100), (111)A, and (111)B InGaAs MOS interfaces. Figure 4(d) shows the energy distribution of the D_{it} values of the (100), (111)A, and (111)B $Al_2O_3/InGaAs$ MOS interfaces with In content of 0.53 with NH_4OH cleaning and S cleaning. They were evaluated from $C - V$ curves with 1 MHz by the Terman method. Here, E_{mid}, E_c, and ψ_s mean the energy at mid gap and the conduction band edge of InGaAs with In content of 0.53, and the surface potential, respectively. We used the N_D value of 3×10^{16} cm^{-3} as the same as the designed doping value. The D_{it} values near the conduction band edge with the S cleaning were lower than those with the NH_4OH cleaning. The energy distribution of D_{it} is similar among the three orientations. As a result, we have found the minimum D_{it} values of $1 - 2 \times 10^{12}$ cm^{-2}eV^{-1} thanks to the S cleaning.

It is interesting that all of the $Al_2O_3/InGaAs$ MOS interfaces show similar $C - V$ characteristics and D_{it} values and that the minimum D_{it} values of the $Al_2O_3/InGaAs$ MOS interfaces are not significantly different between only NH_4OH cleaning and S cleaning. However, the $C - V$ curves of $Al/Al_2O_3/InGaAs$ MOS capacitors with only NH_4OH cleaning (not shown here) showed larger frequency dispersions in the accumulation region than those with S cleaning. Combined with the XPS results, we can consider that this large frequency dispersion is attributable to the large amount of III-V oxides. The characteristic in the accumulation region is basically the same among (100), (111)A, and (111)B InGaAs MOS interfaces. These results strongly suggest that the sulfur passivation and trivalent oxide passivation on InGaAs surfaces is important to realize good InGaAs MOS interfaces. Actually, Robertson *et al.* have theoretically reported the importance of the existence of the trivalent oxides on the III-V semiconductor MOS interfaces [4]-[6], and we have experimentally reported the impact of $ALD-Al_2O_3$ interlayers for InGaAs MOS interfaces [7]. Depositing $ALD-Al_2O_3$ layers can be a key process for trivalent oxide passivation, and annealing treatment after $ALD-Al_2O_3$ layer deposition can make further improvement in trivalent oxide passivation by growing ultrathin As_2O_3 layers by residual hydroxyl groups or water. This interpretation suggests that not only Al_2O_3 but also As_2O_3 could be a good

candidate of trivalent oxides for III-V MOS interface passivation.

Next, we have studied $Al/Al_2O_3/InGaAs$ MOS interface properties with In content of 0.70. Figures 5(a), 5(b), and 5(c) show the frequency dispersion with frequency of 1 kHz, 10 kHz, 100 kHz, and 1 MHz of $C - V$ curves of $Al/Al_2O_3/InGaAs$ MOS capacitors with In content of 0.70 using (100), (111)A, and (111)B orientations, respectively, with S cleaning, respectively. Here, the solid and broken curves indicate the $C - V$ curves measured by sweeping the voltage from V_{acc} to V_{inv} and from V_{inv} to V_{acc}, respectively. The small hysteresis of the $C - V$ curves is identical among the three surface orientations. The obtained $C - V$ characteristics with the small frequency dispersion in the accumulation are basically the same among (100), (111)A, and (111)B InGaAs MOS interfaces. Figure 5(d) shows the energy distribution of the D_{it} values evaluated by the Terman method of (100), (111)A, and (111)B $Al_2O_3/InGaAs$ MOS interfaces with S cleaning, respectively. Here, E_c means the conduction band edge of InGaAs with In content of 0.70. The (111)A interface shows the low D_{it} values of $\sim 10^{12}$ cm^{-2}eV^{-1} near the conduction band edge, whereas the (100) and (111)B interfaces show the D_{it} values of $\sim 10^{13}$ cm^{-2}eV^{-1}. These results indicate that the good MOS interfaces with low D_{it} values can be realized for the (111)A InGaAs surfaces even with high In content by S cleaning and Al_2O_3 ALD.

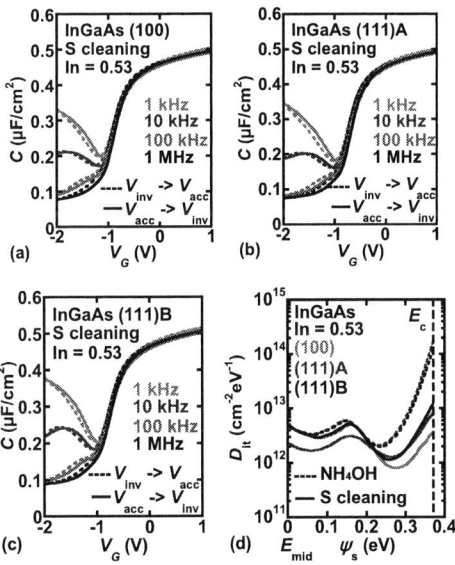

Fig. 4 Surface orientation dependence of frequency dispersion of $C - V$ curves at (a) (100), (b) (111)A, and (c) (111)B $Al_2O_3/InGaAs$ MOS capacitors with In of 0.53 with 1 kHz, 10 kHz, 100 kHz, and 1 MHz, respectively. The solid and broken curves indicate the $C - V$ curves measured by sweeping the voltage from V_{acc} to V_{inv} and from V_{inv} to V_{acc}, respectively. (d) Energy distribution of D_{it} for $Al/Al_2O_3/InGaAs$ MOS interfaces with In content of 0.53 evaluated from $C - V$ curves with 1 MHz by the Terman method. The solid and broken curves are for S cleaning and NH_4OH cleaning, respectively.

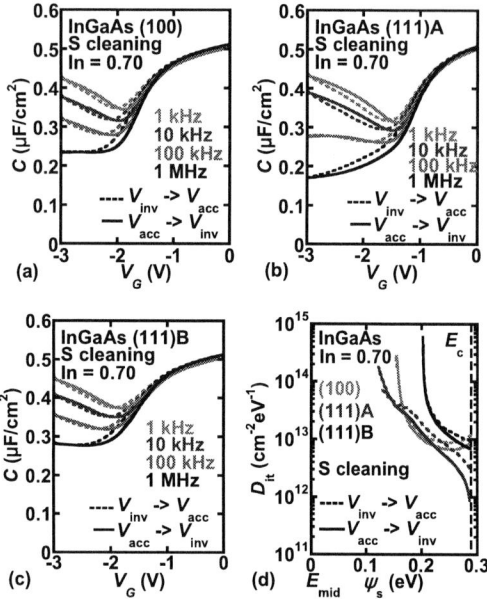

Fig. 5 Surface orientation dependence of frequency dispersion of C – V curves at (a) (100), (b) (111)A, and (c) (111)B Al/Al$_2$O$_3$/InGaAs MOS capacitors with In content of 0.70 with 1 kHz, 10 kHz, 100 kHz, and 1 MHz, respectively. The solid and broken curves indicate the C – V curves measured by sweeping the voltage from V_{acc} to V_{inv} and from V_{inv} to V_{acc}, respectively. (d) Energy distribution of the D_{it} values for (100), (111)A, and (111)B Al/Al$_2$O$_3$/InGaAs MOS interfaces with In content of 0.70 evaluated by the Terman method.

IV. CONCLUSION

We have studied the impact of sulfur cleaning on (100), (111)A, and (111)B InGaAs surfaces with In content of 0.53 and 0.70 by (NH$_4$)$_2$S$_x$ solution and their Al$_2$O$_3$/InGaAs MOS interface properties. We have found that sulfur cleaning can effectively remove the native oxides and that the sulfur passivation and the trivalent oxide passivation during Al$_2$O$_3$ ALD can realize good InGaAs MOS interfaces with the small frequency dispersion of the C – V curves in accumulation region, irrespective of the In content and the surface orientation. As a result, the good InGaAs MOS interfaces with the low D_{it} values near E_C could contribute to realize the high electron mobilities in InGaAs MOSFETs.

ACKNOWLEDGEMENT

This work was supported by New Energy and Industrial Technology Development Organization.

REFERENCES

[1] S. Takagi, T. Irisawa, T. Tezuka, T. Numata, S. Nakaharai, N. Shichijo, R. Nakane, S. Sugahara, M. Takenaka, and N. Sugiyama, "Carrier-Transport-Enhanced Channel CMOS for Improved Power Consumption and Performance," *IEEE Trans. Electron Dev.*, vol. 55, no. 1, pp. 21 – 39, January 2008.

[2] Y. Xuan, H. C. Lin, P. D. Ye, and G. D. Wilk, "Capacitance-voltage studies on enhancement-mode InGaAs metal-oxide-semiconductor field-effect transistor using atomic-layer-deposited Al$_2$O$_3$ gate dielectric," *Appl. Phys. Lett.*, vol. 88, issue 26, p. 263518, June 2006.

[3] E. J. Kim, E. Chagarov, J. Cagnon, Y. Yuan, A. C. Kummel, P. M. Asbeck, S. Stemmer, K. C. Saraswat, and P. C. McIntyre, "Atomically abrupt and unpinned Al$_2$O$_3$/In$_{0.53}$Ga$_{0.47}$As interfaces: Experiment and simulation," *J. Appl. Phys.*, vol. 106, issue 12, p. 124508, December 2009.

[4] L. Lin and J. Robertson, "Defect states at III-V semiconductor oxide interfaces," *Appl. Phys. Lett.* vol. 98, issue 8, p. 082903, February 2011.

[5] J. Robertson and L. Lin, "Bonding principle of passivation mechanism at III-V-oxide interfaces," *Appl. Phys. Lett.*, vol. 99, issue 22, p. 222906, November 2011.

[6] L. Lin and J. Robertson, "Passivation of interfacial defects at III-V oxide interfaces," *J. Vac. Sci. Technol. B*, vol. 30, issue 4, p. 04E101, May 2012.

[7] R. Suzuki, N. Taoka, M. Yokoyama, S. H. Lee, S. H. Kim, T. Hoshii, T. Yasuda, W. Jevasuwan, T. Maeda, O. Ichikawa, N. Fukuhara, M. Hata, M. Takenaka, and S. Takagi, "1-nm-capacitance-equivalent-thickness HfO$_2$/Al$_2$O$_3$/InGaAs metal-oxide-semiconductor structure with low interface trap density and low gate leakage current density," *Appl. Phys. Lett.* vol. 100, issue 13, p. 132906, March 2012.

[8] Y. Xuan, Y. Q. Wu, T. Shen, T. Yang, and P. D. Ye, "High performance submicron inversion-type enhancement-mode InGaAs MOSFETs with ALD Al$_2$O$_3$, HfO$_2$ and HfAlO as gate dielectrics," *IEEE IEDM Tech. Dig.*, pp. 637 – 640, December 2007.

[9] H. Ishii, N. Miyata, Y. Urabe, T. Itatani, T. Yasuda, H. Yamada, N. Fukuhara, M. Hata, M. Deura, M. Sugiyama, M. Takenaka, and S. Takagi, "High Electron Mobility Metal–Insulator–Semiconductor Field-Effect Transistors Fabricated on (111)-Oriented InGaAs Channels," *Appl. Phys. Express*, vol. 2, p. 121101, November 2009.

[10] Y. Urabe, N. Miyata, H. Ishii, T. Itatani, T. Maeda, T. Yasuda, N. Fukuhara, M. Hata, M. Yokoyama, N. Taoka, M. Takenaka, S. Takagi, "Correlation between Channel Mobility Improvements and Negative V_{th} Shifts in III-V MISFETs : Dipole Fluctuation as New Scattering Mechanism," *IEEE IEDM Tech. Dig.*, pp. 142 – 145, December 2010.

[11] M. Xu, Y. Q. Wu, O. Koybasi, T. Shen, and P. D. Ye, "Metal-oxide-semiconductor field-effect transistors on GaAs (111)A surface with atomic-layer-deposited Al$_2$O$_3$ as gate dielectrics," *Appl. Phys. Lett.* vol. 94, issue 21, p. 212104, May 2009.

High-Power InP-based Waveguide Photodiodes and Photodiode Arrays Heterogeneously Integrated on SOI

Andreas Beling[1], Molly Piels[2], Allen S. Cross[1], Yang Fu[1], Qiugui Zhou[1], Jon Peters[2], John E. Bowers[2], and Joe C. Campbell[1]

[1] *ECE Department, University of Virginia, 351 McCormick Road, Charlottesville, VA 22904, E-mail:andreas@virginia.edu*
[2] *ECE Department, University of California at Santa Barbara, Santa Barbara, CA 93106-9560, E-mail: molly@ece.ucsb.edu*

Abstract: For the first time we demonstrate evanescently-coupled modified uni-traveling carrier photodiodes (MUTC PDs) on silicon-on-insulator (SOI) waveguide with an internal responsivity of 0.85 A/W, up to 15 GHz bandwidth, and high RF output power. A novel 2-element MUTC PD array has a saturation current-bandwidth-product of >630 mA*GHz and achieves +9 dBm RF output power at 20 GHz.

1. Introduction

Heterogeneous integration of III-V material on silicon is a promising approach to realize high-performance photodiodes on a silicon photonics platform. Owing to their material properties InP-based photodiodes allow for complex bandgap engineering and have the potential to achieve low dark current, high saturation current and wideband absorption over C- and L-bands. In [1] we demonstrated discrete InP-based MUTC photodiodes that have achieved high saturation current and high linearity. In the present work we use a wafer-bonding technology [2,3] to integrate this type of photodiode on SOI/Si waveguides. Experimental results from both, single MUTC PDs and PD arrays on SOI are reported.

2. Layer design and fabrication

Figure 1 shows a cross section of the MUTC PD on SOI waveguide [4]. The InGaAs absorbing region is comprised of a 250 nm p+ and a 50 nm unintentionally-doped absorber layer. In order to form an abrupt doping profile, Be is used as the dopant in the p+ absorbing layer. A 700-nm n.i.d. InP drift layer is inserted between the absorber and the n+ contact layer to reduce junction capacitance. The structure also incorporates a 10 nm moderately n-type doped InP cliff layer to maintain high electric field at the absorber-drift layer interface. This is crucial to achieving high saturation current. A 10 nm thick bonding layer and an InP/InGaAsP super lattice are integrated below the p+ contact layer to facilitate wafer bonding onto Si and to reduce the propagation of defects into the active region. The PD is evanescently coupled to an underlying SOI waveguide. The fabrication process starts with dry etching of the SOI waveguides followed by transferring the III-V wafer to the patterned SOI/Si substrate through low-temperature oxygen-plasma-assisted wafer bonding [3]. After removal of the InP substrate, we deposited an AuGe/Ni/Au blanket metallization and used self-aligned ICP-RIE to form the n-mesa. Next, the p-mesa was etched, the p-contact metal layers were evaporated, and devices were passivated with a 200 nm-thick SiO_2 film. To reduce RF loss originating from the low resistivity Si substrate we deposited the RF pads on a 2 µm-thick SU8 layer (inset of fig. 2).

3. Experimental results

Using a lensed fiber with a spot diameter of approximately 5 µm we measured typical responsivities of 100 µm-long PDs between 0.16 A/W and 0.21 A/W at 1550 nm wavelength (no anti-reflection coating). Based on this result and taking the simulated fiber-chip coupling loss of 5 dB into account we estimated an internal responsivity as high as 0.85 A/W. We measured dark currents as low as 0.1 nA at 3 V before and after passivation. However, after pad deposition the dark current increased by roughly 3 orders of magnitude due to process-related surface leakage currents.

To measure the bandwidth we used an optical heterodyne setup with an optical modulation depth close to 100 %. Figure 2 shows the measured frequency responses. In good agreement with previous simulations, PDs with active areas of 7x200 µm^2 and 7x100 µm^2 reach 3dB-bandwidths of 9 GHz and 15 GHz, respectively. The saturation current (i.e. average photocurrent at 1-dB RF power compression) was measured on a 200-µm long device at a signal frequency of 9 GHz. At 4 V we obtained a saturation current of 20 mA and a maximum electrical output power of +4.3 dBm.

To further enhance saturation current and RF output power we designed and fabricated PD arrays in which the photocurrents of two PDs are combined (fig. 3). The optical input signal is coupled into a singlemode waveguide and then equally split in a multi-mode interference (MMI) splitter to feed both PDs. The 10x37 µm^2 PDs are located on top of the multi-mode waveguide at positions where optical simulations predicted two-fold self-images. We measured a bandwidth of 15 GHz and a saturation current of 42 mA at 5 V bias and 20 GHz corresponding to a saturation current-bandwidth-product

978-1-4673-1725-2/12 $31.00 © 2012 IEEE

of >630 mA*GHz (fig. 4). The maximum RF output power reached +9.3 dBm at 20 GHz (fig. 5). These results are the highest values reported for heterogeneous integration on Si.

4. Conclusion

High-power InP-based MUTC PDs heterogeneously integrated onto SOI/Si waveguides were fabricated and characterized. A 7x100 μm^2 waveguide PD had 0.85 A/W internal responsivity and 15 GHz bandwidth. A 200 μm-long PD reached +4.3 dBm RF output power at 4 V and 9 GHz. A 2-element PD array achieved +9.3 dBm RF output power at 20 GHz signal frequency.

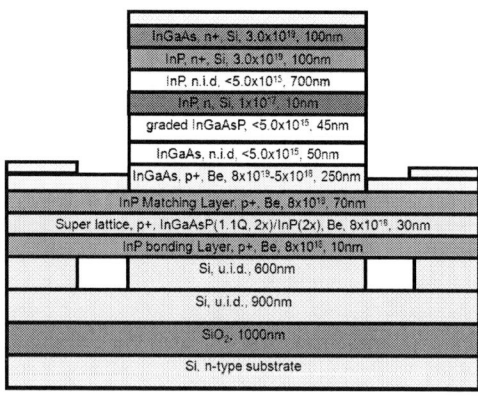

Fig. 1. Layer stack of MUTC PD on SOI. Doping concentrations in cm^{-3}.

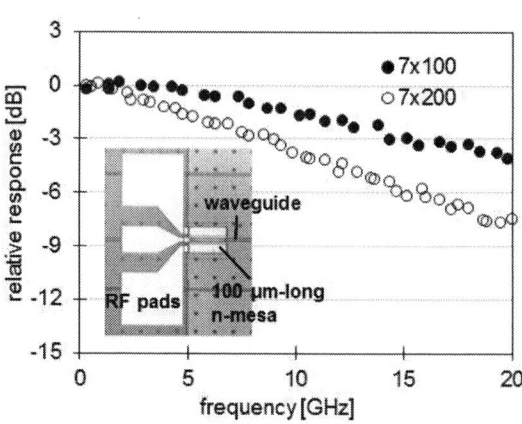

Fig. 2. Frequency responses of waveguide MUTC PDs. Inset: micrograph of fabricated device.

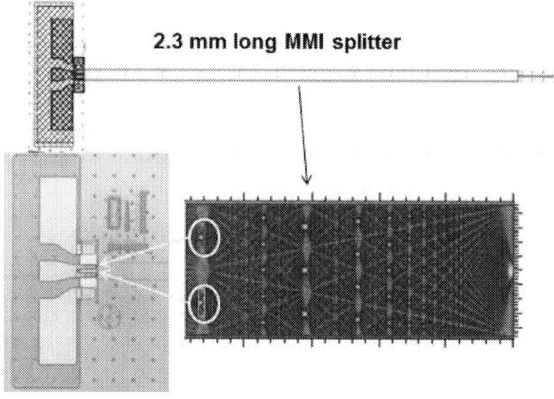

Fig. 3. 2-element PD array: layout (top), PDs with RF pads (bottom left), simulated intensity distribution in multi-mode waveguide (bottom right) with PD locations being circled.

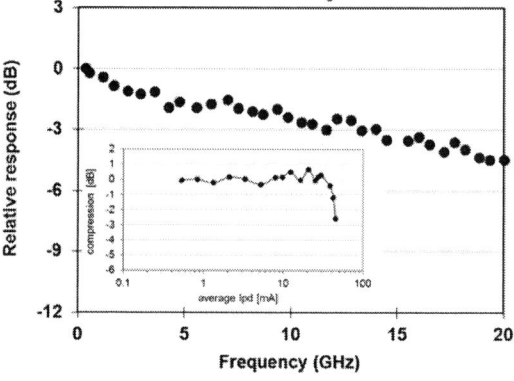

Fig. 4. Frequency response of 2-element waveguide MUTC PD array. Inset: RF power compression vs. average photocurrent (Ipd).

5. References

[1] H. Pan, Z. Li, A. Beling, and J. C. Campbell, "Measurement and modeling of high-linearity modified uni-traveling carrier photodiode with highly-doped absorber," Optics Express **17**(22), pp. 20221-20226 (2009).

[2] H. Park, A. W. Fang, R. Jones, O. Cohen, O. Raday, M. N. Sysak, M. J. Paniccia, J. E. Bowers, "A hybrid AlGaInAs-silicon evanescent waveguide photodetector," Optics Express **15**(10), pp. 6044-6052 (2007).

[3] D. Liang, G. Roelkens, R. Baets, J. E. Bowers, "Hybrid Integrated Platforms for Silicon Photonics," Materials, **3** (3), 1782-1802, March 12, 2010.

[4] A. Beling, Y. Fu, Z. Li, H. Pan, Q. Zhou, A. Cross, M. Piels, J. Peters, J. E. Bowers, J. C. Campbell, "Modified Uni-Traveling Carrier Photodiodes Heterogeneously Integrated on Silicon-on-Insulator (SOI)," Integrated Photonics Research, Silicon, and Nano-Photonics (IPR 2012), Colorado Springs, CO, June 17-20, 2012, paper IM2A.2.

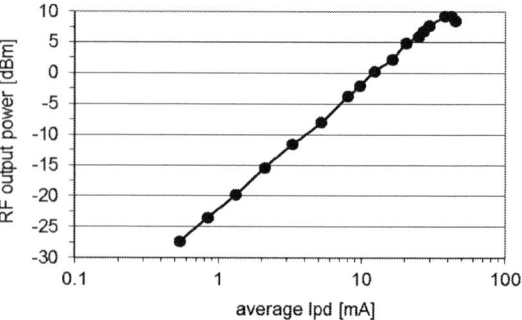

Fig. 5. RF output power at 20 GHz vs. average photocurrent measured at 5 V.

Recent advances in high-speed lasers and amplifiers based on 1.5 μm QD/QDash material

Johann Peter Reithmaier[1], Gadi Eisenstein[2]

[1]*Institute of Nanostructure Technologies and Analytics (INA), CINSaT, University of Kassel, Germany*
[2]*Department of Electrical Engineering, Technion, Haifa 32000, Israel*

Abstract — **A review is given on the recent progress in the epitaxy of 1.55 μm QD/QDash laser material and its application in lasers and semiconductor optical amplifiers. By choosing different growth modes by using either As₄ or As₂, the shape of InAs islands grown on AlGaInAs surfaces can be strongly influenced from dash-like to dot-like geometries. This shape change is accompanied by a strong reduction of the size fluctuation, which increases significantly the modal gain of lasers. This has a large impact on the response time of lasers allowing record values of the modulation speed of InP based QD lasers of 15 GBit/s. On the other hand dash-like structures allow ultra-high speed responses in optical amplifiers due to their local relaxation but remaining coupling to continuum states. By two-photon absorption phenomena instantaneous gain can be obtained on the sub-ps time scale.**

Index Terms — **Quantum dot laser, optoelectronic device, semiconductor optical amplifier, modulation speed, epitaxial growth, InP based material.**

I. INTRODUCTION

The formation of round-shape dots on InP based compounds was, for many years, rather difficult to realize on (001) substrates [1, 2]. Sufficiently high densities for laser operation was mainly obtained by dash formation on (001) surfaces [3] or on strongly tilted surfaces [4]. With dash-like active regions, very impressive device properties, like high speed optical amplifiers and multi-wavelength amplification were demonstrated [5]. However, the laser properties suffered from moderate modal gain values (typically 3 cm⁻¹ per dash layer) and limited modulation speeds [6]. A major improvement was obtained recently when an improved growth mode was developed [7]. The new QD material avails an increase of the modal gain by approximately a factor of three [8] and resulted in record high modulation data rates of 15 GBit/s [9].

In the following, a review on our recent results gained on this new type of high-gain quantum dot (QD) material and its application in optoelectronic devices as well as a prospect for significant further improvement will be given.

II. FORMATION OF CIRCULAR QDS

A standard growth mode of arsenide compounds in a molecular beam epitaxy system is based on the sublimation of As4 molecules, which are cracked to As₂ and, in a second step, to As radicals on the growth surface.

By pre-cracking As₄ into dimers by a high temperature cracker cell, one can surpass a part of the time consuming

surface chemistry, which suppress the migration of group-III elements, like In, on the surface.

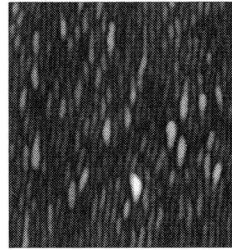

Fig. 1: InAs QDashes (left) and QDs (right) grown in As₄ and As₂ mode, respectively, on InAlGaAs surfaces [7].

As can be seen in Fig. 1, the strong elongation of InAs islands grown on an InAlGaAs surface by using As₄ (left picture) is mostly suppressed by using As₂ and round-shape islands at very high density are formed [7].

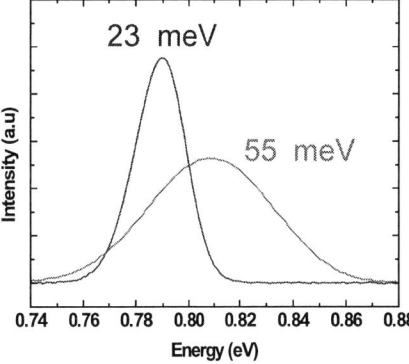

Fig. 2: PL spectra (T = 10 K) for As4 (broadened line) and As₂ growth modes (narrow line) [7].

The suppression of migration leads also to much reduced dot height fluctuations, which is reflected by the much narrower inhomogeneously broadened photoluminescence linewidth of 23 meV shown in Fig. 2.

The change in dot symmetry can also be monitored to a certain extent by polarization dependent PL measurements. Fig. 3 shows clearly a strong reduction in the degree of polarization in structures based on As₂ (left) compared to the As₄ growth mode (right). Although the circular shape is more pronounced, there still exists some residual degree of polarization, which is indicative of a slight elongation.

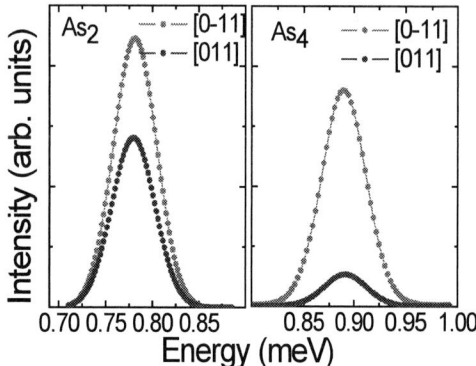

Fig. 3: Polarization dependent PL for dot-like (left) and dash-like (right) structures [10]

III. QD LASER MATERIAL

The QDs with improved optical properties were embedded in a separate confinement heterostructure as illustrated in Fig. 4.

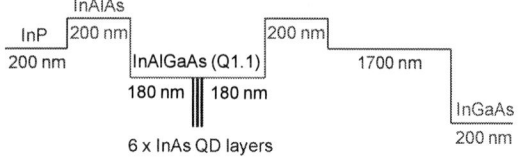

Fig. 4: Schematic layer structure for QD laser starting from the left with the n-doped InP substrate8

Different numbers of QD layers were embedded in the active region using the same layer structure. Broad area (BA) lasers were processed with about 100 μm stripe width and different cavity lengths. In Fig. 5, the light output characteristics of different BA laser structures are shown for a cavity length of 1 mm. By decreasing the number of QD layers, the threshold current reduces and the slope efficiency increases slightly. All the lasers emit near 1.55 μm as can be seen in the inset of Fig. 5.

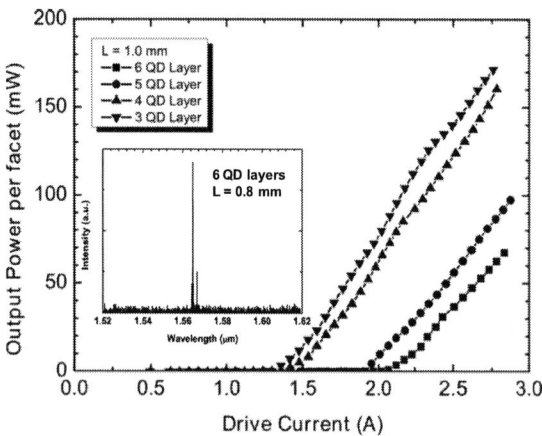

Fig. 5: Light output characteristics of 1 mm long broad area QD lasers with different numbers of QD layers. The inset show the emission spectrum of a 0.8 mm long laser with 6 QD layers [8].

Length dependent measurements are used to determine intrinsic parameters such as the transparent current density and the modal gain. The transparency current density decreases systematically from about 300 A/cm^2 (with 6 QD layers) to 150 A/cm^2 (in the case of 4 QD layers). The modal gain per QD layer is in the range of 10 cm^{-1} and above leading to a very large modal gain of more than 60 cm-1 for the laser with 6 QD layers. Details are given in reference 8.

IV. HIGH-SPEED QD LASERS

The characterization of the dynamic properties of the laser material was performed using ridge waveguide (RWG) lasers. For this purpose lasers with different ridge widths were fabricated and cleaved to 340 μm long cavities. One facet was kept as-cleaved while the other was high reflection (HR) coated to about 95%.

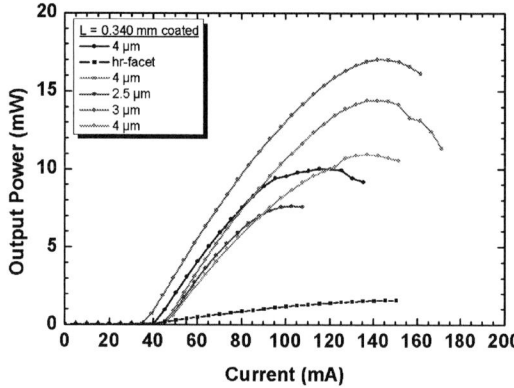

Fig. 6: Output characteristics of a 340 μm RWG laser, HR coating on one facet with different ridge widths [9].

Fig. 6 shows light output (LI) characteristics of QD RWG lasers with different ridge widths. A 3 μm wide ridge yields the highest power of 18 mW, The LI characteristics measured from the HR coated facet is also shown (low power curve).

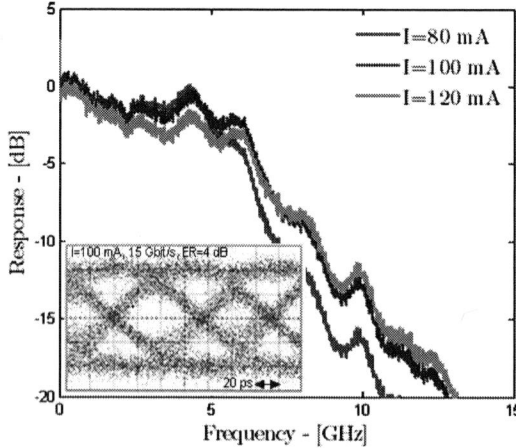

Fig. 7: Small signal modulation at different injection currents of a 340 μm long QD laser. The inset show a clear open eye at 15 GBit/s data rate.9

The small signal modulation properties for different injection currents are shown in Fig. 7. The response is highly damped with a -3 dB bandwidth of 5-6 GHz, with resonances observed up to 9.5 GHz and above. A similar behavior was also observed in GaAs based QD lasers [11]. In contrast to the moderate small signal modulation capabilities, the large signal modulation using a bit pattern generator with a bit rate of 15 GBit/s shows well defined open eyes with an extinction ratio (ER) of 4 dB. This result is indicative of the very complex dynamic governing QD lasers, where digital modulation limits cannot be extrapolated from the small signal characteristics.

V. PROSPECTS TO HIGHER MODULATION SPEEDS

The major limitation of the small signal modulation response in the present QD lasers stems from the transport time in the non-optimized active region. A detailed spatially resolved model which treats electrons and holes separately and includes their mutual Coulomb interaction and the consequent distortion of the energy bands [12] was used to simulate the present structure and variations thereof. Fig. 8 shows calculated responses for an active region with 200 nm barrier widths on both sides of the QD layers (bottom curves) .Measured gain properties were used in the simulations. The simulation fits well to the experimental responses in Fig. 7.

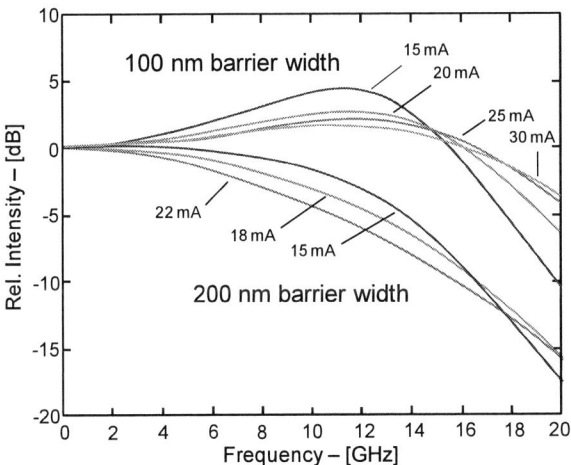

Fig. 8: Simulation of small signal modulation for current (lower curves) and optimized layer design (upper curves) [12].

A reduction of the barrier width to 100 nm has a large impact and can triple the -3 dB bandwidth to more than 20 GHz. This major improvement is mainly caused by the very high modal gain, which allows operating the lasers sufficiently far from QD state saturations. In this condition one can, for the first time, utilize the very high differential gain of quantum dots, which was previously suppressed by state filling effects [13].

Considering this different operation condition of QD laser material, one should be able to push the modulation rates up to 25 GBit/s or more.

VI. QDASH SOAs

QDash SOAs have proven to be effective devices for practical telecom applications [5] where they can offer multi wavelength amplification and signal processing [14] without cross talk, pattern dependence free amplification and regeneration as well as wavelength conversion and switching.

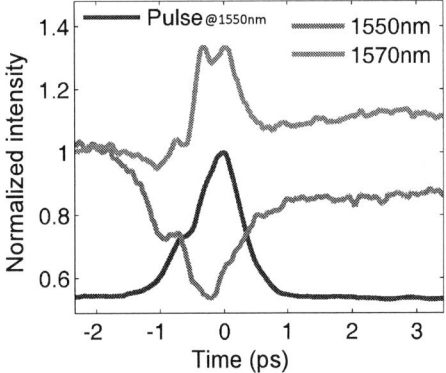

Fig. 9: Multi wavelength pump probe response of a QDash SOA [15]

Recent experiments and modeling concentrated on more basic dynamical properties of these QDash gain media. One new phenomenon is an instantaneous gain response observed in multi wavelength pump probe measurements when intense pump pulses feed a QDash SOA which is driven at a high bias. Pumping at 1550 nm with a 150 fs pulse and probing at 1570 nm yields the dynamic response shown in Fig. 9.

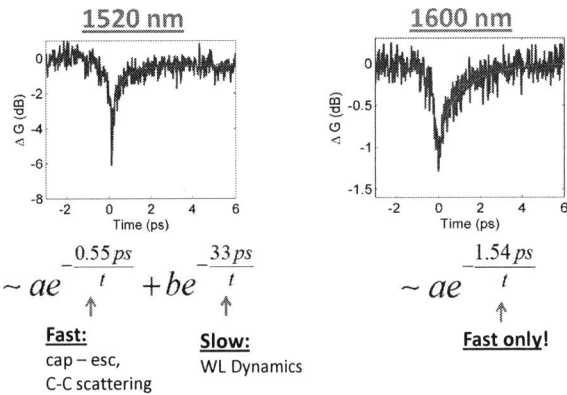

Fig. 10: Wavelength dependent gain relaxation for a 150 fs pump at 1550 nm [15].

The instantaneous gain response is a result of an indirect process by which two-photon absorption generates a large carrier density at high energy levels where the high energy tails of the gain spectra characterizing all Dashes mix. Since the bias is high, carrier – carrier scattering is efficient and the high energy carriers relax fast to the ground states of all dashes where they contribute to gain within less that 100 fs after the perturbation. This observation sheds light on many ultrafast phenomena in such structures and can also be the

978-1-4673-1725-2/12 $31.00 © 2012 IEEE   175

basis for a new generation of new modulation schemes at unprecedented rates.

For lower excitation energies, the instantaneous gain vanishes and it is possible to extract energy dependent relaxation rates [15]. Using the multi wavelength pump probe set up we measured the responses described in Fig. 10.

The difference in the responses is very clear and is related to the energetic distance from the gain region of interest to the high energy continuum states. The high energy side saturates much deeper and the relaxation is governed by two time constants while the saturation on the long wavelength side is much smaller and the relaxation obeys a single exponential with a relatively long time constant. These observations are consistent with theoretical expectations where saturation depends on carrier capture / escape times and the relaxation is determined by the principle of detailed balance and can be quantified by a Boltzmann factor.

VII. Conclusions

Based on improved growth conditions for the InAs QD formation on (001) InAlGaAs lattice matched to InP, high modal gain 1.55 µm QD lasers were realized, which exhibit vastly improved dynamic properties allowing modulation rates of up to 15 GBit/s. Simulation results show that further optimizing the laser design by suppressing parasitic transport effects, may allow to realize directly modulated 1.55 µm high speed QD lasers approaching 25 GBit/s, which could combine in future different very favorite QD laser properties, like low threshold currents, high temperature stability and reduced chirp with high speed direct modulation. This would make this new type of device an ideal candidate for low-cost uncooled light sources for fiber-to-the-home or 100 GBit/s Ethernet systems.

On the other hand in wire-like quantum structures (i.e. QDashes) ultra-fast phenomena could be observed in semiconductor optical amplifiers on a sub-ps time scale. This observation sheds light on many ultrafast phenomena in such structures and can also be the basis for a new generation of new modulation schemes at ultra-high data rates.

VIII. Acknowledgements

The financial support by the EU project DeLight and Gospel is gratefully acknowledged. We would like to thank the technical assistance of F. Schnabel, A. Rippien, K. Fuchs and D. Albert.

IX. References

[1] J. Brault, M. Gendry, G. Grenet, G. Hollinger, J. Olivares, B. Salem, T. Benyattou, and G. Bremond "Surface effects on

shape, self-organization and photoluminescence of InAs islands grown on InAlAs/InP(001)", *J. Appl. Phys.*, vol. 92, pp. 506-510, July 2002.

[2] A. Stintz, T. J. Rotter, and K. J. Malloy, "Formation of quantum wires and quantum dots on buffer layers grown on InP substrates", *J. Cryst. Growth*, vol. 255, pp. 266-277, August 2003.

[3] R. Schwertberger, D. Gold, J.P. Reithmaier, and A. Forchel, "Long wavelength InP based quantum dot lasers", *IEEE Phot. Technol. Lett.*, vol. 14, pp. 735-737, June 2002.

[4] H. Saito, K. Nishi, A. Kamei, and S. Sugou, "Low chirp observed in directly modulated quantum dot lasers," *IEEE Photon. Technol. Lett.*, vol. 12, pp. 1298–1300, Oct. 2000.

[5] J.P. Reithmaier, A. Somers, S. Deubert, R. Schwertberger, W. Kaiser, A. Forchel, M. Calligaro, P. Resneau, O. Parillaud, S. Bansropun, M. Krakowski, R. Alizon, D. Hadass, A. Bilenca, H. Dery, V. Mikhelashvili, G. Eisenstein, M. Gioannini, I. Montrosset, T.W Berg, M. van der Poel, J. Mørk, B. Tromborg, "InP based lasers and optical amplifiers with wire-/dot-like active regions", *J. Phys. D*, vol. 38, pp. 2088-2102, July 2005.

[6] C. Chen, Y. Wang, H.S. Djie, B.S. Ooi, L. Lester, T.L. Koch, J.C:W. Hwang, "Dynamics of Quantum-Dash Lasers", *IEEE J. Sel. Top. QE*, vol. 17, pp. 1167-1174, Sept. 2011.

[7] C. Gilfert, E.M. Pavelescu; J.P. Reithmaier, "Influence of the As_2/As_4 growth modes on the formation of quantum dot like InAs islands grown on InAlGaAs/InP (100)", *Appl. Phys. Lett.*, vol. 96, art. 191903, May 2010.

[8] C. Gilfert, V. Ivanov; N. Oehl, M. Yacob, J.P. Reithmaier, "High gain 1.55 µm diode lasers based on InAs quantum dot like active regions", *Appl. Phys. Lett.*, vol. 98, art. 201102, May 2011.

[9] D. Gready, G. Eisenstein, C. Gilfert, V. Ivanov, J.P. Reithmaier, "High Speed Low Noise InAs/InAlGaAs/InP 1.55 µm Quantum Dot Lasers", *IEEE Phot. Technol. Lett.*, vol. 24, pp. 809-811, May 2012.

[10] M. Yacob, M. Benyoucef, J.P. Reithmaier, unpublished data, 2011.

[11] Y. Tanaka et al., CLEO, paper CTuZ1, 2011.

[12] D. Gready, G. Eisenstein, unpublished data, 2012.

[13] H. Dery, G. Eisenstein, "The impact of energy band diagram and inhomogeneous broadening on the optical differential gain in nanostructure lasers", *IEEE J. of Quantum Electron.*, vol. 41, pp. 26-35, Jan. 2005.

[14] A. Capua, G. Eisenstein, J.P. Reithmaier, "A nearly instantaneous gain response in quantum dash based optical amplifiers", *Appl. Phys. Lett.*, vol. 96, art. 131108, Sept. 2010.

[15] A. Capua G. Eisenstein, J.P. Reithmaier, "Ultrafast cross saturation dynamics in inhomogeneously broadened InAs/InP quantum dash optical amplifiers", *Appl. Phys. Lett.*, vol. 98, art. 101108, March 2011.

1550nm InAs/InP Quantum Dash Based Directly Modulated Lasers for Next Generation Passive Optical Network

N. Chimot, [1] S. Joshi, [1] G. Aubin,[2] K. Merghem, [2] S. Barbet, [1] A. Accard, [1] A. Ramdane[2] and F. Lelarge[1]

[1] III-V Lab, a joint Laboratory of "Alcatel Lucent Bell Labs", "Thales Research & Technology" and CEA-LETI, Route de Nozay, 91460 Marcoussis, France

[2] CNRS, Laboratory for Photonics and Nanostructures, Route de Nozay, 91460 Marcoussis, France

Abstract — High speed transmissions up to 20Gb/s are reported using 1.55µm InP-based Quantum Dashes distributed feedback lasers. Combining a directly modulated laser with an etalon filter, we demonstrate uncompensated and non-amplified SMF 10Gb/s transmissions at a constant bias current from back-to-back up to 65km with extinction ratio as large as 8dB. This result opens the way to the implementation of low cost high bit rate sources for access network applications.

I. INTRODUCTION

The next generation passive optical network (NGPON) requires green optical sources for low cost, low consumption, athermal, access network infrastructure that will support a number of access network topologies. For that purpose, the industry hence aims at longer reach (40-100km), higher modulation bandwidth and increased split ratio [1].

The standard 10Gb/s transmission distance without dispersion compensation obtained with 1.55µm multi-quantum-well directly modulated lasers (DML) is around 25 km. For longer distances, new limitations arise from linewidth enhancement factor -induced chirp. Long reach transmission (>40 km) using DML has been demonstrated by compensating the fiber dispersion to avoid the chirp induced signal degradation. These methods use for instance dispersion compensation fibre [2], specific filtering [3], reduction of spectral broadening [4] or electronic compensation [5]. These solutions require tightly stabilized optical filter, accurate temperature control or specific power consuming electronic detection schemes. The development of a low cost DML allowing to cover a large range of transmission distance (typically 10:80km) without using complex compensation requiring specific in-line adjustment or power consuming electronic cards is therefore of paramount importance.

For that purpose, Quantum Dots (QDots) and Quantum Dashes (QDashes) based lasers emitting in the 1.55µm window are very promising for access networks requiring low-cost devices. The expected higher differential gain and lower chirp in low dimensional heterostructure [6] should indeed lead to DML with high modulation bandwidth and should enable long transmission distances. QDashes lasers with modulation bandwidth in continuous wave (CW) mode operation up to 10GHz [7-8] have been demonstrated. More recently, we demonstrated very low chirp distributed feedback (DFB) lasers, even at high injected current using optimized

QDashes devices with p-doped active layers [9]. In this contribution, we show that the combination of a QDash based laser with a commercially available etalon filter leads to SMF 10Gb/s transmissions at a constant bias current from back-to-back up to 65km with extinction ratio as large as 8dB. In addition, the potential of QDash lasers for higher bit rate is also investigated.

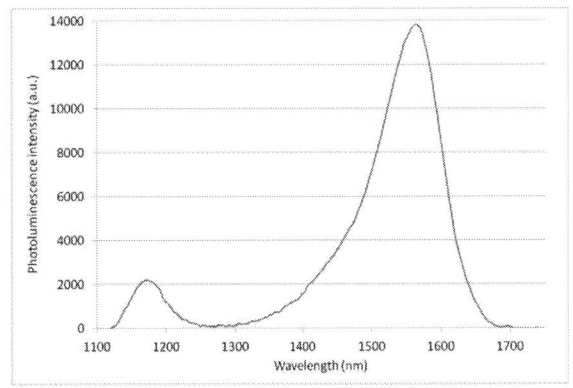

Fig.1: Room temperature photoluminescence spectra of the p-doped DWELL epiwafer.

II. DEVICE CHARACTERISTICS

The dashes-in-a-well (DWELL) laser structure was grown by gas source molecular beam epitaxy on a S-doped (001) InP substrate. It comprises 6 layers of InAs QDs embedded within a 6nm-thick InGaAsP quantum well (λg=1.45µm) each, separated by 20nm-thick InGaAsP p-doped barriers (λg=1.17µm). Optimized growth condition leads to the formation of high-density QDashes with significantly reduced photoluminescence (PL) linewidth (down to ~50meV) owing to homogeneous dash size. As shown in Fig.1, the room temperature (RT) spectra evidences a well defined signature of the DWELL emission at 1.56μm and a signature at 1.17μm of the quaternary layer used for the grating fabrication.

For transmission evaluation, we fabricated Buried Ridge Stripe (BRS) Distributed Feedback lasers based on a DWELL active layer. A 15nm laser detuning with respect to maximum of gain has been chosen as a good compromise between optimum dynamic properties and temperature behavior allowing the investigation of 10Gb/s transmission both at room temperature and in semi-cooled operation. The length of DFB lasers is 500μm. A λ/4 phase-shift is inserted for single mode operation. An anti reflection coating has been deposited on the front facet.

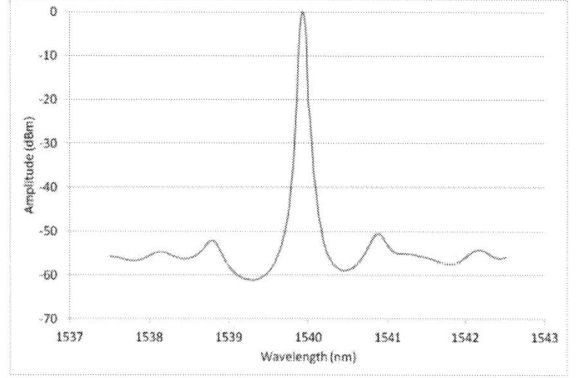

Fig.2: Optical spectra of the 500μm-long BRS DFB laser measured under CW operation at 25°C for an injected current of 80mA.

We studied 500μm-long DFB lasers emitting at 1540nm with side mode suppression ration as large as 50dB (Fig.2). As shown in Fig.3, we obtained a threshold current of 30mA and an external efficiency of 0.15W/A per facet despite the introduction of p-doping within the barriers. An output power up to 16mW is achieved at 160mA in CW operation.

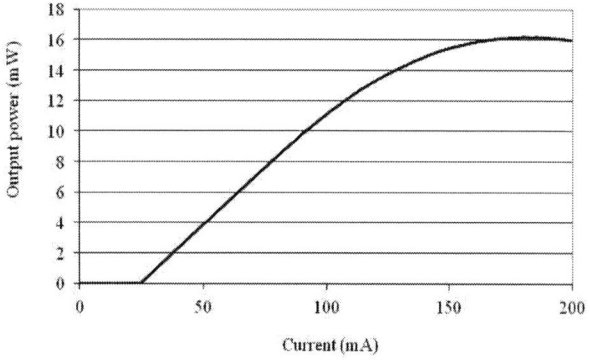

Fig.3: Output power versus injection current of the 500μm-long BRS DFB laser measured under CW operation at 25°C.

III. DYNAMIC PROPERTIES

As shown in Fig.4, the small signal -3dB modulation bandwidth of the DWELL-based BRS laser is about 10GHz at 140mA. The typical capacitance of the BRS laser is about 10pF, leading to a strong RC parasitic limitation (~5GHz). From the electrical response, we can extract the intrinsic response of the DFB laser and derive a theoretical modulation bandwidth as large as 16GHz.

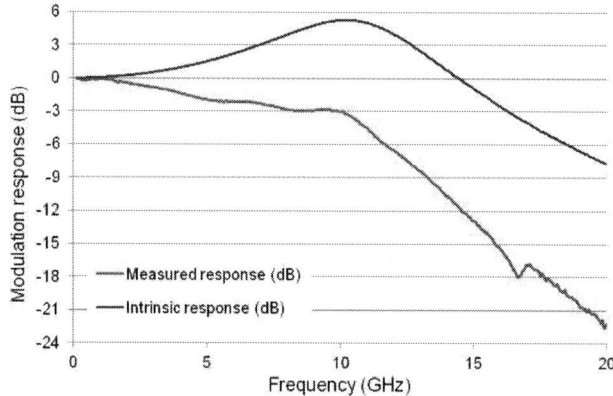

Fig.4: Small signal modulation response of the DFB laser.

The large signal chirp induced by the modulation of injected current is a parameter of paramount importance for long distance transmission in a fibre. We therefore characterized the large signal chirp induced by 1ns long square pulses. In the experiment, the optical signal is analyzed through a Mach-Zehnder interferometer, in order to measure the time-resolved amplitude and frequency response. Fig.5 displays these variations at 25°C. The biasing current is 100mA and the peak-to-peak modulating current is 50mA. The adiabatic chirp is about 7GHz whereas the transient chirp is kept at a low value around 1GHz. The strong damping of resonance frequency and the RC parasitic limitation related to the BRS process attenuates the transient chirp. This reduction is valuable for long reach transmissions as it compensates for the highly detrimental effect of transient chirp.

The large fibre-coupled power, the low transient chirp factor and the 10GHz-modulation bandwidth enable to properly evaluate the potential for transmission of such low dimensional material sources.

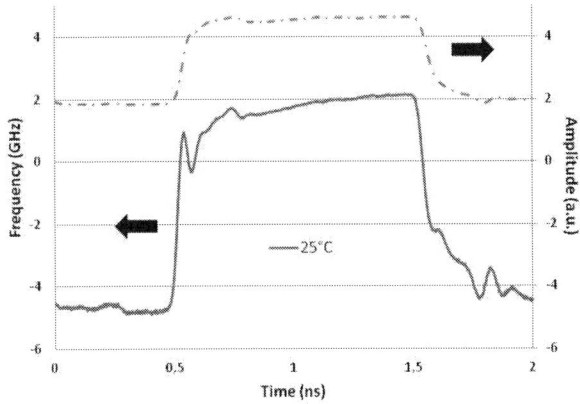

Fig.5: Large signal chirp of the DFB laser.

IV. LARGE SIGNAL TRANSMISSION

The transmission evaluation is performed at 10Gb/s and at 20 Gb/s using a PRBS signal with two different receivers where the output signal is sent, before error detection, either directly after the laser (back-to-back measurement) or after transmission on a standard single-mode fiber. As shown in Fig.6, open eyes are observed up to 20Gb/s, which confirms the potential of QDash lasers for very high speed applications. As shown in the bit error rate (BER) reported in back-to-back configuration and after fibre transmissions, error free 10Gb/s transmission is achieved up to 65km at 10Gb/s and to 3km at 20Gb/s (Fig.7). The fabrication of low capacitance ridge waveguide lasers based on similar active layer should significantly improve the 20Gb/s transmission performances.

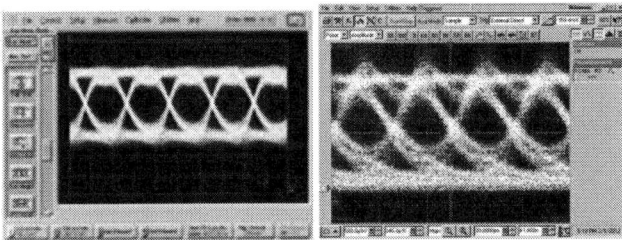

Fig.6: 10Gb/s (Left) and 20Gb/s (Right) eye diagrams in back to back of a 500µm-long DFB laser.

V. 10GB/S TRANSMISSION USING ETALON FILTER

Full exploitation of such 10Gb/s transmissions over 65km is still restricted by the ~3dB extinction ratio whereas the access network standards require 6 or 8dB. In order to increase the extinction ratio, we combined such QDash DML with a commercially available free space etalon filter.

Using a 500µm-long cavity DWELL based laser, the average DC current (100mA) and the modulation depth (+/- 50mA) were chosen to optimize the bit error rate (BER) at

10Gb/s for 40km transmission and these bias conditions were kept constant for all distances. As shown in Fig.8, error free 10 Gb/s transmission is achieved over 65 km with an extinction ratio as large as 8dB, in full compliance with access network standards.

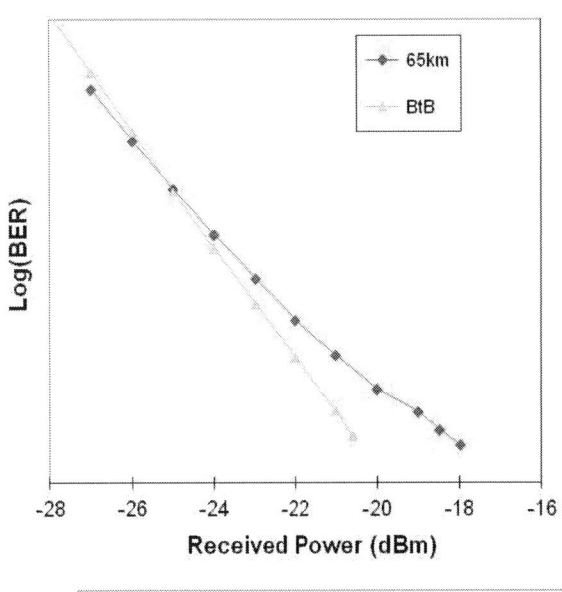

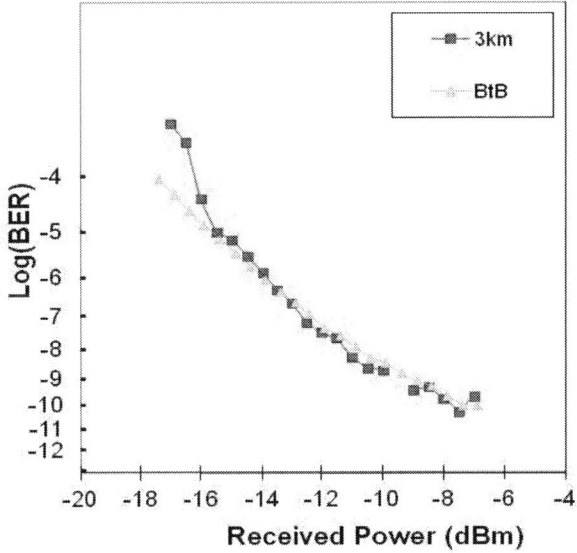

Fig.7: 20Gb/s and 10Gb/s bit error rate at 25°C in back to back and after SMF transmission over 3 km and 65 km, respectively.

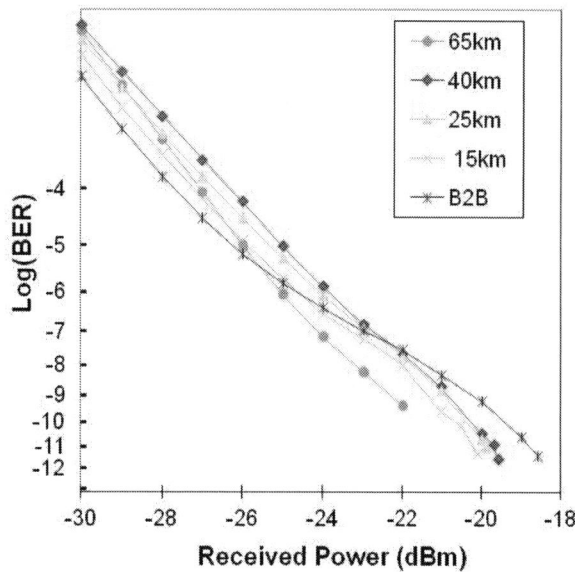

Fig.8: 10Gb/s bit error rate at 25°C in back to back and after 15, 25, 40 and 65km SMF transmission of a 500μm-long DFB laser combined with an etalon filter.

VI. CONCLUSION

We report on transmission performances of InP-based Quantum Dashes 1.55μm distributed feedback buried ridge stripe lasers. The high intrinsic dynamic performances of p-doped DWELL design lead to error free transmissions up to 20Gb/s. Combining the directly modulated laser with an etalon filter, we demonstrate uncompensated and non-amplified SMF 10Gb/s transmissions at a constant bias current from back-to-back up to 65km with extinction ratio as large as 8dB. This result paves the way to the implementation of low cost sources for access network applications.

ACKNOWLEDGMENTS

This work has been partially supported by the French ANR-DIQDOT, the ITN-PROPHET projects and the System@tic Paris-region competitiveness cluster.

REFERENCES

[1] Ericsson Review No. 2, p.64 (2008)
[2] O. Mizuhara et al., Electron.Lett. 31, 660 (1995)
[3] P.A. Morton et al., Proc. OFC'96, TuH6 (1996)
[4] B. Wedding, Electron. Lett. 28, 1298 (1992)
[5] Y. Matsui et al., IEEE Photon. Technol. Lett. 18, 385 (2006)
[6] Y. Arakawa et al., IEEE J. of Quantum Electronics, 22, 1887 (1986)
[7] B. Dagens et al., IEEE Photon. Technol. Lett., 20, 903 (2008)
[8] S. Hein et al., Electronics Letter, 43, 1093 (2007)
[9] N. Chimot et al., Optical Communication (ECOC), 2010, Th.9.D

20 GHz to 83 GHz single section InAs/InP quantum dot mode-locked lasers grown on (001) misoriented substrate

K. Klaime[1], R. Piron[1], C. Paranthoen[1], T. Batte[1], F. Grillot[1], O. Dehaese[1], S. Loualiche[1], A. Le Corre[1], R. Rosales[2], K. Merghem[2], A. Martinez[2] and A. Ramdane[2]

[1]UEB INSA-RENNES, CNRS UMR6082 FOTON, FRANCE

[2] CNRS LPN MARCOUSSIS, FRANCE

ABSTRACT — **We report original results on GSMBE grown InAs/InP QD structures. Three single section devices show passive mode locking from 20 GHz to 83 GHz with low RF spectral width (32 kHz) and low pulse duration of 1.3 ps. We report also a double wavelength emission at high injection current, associated with degradation of mode locking properties. The real cause of these phenomena is still unclear.**

Index Terms — **Mode-locking, quantum dots (QDs), semiconductors laser.**

I. INTRODUCTION

Mode-locked semiconductor laser (MLL) diodes have for many years been regarded as potentially suitable for a large range of photonic applications. According to their short pulse and high frequency, These short pulse sources are well suited for ultra high bit rate optical telecommunications for the achievement of all optical systems for clock recovery, clock distribution, optical sampling, radio over fiber signal generation so as to overcome the bandwidth limitations of electronic systems at low cost [1]. According to their ultra broad bandwidth, ultra fast gain dynamics, low threshold current density, high characteristic temperature and easily saturated gain, quantum dots (QDs) materials, are ideal candidate for semiconductor MLLs. Many results are demonstrated on GaAs material, however this material system does not allow reaching the C and L bands. InP material system has shown some excellent performances on InAs quantum dashes (QDashes) and QDs mono-section laser devices [2][3], and InAs QDashes multi-section laser devices [4]. InAs QDashes are currently observed on (001) InP by Molecular Beam Epitaxy (MBE). Recently, we demonstrated the growth of real QD nanostructures on conventional (001) InP substrate [5]. Our approach consists to reduce the elongation of QDashes by using misoriented substrates. Here, we report the growth, the fabrication and characterization of a single section InAs/InP QD mode locked laser (QDMLL) on (001) misoriented substrate emitting at 1.65 μm.

II. FABRICATION AND STATIC CHARACTERISATIONS

The active zone contains 6 QDs layers embedded in a 360 nm of GaInAsP (1.18 μm) waveguide grown by Gas source MBE on n doped (001)InP misoriented 2° toward (111)B orientation

using the double cap technique [6]. The top cladding is a 2 μm of p-InP with a 200 nm p-InGaAs top contact layer. Fig. 1. shows the epitaxial structure described in this section.

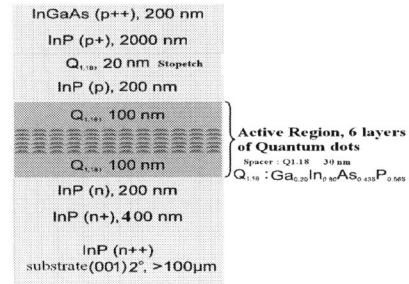

Fig. 1. Laser structure

As shown on Fig. 2. QD exhibits a photoluminescence spectrum centred at 1.6 μm and a 160 nm Full Width Half Maximum resulting from the large size dispersion of quantum dots. This inhomogeneous enlargement of the optical spectrum increases the number of modes involved in the laser mode locking emission, and then if all these modes have the same phase, the ML pulse width was reduced to the minimum. The optical polarisation ratio appears to be 11%, smaller in comparison with conventional QDashs (30%), and close to the polarisation ratio of the QDs elaborated on (311)B substrate (9%). This low polarisation ratio is an indication of the presence of real dots in the active zone. Fabricated devices are single section Fabry Perot (FP) mono-mode ridge lasers. The width of the ridge is 3 μm. BCB processing has been used for insulation and planarization between lasers. To form the top p-contact metal, we deposit a Ti/Au using e-beam evaporator. After thinning and polishing the substrate, an Au/Ge n-contact metal was deposited on the backside of the InP substrate. The wafer was cleaved into 3 different laser FP cavity lengths. The transversal single mode emission has been demonstrated by far field measurements. Three FP laser with cavity lengths of 2 mm, 1 mm and 0.5 mm are realized. The lasers without highly reflective facet treatment exhibit relatively low threshold currents of 100 mA, 57 mA and 46 mA respectively for these lengths. The corresponding slope efficiencies of these lasers are 0.04 W/A, 0.07 W/A and 0.088 W/A per facet. Modal gain and internal losses were estimated to be 24 cm^{-1} and 15 cm^{-1} respectively.

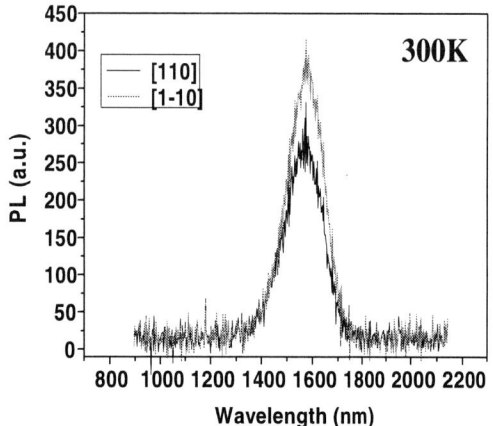

Fig. 2. Polarization resolved photoluminescence spectra.

III. MEASUREMENTS AND DISCUSSIONS

In this section, we present the results of a study on passively single section mode-locked lasers for the 3 previous sets of lasers FP cavity length. The repetition rates are 21.6 GHz, 41.6 GHz and 83 GHz. The used measurement setup is described in Fig. 3. The temperature is regulated at 20°C using a Peltier cooler. The laser signal is collected by an antireflection coated lensed fiber while an optical isolator is used to prevent optical feedback from reflections into the laser cavity. An optical spectral analyser, an RF electrical spectral analyser with a 50 GHz InGaAs photodiode and a home-made autocorrelator based on second harmonic generation are also used. The typical operating condition gain current is about 270 mA for the 2 mm device, 150 mA for the 1 mm device and 210 mA for the 0.5 mm device, corresponding to a minimum of background and a smaller pulse width. Measurements were taken after propagation in 170 m, 120 m and 60 m of a Single mode fiber (SMF) respectively for the 2 mm, 1 mm and 0.5 mm devices. These fiber lengths are used to compensate the group delay dispersion of the laser and thus shorten the optical pulses. Fig. 4. shows the RF spectrum of the 1 mm and 2 mm cavity length devices. The frequency is respectively 41.6 GHz and 21.6 GHz. The RF linewidth at -3 dB is 32 kHz and 130 kHz indicating a ML regime with a low phase noise.

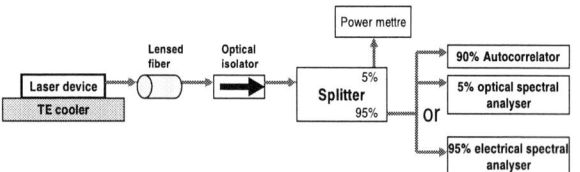

Fig. 3. Measurement setup of mode-locking lasers.

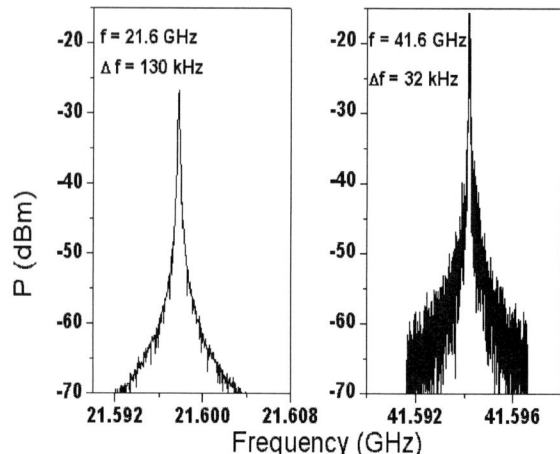

Fig. 4. RF spectrum of the 2 mm (left) and 1 mm (right) devices at 300 mA and 160 mA, the resolution is 30 and 17 kHz respectively.

The autocorrelation trace and the optical spectrum (in inset) are represented in Fig. 5 for the three devices. The optical spectrum width at -3 dB is calculated to be 3.5 nm, 5.8 nm and 6 nm respectively for the 2 mm, 1 mm and 0.5 mm cavity length. The autocorrelation maximum value over the background ratio called the on/off extinction ratio is about 15 dB. This value indicates the absence of a secondary pulse. We obtain pulse width values between 1.3 ps and 2 ps after deconvolution of the Gaussian line shape, which confirm a well dispersion compensation by using a single mode fibre. The lowest value is measured for the shorter FP cavity which requires lower external dispersion compensation. Then we can calculate the time bandwidth product (TBP) to be 0.6, 1.3 and 0.9. These values are lightly above the 0.44 Gaussian Fourier limit indicating some residual frequency chirp being present in the pulses. All these measurements were taken without optical amplification. Table 1 presents the pulses width, the optical spectrum width at -3dB and the time bandwidth product (TBP) for different devices.

FP cavity length (mm)	$\Delta\tau$(ps)	$\Delta\lambda$(nm) ($\Delta\nu$(THz))	$\Delta\tau*\Delta\nu$ (TBP)
2	1.6	3.5 (0.38)	0.6
1	2.0	5.8 (0.63)	1.3
0.5	1.3	6.0 (0.67)	0.9

Table. 1. Pulse and optical spectrum widths of the devices.

We note that the ML pulses are more easily generated at higher repetition rates for shorter cavity lengths and require lower external dispersion compensation by SMF. This behaviour has been also reported by other research groups for QD or QDash lasers [2][3]. 300 fs pulses with 0.86 mm long single section QDMLL has been reported using a Chemical Beam Epitaxy (CBE). R. Rosales et al have showed that the Group Delay Dispersion (GDD) is always smaller for shorter laser. The decrease of the pulse width with increasing the gain current is also reported and has been attributed to a compensation of the dispersion by non linear effects enhanced by higher carrier density and larger intracavity laser fields. The optimal single fibre length decreases with the cavity length. The high optimal current density for the 0.5 mm cavity length device (15 kA/cm^2) increases the effects of non-linear processes which compensates a part of the group delay dispersion [7]. Different non linear processes such as four-wave mixing, self-phase modulation and cross phase modulation are proposed but at present the exact origin of this effect remains unclear.

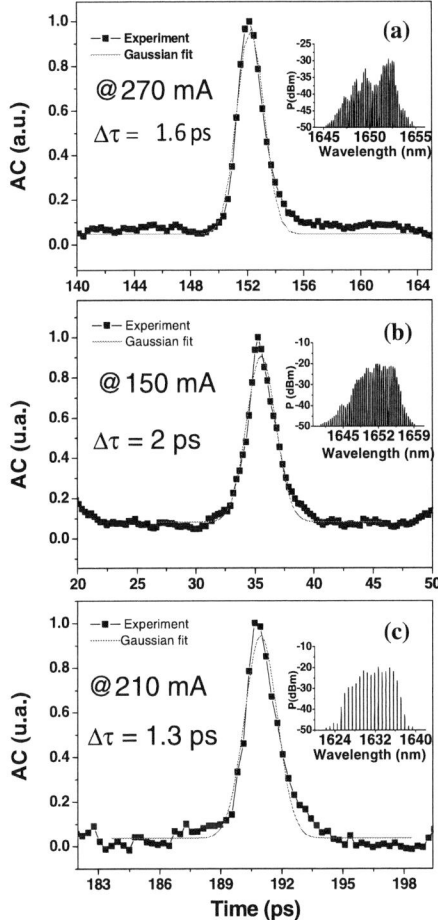

Fig. 5. Autocorrelation trace of the (a) 2mm, (b) 1 mm and (c) 0.5 mm cavity length device after propagation in 170 m, 120 m and 60 m of SMF respectively, (inset) the optical spectrum.

During mode locking process, we observe two different regimes. First, the pulse width continuously decreases as the current increases, reaching its minimum value reported in table 1. Then, for higher current, we observe simultaneously an increase of the pulse width and a peculiar two color emission separated by 7-23 nm. Fig. 6. shows the wavelength spacing change versus current for the 1 mm long device, we observe that over 170 mA, the laser emission splits in two sub-bands. These two optical sub-bands moves away from each other when we increase the injection current. The insets of the figure show the RF spectra for different currents. We can see the degradation of the RF spectra when the double emission appears. These sub-

band splitting cannot be related only to the ground and exited states (GS,ES), indeed the spectral separation is less than GS and ES transition shift . This phenomenon was observed by many research groups[8][9] and there are two recently proposed interpretations. The first one is related to the optical field and involves the interaction between the light and the excitonic based materials, like Rabi-oscillations. In this model, the single emission peak due to recombination of an exciton trapped by a QD will split into two peaks when the exciton state interacts with the strong electromagnetic field. In this case, the two peaks separation frequency is proportional to the square root of the power density. The second interpretation takes into account mode competition effects through the homogeneous and inhomogeneous broadening and has been demonstrated for QD lasers in cw and pulsed operation[9][10]. The sub-band splitting at high current injection is attributed to the enhancement of spectral hole burning at the center of the laser spectrum. Regardless of the origins of this phenomenon, and considering our experimental observations, we believe that the pulse width increasing and the degradation of the RF spectra are mainly related to the apparition of this second emission peak. Effects of temperature and laser cavity length on MLL properties are currently under investigations.

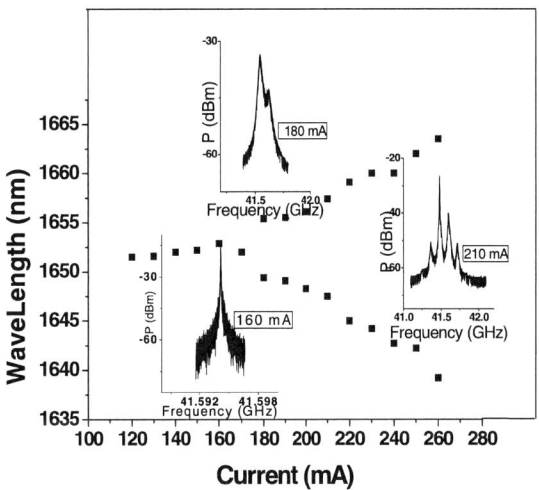

Fig. 6. Current dependence of the maximum power wavelength of the lasing peaks. Inset : RF spectra for different currents.

IV. CONCLUSION

Here we report, for the first time a single section QDMLL using GSMBE growth with excellent control of QD lateral dimension and density using diffusion properties of In adatoms along precisely oriented terraces generated on misoriented substrates. Mode-locking is demonstrated for 21.6, 41.6 and 83 GHz of repetition rates on a single section 6 QD layers laser grown on InP (001) misoriented. We observed at high current injection, a two colours emissions, these have a detrimental effects on the mode-locking process. For the optimal current condition, non-background pulses can be measured using the autocorrelator after dispersion compensation by propagation in SMF. Pulse width as low as 1.3 ps is reported with a low TBP factor close to

978-1-4673-1725-2/12 $31.00 © 2012 IEEE   183

the Fourier limit. It is interesting to note that the lower pulse width is obtained for the shorter cavity length.

Acknowledgement: This work was supported by the French National Research Agency through the project TELDOT.

REFERENCES

[1] T. Ohno, K. Sato, R. Iga, Y. Kondo, I. Ito, T. Furuta, K Yoshino, and H. Ito, "Recovery of 160 GHz optical clock from 160 Gbit/s data stream using mode locked laser diode" Electron. Lett., vol. 40, 265-267, 2004.

[2] R. Rosales, S.G. Murdoch, R.T. Watts, K. Merghem, A. Martinez, F. Lelarge, A. Accard, L.P. Barry, A. Ramdane, "High performance mode locking characteristics of single section quantum dash lasers", Opt. Express, vol. 20, 8649-8657 (2012)

[3] Z.G. Lu, J.R. Liu, P.J. Poole, Z.J. Jiao, P.J. Barrios, D. Poitras, J. Caballero, X.P. Zhang, "Ultra-high repetition rate InAs/InP quantum dot mode-locked lasers, Optics Communications 284, 2323-2326 (2011)

[4] M. Dontabactouny, R. Piron, K. Klaime, N. Chevalier, K. Tavernier, S. Loualiche, A. Le Corre, D. Larsson, C. Rosenberg, E. Semenova, K. Yvind. « 41 GHz and 10.6 GHz low threshold and low noise InAs/InP quantum dash two-section mode-locked lasers in L band » Journal of Applied Physics, 111, 023102 (2012)

[5] G.Elias, A. Letoublon, R. Piron, I. Alghouraibi, A Nakkar, N. Chevalier, K. Tavernier, A. Le Corre, N. Bertru and S. Loualiche « Achievement of high density InAs/GaInAsP quantum dots on misoriented InP(001) substrates emitting at 1.55µm »,Jpn. J. of Appl. Phys. 48, 070204 (2009).

[6] C. Paranthoen, N. Bertru, O. Dehaese, A. Le. Corre, S. Loualiche, B. Lambert, and G. Patriarche, "Height dispersion control of InAs/InP quantum dots emitting at 1.55µm", *Applied Physics Letters*, **78,** 1751 (2001).

[7] R. Maldonado-Basilio, J. Parra-Cetina, S. Latkowski, and P. Landais, "Timing-jitter, optical, and mode-beating linewidths analysis on subpicosecond optical pulses generated by a quantum-dash passively mode-locked semiconductor laser," Opt. Lett. 35, 1184 (2010).

[8] E. Bente, S. Tahvili, B. Tilma, J. Kotani, M. Smit and R. Notzel, "Modelocked and tunable InAs/InP (100) quantum dot lasers in the 1.55 µm region" ICTON 2010

[9] C. Mesaritakis, C. Simos, H. Simos, S. Mikroulis, I. Krestnikov and D. Syvridis, "Pulse width narrowing due to dual ground state emission in quantum dot passively mode locked lasers" Applied physics letters, 96, 211110, 2010

[10] C. G. Leburn, N. K. Metzger, A. A. Lagatsky, C. T. Brown, W. Sibbett, M. Lumb, E. Clarke and R. Murray, advanced solid-state photonics, OSA technical digest series (CD) (optical society of America, Washington, D.C. 2009).

InAs/InP Quantum Dash based mode locked lasers for 60 GHz radio over fiber applications

R. Rosales[1], B. Charbonnier[2], K. Merghem[1], F. Van Dijk[3], F. Lelarge[3], A. Martinez[1] & A. Ramdane[1]

1 : CNRS, Laboratory for Photonics and Nanostructures, Route de Nozay, 91460 Marcoussis, France,
2 : France Telecom Orange Labs, Advanced Studies on Home and Access networks, Lannion, France
3 : III-V Lab, a joint Laboratory of "Alcatel Lucent Bell Labs", "Thales Research &Technology" and "CEA-LETI" Route de Nozay, 91460 Marcoussis, France

Abstract — **Material growth optimization of InAs/InP quantum dashes has allowed the achievement of single section mode-locked lasers with improved performance in terms of modal gain > 40 cm⁻¹, average output power > 40 mW, radio frequency linewidth as narrow as 10 kHz together with a direct modulation bandwidth over 7 GHz. These devices were subsequently investigated as potential optical transmitters for mm-wave generation, namely 60 GHz radio-over-fiber applications. Relatively low error vector magnitude errors of ~ 11 % as well as high extinction ratio > 25 dB shows the potential of these devices for > 2.5 Gb/s short distance indoor communications.**

I. INTRODUCTION

Semiconductor monolithic mode locked lasers are very attractive for many applications including short pulse generation for optical time division multiplexing (OTDM), high bit rate all-optical clock recovery, coherent frequency comb generation for wavelength division multiplexing (WDM) or optical sampling [1]. Recent work on InAs/InP Quantum Dash based Fabry-Perot devices has exploited their relatively low phase noise for the implementation of an all-optical oscillator with record noise levels [2]. Specific bandgap engineering of the Quantum Dashes also allowed the achievement of high modal gain structures [3] and the demonstration of extremely high repetition rates based on very short laser cavities [4]. Wireless systems operating at higher frequencies, in the millimeter wave region, are quite attractive for the increased available bandwidth. Recent standardization efforts are directed towards the use of a frequency around 60 GHz for which a bandwidth ~ 7 GHz is allocated for wireless communications. The target is then the provision of > 1 Gbit/s data rate for e.g. indoor short reach. Mode-locked lasers oscillating within the 60 GHz frequency range are hence natural candidates for this novel application as they are capable of generating the millimeter wave tone without the need of any external reference oscillator [5]. The 60 GHz radiofrequency signal linewidth is moreover sufficiently narrow (a few tens of kHz) to serve as the optical carrier. Direct modulation of the mode locked laser should also allow data modulation without resorting to an external modulator, which is another advantage of this approach.

In the following we report on the optimization of a high modal gain single section mode locked laser oscillating at ~60 GHz for radio-over-fiber broadband wireless communications. Quadrature Amplitude Modulation (QAM) transmission is successfully demonstrated at 3 Gbit/s for indoor multimedia applications.

II. LASER DESIGN, FABRICATION AND CHARACTERISTICS

The laser heterostructure was grown by Gas Source Molecular Beam Epitaxy (GSMBE) on an S-doped (001) InP substrate. The active region consists of a dashes-in-a-barrier design where 9 layers of InAs quantum dashes separated by InGaAsP barriers [3]. Growth conditions were adjusted in order to keep a full width half maximum (FWHM) of PL spectrum almost constant when the number of stack is increased from 4 to 9, leading to a value of ~ 80 nm at room temperature.

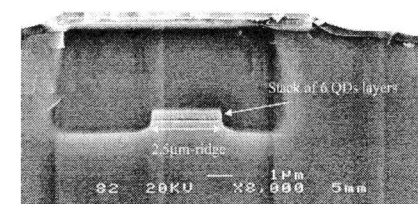

Fig. 1. SEM view of a single mode BRS

From this dash-in-a-barrier structure, broad area lasers were processed. Modal gain and internal losses were determined from threshold current vs. reciprocal length and slope efficiency vs. length curves from broad area lasers. They have been estimated at 50 cm⁻¹ and 18 cm⁻¹ respectively. Owing to a high modal gain, laser emission is observed for cavity length as short as 150 μm. Single mode buried ridge stripe lasers (ridge width of 1.5 μm) were then fabricated from the as-grown structures and devices were as cleaved (fig.1). Devices were soldered p-side up on copper submounts for thermal heat management purposes. The lasers and a temperature electrical cooler was used to monitor the temperature. All experiments were done at room temperature.

978-1-4673-1725-2/12 $31.00 © 2012 IEEE

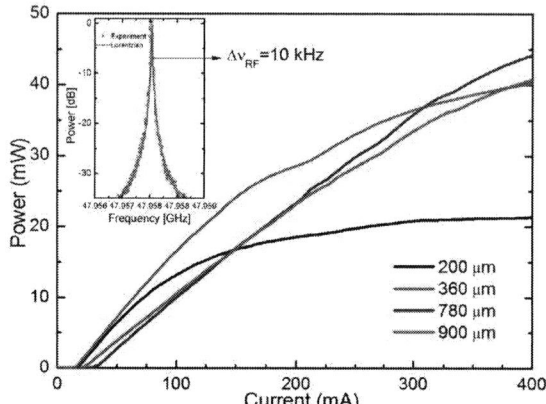

Fig.2.Typical light-current characteristics of 9 DBAR single mode Fabry-Perot lasers at room temperature. Insert: RF spectrum of the 900 µm laser at 250 mA.

Lasing occurs in continuous regime for cavities as short as 200 µm, with a threshold current of 17 mA and a maximum output power of ~ 20 mW (Fig. 2). The highest slope efficiency is obtained for a 360 µm long cavity and amounts to 0.2 W/Λ per facet and 16 mW of power in the linear regime. All these Fabry-Perot type lasers exhibit narrow radio frequency (RF) spectra at measured repetition frequencies from ~ 24 GHz to 209 GHz for corresponding cavity lengths of 1820 µm down to 205 µm. The potential of these devices for radio-over fiber applications is preliminary assessed by means of measurement of the RF linewidth. At repetition frequencies of 50-60 GHz, the obtained linewidths are ~10 kHz (insert Fig. 2), which implies very low phase noise.

Subsequent direct modulation bandwidth experiments were carried out. The -3 dB direct modulation bandwidth of 780 µm-long lasers exceeds 7 GHz owing to a relatively high differential gain. This value was high enough to accommodate the 2.5 GHz bandwidth signal centred at ~ 4 GHz for radio-over-fiber experiments as described in the next paragraph.

III. RADIO-OVER-FIBER HIGH DATA RATE TRANSMISSION

The 780 µm-long lasers are as used as optical transmitter and for direct generation of a millimeter wave tone at ~ 55 GHz without any external reference oscillator. The device is biased at about 100 mA. The experimental set-up is illustrated in fig. 3. A lensed fibre is used to couple light from the device but the coupling is willingly detuned in order to avoid detrimental back-reflections into the chip. The test signal is compliant with the IEEE802.15.3c standard and carries up to 3 Gbps for indoor multimedia applications. The base band signal is generated by a dual-output Arbitrary Waveform Generator and up-converted to 4.4 GHz intermediate frequency by an In phase - Quadrature mixer. The optical modulation creates two side bands around the natural RF oscillation frequency of the laser (55.27 GHz). Two up-

converted copies of our original 4.4 GHz signal are obtained: the first one at 50.87 GHz and the second one at 59.67 GHz as well as a very strong RF peak at 55.27 GHz. To obtain a proper up-converted signal, we want to keep only the upper modulation sideband and remove the two other signals (RF peak and lower modulation sideband). The filtering induced by the LNA gain (55-65 GHz) is not sufficient to remove those unwanted signals so an appropriate band-pass filter is used (57-62 GHz). At this point the up-converted radio signal would be ready for broadcasting into the air.

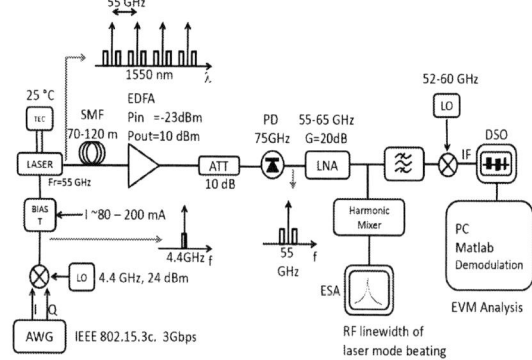

Fig. 3. High-Data Rate Transmission Experiment for direct modulation.

In order to assess the quality of this signal, we perform its down-conversion to around 5 GHz using a mixer fed with a Local Oscillator at 55 GHz.

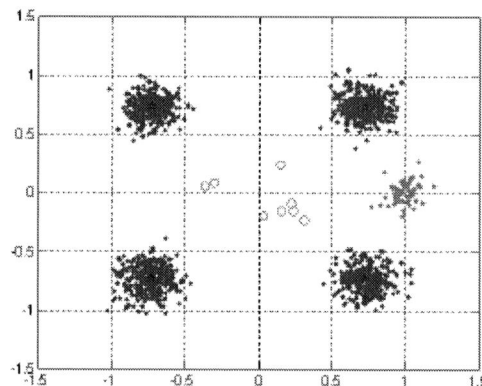

Fig. 4. Error Vector Magnitude of the test signals after optical up-conversion and electrical down-conversion from a 55 GHz InAs/InP quantum dash based mode-locked laser.

The intermediate frequency signal is then captured by a 40 GS/s Digital Sampling Oscilloscope and demodulated "offline" on a PC running Matlab®. To quantify the performance of a digital radio transmitter, we measured the error vector magnitude (EVM). The EVM equals to the ratio of the power of the error vector to the power of the reference.

This measurement tells how far the constellation points are from an ideal transmitter, a constellation diagram being a two-dimension diagram that consists of a real axis (in phase, I-axis) and imaginary axis (quadrature phase, Q-axis). Typical EVM values of 11% with a high signal to noise radio of 25 dB are obtained, demonstrating the high performance of the devices (Fig. 2). Current experiments aim at decreasing the phase noise and increasing the differential gain of the laser in order to attain data rates as high as 10 Gb/s for future indoor applications.

IV. CONCLUSION

We have shown the potential of quantum dash based mode locked lasers for radio-over-fiber applications at ~60 GHz. The main important parameters are the small RF linewidth together with high RF power thanks to the high average output power achievable with the optimized structures. Improved performance is further expected with newly designed and optical feedback immune chips.

ACKNOWLEDGEMENTS

This work has been partially funded by the ITN PROPHET and French ANR TELDOT projects.

REFERENCES

[1] E. A. Avrutin et al., 'Monolithic and multi-gigahertz mode-locked semiconductor lasers : constructions, experiments, models and applications', **IET Proc. Optoelectron., vol. 147, 251 (2000).**

[2] A. Akrout, et al., 'Low phase noise all-optical oscillator using quantum dash modelocked laser' **Electron. Lett. 46**, 73 (2010)

[3] F. Lelarge, et al., "Recent advances on InAs/InP quantum dash based semiconductor lasers and optical amplifiers operating at 1.55 μm," **IEEE J. Sel.Top. Quantum Electron. 13(1), 111–124 (2007).**

[4] K. Merghem, et al., 'Pulse generation at 346 GHz using a passively mode locked quantum-dash-based laser at 1.55 μm' **Appl. Phys. Lett. 94, 021107 (2009).**

[5] A. Stöhr, et al., '60 GHz radio-over-fiber technologies for broad band wireless services' **J. Optical Netw. Vol. 8, 471 (2009).**

InP HBT with 55-nm-wide Emitter and Relationship between Emitter Width and Current Density

Keishi Tanaka[1], and Yasuyuki Miyamoto[1]

[1]*Department of Physical Electronics, Tokyo Institute of Technology,*
2-12-1-S9-2, O-okayama, Meguro-ku, Tokyo 152-8552, Japan

Abstract—We fabricated an InP heterojunction bipolar transistor (HBT) with a 55-nm-wide emitter. For an emitter width of 55 nm and that greater than 300 nm, the maximum gain was around 15 and around 120, respectively. To confirm the relationship between the acceptable current density and the emitter width, we measured the current density when the current gain was half its maximum value. The measured current density J_{half} increased with a decrease in the emitter width. The highest observed current density was approximately 5 MA/cm² and was nearly equal to the highest reported current density of InP HBTs.

Index Terms—heterojunction bipolar transistor, InP, electron beam lithography, self-heating effect

I. INTRODUCTION

Scaling laws for heterojunction bipolar transistors (HBTs) have been proposed for their high-speed operation [1]. The important factors in realizing the scaling are the emitter width and current density. In general, it is believed that a narrower emitter width allows for a higher current density; in this regard, a current density of 3.6 MA/cm² has been expected for an emitter width of 64 nm [1]. However, to the best of our knowledge, the narrowest reported emitter width was 85 nm [2]. When the emitter width is narrow, the current gain can be reduced owing to electron-hole recombination at the surface periphery.

In our study, we fabricated an InP HBT with a 55-nm-wide emitter. On the basis of the definition of the acceptable current density from the degradation of the current gain, we evaluated the relationship between the acceptable current density and the emitter width.

II. STRUCTURE AND PROCESS

To obtain increased current density, the carrier concentration of the 30-nm-thick InP emitter adjacent to the InGaAs base was increased to 5×10^{18} cm⁻³. In general, the maximum current density is calculated as a product of the number of carriers and their Fermi velocity. The calculated maximum current density at the InP emitter with a carrier concentration of 1×10^{18} cm⁻³ is limited to 2.8 MA/cm² whereas that with a carrier concentration of 5×10^{18} cm⁻³ is around 18.6 MA/cm². The thickness and carrier concentration of the InGaAs collector were 75 nm and 7×10^{16} cm⁻³, respectively, to reduce the Kirk effect. The 15-nm-thick InGaAs base layer had a composition grading of indium from 0.46 to 0.53. As a result of carbon doping, the carrier concentration of the base layer was 5×10^{19} cm⁻³.

A schematic of the cross-sectional view of the fabricated device is shown in Fig. 1. In the fabrication process, a 350-nm- thick TiW emitter was first deposited by sputtering. A dry-etching mask (Cr/Au/Cr = 10 nm/40 nm/30 nm) was fabricated by electron beam lithography and a lift-off process. The designed widths of the mask were 80 nm, 100 nm, 200 nm, 400 nm, and 800 nm. The emitter length was 5 μm. To confirm the dependence on the emitter length, we also fabricated HBTs with emitter lengths of 10 μm, 20 μm, and 40 μm, when the designed width of the mask was 200 nm. The TiW emitter metal was formed via CF_4/O_2 reactive-ion etching (RIE), followed by etching of the InGaAs emitter cap by citric acid solution. Next, the InP emitter layer was etched by a hydrochloric acid solution. In this experiment, the base mesa was fabricated by conventional optical lithography because the purpose of this experiment was limited to DC measurements. Thus, the area of the base mesa was 50 μm × 100 μm. The base and collector contacts were obtained simultaneously by the single evaporation of a layer of Ti/Pd/Au = 15 nm/15 nm/30 nm. Self-alignment was used to achieve isolation between the emitter mesa and the base. Next, a benzocyclobutene (BCB) layer was coated on the sample, and large openings for the direct contact of probes were made via optical lithography and RIE. The top of the emitter metal was exposed using the etch-back process. Finally, an emitter contact pad was formed on the emitter via optical lithography and the lift-off process. Pads for the collector and the base were added simultaneously. Figure 2 shows an optical microscope image of the fabricated device.

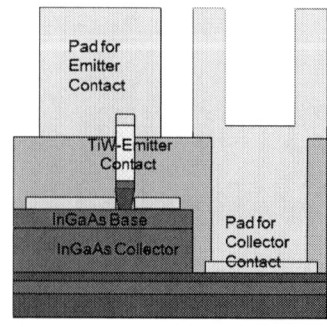

Fig. 1.   Schematic cross-sectional view of the fabricated device.

Fig. 2. Optical microscope image of the fabricated devices. Large contact pads for emitter, base, and collector can be observed.

The cross-sectional scanning electron microscope (SEM) images of the fabricated device are shown in Figs. 3 and 4. Figure 3 shows the device with an 80-nm-wide dry-etching mask in the design, and Fig. 4 shows the device with a 100-nm-wide dry-etching mask in the design. Owing to the slanted etched facet of InP, the actual widths at the emitter-base interface were 55 nm in Fig. 3 and 75 nm in Fig. 4. Thus, the actual width was 25 nm less than the designed width.

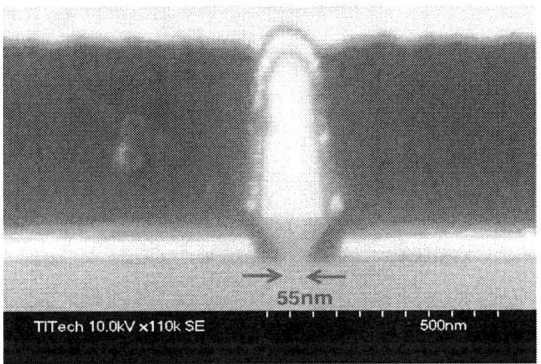

Fig. 3. Cross-sectional SEM view of the fabricated device. The designed width of the Cr/Au/Cr dry-etching mask was 80 nm.

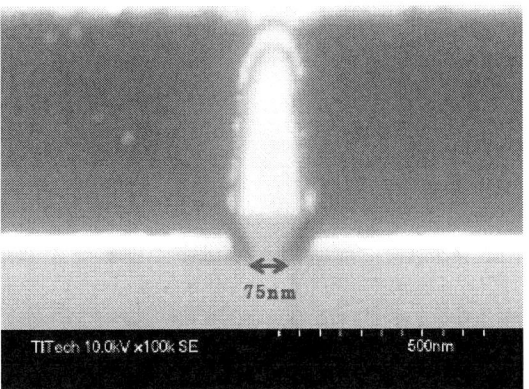

Fig. 4. Cross-sectional SEM image of the fabricated device. Designed width of Cr/Au/Cr dry-etching mask was 100 nm.

III. RESULTS AND DISCUSSIONS

The Gummel plot of the device with a 55-nm-wide emitter is shown in Fig. 5. The base ideality factor is around 2.2, whereas the collector ideality factor is around 1.1. The poor ideality factor of the base current can be attributed to the high carrier concentration of the emitter region (5×10^{18} cm^{-3}).

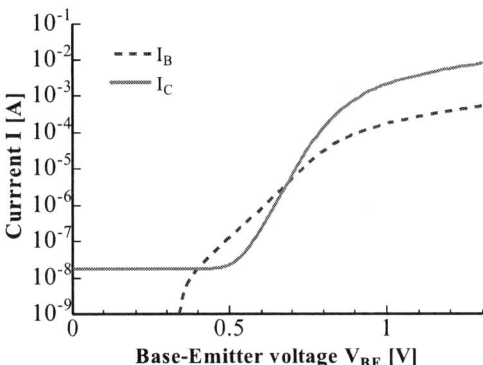

Fig. 5. Gummel plot of the device with 55-nm-wide emitter.

Figures 6 and 7 show the common-emitter current-voltage (I-V) characteristics of the device with a 55-nm-wide emitter and a 775-nm-wide emitter, respectively. From a comparison between Fig. 6 and Fig. 7, we find that the current gain increases with the emitter width. Figure 8 shows the relationship between the emitter width and the current gain. When the emitter width was 55 nm, the maximum gain was around 15, whereas it was around 120 when the emitter width was greater than 300 nm, as shown in Fig. 8. A drastic decrease in the current gain was observed for reduced emitter widths. The observed reduction was so drastic that it cannot be explained as owing to the conventional recombination current at the surface periphery [3].

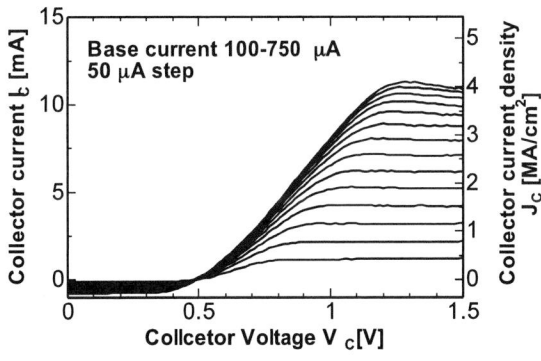

Fig. 6. Common-emitter characteristics of device. a) Device with 55-nm-wide emitter. Integration time for each point of measurement was 20 ms.

978-1-4673-1725-2/12 $31.00 © 2012 IEEE

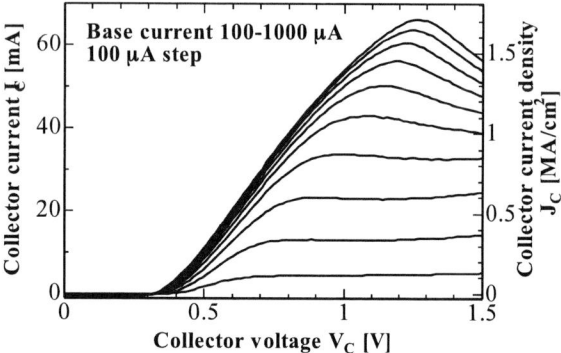

Fig. 7. Common-emitter characteristics of the fabricated device. Device with a 775-nm-wide emitter. Integration time for each point of measurement was 20 ms.

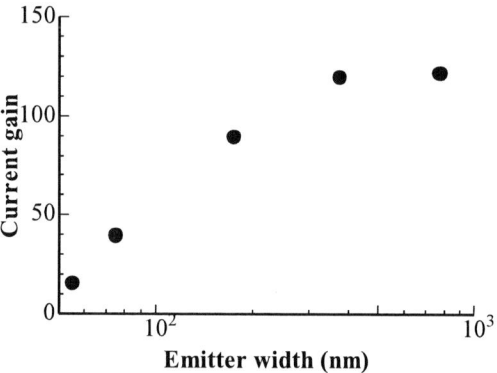

Fig. 8. Relationship between the acceptable current density J_{half} and the emitter width. Integration time for each point of the measurement was 20 ms.

When the base current was increased, the current gain decreased, as shown in Figs. 6 and 7. Strong negative output conductance was also observed. The negative output conductance can be explained by a self-heating effect [4]. Because the current density for a change in the current gain and output conductance varied depending on the emitter width, we believe that the observed change is attributed to the self-heating effect, although the reported self-heating effect in InP was less pronounced than that of GaAs [5]. As shown in Fig. 8, the current gain decreases drastically with the emitter width. In order to estimate the relationship between the emitter width and the current density, we defined the acceptable current density J_{half} for a current gain that was half its maximum gain. The relationship between J_{half} and the emitter width is shown in Fig. 9. This figure shows that J_{half} increases with a decrease in the emitter width. This improvement can be explained on the basis of a decrease in the total heat in the device.

Because the maximum current is limited by the self-heating effect of the device, a decrease in the total heat might increase the acceptable current level. In our measurement, the usual integration time for one point of measurement was 20 ms. One curve for one constant value of the base current consists of 101 points. Thus, the self-heating effect might be reduced for a short integration time. An integration time of 0.64 ms showed an increase in J_{half} when the emitter width was less than 100 nm (Fig. 9).

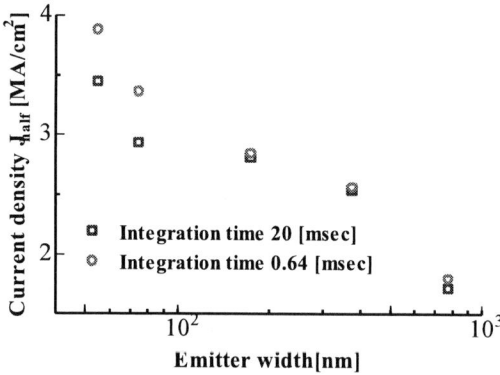

Fig. 9. Relationship between the acceptable current density J_{half} and the emitter width. Integration time for each measurement was 20 ms and 0.64 ms.

Figure 10 shows the obtained common-emitter characteristics of the device with the 55-nm-wide-emitter when the integration time of the measurement is 0.64 ms. The observed highest current density was approximately 5 MA/cm^2, and this was nearly equal to the highest reported current density of InP HBTs [6].

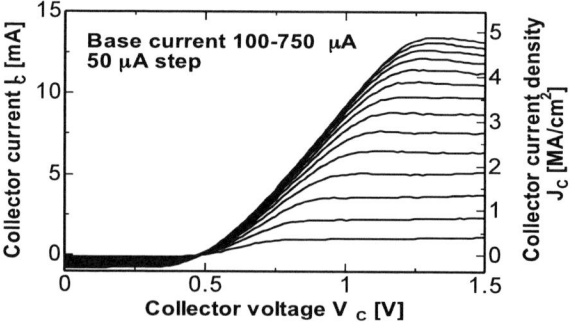

Fig. 10. Common-emitter characteristics of device with 55-nm-wide emitter. Integration time for each point of the measurement was 0.64 ms.

We also measured the relationship between J_{half} and the emitter length when the emitter width was 175 nm. The current gain is shown in Fig. 11. J_{half} increases with a decrease in the emitter length. Thus, the change in the thermal resistance by emitter area [7] was confirmed.

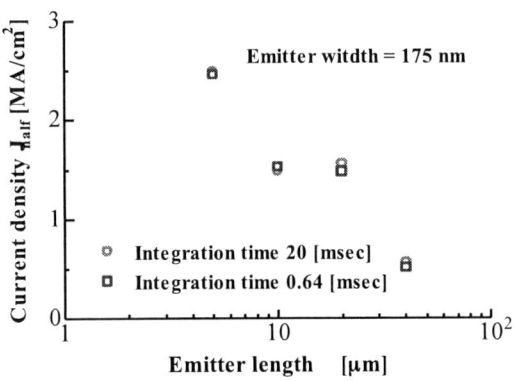

Fig. 11. Relationship between the acceptable current density J_{half}, and the emitter length. Integration time for each point of the measurement was 20 ms and 0.64 ms.

IV. CONCLUSIONS

We fabricated an InP HBT with a 55-nm-wide emitter. For an emitter width of 55 nm, the maximum gain was around 15, and for an emitter width greater than 300 nm, the maximum gain was around 120. The relationships between the acceptable current density, emitter width, and emitter length were measured on the basis of the current density when the current gain was half its maximum value. The measured current density J_{half} increased with a decrease in the emitter area. The highest observed current density was approximately 5 MA/cm^2 and was nearly equal to the highest reported current density of InP HBT.

ACKNOWLEDGMENTS

This work was supported by the Ministry of Education, Culture, Sports, Science and Technology through a Grant-in-Aid for Specially Promoted Research and Scientific Research (S) and Strategic Information and Communications R&D Promotion Programme (SCOPE) of the Ministry of Internal Affairs and Communication, Japan.

REFERENCES

[1] M. J. W. Rodwell, M. Le, and B. Brar, "InP Bipolar ICs: Scaling Roadmaps, Frequency Limits, Manufacturable Technologies," *IEEE Proc.*, vol. 96, no. 2, pp. 271-286, February 2008.

[2] M. Urteaga, M. Seo, J. Hacker, Z. Griffith, A. Young, R. Pierson, P. Rowell, A. Skalare, V. Jain, E. Lobisser, and M. J. W. Rodwell, "InP HBTs for THz Frequency Integrated Circuits," *InP related Conference*, Mo. 1.2.1, May 2011.

[3] K. Kurishima, H. Nakajima, S. Yamahata, T. Kobayashi, and Y. Matsuoka, "Growth, Design and Performance of InP-Based Heterostructure Bipolar Transistors," *Trans. IEICE*, vol. E-78, no. 9, pp. 1171-1181, 1995.

[4] P. M. Asbeck, M. F. Chang, K.-C. Wang, D. L. Miller, G. J. Sullivan, N. H. Sheng, E. Sovero, and J. A. Higgins, "Heterojunction Bipolar Transistors for Microwave and Millimeter-wave Integrated Circuits," *IEEE Trans. Microwave Theory and Techniques,* vol. MTT-35, no. 12 pp. 1462-1470 1987.

[5] W. Liu, H.-F. Chau, and E. Beam, III, "Thermal Properties and Thermal Instabilities of InP-Based Heterojunction Bipolar Transistors," *IEEE Trans. Electron Devices*, vol. 43, no. 3, pp. 388-395, 1996.

[6] V. Jain, J. C. Rode, H.-W. Chiang, A. Baraskar, E. Lobisser, B. J. Thibeault, M. Rodwell, M. Urteaga, D. Loubychev, A. Snyder, Y. Wu, J. M. Fastenau, and W. K. Liu, "1.0 THz f_{max} InP DHBTs in a Refractory Emitter and Self-Aligned Base Process for Reduced Base Access Resistance," *Device Research Conference*, pp. 271-272, June 2011.

[7] B. Grandchamp, V. Nodjiadjim, M. Zaknoune, G. A. Kone, C. Hainaut, J. Godin, M. Riet, T. Zimmer, and C. Maneux, "Trends in Submicrometer InP-Based HBT Architecture Targeting Thermal Management," *IEEE Trans. Electron Dev.*, vol. 58, no. 8, pp. 2566-2572, 2011.

InP/GaInAs DHBT with TiW emitter demonstrating f_T/f_{max} ~340/400GHz for 100 Gb/s circuit applications

V. Nodjiadjim, S. Cros-Chahrour, J.-Y. Dupuy, M. Riet, P. Berdaguer, J.-L. Gentner, B. Saturnin, J. Godin

III-V Lab (Bell Labs, TRT and CEA/LETI joint Lab), Route de Nozay, 91460 Marcoussis, France

Abstract — **In this work we report the performances of an InP/GaInAs DHBT developed in III-V Lab with a TiW metal emitter. 0.5 μm effective emitter size HBTs demonstrate f_T and f_{max} above 320 and 430 GHz respectively. Very high yield 0.7 μm emitter width HBTs showing f_T/f_{max} ~ 340/400 GHz have been used to fabricate a trans-impedance amplifier with single-ended input and differential output for 100-Gb/s optical communications.**

Index Terms — **Heterojunction bipolar transistor, TiW, InP, trans-impedance amplifier**

I. INTRODUCTION

InP DHBT technology is well suited for applications demanding both high speed and high swing thanks to its capability to provide simultaneously high cut-off frequencies and high breakdown voltage. In order to maximize the frequency performances, the device operates at high current density, hence at an elevated junction temperature which is detrimental to the device reliability. For these reasons the optimization of the device epitaxial layers to reduce the device thermal resistance as well as the use of refractive metals (W or Mo) for the emitter have been investigated for some years [1]-[2]. For instance, W acts as a barrier to Ti and Au diffusion into the emitter and base which is one of the causes for current gain degradation in InP HBTs.

In this paper, we present an InP DHBT with a 0.7 μm TiW emitter which demonstrates f_T and f_{max} of 340 and 400 GHz respectively. This device served as a key component for the fabrication of trans-impedance amplifiers for 100Gb/s transmission.

II. STUCTURE AND PROCESS

HBT heterostructures were home-grown by Gas Source MBE on a 3 inches semi-insulated substrate. The structure includes a 130 nm thick composite collector with an InGaAs spacer, a 20 nm highly doped InP region and a low-doped InP layer (2.5×10^{16} cm^{-3}). A very thin (5 nm) sub-collector InGaAs layer is also used for lower thermal resistance. The base (~ 30nm) is compositionally graded. The highly doped level of this layer (C-doping: 8×19 cm^{-3}) leads to a sheet resistance of 800Ω/□ and a low contact resistivity about 2 Ω.μm². In order to reduce the emitter contact resistance a very highly doped In$_{0.85}$Ga$_{0.15}$As is used in the emitter cap layer.

Devices with an emitter width of 0.5 and 0.7μm were processed using a self-aligned triple mesa technology. Most of

steps are realized using stepper lithography. The emitter and base contacts are defined by e-beam lithography. TiW was used for the emitter metallic contact because of its refractory metal properties. This alloy is characterized by a low dilatation coefficient, a high thermal conductivity as well as a good resistance to thermal variations. The deposition and the dry etching process of TiW have been validated and are reported in [3]. The devices were fully encapsulated with SiN before polyimide planarization. A cross-sectional view of the device is presented in Fig. 1.

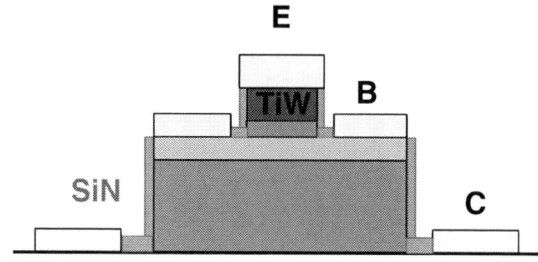

Fig. 1. Schematic cross-sectional view of HBT

Additional steps for circuit fabrication included thin film resistors and capacitors as well as 3 metal interconnect levels.

III. DEVICE PERFORMANCES

DC characteristics show a maximum current gain of 25 and a breakdown voltage above 5V @I_C=100μA. At J_C ~7mA/μm² f_T and f_{max} peak at 320 GHz and 430 GHz respectively for the 0.5μm emitter width HBTs, whereas they exhibit 340 GHz and 400 GHz respectively for the 0.7 μm emitter width HBTs (Fig. 2). These frequency performances have been improved compared to our baseline submicron technology showing f_T and f_{max} in the 250-300 GHz range [4].

Based on the simplified equivalent circuit presented in Fig. 3, the small-signal model parameters of 0.7x5μm² HBT were extracted from RF data at a bias condition of V_{CE}=1.6V and J_C=7mA μm² (peak f_T). Table I shows a comparison of the parameter values between our baseline technology and the new TiW technology. Corresponding f_T and f_{max} values are also presented. They are quite similar to the ones extrapolated from H_{21} and U respectively, determined from S parameter measurements up to 110 GHz.

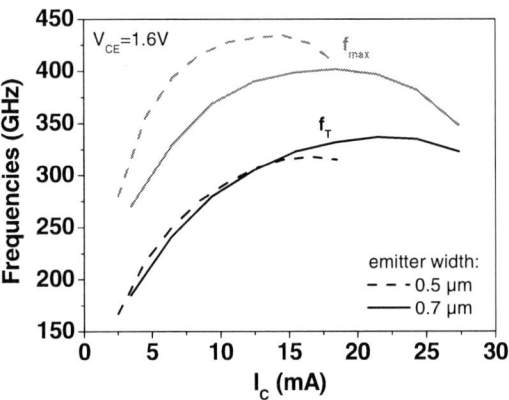

Fig. 2. Ft and f_{max} versus collector current of 0.5x5µm² and 0.7x5 µm² InP DHBTs at an emitter-collector voltage of 1.6V

TABLE I
SMALL-SIGNAL PARAMETER VALUES EXTRACTED AT PEAK F_T

	Baseline technology	Present work
R_E (Ω)	5.4	2.5
C_{BE} (fF)	20.8	21.9
r_E (Ω)	2.8	2.0
R_{Bi} (Ω)	29.8	13.6
R_{Bx} (Ω)	13.7	7.9
C_{BCi} (fF)	1.6	1.2
C_{BCx} (fF)	8.4	8.2
R_{Cx} (Ω)	3.0	3.8
τ_F (ps)	0.38	0.35
$f_{T\,calculated}$ (GHz)	288	337
$f_{max\,calculated}$ (GHz)	249	387

We note that f_T has been increased by 17%. This is essentially due to the emitter resistance decrease from 5.4 to 2.5 Ω resulting from the emitter cap layer optimization and the shrinking of the emitter semiconductor allowed by the introduction of a dry-etched TiW contact.

Additionaly, f_{max} has been enhanced by 55%. This results not only from the f_T improvement but also to the total base resistance decrease from 43 to 21 Ω which is mainly explained by a base sheet resistance reduction from 1200 to 800 Ω/□.

Thanks to the InGaAs layers shrinking in the emitter and subcollector as well as the replacement of a part of the emitter mesa by TiW metal, heat evacuation has also been improved compared to previous HBTs presented in [5].

Hence, the thermal resistance of 0.7x5 µm² HBTs has been reduced from 4K/mW to 2.5 K/mW at low power (~15mW). This corresponds to a decrease about 40%. As a result, the safe operating area of the device has been improved. As shown on Fig. 4, the negative slope on Jc(Vce) characteristics appears only for a high current density of 8 mA/µm².

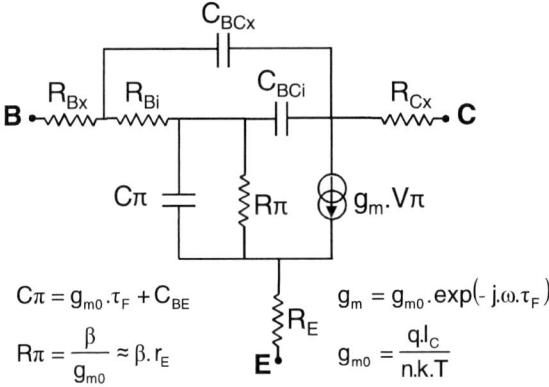

Fig. 3. Small-signal Π equivalent circuit

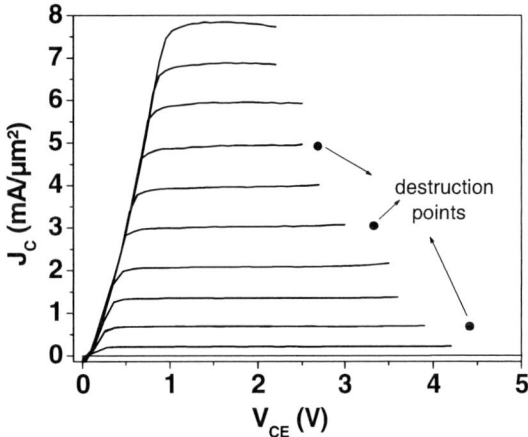

Fig. 4.   Safe operating area of 0.7x5µm² DHBT

IV. CIRCUIT RESULTS

A trans-impedance amplifier with automatic offset compensation loop dedicated to 100-Gb/s optical communications has been fabricated using the technology presented above. Except a modification of the chip length, as depicted in Fig. 5, the circuit design is identical to that reported in [6]. This circuit contains 20 HBTs with a $0.7 \times 5 \mu m^2$ emitter area for the RF part and 4 HBTs with a $0.7 \times 10 \mu m^2$ emitter area for the DC loop.

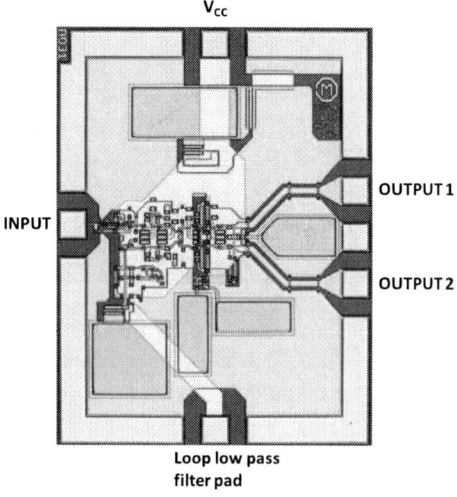

Fig. 5. Microphotograph of the fabricated trans-impedance amplifier. The chip size is $1.2 \times 1.5 mm^2$.

Thanks to technological improvements, a 3-dB bandwidth of 94 GHz and a gain of 56 dB-Ohm are obtained for the differential trans-impedance gain, as depicted in Fig. 6. The common mode trans-impedance gain is kept below 40 dB-Ohm up to 110 GHz.

As depicted in Fig. 7 and Fig. 8, S_{11} is below -10 dB up to 110 GHz and S_{22} is below -10 dB up to 37 GHz and below -5dB up to 55GHz, respectively. As depicted in Fig. 9 S_{12} is below -35 dB up to 110 GHz.

Consuming 170 mA from a single +3.3 V power supply, this is, to our knowledge, the highest bandwidth reported for a trans-impedance amplifier with single-ended to differential conversion integrating an automatic offset compensation loop, as needed when used with a DC-coupled photodiode.

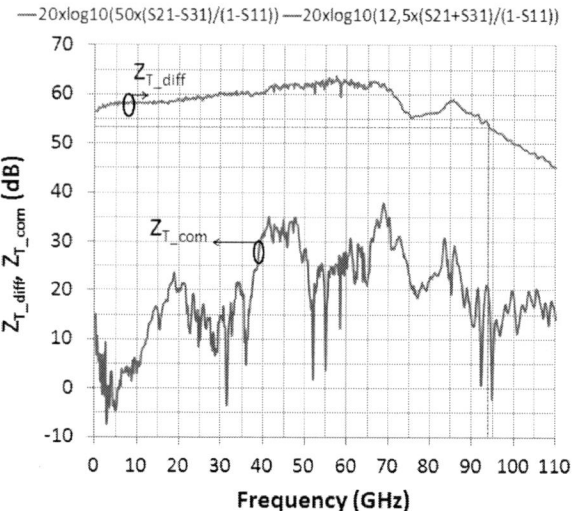

Fig. 6. Differential (Z_{T_diff}) and common mode (Z_{T_com}) trans-impedance gain extracted from S-parameter measurements.

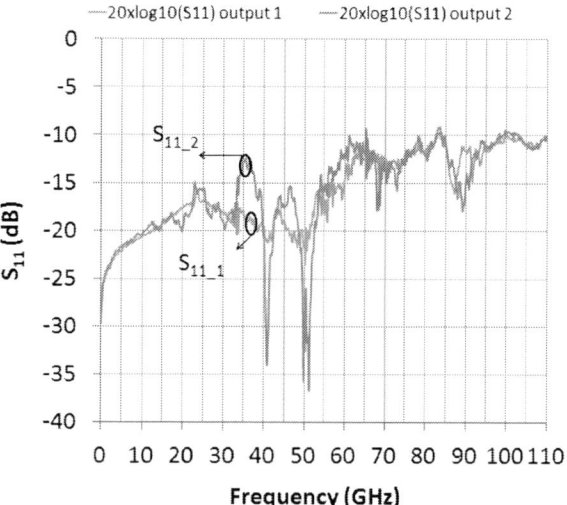

Fig. 7. Measured S_{11} (S_{11_1} or S_{11_2} correspond to S_{11} with port 1 at the input and port 2 at the output 1 or 2, respectively).

978-1-4673-1725-2/12 $31.00 © 2012 IEEE

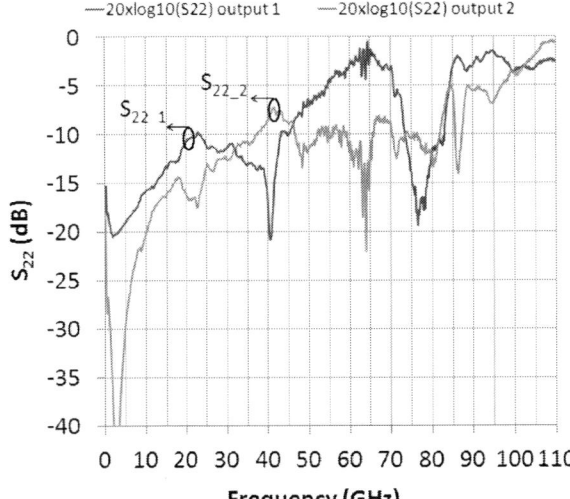

Fig. 8. Measured S_{22} (S_{22_1} or S_{22_2} correspond to S_{22} with port 1 at the input and port 2 at the output 1 or 2, respectively)

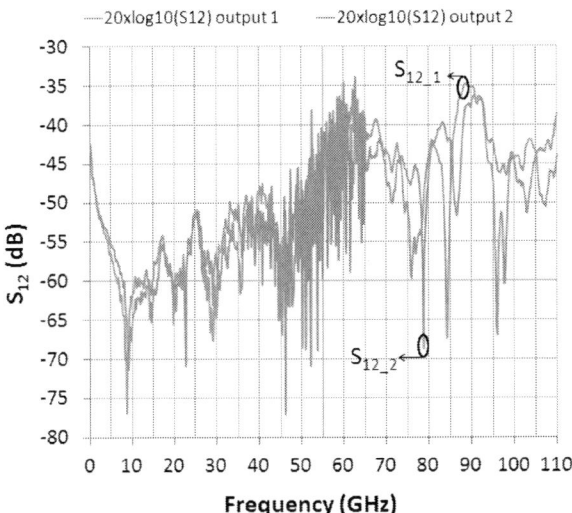

Fig. 9. Measured S_{12} (S_{12_1} or S_{12_2} correspond to S_{22} with port 1 at the input and port 2 at the output 1 or 2, respectively).

V. CONCLUSION

We have presented a TiW-emitter-based submicron InP DHBT with f_T and f_{max} of 340 GHz and 400 GHz respectively. Thanks to epitaxial and technological improvements the emitter and base resistances have been considerably reduced and the thermal behavior has been improved compared to our baseline technology previously reported. This device has been used to build a trans-impedance amplifier for 100 Gb/s transmission including a single-ended input and differential output TIA with state-of-the-art performances.

ACKNOWLEDGMENT

The authors are grateful to T. K. Johansen for transistor modeling, J. Moulu for CAD support, O. Drisse and E. Derouin for e-beam lithography, S. Bernard and C. Jany for process support, and André Scavennec for fruitful discussions. This work was partly supported by the French National Research Agency (ANR) through ROBUST program.

REFERENCES

[1] Y. K. Fukai, K. Kurishima, N. Kashio, S. Yamahata, "Reliability study on InP/InGaAs emitter-base junction for high-speed and low-power InP HBT", *22nd International Conference on Indium Phosphide and Related Materials*, pp. 119-122, 2010.

[2] V. Jain, J. C. Rode, H.-W. Chiang, A. Baraskar, E. Lobisser, B. J. Thibeault, M. Rodwell, M. Urteaga, D. Loubychev, A. Snyder, Y. Wu, J. M. Fastenau, W. K. Liu, "1.0 THz f_{max} InP DHBTs in a refractory emitter and self-aligned base process for reduced base access resistance", *69th annual Device Research Conference*, pp. 271-272, 2011.

[3] S. Cros-Chahrour, M. Riet, V. Nodjiadjim, P. Berdaguer, J.-L. Gentner, O. Drisse, J. Godin, "Performances of InP/InGaAs HBTs with a dry-etched TiW metal emitter process", *20th European Workshop on Heterostructure Technology (HeTech)*, Nov. 2011.

[4] J. Godin, V. Nodjiadjim, M. Riet, P. Berdaguer, O. Drisse, E. Derouin, A. Konczykowska, J. Moulu, J.-Y. Dupuy, F. Jorge, J.-L. Gentner, A. Scavennec, "Submicron InP DHBT technology for high-speed high-swing mixed-signal ICs", *CSICS08*, Oct. 2008.

[5] B. Grandchamp, V. Nodjiadjim, M. Zaknoune, G. A. Kone, C. Hainaut, J. Godin, M. Riet, T. Zimmer, C. Maneux, "Trends in submicrometer InP-based HBT architecture targeting thermal management, *IEEE Transaction On Electron Devices*, Vol. 58, n° 8, pp. 2566-2572, August 2011.

[6] J.-Y. Dupuy, F. Jorge, M. Riet, A. Konczykowska, and J. Godin, "InP DHBT transimpedance amplifiers with automatic offset compensation for 100 Gbit/s optical communications", in *Proc. 2010 Eur.Microwave Integrated Circuits Conference (EuMIC)*, pp. 341–344, Sep. 2010.

Lower Limits To Specific Contact Resistivity

Ashish Baraskar[1], Arthur C. Gossard[2,3], Mark J. W. Rodwell[3]

[1]*GLOBALFOUNDRIES, Yorktown Heights, NY*

Depts. of [2]Materials and [3]ECE, University of California, Santa Barbara, CA

Abstract — We calculate minimum feasible contact resistivities to n-type and p-type InAs and $In_{0.53}Ga_{0.47}As$. Resistivities were calculated for a range of Schottky barrier heights as well as for the case where the transmission probability is unity (Landauer limit). Calculations are compared with recent experimental data. Experimental contact resistivities for n-$In_{0.53}Ga_{0.47}As$ and n-InAs lie within 2.5:1 of calculated resistivities given generally accepted values of Schottky barrier potential. Computed resistivities in the presence of a barrier are only 3.5:1 to 4:1 above Landauer limits.

Index Terms — Contact resistivity, Landauer limit, metal semiconductor junctions, Schottky barrier, transmission probability.

I. INTRODUCTION

Low-resistivity metal-semiconductor contacts are fundamental to the scaling of transistors in both nm VLSI and sub-mm-wave/THz applications [1, 2]. In high-frequency transistors, the conductivity and operating current densities of contacts must both increase in proportion to the square of operating frequency. Similar scaling is required of MOSFET source/drain contacts in VLSI because of decreasing S/D contact pitch and of increasing drain current per unit gate width. Improved contacts are under development for both group IV and III-V compound semiconductors. Degenerate active carrier concentration, Schottky barrier height, and semiconductor surface preparation are the primary factors that determine contact resistivity [3, 4, 5]. Contact resistivity is determined by finite values of transmission probability T, electron velocity and density of available conduction states. For $T = 1$, this lower limit is known as the Landauer quantum conductivity limit. Here we compare published InGaAs and InAs contact resistivity data with calculations of contact resistivity both in the presence of an interfacial Schottky barrier and in the Landauer limit. We find that experimental contact resistivities for n-$In_{0.53}Ga_{0.47}As$ and n-InAs lie within 2.5:1 of calculated resistivities given generally accepted values of Schottky barrier potential. Further, computed resistivities in the presence of a barrier are only 3.5:1 to 4:1 above Landauer limits.

II. CURRENT DENSITY AND CONTACT RESISTIVITY CALCULATIONS

We first present the methods used here to calculate contact resistivity. Assuming conservation of transverse momentum and total energy, the net current density crossing the metal-semiconductor interface is [6]

$$J = \frac{2q}{(2\pi)^3} \int_{k_{sx}=-\infty}^{k_{sx}=\infty} \int_{k_{sy}=-\infty}^{k_{sy}=\infty} \int_{k_{sz}=0}^{k_{sz}=\infty} v_{sz} \cdot (f_s - f_m) \cdot T \cdot dk_{sx} dk_{sy} dk_{sz} , \quad (1)$$

where k_{sx}, k_{sy} and k_{sz} are the wave vectors in the semiconductor in x, y and z (transport) directions, v_{sz} is the z component of the electron group velocity in the semiconductor, and T is the interface transmission probability. The Fermi function in the semiconductor is $f_s = (1 + \exp((E - E_{fs})/kT))^{-1}$, while that of the metal is $f_m = (1 + \exp((E - E_{fm})/kT))^{-1}$, where E is the total electron energy, E_{fs} and E_{fm} are the Fermi energies in the semiconductor and in the metal, and $E_{fs} - E_{fm} = qV$, where V is the applied bias voltage.

The contact resistivity ρ_c is $(dV/dJ)|_{V=0}$ and hence is

$$\frac{1}{\rho_c} = \frac{2q^2}{(2\pi)^3 kT}$$

$$\times \int_{k_{sx}=-\infty}^{k_{sx}=\infty} \int_{k_{sy}=-\infty}^{k_{sy}=\infty} \int_{k_{sz}=0}^{k_{sz}=\infty} \frac{v_{sz} \cdot T \cdot \exp\left(\dfrac{E - E_{fs}}{kT}\right)}{\left(1 + \exp\left(\dfrac{E - E_{fs}}{kT}\right)\right)^2} dk_{sx} dk_{sy} dk_{sz} . \quad (2)$$

We approximate the metal's $E-k$ dispersion relationship as a single parabolic band with conduction band energy E_{cm}. For a semiconductor with parabolic energy dispersion, the total electron energy is

$$E = q\phi_R + \frac{\hbar^2}{2m_s}(k_{sx}^2 + k_{sy}^2 + k_{sz}^2), \quad (3)$$

where $q\phi_R = E_{cs} - E_{cm}$ is the difference between the metal and semiconductor conduction band energies, m_s is the electron mass in the semiconductor. Energies are computed relative to E_{cm}.

The electron group velocity is given by

$$v_{sz} = \frac{1}{\hbar} \frac{\partial E}{\partial k_{sz}} = \frac{\hbar k_{sz}}{m_s} . \quad (4)$$

From equations (2), (3) and (4),

$$\frac{1}{\rho_c} = \frac{2q^2 \hbar}{(2\pi)^3 m_s kT}$$

$$\times \int_{k_{st}=-\infty}^{k_{st}=\infty} \int_{k_{sz}=0}^{k_{sz}=\infty} \int_{\theta=0}^{\theta=2\pi} \frac{T \cdot \exp\left(\dfrac{E - E_{fs}}{kT}\right)}{\left(1 + \exp\left(\dfrac{E - E_{fs}}{kT}\right)\right)^2} k_{st} dk_{st} k_{sz} dk_{sz} d\theta , \quad (5)$$

where $k_{st}^2 = k_{sx}^2 + k_{sy}^2$ and $dk_{sx} dk_{sy} = k_{st} dk_{st} d\theta$,

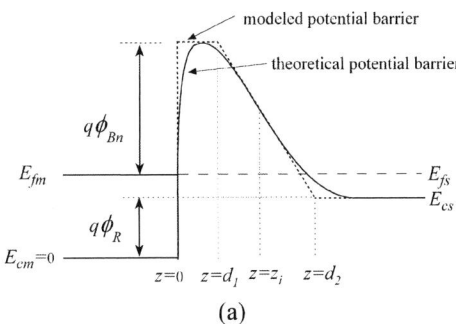

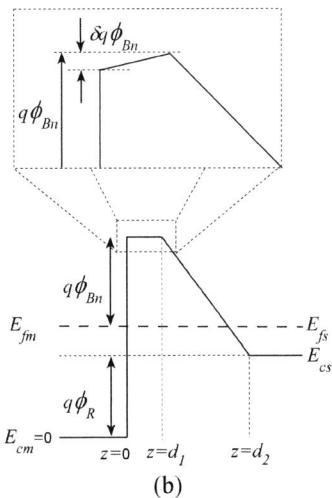

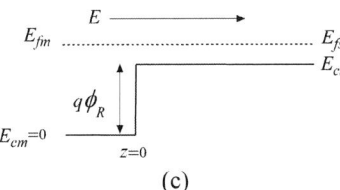

Figure 1: (a) Schematic of the theoretical and modeled potential barrier. (b) Detailed schematic of the modeled potential barrier. (c) Schematic of a step potential energy barrier.

$$\frac{1}{\rho_c} = \frac{q^2\hbar}{2\pi^2 m_s kT} \int_{k_{st}=-\infty}^{k_{st}=\infty} \int_{k_{sz}=0}^{k_{sz}=\infty} \frac{T \cdot \exp\left(\dfrac{E-E_{fs}}{kT}\right)}{\left(1+\exp\left(\dfrac{E-E_{fs}}{kT}\right)\right)^2} \, k_{st} dk_{st} k_{sz} dk_{sz} \ . \quad (6)$$

Equation (6) gives the contact resistivity for the case of parabolic bands.

For a semiconductor with non-parabolic energy dispersion, the total electron energy is approximately [7]

$$(E-q\phi_R)(1+\alpha(E-q\phi_R)) = \frac{\hbar^2}{2m_s}(k_{st}^2 + k_{sz}^2) \ , \quad (7)$$

from which we find

$$E = q\phi_R + \frac{1}{2\alpha}\left[\sqrt{1+\frac{2\alpha\hbar^2(k_{st}^2 + k_{sz}^2)}{m_s}} - 1\right], \quad (8)$$

where α is the non-parabolicity factor. The group velocity is then

$$v_{sz} = \frac{1}{\hbar}\frac{\partial E}{\partial k_{sz}} = \frac{\hbar k_{sz}}{m_s}\left(1+\frac{2\alpha\hbar^2(k_{st}^2 + k_{sz}^2)}{m_s}\right)^{-1/2}, \quad (9)$$

from which we find the contact resistivity given non-parabolic bands,

$$\frac{1}{\rho_c} = \frac{q^2\hbar}{2\pi^2 m_s kT}$$

$$\times \int_{k_{st}=-\infty}^{k_{st}=\infty} \int_{k_{sz}=0}^{k_{sz}=\infty} \frac{T \cdot \exp\left(\dfrac{E-E_{fs}}{kT}\right)}{\left(1+\exp\left(\dfrac{E-E_{fs}}{kT}\right)\right)^2} \frac{k_{st} dk_{st} k_{sz} dk_{sz}}{\left(1+\dfrac{2\alpha\hbar^2(k_{st}^2+k_{sz}^2)}{m_s}\right)^{1/2}} \ . \quad (10)$$

A. Calculation Of Transmission Probability,

The Wentzel-Kramers-Brillouin (WKB) approximation is frequently used to calculate metal-semiconductor interface transmission probability [8]. This approximation breaks down at the regions close to the maximum of the potential energy barrier and neglects quantum mechanical reflection at the metal-semiconductor interface. These limitations are important in modern heavily-doped junctions where the tunneling probability approaches unity.

Here we have calculated the exact transmission coefficient, including quantum mechanical reflection and valid in all energy ranges. From the combined effects of depletion region electrostatics and image-force, the semiconductor band energy E_{cs} is first computed as a function of position. This is then fit to a piecewise-linear approximation (fig. 1(a)). For $0 < z < d_1$, E_{cs} is fit to the peak barrier potential. In the region $d_1 < z < d_2$, the magnitude $\mathcal{E}_{max}$ and location ($z = z_i$) of the point of maximum field in the depletion region is first calculated. E_{cs} is then approximated for $d_1 < z < d_2$ with a first-order linear potential fit of field $\mathcal{E}_{max}$ passing through the point $z = z_i$.

Schrodinger's equation is then solved using Airy functions. An infinitesimal gradient $\delta q\phi_{Bn}$ is introduced in the $0 < z < d_1$ region (fig. 1(b)) as it facilitates the use of Airy functions in this region. The Airy function solutions are valid in all the energy ranges [9] i.e. $q\phi_R < E < \infty$ making the calculations less cumbersome. If a barrier with constant potential energy was chosen for this region ($0 < z < d_1$), it would require solutions of Schrodinger equations for $q\phi_R < E < q\phi_{Bn}$, $E = q\phi_{Bn}$ and $E > q\phi_{Bn}$, making the calculations tedious. The detailed calculations of transmission probability are presented in [10].

It must be noted that the present calculations neglect band gap narrowing arising from heavy doping. Treatment of the metal $E - k$ dispersion relationship as a single parabolic band introduces errors in the metal-semiconductor interface reflection probability. This limitation should be addressed in future work.

978-1-4673-1725-2/12 $31.00 © 2012 IEEE

B. Landauer Contact Resistivity

In modern III-V transistors serving high-frequency applications, under the contacts a semiconductor doping of 5×10^{19} cm^{-3} to 1×10^{20} cm^{-3} is typical. For In$_{0.53}$Ga$_{0.47}$As and InAs, where ϕ_{Bn} is small (<0.2 V), the associated depletion depths are 1 nm or less, and tunneling probability through the barrier potential--even if the barrier energy is positive-- is high. In such cases, contact resistivity remains nonzero because of quantum mechanical reflection at the interface, finite electron velocity and finite density of energy states available for electron transport.

Contact resistivity can be expressed as the inverse of the product of the density per unit area, conductivity $q^2/\pi\hbar$, and transmission probability T of available 1-D Landauer conduction channels. In the Landauer limit, $T = 1$ and $\sigma_c = (q^2/\hbar)(3/8\pi)^{2/3} \cdot \sum_{i=1}^{g}(m_{xi}m_{yi}/m_{zi}^2)^{1/6} n_{ei}^{2/3}$, where g is the # of valleys, m_{xi}, m_{yi}, m_{zi} the masses in the x, y and z (transport) directions, and n_{ei} the electron concentration in the ith valley. $\sigma_c = (q^2/\hbar)(3/8\pi)^{2/3} n^{2/3}$ for Γ-band (III-V) semiconductors, while for contacts to (100) Si, $\sigma_c = (q^2/\hbar)(3/8\pi)^{2/3}(n/6)^{2/3}(4(m_t/m_l)^{1/6} + 2(m_t/m_l)^{1/3})$.

III. RESULTS AND DISCUSSION

Figure 2 compares calculated contact resistivities for parabolic and non-parabolic bands for n-InAs. A step-potential barrier (fig. 1(c)) was assumed. Resistivities lie slightly above Landauer limits because of interface quantum reflectivity; parabolic and non-parabolic bands show differing $(E_{fs} - E_{cs})$ and hence differing interface reflectivity. At a given electron concentration, Landauer contact resistivities are slightly lower in Si than in Γ-minima InAs because of the multiple band minima and the anisotropic bands.

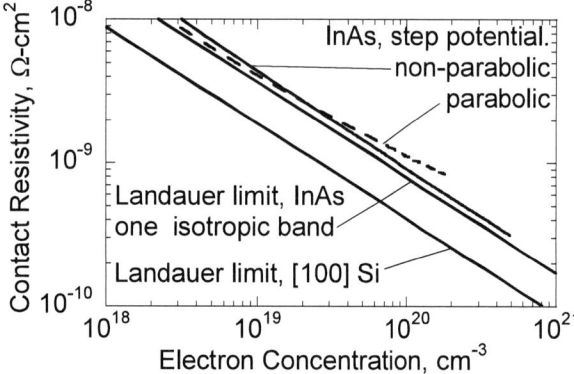

Figure 2: Landauer contact resistivity: for a single isotropic band and for (100) Si, and resistivity of InAs contact with a step-potential interface.

For Schottky tunnel barriers (fig. 1(b)), T was calculated assuming parabolic bands. Contact resistivities were calculated (fig. 3) and compared with published experimental data for n-In$_{0.53}$Ga$_{0.47}$As, p-In$_{0.53}$Ga$_{0.47}$As, n-InAs and p-InAs, as a function of electron/hole concentration and barrier height ϕ_B. We had earlier reported ultra-low contact resistivities obtained for n-InAs, n-In$_{0.53}$Ga$_{0.47}$As and p-In$_{0.53}$Ga$_{0.47}$As [11, 12, 13]. The contact resistivities were $(0.6 \pm 0.4) \times 10^{-8}$ Ω-cm^2, $(1.1 \pm 0.5) \times 10^{-8}$ Ω-cm^2 and $(0.6 \pm 0.5) \times 10^{-8}$ Ω-cm^2 for n-InAs, n-

In$_{0.53}$Ga$_{0.47}$As and p-In$_{0.53}$Ga$_{0.47}$As, respectively, which are the lowest contact resistivities reported to date for these semiconductors.

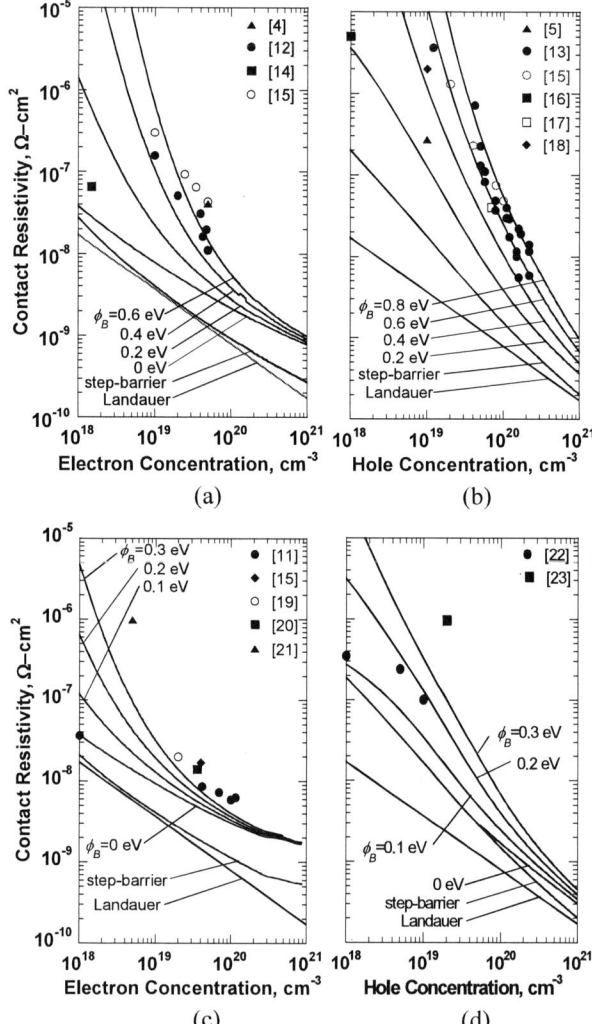

Figure 3: Calculated dependence (represented by lines) of contact resistivities (ρ_c) on bulk electron/hole concentration and Schottky barrier height (ϕ_B) for (a) n-In$_{0.53}$Ga$_{0.47}$As (b) p-In$_{0.53}$Ga$_{0.47}$As (c) n-InAs and (d) p-InAs. Experimental data from the literature is shown for comparison.

Computed contact resistivities show the expected strong dependence on electron/hole concentration and on ϕ_B. Even for contacts formed by *in-situ* deposition of refractory metals [11, 12, 13], where interfaces are expected to have an oxide/contaminant free metal-semiconductor interface, experimental resistivities of n-type contacts lie above theory given generally reported values of barrier potential. Measured contact resistivity to n-In$_{0.53}$Ga$_{0.47}$As at 5×10^{19} cm^{-3} electron concentration is 2.3:1 higher than calculated assuming $\phi_B = 0.2$ eV, while measured contact resistivity to n-InAs at 10^{20} cm^{-3} electron concentration is 1.9:1 higher than calculated assuming $\phi_B = 0$ eV. In contrast, measured contact resistivity to p-

$In_{0.53}Ga_{0.47}As$ at 2.2×10^{20} cm^{-3} hole concentration correlates well with theory if $\phi_B = 0.6$ eV is assumed.

Calculations also show the degree to which the Schottky barrier increases contact resistivity. Computed contact resistivity for n-$In_{0.53}Ga_{0.47}As$ at 5×10^{19} cm^{-3} electron concentration and $\phi_B = 0.2$ eV is only 3.9:1 larger than the Landauer limit, while computed resistivity of n-InAs at 10^{20} cm^{-3} electron concentration and $\phi_B = 0$ eV is only 3.6:1 larger than Landauer limit. For p-$In_{0.53}Ga_{0.47}As$ at 2.2×10^{20} cm^{-3} hole concentration and $\phi_B = 0.6$ eV, computed resistivity lies 13:1 above the Landauer limit; the tunneling probability remains low.

Assuming that such electron/hole concentration levels can be made feasible, contact resistivity will approach 10^{-9} Ω-cm^2 as electron/hole concentration is increased to $\sim 10^{21}$ cm^{-3}, both because of increased Landauer conductivity and increased contact transmission probability. Noting the curves for $\phi_B = 0$ eV in fig. 3(a) and 3(c), it is seen that for n-$In_{0.53}Ga_{0.47}As$ and n-InAs contacts, because of interface quantum reflectivity, contact resistivities do not drop far below 10^{-9} Ω-cm^2 even for electron concentration approaching 10^{21} cm^{-3}.

REFERENCES

[1]. M. J. W. Rodwell *et al.*, "InP bipolar ICs: scaling roadmaps, frequency limits, manufacturable technologies," *Proceedings of the IEEE, Special Issue, "A Future of Integrated Electronics: Moving Beyond Moore's Law and Off the Roadmap,"* vol. 96, no. 2, pp. 271-286, February 2008.

[2]. M. J. W. Rodwell *et al.*, "THz bipolar transistor circuits: technical feasibility, technology development, integrated circuit results," *IEEE Compound Semiconductor IC Symposium*, Monterey, CA, October 2008.

[3]. S. M. Sze and K. K. Ng., *Physics of Semiconductor Devices*, John Wiley and Sons, Inc., New Jersey, 2007.

[4]. V. Jain *et al.*, "Effect of surface preparations on contact resistivity of TiW to highly doped n-InGaAs," *IEEE/LEOS Intern. Conf. Indium Phosphide and Related Materials*, Newport Beach, CA, May 2009.

[5]. E. F. Chor *et al.*, "Electrical characterization, metallurgical investigation, and thermal stability studies of (Pd, Ti, Au)-based ohmic contacts," *J. Appl. Phys.*, vol. 87, no. 5, pp. 2437-2444, March 2000.

[6]. J. W. Conley *et al.*, "Electron tunneling in metal-semiconductor barriers", *Phys. Rev.*, vol. 150, no. 2, pp. 466-469, October 1966.

[7]. E. O. Kane, "Band structure of indium antimonide," *J. Phys. Chem. Solids,* vol. 1, no. 4, pp. 249-261, January 1957.

[8]. C. R. Crowell and V. L. Rideout, "Normalized thermionic-field (T-F) emission in metal-semiconductor (Schottky) barriers," *Solid-State Electron.*, vol. 12, no. 2, pp. 89-105, February 1969.

[9]. D. N. Christodoulides *et al.*, "Analytical calculation of the quantum mechanical transmission coefficient for a triangular, planar-doped potential barrier," *Solid-State Electron.*, vol. 28, no. 8, pp. 821-822, August 1985.

[10]. A. Baraskar, "Development of ultra-low resistance ohmic contacts for InGaAs/InP HBTs," PhD dissertation, Dept. Elec. Eng., Univ. California, Santa Barbara, 2011.

[11]. A. Baraskar *et al.*, "High doping effects on in-situ ohmic contacts to n-InAs," *22nd IEEE Intern. Conf. Indium Phosphide and Related Materials*, Kagawa, Japan, May 2010.

[12]. A. Baraskar *et al.*, "Ultralow resistance, nonalloyed Ohmic contacts to n-InGaAs," *J. Vac. Sci. Tech. B*, vol. 27, no. 4, pp. 2036-2039, July 2009.

[13]. A. Baraskar *et al.*, "In-situ and ex-situ ohmic contacts to heavily doped p-InGaAs," *16th Int. Conf. Molecular Beam Epitaxy*, Berlin, Germany, Aug 2010.

[14]. Y. H. Yeh *et al.*, "Low contact-resistance and shallow Pd/Ge ohmic contacts to n-$In_{0.53}Ga_{0.47}As$ on InP substrate formed by rapid thermal annealing," *Jap. J. Appl. Phys.*, vol. 35, no. 12A, pp. L1569-L1571, December 1996.

[15]. G. Stareev *et al.*, "A controllable mechanism of forming extremely low-resistance nonalloyed ohmic contacts to group III-V compound semiconductors," *J. Appl. Phys.*, vol. 74, no. 12, pp. 7344-7356, December 1993.

[16]. A. Katz *et al.*, "W(Zn) selectively deposited and locally diffused ohmic contacts to p-InGaAs/In formed by rapid thermal low pressure metalorganic chemical vapor deposition," *Appl. Phys. Lett.*, vol. 62, no. 21, pp. 2652-2654, May 1993.

[17]. V. Jain *et al.*, "High performance 110 nm InGaAs/InP DHBTs in dry-etched in-situ refractory emitter contact technology," *IEEE Dev. Res. Conf.*, South Bend, IN, June 2010.

[18]. P. Jian *et al.*, "Microstructural study of Au-Pd-Zn ohmic contacts to p-type InGaAsP-InP," *J. Matl. Sci.: Matl. Elec.*, vol. 7, no. 2, pp. 77-83, April 1996.

[19]. Y. Shiraishi *et al.*, "Influence of metal-n-InAs-interlayer-n-GaAs structure on nonalloyed ohmic contact resistance," *J. Appl. Phys.*, vol. 76, no. 9, pp. 5099-5110, November 1994.

[20]. U. Singisetti *et al.*, "Ultralow resistance in situ Ohmic contacts to InGaAs/InP," *App. Phys. Lett.*, vol. 93, no. 18, pp. 183502 1-3, November 2008.

[21]. C. T. Lee *et al.*, "Thermal stability of Ti/Pt/Au ohmic contacts on InAs/graded InGaAs layers," *Solid-State Elec.*, vol. 42, no. 5, pp. 871-875, May 1998.

[22]. A. Katz *et al.*, "Electrical and structural properties of Pt/Ti/p+-lnAs ohmic contacts," *J. Vac. Sci. Tech. B*, vol. 8, no. 5, pp. 1125-1127, September/October 1990.

[23]. E. M. Lysczek *et al.*, "Ohmic contacts to p-type InAs," *Matl. Sci. Eng. B*, vol. 134, no. 1, pp. 44-48, September 2006.

Study of the Ni-InGaAs alloy as an ohmic contact to the p-type base of InP/InGaAs HBTs

Shlomo Mehari, Arkady Gavrilov, Shimon Cohen, and Dan Ritter

The authors are with the Department of Electrical Engineering Technion- Israel Institute of Technology, Haifa 3200, Israel (email:smehari@tx.technion.ac.il)

Abstract — Following the silicide to silicon contact approach, Ni-InGaAs alloy was studied as an ohmic contact to p-type InGaAs. The Schottky barrier height of this system is similar to that of conventional metals such as Ti, but the interface is oxide free. The obtained specific contact resistivity to p-type material was substantially lower than that of standard Pt based metallic contacts. However, if employed for the fabrication of the base contact to HBTs, nickel thickness variations may degrade the performance of the base collector junction.

Index Terms — HBT, Ni-InGaAs, ohmic contacts, TLM.

I. INTRODUCTION

Silicides are universally used as high quality ohmic contacts to silicon devices with contact resistivity values as low as $0.6\ \Omega\mu m^2$ [1]. There are many stable silicides, and due to the self limiting nature of their formation the metal does not diffuse appreciably into the silicon [2]. By contrast to wide spread use of silicides, metal-semiconductor alloy contacts to III/V semiconductors have not yet been extensively studied. Previous literature on this subject described mainly the interfacial reactions and phase formation of the Ni/GaAs material system [3]-[4]. Recently, self-aligned Ni-InGaAs alloy was used to create the source and drain of metal-oxide-semiconductor field effect transistors [5]-[7]. Metallic source and drain for n-MOSFETs require a low Schottky barrier height (SBH) to the conduction band for low source to drain resistance and for high I_{on}/I_{off} ratio.

We recently evaluated the SBH between Ni-InGaAs alloy and n-InGaAs using the temperature dependence of the current-voltage characteristics [8]. The value we obtained was similar to the SBH of standard metal contacts to InGaAs such as Ti, and larger than reported in [7] and [9]. Our experimental results also indicated that the interface quality between the Ni-InGaAs alloy and InGaAs was superior to that of the interface between various metals and InGaAs. Thus, in spite of the similar SBH, it is of interest to find out if the contact resistivity of Ni-InGaAs is lower than that of standard metallic contacts. This work investigates only contacts to p-type material, crucial for heterojunction bipolar transistors (HBTs). Similar contacts to n-type material may also be of interest.

II. SCHOTTKY DIODES

Fig. 1 presents the Richardson plots of Ni-InGaAs Schottky diodes and of conventional metal Schottky diodes fabricated on the same wafer. The saturation current and ideality factor were obtained from a plot of $I/\{1 - exp(-qV/k_BT)\}$ at reverse and forward bias [10]. The Richardson plots of the Ni-InGaAs diodes are perfectly linear across a wide range of temperatures, while those of the conventional metallic diodes are not. The non-linear nature of the Richardson plots of the metal diodes is an indication for the presence of a thin oxide layer between the metal and the semiconductor [11]. The linear Richardson plot of the Ni-InGaAs Schottky diodes allowed an accurate extraction of the energy barrier of $0.239 + 0.01\ eV$. The ideality factor values of Ni-InGaAs diodes, shown in Fig. 2, are very close to unity, and agree well with the theoretical value related to the image force effect and thermionic assisted tunneling [10]. The ideal nature of the Ni-InGaAs Schottky diodes thus indicates that their interface is oxide free and clean.

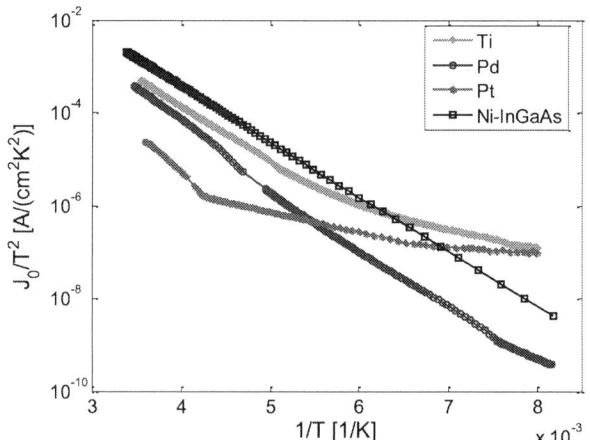

Fig. 1   Richardson plots of Ni-InGaAs alloy and of conventional metals on n-InGaAs. Note the linear behavior of the Ni-InGaAs curve across the 80-300 K temperature range, which indicates an oxide free interface, and allows an accurate extraction of the Schottky barrier height.

III. TRANSFER LENGTH METHOD MEASUREMENTS

Two types of transfer length method (TLM) structures shown schematically in Fig. 3 were fabricated to evaluate the relevant contact resistivities. The purpose of the structure shown in Fig. 3(a) was to measure the resistivity of the Ni-InGaAs alloy, and the specific contact resistivity of the metal pads to

978-1-4673-1725-2/12 $31.00 © 2012 IEEE

the Ni-InGaAs layer. It was prepared by depositing 10 nm of Ni on 25 nm thick p-type $In_{0.53}Ga_{0.47}As$ layer carbon doped to $5x10^{19}cm^{-3}$ grown on semi-insulating InP (100) substrate by metal organic molecular beam epitaxy system [12]. After Ni deposition the sample was rapid thermal annealed (RTA) at 250°C for 1 minute in forming gas atmosphere and the un-reacted Ni was etched away by an HCl solution. The TLM metal pad stack deposited after the excess Ni etch was Ti/Pt/Au 20/20/200 nm.

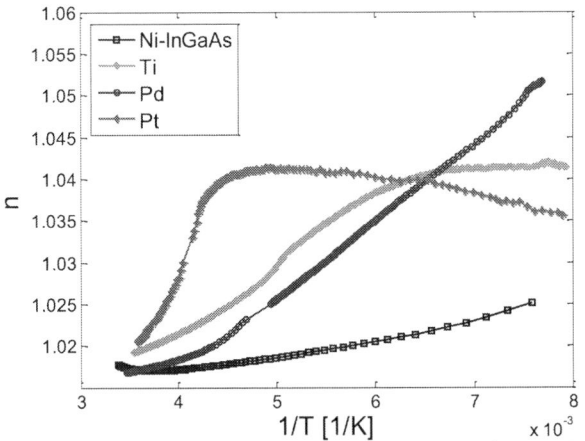

Fig. 2   Temperature dependence of the ideality factor of Ni-InGaAs and conventional metal Schottky diodes. Values obtained for the Ni-InGaAs Schottky diodes are close to the theoretical value of 1.02 [10], indicating an oxide free interface.

The purpose of the TLM structure shown in Fig. 3(b) was to measure the specific contact resistivity of Ni-InGaAs to the p-InGaAs layer. It was prepared by first covering the sample by a 20 nm thick of Si_3N_4 insulator. Opening were etched using the TLM mask pattern, and 10 nm thick Ni was evaporated into the openings, followed by 1 min RTA at 250°C, and un-reacted Ni etch by HCl. Finally the TLM pads were evaporated into the same opening. Note that the contact area in the structure shown in Fig. 3(b) is determined by the thickness of the Ni-InGaAs layer rather than by the transfer length.

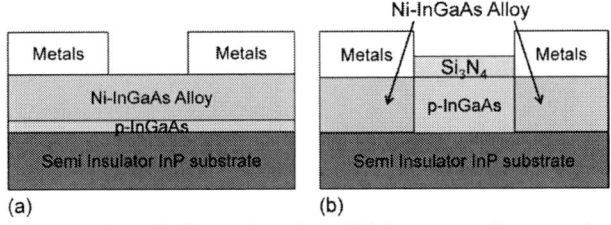

Fig. 3   Schematic illustration of two TLM structures for measuring the metal/Ni-InGaAs interface contact resistance (a) and the contact resistance of the Ni-InGaAs alloy to p-InGaAs layer (b).

Since the experiment shown in Fig. 3(b) is somewhat non-conventional, we present here a brief analysis of this structure. Its equivalent circuit model is shown in Fig. 4. A standard solution yields a total resistance of

$$R_T = \frac{2\rho_{C2}}{wt} + \frac{2\rho_{C1}}{wL_{T1}} + \frac{R_{SH2}}{w}d \qquad (1)$$

where d is the contact separation, t is the thickness of the Ni-InGaAs layer, L_{T1} is the transfer length between the metal pad and Ni-InGaAs layer, ρ_{C1} is the specific contact resistivity of the metal/Ni-InGaAs interface, ρ_{C2} is the specific contact resistivity of the Ni-InGaAs/p-InGaAs interface, R_{SH2} is the sheet resistance of the p-InGaAs layer, and w is the width of the TLM structure.

Error estimation as in [13] for this structure yields

$$\sigma_{\rho_{C2}} = \frac{wt}{\sqrt{N}}\left\{\sqrt{\left(\frac{R_{SH2}}{w}\right)^2 \sigma_d^2 + \sigma_{R_T}^2} + \frac{1}{w^2}\left(\frac{\rho_{C2}}{t} + \frac{\rho_{C1}}{L_{T1}}\right)\sigma_w\right\} +$$
$$\left(\frac{\rho_{C2}}{t}\right)\sigma_t + \frac{1}{w}\left(\frac{\rho_{C2}}{t} + \frac{\rho_{C1}}{L_{T1}}\right)t\sigma_w + \frac{t}{L_{T1}}\sigma_{\rho_{C1}} + \frac{\rho_{C1}t}{L_{T1}}\sigma_{L_{T1}} \qquad (2)$$

where σ_i is the uncertainty in the parameter i, and N is the number of gaps in the TLM structure.

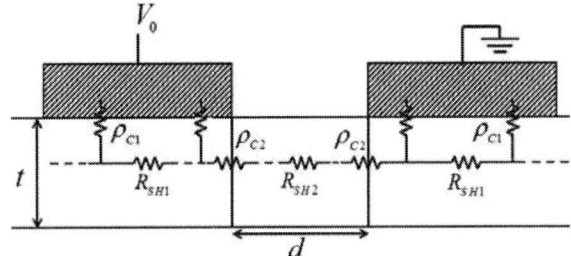

Fig. 4   Equivalent circuit model of the structure shown in Fig.3(b).

The experimental results are shown in Fig. 5. The specific contact resistivity between the metals and the Ni-InGaAs layer obtained from the structure shown in Fig. 3(a) was 4.1 ± 0.5 $\Omega\mu m^2$. This value is large compared to the contact resistivity between silicides and metals, but sufficiently low for this study. Future work to improve this contact (or possibly just to measure it more accurately) should be carried out. The obtained resistivity of the ~20 nm Ni-InGaAs layer (10 nm of Ni were deposited) was 64 $\mu\Omega cm$. This result is lower than the value of 137 $\mu\Omega cm$ reported in [14], and comparable to that of various silicides [15].

The specific contact resistivity between Ni-InGaAs layer and p-type InGaAs obtained from the structure shown in Fig. 3(b) was 6.2 ± 1.23 $\Omega\mu m^2$. It is significantly smaller than the specific contact resistivity of Pt based contacts that were prepared on the same p-type InGaAs layer which was of the order of 50-80 $\Omega\mu m^2$. The spread in contact width, σ_w, and contact separation, σ_d, was 0.05μm. The error in the measured resistance by four point probe was 0.01Ω.

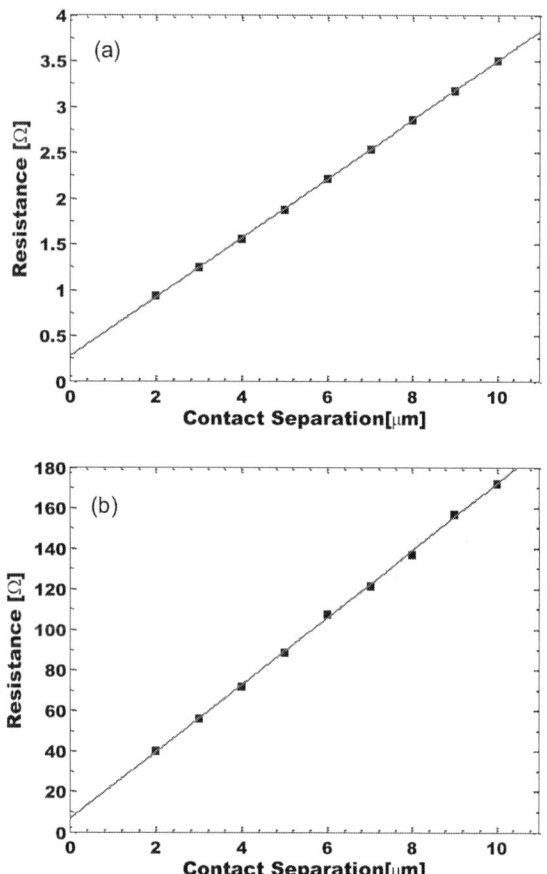

Fig. 5   Experimental results obtained from the structure shown in Fig. 3. (a) Ni-InGaAs alloy under and between metal pads as illustrated in Fig. 3(a), and (b) Ni-InGaAs alloy only under metal pads as illustrated in Fig. 3(b).

IV. CONTACT TO THE BASE COLLECTOR JUNCTION

A. Electrical characteristics

Large area p-InGaAs/n-InP diodes that emulate base collector layers of HBTs were used to test the influence of Ni-InGaAs alloy on junction performance. Fig. 6 shows the I-V plots of large area p-InGaAs/n-InP diodes with standard Pt based metallic contacts and Ni-InGaAs contacts to the p-type InGaAs layer. The p-type layer was 25 nm thick and carbon doped to $5 \times 10^{19} cm^{-3}$, grown on an unintentionally doped n-type InP layer. The Ni-InGaAs contact prepared by deposition of 6 nm thick layer of Ni did not degrade the performance of the diode, but the Ni-InGaAs layer prepared by depositing 10 nm of Ni degraded the diode performance. Heat treatment of both samples was identical: 1 min RTA at 250°C in forming gas atmosphere.

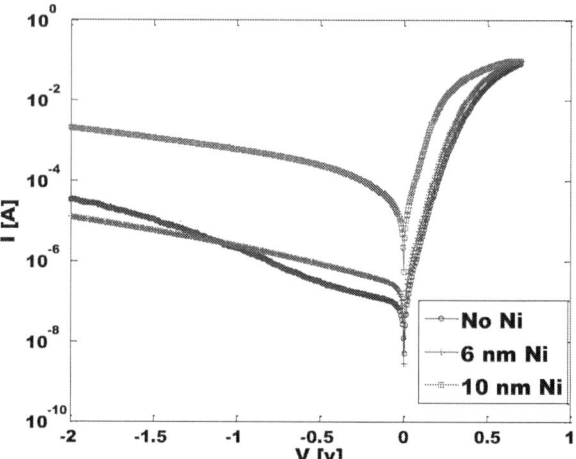

Fig. 6   I-V plots of large area p-InGaAs/n-InP diodes made with different deposited Ni thicknesses on p-InGaAs layer. For the deposition of 10 nm thick Ni a Ni-InP/n-InP Schottky diode I-V characteristic is obtained.

B. Structural analysis

A HRTEM cross section micrograph of the large area p-InGaAs/n-InP diodes with Ni-InGaAs alloy prepared by deposition 6 nm of Ni on 25 nm thick p-type InGaAs layer is shown in Fig. 7. The thickness of the Ni-InGaAs layer is about 12 nm, roughly twice the evaporated Ni thickness. The interface between the Ni-InGaAs layer and the InGaAs layer is fairly abrupt, similar to the results obtained in [7].

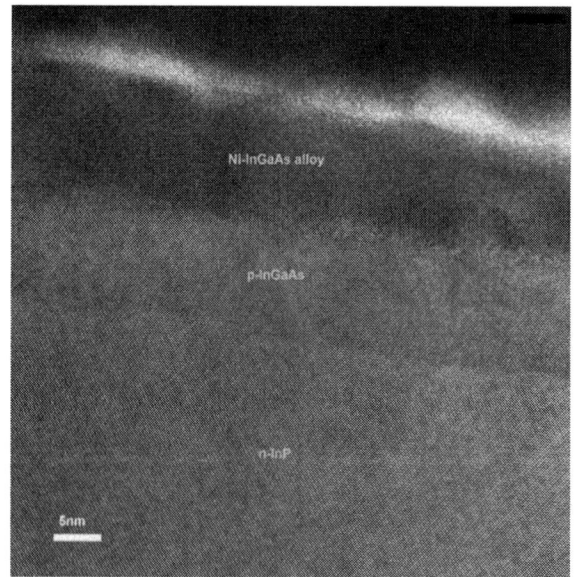

Fig. 7   HRTEM micrograph of reacted 6 nm Ni film with 25 nm p-InGaAs layer. About 12 nm Ni-InGaAs layer was formed.

Fig. 8 shows the HRTEM cross section micrograph of the sample prepared by deposition of 10 nm thick of Ni on a 25 nm thick p-InGaAs layer grown on InP. The Ni reacted with the entire p-InGaAs layer, penetrated into the InP layer, and generated an alloy of Ni-InP. This TEM explains the degraded I-V characteristic shown in Fig. 6. Presently it is not clear why a Ni alloy was obtained through the full thickness of the InGaAs layer and the first 15 nm of the InP layer.

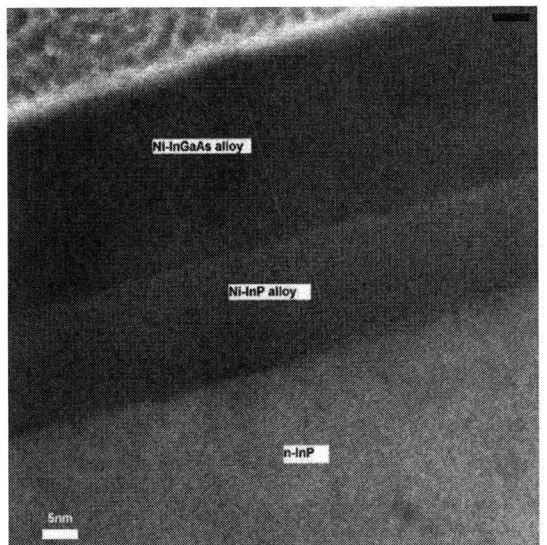

Fig. 8   HRTEM micrograph of reacted 10 nm Ni with 25 nm p-InGaAs layer. Ni reacted with the underling unintentionally doped n-InP layer to form Ni-InP alloy.

IV. CONCLUSIONS

The nearly-ideal, oxide free interface of the Ni-InGaAs alloy to InGaAs provides an order of magnitude lower ohmic contact resistance to p-type InGaAs than conventional Pt based contacts. If used for HBT base contacts, care must be taken to prevent Ni diffusion or dopants redistribution that may degrade the base-collector junction.

We thank Dr. Guy Cohen of IBM Yorktown Heights Research Center for proposing the research of the Ni-InGaAs material system and for illuminating discussions, and Dr. Yaron Kauffmann for the TEM analysis.

REFERENCES

[1]   Z. Zhang, F. Pagette, C. D'Emic, B. Yang, C. Lavoie, Y. Zhu, M. Hopstaken, S. Maurer, C. Murray, M. Guillorn, D. Klaus, J. Bucchignano, J. Bruley, J. Ott, A. Pyzyna, J. Newbury, W. Song, V. Chhabra, G. Zuo, K.L. Lee, A. Ozcan, J. Silverman, Q. Ouyang, D.G. Park, W. Haensch, and P. M. Solomon, "Sharp reduction of contact resistivities by effective Schottky barrier lowering with silicides as diffusion sources," *IEEE Electron Device Lett.*, vol 31, no. 7 July 2010.

[2]   Shyam P. Murarka, "Silicide thin films and their applications in microelectronics," *Intermetallics*, vol. 3, pp.173-186, 1995.

[3]   A. Lahav, M. Eizenberg, and Y. Komem "Interfacial reactions between Ni films and GaAs," *J. Appl. Phys.*, vol. 60, no. 3, pp.991-1001, April 1986.

[4]   Yu-Lun Chueh, Alexandra C. Ford, Johnny C. Ho, Zachery A. Jacobson, Zhiyong Fan, Chih-Yen Chen, Li-Jen Chou, and Ali Javey "Formation and characterization of $Ni_X InAs/InAs$ nanowire heterostructures by solid source reaction," *Nano Lett.*, vol. 8, no. 12, pp.4528-4533, April 2008.

[5]   L. Czornomaz, M. El Kazzi, M. Hopstaken, D. Caimi, P. Machler, C. Rossel, M. Bjoerk, C. Marchiori, H. Siegwart, J. Fompeyrine, "CMOS compatible self-aligned S/D regions for implant-free InGaAs MOSFETs," *Solid-State Electron.*, vol. 74, pp. 71-74, April 2012.

[6]   Xingui Zhang, Huaxin Guo, Hau-Yu Lin, Ivana, Xiao Gong, Qian Zhou, You-Ru Lin, Chih-Hsin Ko, Clement H. Wann, and Yee-Chia Yeoa, "Reduction of Off-State leakage current in $In_{0.7}Ga_{0.3}As$ Channel n-MOSFETs with self-aligned Ni-InGaAs contact metallization," *Electrochem. And Solid-state Lett.*, vol. 14, no. 5, pp. H212-H215, March 2011.

[7]   SangHyeon Kim, Masafumi Yokoyama, Noriyuki Taoka, Ryo Iida, Sunghoon Lee, Ryosho Nakane, Yuji Urabe, Noriyuki Miyata, Tetsuji Yasuda, Hisashi Yamada, Noboru Fukuhara, Masahiko Hata, Mitsuru Takenaka, and Shinichi Takagi, "Self-Aligned metal source/drain $In_X Ga_{1-X}As$ n-MOSFET transistors using Ni-InGaAs alloy," *Appl. Phys. Express*, vol. 4, 024201, January 2011.

[8]   Shlomo Mehari, Arkady Gavrilov, Shimon Cohen, Pini Shekhter, Moshe Eizenberg, and Dan Ritter, "Measurement of the Schottky Barrier Height between Ni-InGaAs alloy and $In_{0.53}Ga_{0.47}As$," unpublished.

[9]   Ivana, Jisheng Pan, Zheng Zhang, Xingui Zhang, Huaxin Guo, Xiao Gong and Yee-Chia Yeo, "Photoelectron spectroscopy study of band alignment at interface between Ni-InGaAs and $In_{0.53}Ga_{0.47}As$," *Appl. Phys. Lett,.* vol. 99, no. 1, 012105, July 2011.

[10] E. H. Rhoderick and R. H. Williams, *Metal-Semiconductor contacts*, second edition, Clarendon press, Oxford 1988, pp. 89-129.

[11] J. Werner, A. F. J. Levi, R. T. Tung, M. Anzlowar and M. Pinto, "Origin of the excess capacitance at intimate Schottky contacts," *Phys. Rev. Lett.,* vol. 60, no. 1, pp.53-56, January 1988.

[12] R. A. Hamm, D. Ritter, and H. Temkin, "Compact metalorganic molecular-beam epitaxy growth system," *J. Vac. Sci. Technol. A*, vol. 12, no. 5,pp.2790-2794 (1994).

[13] Keh-Ching Huang, David B. Janes, Kevin J. Webb, and Michael R. Melloch, "A Transfer Length Model for contact Resistance of Two-Layer systems with arbitrary interlayer coupling under the contacts," *IEEE Trans. On Electron Devices*, vol. 43, no. 5 May 1996.

[14] Sujith Subramanian, Ivana, Qian Zhou, Xingui Zhang, Mahendran Balakrishnan, and Yee-Chia Yeo, "Selective wet etching process for Ni-InGaAs Contact formation in InGaAs N-MOSFETs with self-aligned source and drain," *Journal of The Electrochemical Society*, vol. 159, no. 1, pp.H16-H21, December 2011.

[15] J. P. Gambino, E.G. Colgan, "Silicides and ohmic contacts," *Materials Chemistry and Physics,* vol. 52, pp.99-146, June 1998.

Multi-finger 250nm InP HBTs for 220GHz mm-Wave Power

Zach Griffith, Miguel Urteaga, Petra Rowell, Richard Pierson, and Mark Field

Teledyne Scientific Company, 1049 Camino Dos Rios, Thousand Oaks, CA 91360

zgriffith@teledyne.com, 805-453-8011

Abstract — We present here measured DC and RF performance of multi-finger 250nm InP HBTs intended for power amplifier design at high-mm, sub-mm-wave frequencies. The designs presented are in common-emitter and common-base configuration, having 24um periphery. Performance limitations for the PA cell have been identified and mitigated through novel design and layout – they include HBT thermal impedance, RF bandwidths f_t and f_{max}, MAG/MSG @ 220GHz, and reduced common-base stability from parasitic base inductance L_b and/or collector-emitter capacitance C_{ce}. The PA cells are realized using substrate-shielded non-inverted thin-film microstrip wiring to minimize L_b and C_{ce}, make small the feed lines to the multi-finger devices, and prevent parasitic substrate-mode excitation in the 12.8-ε_r InP substrate.

I. INTRODUCTION

State-of-the-art InP HBTs at the 250nm technology node have demonstrated the requisite device bandwidths and gains for power amplifiers (PA) at high-mm-, sub-mm-wave frequencies [1]. While 250nm InP DHBTs have higher off-state breakdown compared to state-of-the-art InP HEMTs [2], the usable large-signal RF voltage and current handling (toform the load-line of a power amplifier cell) of these technologies, where the MAG/MSG is appreciable, is similar. MAG/MSG of a cascode cell formed with these technologies shows that the HBT cell yields many more dB of gain compared to its HEMT counterpart. When the quiescent DC bias for a multi-finger PA cell is considered, the higher breakdown voltage and higher sustainable operating voltage at a given I_c for the HBT permits the quiescent PA bias to be satisfied with at least 15% backoff from destructive bias; this backoff margin is critically important when a PA is packaged, and/or needs to operate in a higher temperature environment. Low-loss interconnect is vital to realize the power splitters/combiners necessary for higher P_{out}. Substrate-shielded, thin-film microstrip is used to prevent parasitic substrate mode excitation in the InP substrate. Characterization of passive structures by EM simulation in this interconnect is less complicated and more accurately predicted compared to GCPW-waveguide on 2-mil InP having a complicated ground return network.

II. TSC 250NM INP HBT TECHNOLOGY

The multi-finger HBT cells are formed using a 250nm technology, having ~4.5V breakdown voltage. The TSC InP HBT technology is described in detail in [3,4]. The peak

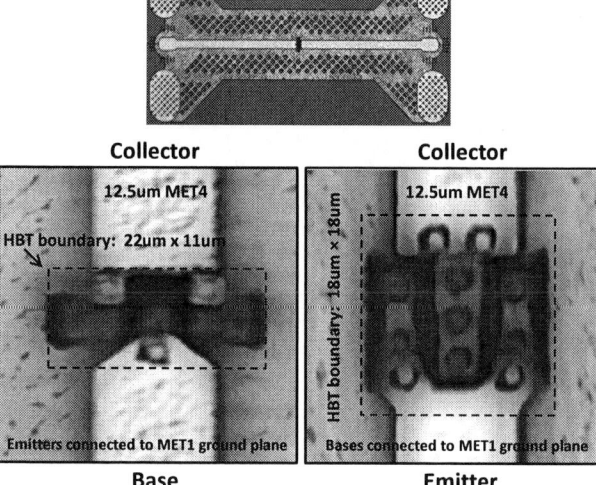

Common-Emitter HBT **Common-Base HBT**

Fig. 1 IC Micrograph of a 4-finger common-emitter and 2-finger common-base configured 250nm InP HBT, both having 24um periphery. Compact stacked vias present signal from MET4 to the base (or emitter) and collector. The HBT size (to the edge of signal reference plane) is $22 \times 11um^2$ for the common-emitter HBT and $18 \times 18um^2$ for the common-base HBT.

bandwidth of a single HBT (6um L_e) was 578GHz f_{max} and 355GHz f_τ. Figure 2 shows a cross-section of the interconnect used for this work, as well as for the 220GHz solid-state power amplifiers (SSPA) demonstrating 48, 58, and 90mW P_{out} using the PA cells presented here [5-7].

III. MULTI-FINGER HBT DESIGN AND RESULTS

Multi-finger HBTs having 12, 24, and 44um periphery were examined for 220GHz PA cell design. Here, only the 4-finger, 24um devices are presented, as they are the only HBTs used in the aforementioned 220GHz SSPA designs. Figure 1 shows an IC micrograph at high magnification of common-emitter (CE) and common-base (CB) devices, showing how connection from the base (or emitter) and collector feeds are made to the devices. The emitters are in a 2×2 for the CE and 2×1 for the CB configuration – this approach permits a multi-finger HBT having lower thermal impedance and higher operating bias before destruction, which in-turn allows the PA

978-1-4673-1725-2/12 $31.00 © 2012 IEEE

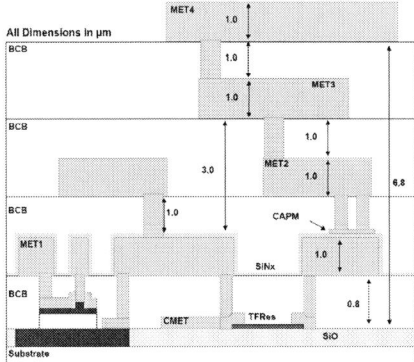

Fig. 2 Cross-section of the TSC mixed-signal InP HBT process with four metal layers separated by BCB (2.7 ε_r) as the interlayer dielectric. This interconnect and the use of stacked interconnect vias permits compact, low parasitic layout for the multi-finger HBT.

cell to operate at a higher DC voltage, increasing the P_{out} per cell. The CE base feed using MET4 is presented to the center of the multi-finger structure and vias connect to MET1 where this signal is distributed to the four fingers. The collector feed branches from MET4 before connection is made to MET1 for connection downward to the device. For each side of the device, a collector Ohmic is shared between two fingers. For the CE HBT, emitters are connected directly to the MET1 ground plane; this makes zero parasitic ground-return inductance, which in mm-, sub-mm-wave amplifier design

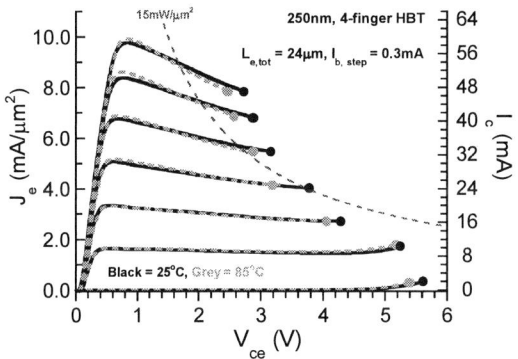

Fig. 3 Common-emitter DC characteristics of the 4-finger PA cell HBT (24um periphery) swept to destruction at 25, and 85°C

degrades gain and bandwidth.

The common-base device is a two-finger device having 24um periphery. Each finger has a base terminal on both ends (aka, dual base-post HBT) creating a plane of symmetry in the x- and y-directions. Due to this symmetry, this is effectively a four-finger HBT of emitter length 6um per finger. The base terminals are connected directly to the MET1 ground plane, minimizing the parasitic feed inductance L_b of such connections. For common-base and cascode operation, amplifier stability is greatly reduced when L_b is present, and even when it is compensated, amplifier bandwidth is made narrower. The collector Ohmic is shared between the two fingers. This arrangement allows the emitter signal

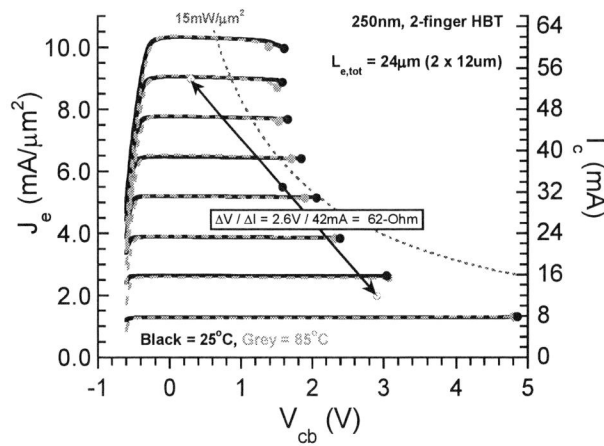

Fig. 4 Common-base DC characteristics of the 2-finger PA cell HBT (24um periphery) swept to destruction at 25 and 85°C. The PA cell load-line is superimposed showing the usable RF voltage where the HBT has sufficient gain for 220GHz SSPAs, specified by the operating endpoints. The projected P_{out} / cell is 13-15mW.

interconnect (in MET4) to 'fork' around the center of the device where collector Ohmic resides. This arrangement removes overlap capacitance and makes small the interconnect fringing capacitance between the emitter and collector terminals C_{ce}, which acts as positive feedback and degrades amplifier stability and bandwidth as well.

Figure 3 shows the CE characteristics of a 24um, 4-finger InP HBT at 25°C and 85°C wafer chuck temperature, measured on the 25-mil InP substrate. The device shows an acceptable reduction of its maximum power density before destruction when compared to a single 6um device.

Figure 4 shows the CB characteristics for a 24um, 2-

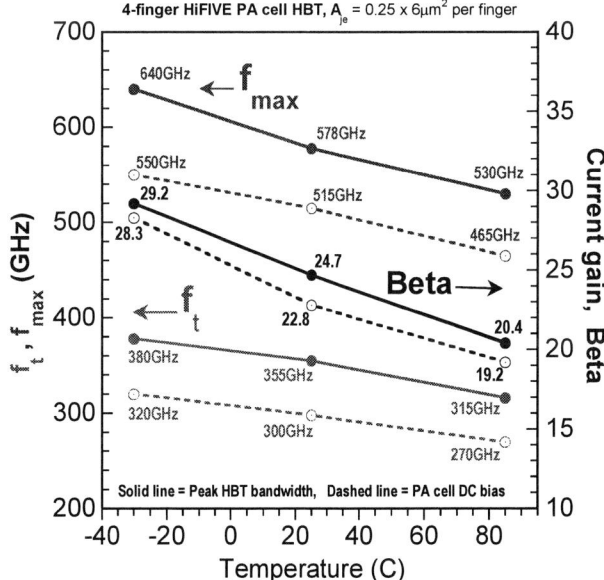

Fig. 5 Measured f_t, f_{max}, and beta of the 4-finger CE PA cell HBT (24um periphery) biased at peak RF bandwidth and at the quiescent bias of a 220GHz PA cell over temperature -30, 25, and 85°C.

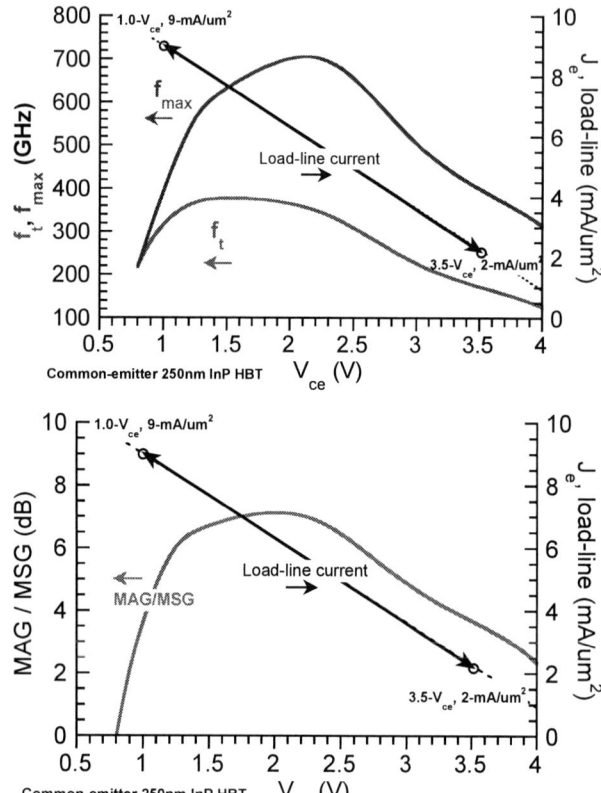

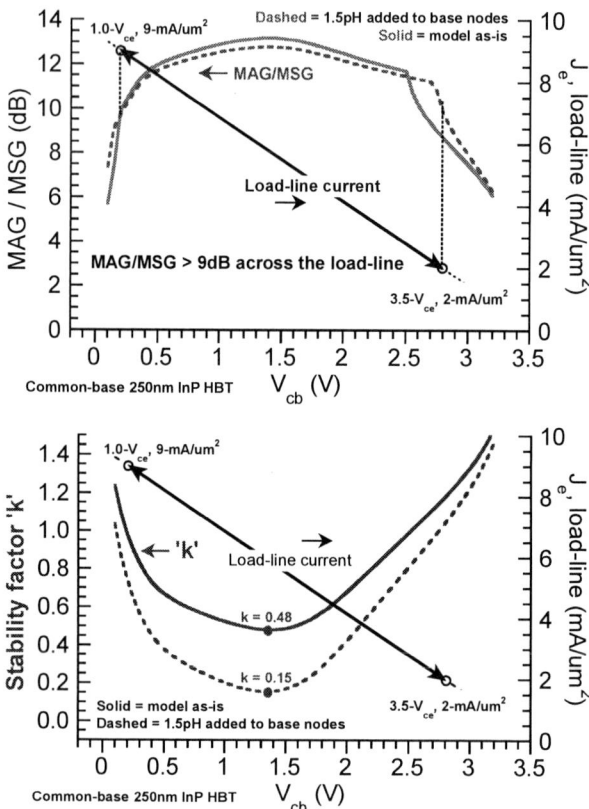

Fig. 6 Simulations f_t, f_{max} (extrapolated from 90GHz), and MAG/MSG at 220GHz for the 4-finger, 24um common-emitter PA cell HBT, plotted against the V_{ce} and I_c pairs along the PA load-line for 220GHz operation.

Fig. 7 Simulations MAG/MSG and stability factor 'k' at 220GHz operation for the 2-finger, 24um PA cell common-base HBT, whose values are plotted against the V_{cb} and I_c pairs along the PA load-line at 220GHz operation. The solid line represents the HBT model as-is, whereas the dashed line shows how by adding only 1.5pH inductance at the base node, stability is reduced, modifying MAG/MSG.

finger InP HBT at 25°C and 85° wafer chuck temperature, measured on the 25-mil InP substrate. Superimposed on the curves is the specified load-line used in [5-7] where the HBT has sufficient gain at 220GHz operation over the current and voltage range shown. The quiescent PA cell bias is 33mA (5.5mA/um²) and 2.3V_{ce}, where there is at least 15% back-off margin from destruction.

Figure 5 plots the f_t, f_{max}, and beta for the 24um, 4-finger common-emitter InP HBT when it is biased at peak RF bandwidth and at the quiescent 220GHz PA cell bias listed above, at temperatures -30, 25, and 85°C. Based on this data, we expect an acceptable reduction to the saturated SSPA P_{out} @ 85°C by 10-15%, as the MAG/MSG will be lower at the high-voltage, low current region of the load-line (fig. 4), in-turn reducing the usable voltage swing at 220GHz operation.

Figures 6 and 7 present simulated device figures-of-merit at the voltate-current pairs V_{ce} and I_C associated with the operating PA load-line for the PA's in [5-7]. A large signal Agilent ADS model for the TSC 250nm InP HBT technology was used; this same proven model was used for the 220GHz SSPAs. For the CE HBT, the gain and usable voltage is not sufficiently high for 220GHz SSPA design. While the f_{max} between the CE and CB PA cell HBTs we know is identical, figure-7 shows the CB HBT has considerably more gain.

More importantly, the CB HBT maintains sufficiently high gain and usable voltage across the span of the load-line specified for the 220GHz SSPA work for highest mW/um HBT operation, even when the HBT f_{max} is reduced by ~ 40% at the high voltage, low current regime of the load-line.

Figure 7 also shows the simulated stability factor 'k' for the CB PA cell HBT. The purpose of the plot is to show that at 220GHz operation in this HBT technology, 'k' is already < 1. When only 1.5pH of inductance is added to the base node of the HBT, the stability factor is modified significantly. With this understanding, the CB HBT design here purposely has its DC and AC base nodes connected to the large metal ground plane to minimize unwanted inductance at these nodes.

ACKNOLEDGEMENTS

The authors wish to recognize the 220GHz SSPA designs performed by Thomas Reed and Prof. Rodwell at UC Santa Barbara.

This work was supported under the DARPA HiFIVE program. The views, opinions, and/or findings contained in this article are those of the authors and should not be interpreted as representing the official views or polices, either expressed or implied, of the Defense Advanced Projects Agency or the Department of Defense.

IV. REFERENCES

[1] M. Rodwell. M. Le, B. Brar, *Proceedings of the IEEE*, vol. 96, no. 2, Feb. 2008.

[2] V. Radisic, K. Leong, S. Sarkozy, X. Mei, W. Yoshida, P-H Liu, R. Lai, "A 75mW 210GHz Power Amplifier Module", *Proc. IEEE Compound Semiconductor IC Symposium*, Honolulu, HI, Oct. 16-19, 2011.

[3] M. Urteaga, R. Pierson, P. Rowell, M. Choe, D. Mensa, B. Brar, *Proc. IEEE/LEOS Indium Phosphide and Related Materials Conference*, Versailles, France, May 25-29, 2008.

[4] M. Urteaga. M. Seo, J. Hacker, Z. Griffith, A. Young, R. Pierson, P. Rowell, A. Skalare, M. Rodwell, *Proc. IEEE Compound Semiconductor IC Symposium, Monterey, CA,* Oct. 3-6, 2010.

[5] T. Reed, M. Rodwell, Z. Griffith, P. Rowell, M. Urteaga, M. Field, J. Hacker, "48.8mW Multi-cell InP HBT Amplifier with On-Wafer Power Combining at 220GHz", *Proc. IEEE Compound Semiconductor IC Symposium*, Honolulu, HI, Oct. 16-19, 2011.

[6] T. Reed, M. Rodwell, Z. Griffith, P. Rowell, M. Field, M. Urteaga, "A 58.4mW Solid-State Power Amplifier at 220GHz using InP HBTs", *Proc. IEEE MTT-S IMS*, June 17-22, 2012.

[7] T. Reed, M. Rodwell, Z. Griffith, P. Rowell, A. Young, M. Urteaga, M. Field, "A 220GHz InP HBT Solid-State Power Amplifier MMIC with 90mW P_{out} at 8.2dB Compressed Gain", *to be presented at the IEEE Compound Semiconductor IC Symposium*, La Jolla, CA, Oct. 14-17, 2012.

Analysis of InP/GaAsSb DHBT failure mechanisms under accelerated aging tests

G.A. Koné[a], C. Maneux[a], N. Labat[a], T. Zimmer[a], B. Grandchamp[b], P. Frijlink[b] H. Maher[b].

[a] IMS, Université Bordeaux 1, 351 cours de la libération, 33405 Talence, FRANCE
[b] OMMIC, 2 Chemin du Moulin B.P. 11, 94453 Limeil-Brévannes Cedex, FRANCE

Abstract — **We report on the reliability of InP/GaAsSb/InP DHBTs dedicated to very high-speed ICs applications. The devices under tests were fabricated by OMMIC [1]. Accelerated aging tests under thermal stress were previously performed on the same technology and the results are detailed in [2]. In this paper, we present the accelerated aging tests under bias stress performed on 15 DHBTs up to 2000 hours. The collector current density was fixed at 400 kA/cm² with various collector-emitter voltages V_{CE} (from 1.5 V to 2 V) and various base plate temperature T_a (from 30°C to 120°C). The associated junction temperature range is 80-180°C. From the Gummel characteristic, we observe that the major degradation mechanism is the base current decrease for $V_{BE} < 0.8$ V. The failure mechanism leading to the base current I_B decrease was analysed by physical simulation using TCAD. The degradation mechanism leading to the gradual decrease of the base current is linked to the emitter sidewall trap density decrease. To correctly simulate the base current during aging tests, a unique donor trap level at $E_T - E_V = 1.15$ eV were modified. The traps density evolution suggests a surface state improvement. Associated activation energy E_a of 0.63 eV has been extracted.**

Keywords: Semiconductors, DHBTS, GaAsSb, InP, type II heterojunction, accelerated aging tests, life tests, reliability.

I. INTRODUCTION

In the race for towards the Terahertz domain, InP/GaAsSb/InP double heterojunction bipolar transistor (DHBT) is one of the most suitable devices. In the literature, high measured f_T and f_{MAX} published are between 365 and 603 GHz with the associated breakdown voltage BV_{CE0} range 3.5-6 V [3][4][5]. In [6], the authors show measured f_T and f_{MAX} dependences on J_C for a 2 x 0.13 µm² of respectively 521 GHz and 1150 GHz. InP/GaAsSb/InP (type II heterojunction) have been extensively proposed as an alternative of InP/InGaAs/InP DHBTs (type I heterojunction) [7]. The main advantage of the type II heterojunction is the absence of collector current blocking effect [8] at the base-collector junction and the enhancement of electron conduction from base into collector [9]. Therefore, InP/GaAsSb/InP DHBT does not require base-collector junction grading. Besides, this base-collector junction presents a higher breakdown voltage than the InGaAs/InP junction one. Moreover, in carbon doped InP/GaAsSb/InP DHBT grown by MOCVD the suppression of hydrogen passivation has been assessed [10]. Besides all these mentioned advantages, if InP/GaAsSb technology can enjoy a high reliability and

robustness, it will appear as the most suitable device for mixed-signal ICs.

II. DEVICES UNDER TESTS AND TEST CONDITIONS

A triple-mesa self-aligned process was used to realize the DHBTs. The epitaxial material was grown by MOCVD. Table 1 shows the epitaxial structure of the devices under tests which are passivated by SiN and planarized by Benzocyclobutene (BCB). Besides, The DHBT has Pt/Ti/Pt/Au base ohmic contact metallization and Ti/Pt/Au for the emitter and collector contacts. The transistors under tests have a hexagonal geometry and three emitter areas: 5 x 0.65 µm², 7 x 0.65 µm² and 10 x 0.65 µm² respectively noted T5, T7 and T10. The frequencies f_T and f_{MAX} are respectively 300 and 360 GHz with BV_{CE0} close to 4V. The process fabrication is detailed in [1].

TABLE I
Epitaxial structure.

Name	Material	Doping Level (cm⁻³)	Thickness (nm)
Cont Em	InGaAs	Si : 4×10^{19}	110
Emitter 2	InP	Si : 5×10^{18}	100
Emitter 1	InP	Si : 1×10^{17}	35
Base	GaAsSb	C : 5×10^{19}	25
Collector 2	InP	Si : 1×10^{17}	130
Collector 1	InP	Si : 2×10^{18}	10
Cont Co 2	InGaAs	Si : 4×10^{19}	50
Cont Co 1	InP	Si : 5×10^{18}	250
Etch-Stop	InGaAs	nid	20

For all stress conditions the collector current density is fixed at 400 kA/cm². The collector emitter voltage V_{CE} is fixed at 1.5V, 2V, 2V and 1.75V respectively for bias points named P1, P2, P3 and P4. The base plate temperature T_a is equal to 30°C for P1 and P2, 70°C for P3 and 120°C for P4. The associated junction temperature range is 80-180°C. Fig. 1 shows bias points with respect to the output characteristics and the corresponding isothermal curves for T5 devices. In this graph, are shown the isothermal curves for bias points with base plate temperature T_a of 30°C. For each aging bias point, the junction temperature T_J is determined by the relation between T_J and the thermal resistance R_{TH} [11]. The thermal resistances are 3250, 2300 and 1700 K/W respectively for T5, T7 and T10 DHBTs. From Fig. 1, one observed that the stress conditions are located in the device safe operating area.

978-1-4673-1725-2/12 $31.00 © 2012 IEEE

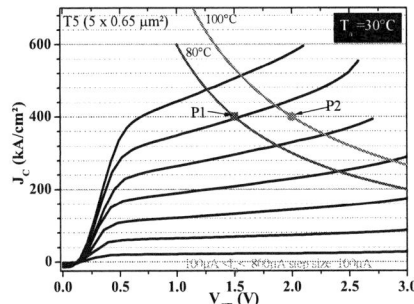

Fig. 1. Output characteristics of 5 x 0.65 μm² devices with bias points conditions location and associated junction temperature isothermal curves.

III. RESULTS

In previous works [2], the accelerated aging tests under thermal stress have been performed at 180°C, 210°C and 240°C oven temperature. The main degradation observed was the base and collector currents (respectively I_B and I_C) decrease during the tests for $V_{BE} > 0.8V$ (Fig. 2). This failure mode was slight for DHBTs submitted at 180°C and more important for the ones at 210°C and 240°C. The extraction of compact model parameters has lead to associate this failure mode to the emitter access resistance R_E. We assumed that the R_E increase is due to the metal diffusion in the emitter contact layer.

This time we present the results of aging tests performed under bias. DHBTs DC characterizations are performed during aging tests at 0, 1, 2, 4, 8, 24, 72, 144, 250, 750, 1000, 1500 and 2000 hours. 15 HBTs have been tested. Fig. 3 shows the typical Gummel characteristic evolutions for DHBT under bias tests. Before the aging tests, the base current ideality factor η_B is around 1.5. This high η_B value can be due to the high density of surface states located in the base surface and/or at the emitter periphery as stated in [12] [13]. The high η_B can be also associated with the tunnelling recombination current, which takes place in the emitter-base space charge region. This recombination is due to the electron pile up at the InP/GaAsSb emitter-base junction [14].

In bias tests, as the junction temperature did not exceed 180°C, the observed R_E evolution is neglected. Fig. 3 (a) shows T5 forward Gummel characteristic behaviour during aging tests under P1 ($V_{CE} = 1.5$ V, $J_C = 400$ kA/cm², Ta = 30°C). Due to low junction temperature (80°C) associated to P1, no changes of base and collector currents are observed. On the contrary of DHBTs submitted to P1 and P2 bias tests, the ones submitted to P3 and P4 bias tests show a base current variation. Fig. 3 (b) shows T7 forward Gummel characteristic behaviour during aging tests under P3 ($V_{CE} = 2V$, $J_C = 400$ kA/cm², $T_a = 70°C$, $T_J = 137°C$). Its T_J being greater than the one of P1 and P2, one observed the base current decrease for $V_{BE} < 0.8$ V. The currents evolution observed for T7 submitted to P3 bias tests are similar to the

one at P4. For $V_{BE} > 0.8$ V a slight decreases of base current (I_B) and the collector current (I_C) leads to the current gain decrease.

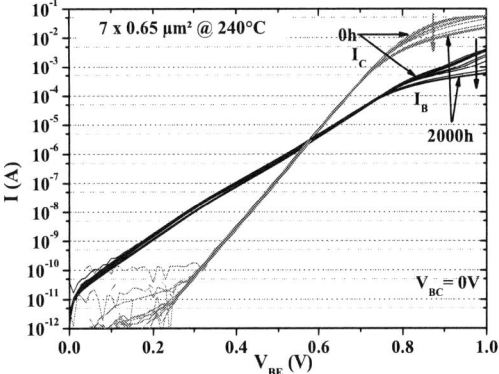

Fig. 2. Gummel characteristic for 7 x 0.65 μm² DHBT during aging tests at 240°C

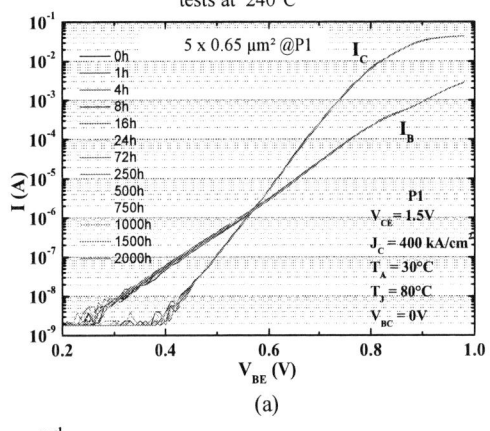

(a)

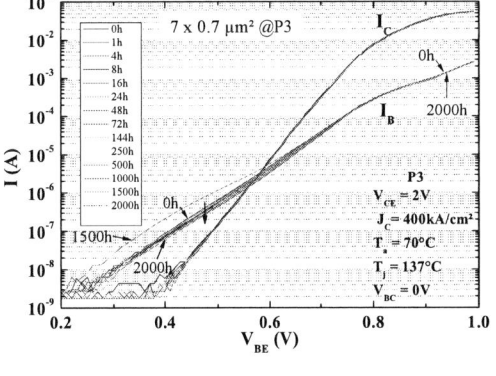

(b)

Fig. 3. Gummel characteristic evolutions for DHBTs under bias tests.
(a) 5 x 0.65μm² DHBT aged at P1 bias point.
(b) 7 x 0.65μm² DHBT aged at P3 bias point.

Fig. 4 shows normalized current gain evolution of four T7 DHBTs aged at P1, P2 and P3 bias tests. At 2000 hours, current gain decreases by 4% for DHBTs under P1 and P2 against 15% for HBTs submitted to P3 bias tests. The observed failure modes are analysed using physical simulation.

IV. ANALYSIS

A. TCAD Calibration

The electrical signature of failure mechanisms observed from bias tests were analysed by physical simulation using TCAD. The gummel characteristic of the DHBTs under tests was simulated before accelerated aging tests. The device is simulated using 2D HBT structure considering the metal layer, the substrate and the passivation. Physical and hydrodynamic transport model equations using the Stratton formulation have been used to perform the simulation [15][16]. To perfectly simulate the base and the collector current, we introduce the surface traps with different energy level located on the emitter sidewalls and the extrinsic base surface (Fig. 5). On the contrary of type I heterojunction, with type II no need to introduce traps at the emitter-base interface [17]. The table II shows the location, the energy level in the bandgap and the type of traps (acceptor or donor) used at 0 hour (before aging tests). At the extrinsic base surface, we introduce traps with energy level E_T-E_V = 0.2 eV. In addition, it was necessary to introduce traps at the emitter sidewall with energy level at E_T-E_V = 1.15 eV and 1.20 eV. The traps location and type were also used in [18] with closed energy level.

TABLE II
SUMMARIZATION OF TRAPS USED IN 2D PHYSICAL SIMULATION.

Traps location	Type	E_T-E_V (eV)
Extrinsic base surface	Acceptor	0.2
Emitter sidewall	Donor	1,15 1,29

B. TCAD Results

TCAD simulations have been performed for each DHBT under tests and for each characterization step during the aging. Fig. 6 shows simulated and measured T5 DHBTs Gummel characteristic before aging tests and after 1500 hours submitted to P4 bias tests. Perfect agreement between measurement and simulation is observed. To simulate the degradation observed during the tests, only the traps density located at E_T-E_V = 1.15 eV were modified to correctly simulate the base current. Consequently, the base current decrease is associated with the emitter sidewall traps density decrease. Fig. 7 shows the P3 and P4 the emitter sidewall donor traps density evolution with stress time. For each bias

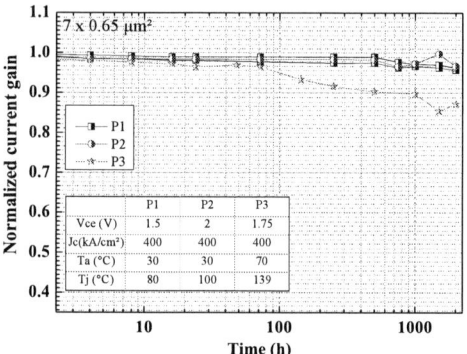

Fig. 4. T7 HBTs normalized current gain evolution at P1, P2 and P3 stress conditions.

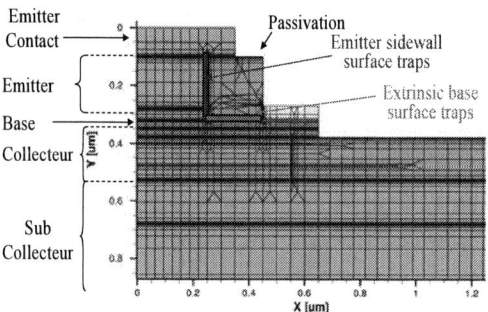

Fig. 5. Simulated 2D structure with surface traps locations.

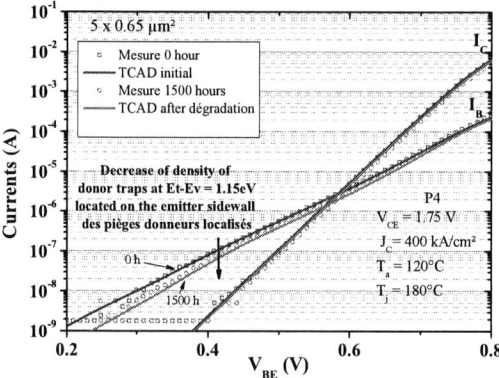

Fig. 6. Comparison between measured and simulated curves performed at 0 and 1500 hours from HBT submitted to P4 bias tests.

point, the emitter sidewall donor traps density is plotted for two DHBTs The traps density decrease, then tends to a saturation value. On the contrary of DHBTs submitted to P3 bias tests, the saturation value occurs earlier for DHBTs submitted to P4 bias tests. Moreover, DHBTs submitted to P4 bias tests have higher amplitude of donor traps density decrease. The traps density evolution suggests a surface state improvement. From the extraction of the mean time constant

978-1-4673-1725-2/12 $31.00 © 2012 IEEE   210

for each temperature as a function of the inverse of temperature a activation energy E_a of 0.63 eV has been extracted. Concerning storage aging tests, the degradation mechanism leading to the gradual decrease of the base and collector currents for $V_{BE} > 0.8$ V is due to the Au and/or Ti diffusion in the semiconductor [19] with a associated E_a of 1.8 eV [2].

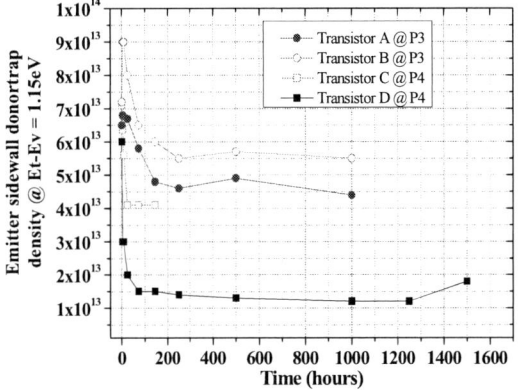

Fig. 7. Emitter sidewall trap density evolution of HBTs submitted to P3 and P4 during aging tests

V. CONCLUSION

Accelerated aging tests under thermal stress and bias stress were carried out on InP/GaAsSb/InP DHBTs from OMMIC. In storage accelerated aging tests, we pointed out the base and collector currents decrease for $V_{BE} > 0.8$ V. This failure mode is associated with the emitter access resistance evolution during aging tests. This degradation is due to the metal (Au and/or Ti) diffusion in the emitter layer ohmic contact. Moreover it is more important for higher aging temperature. Accelerated aging tests under bias stress pointed out the emitter sidewall donor traps density decrease for $V_{BE} < 0.8$ V. This degradation mechanism suggests a surface state improvement. Disregarding the high η_B factor value of the InP/GaAsSb/InP DHBTs, we showed that this technology is very robust. After 2000 hours, the observed failure modes are not critical for the devices.

Acknowledgements

The authors would like to thank ANR for funding ROBUST program research (http://extranet.imsbordeaux. fr/ROBUST/) and all partners of this project for fruitful discussions.

REFERENCES

[1]   H. Maher et al. « A 300 GHz InP/GaAsSb/InP HBT for high data rate applications », in *Compound Semiconductor Week (CSW/IPRM), 2011 and 23rd International Conference on Indium Phosphide and Related Materials*, 2011.

[2]   G. A. Koné et al. « Preliminary results of storage accelerated aging test on InP/GaAsSb DHBT », in *Compound Semiconductor Week (CSW/IPRM), and 23rd International Conference on Indium Phosphide and Related Materials*, Berlin, 2011.

[3]   R. Lovblom et al. « InP/GaAsSb DHBTs With 500-GHz Maximum Oscillation Frequency », *Electron Device Letters, IEEE*, mai 2011.

[4]   H. G. Liu et al. « 600 GHz InP/GaAsSb/InP DHBTs Grown by MOCVD with a Ga(As,Sb) Graded-Base and f_T x $BV_{CEO} > 2.5$ THz-V at Room Temperature », in *Electron Devices Meeting, IEDM 2007*.

[5]   H. G. Liu et al. « 15-nm base type-II InP/GaAsSb/InP DHBTs with FT=384 GHz and a 6-V BVCEO », *IEEE Transactions on Electron Devices*, mars 2006.

[6]   M. Urteaga1 et al. « 130nm InP DHBTs with *ft* >0.52THz and *fmax* >1.1THz », IEEE 2011.

[7]   C. R. Bolognesi et al. « InP/GaAsSb/InP double HBTs: a new alternative for InP-based DHBTs », *IEEE Transactions on Electron Devices*, nov. 2001.

[8]   C. R. Bolognesi et al. « Non-blocking collector InP/GaAs$_{0.51}$Sb$_{0.49}$/InP double heterojunction bipolar transistors with a staggered lineup base-collector junction », *IEEE Electron Device Lett.*, avr. 1999.

[9]   Y. Tian et al. « Temperature dependence of DC characteristics of NpN InP/GaAsSb/InP double heterojunction bipolar transistors: an analytic study », *Microelectronics journal*, 2006.

[10]   Y. Oda et al. « C-doped GaAsSb base HBT without hydrogen passivation grown by MOVPE », *Journal of Crystal Growt*, dec. 2004.

[11]   B. Grandchamp et al. « Study of Failure Mechanisms in InP/GaAsSb/InP DHBT Under Bias and Thermal Stress », in *IEEE 19th International Conference on Indium Phosphide & Related Materials*, 2007.

[12]   N. G. Tao et al. «Impact of surface state modeling on the characteristics of InP/GaAsSb/InP DHBTs », *Solid-State Electronics*, 2007.

[13]   Z. Jin et al. « Surface recombination mechanism in graded-base InGaAs-InP HBTs », *IEEE Transactions on Electron Devices*, 2004.

[14]   S. W. Cho et al. « High performance InP/InAlAs/GaAsSb/InP double heterojunction bipolar transistors », *Solid-State Electronics*, juin 2006.

[15]   B. Grandchamp et al. « Trends in Submicrometer InP-Based HBT Architecture Targeting Thermal Management », *IEEE Transactions on Electron Devices*, août 2011.

[16]   G. A. Koné et al. « Preliminary results of storage accelerated aging test on InP/InGaAs DHBT », *Microelectronics Reliability*, sept. 2010.

[17]   G. A. Koné et al. « Reliability of submicron InGaAs/InP DHBT under thermal and electrical stresses », *Microelectronics Reliability*, 2011.

[18]   J. M. Ruiz-Palmero et al., « Hydrodynamic 2D model for simulation and scaling of InP/InGaAs(P) DHBTs and circuits with limited complexity », *Solid State Electron*, 2006.

[19]   Y. K. Fukai et al. « Emitter-metal-related degradation in InP-based HBTs operating at high current density and its suppression by refractory metal », *Microelectronics Reliability*, avr. 2009.

MICROWAVE PHOTONICS

Keith J. Williams

Naval Research Laboratory, Photonics Technology Branch, Code 5650, Washington, D.C. 20375

Abstract

An overview of analog microwave photonics will be presented. The performance requirements for externally-modulated analog microwave photonic links will be reviewed with specific emphasis placed on modulator efficiency, laser noise, detected photocurrent, and link linearity.

The ability to remote microwave antennas with optical fiber offers attractive possibilities for new broadband semi-nonintrusive RF systems. These systems must however compete with electrically amplified coax systems which, for 100 meter or shorter links, can exhibit noise figures below 5 dB and dynamic ranges above 170 dB-Hz (compression) and 120 dB-Hz$^{2/3}$ (single-octave spur-free) with multi-gigahertz instantaneous bandwidths. To achieve low noise figures and maintain dynamic range, externally modulated photonic links must operate with efficient modulators, low noise lasers and high output photocurrents to minimize the electronic gain required from the preamplifier. If a lower quality optical link is chosen, the preamplifier gain requirements increase to maintain low noise figure which causes the overall dynamic range of the system to decrease. For long distance links where optical fiber losses become important, the use of one or more optical amplifiers can be used to minimize the degradation caused by the power lost to help maintain a high dynamic range and low noise figure.

While analog photonic systems are beginning to find application in some radar, electronic warfare, and other RF systems, the widespread utilization of photonics has not occurred because of the technical challenges facing component technology. Digital fiber communications applications have been instrumental in developing many of the components needed, however the remaining technical hurdles will have to born by the analog applications due to their unique and demanding performance requirements. In addition, the low volume of specialized optical components for analog applications can inhibit the cost-competitiveness of the technology.

As an example, figure 1 plots the noise figure of a quadrature-biased non-preamplified externally-modulated link as a function of detected photocurrent for various levels of laser relative intensity noise (RIN). Generally as the photocurrent increases, the noise figure decreases, however, if the laser RIN is too large, the noise figure reaches a minimum. For a 1 Volt modulator Vpi, a -180 dBc/Hz laser RIN and 100 mA of photocurrent are required to achieve a link noise figure (NF) of less than 6 dB. Note how the photocurrent requirements decrease to 25 mA for a 6 dB NF if a Vpi of 0.5V could be achieved. While modulator Vpi remains one of

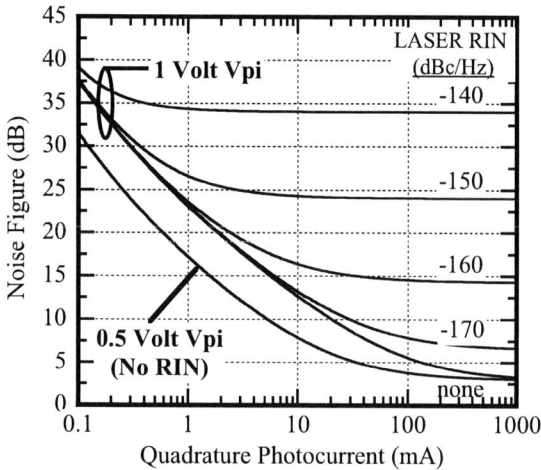

Figure 1: Calculated noise figure versus photocurrent for various levels of laser relative intensity noise (RIN) for an externally modulated link utilizing a 1 volt Vpi modualtor. Also included is the calculation for a 0.5 V Vpi modulator excluding laser RIN.

the most difficult challenges to make amplifierless low noise analog photonic links a reality, there has been progress. The most promising results are in the industry standard LiNbO3 material system used for most COTS digital transmission systems. Results in LiNbO3 [1] have demonstrated a modulator Vpi of below 1.5V at 6 GHz with further room for improvement possible.

With modulator efficiencies improving, decreasing laser RIN will become important. For frequencies greater than 10 MHz, low RIN [2] is easily achieved in diode pumped solid state lasers due to the slow gain dynamics of the Erbium ion in glass or the Neodymium ion in YAG. Figure 2 plots the RIN from a Erbium doped glass laser at 1.55μm and a Nd:YAG laser at 1.32μm. As can be seen, RIN levels near –170 dBc/Hz and below are achievable while at the same time providing for moderate oscillating powers of 50 to 1000 mW. RIN levels of DFB diode lasers have also made recent progress in achieving lower RIN. At frequencies below 10 MHz, DFBs often outperform solid state lasers, however, achieving levels below 165 dBc/Hz at microwave frequencies remains a challenge.

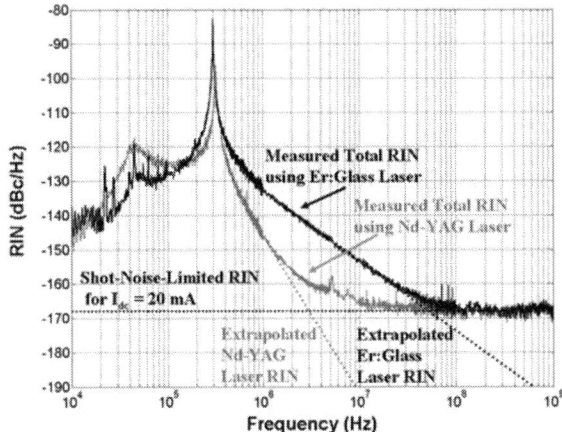

Fig. 2. Measured relative intensity noise spectra for I_{dc} = 20 mA using the Er:glass laser (solid black) and the Nd-YAG laser (solid gray). Also shown are the shot-noise-limited RIN (dashed black), the extrapolated Er:glass laser RIN (dotted black), and the extrapolated Nd-YAG laser RIN (dotted gray).

As optical powers increase to increase link dynamic range, photocurrents must also increase. Increasing photocurrents to above 100 mA has been accomplished for frequencies above 10 GHz [3]. In addition to allowing for an increase in link dynamic range and lower link noise figures, high photocurrents also allow for high RF output power directly from the photodetector where powers above +25 dBm are possible [4,5].

As photocurrents increase, maintaining high SFDR places demands on the photodetector linearity. For 100 mA Mach-Zehnder modulator links (IMDD), photodetector linearity has already been shown to limit the achievable linearity [6] as compared to the distortion intrinsic to the sinusoidal tranfer function of the modulator. If the transfer function if further linearized or if techniques to linearize the modulation [7] transfer function are utilized, additional improvements to photodetector linearity are needed [8] to improve overall link dynamic range.

In longhaul links or high loss links where optical amplifiers are utilized, careful design rules must be used to minimize the degradation due to amplifier noise. The concept of noise penalty [2] was introduced as a simple metric to directly access the penalty associated with adding an optical amplifier to an otherwise shot-noise-limited link design. The noise penalty can simply be stated as the penalty, in dB, for using an optical amplifier as compared to a shot-noise-limited link having the same output photocurrent. The use of the noise penalty allows for a direct modification to the link gain and noise figure equations for quick application of optical amplifiers into the basic link design. Figure 3 plots [from ref. 2] the noise penalty (along with the gain, noise figure and

resulting link SFDR) of a Erbium-doped fiber amplifer designed for low noise penalty. As can be seen, even though the amplifier has a high optical noise figure of 8 to 10 dB in hard compression, the resulting noise penalty can be quite low (few dB). This is due to the high-injected, low-noise laser power into the amplifier. In links where more than a single optical amplifier is needed, cascading rules for amplifier noise must be used [9].

This invited talk will discuss these and other component requirements for analog photonic links.

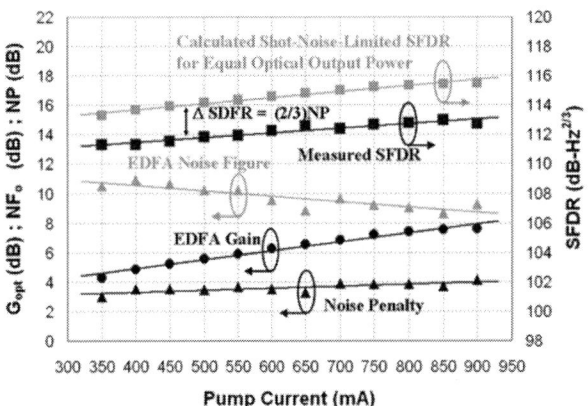

Fig. 3. Measured *SFDR* for the amplified Er:glass link as a function of pump current for the EDFA (black squares). For comparison, the calculated shot-noise-limited *SFDR* for the same optical power is shown (gray squares). In addition, measured EDFA noise figure (gray triangles), EDFA gain (black circles), and noise penalty (black triangles) are also shown.

1. S. Thaniyavarn, G. Abbas and W. Charczenko, "Very-Low-Loss Electro-Optic LiNbO$_3$ Components for RF/Analog Transmission," IEEE Avionics and Fiber Optics Conference, 2005, Paper ThC3.
2. V. J. Urick, M. S. Rogge, F. Bucholtz, and K. J. Williams, "The performance of analog photonics links employing highly-compressed erbium-doped fiber amplifiers," *IEEE Trans. Microwave Theory Tech.*, vol. 54, no. 7, pp. 3141, 3145, July 2006.
3. X. Li, N. Li, S. Demiguel, J. Campbell, D. Tulchinsky, and K. Williams, "A Comparison of Front- and Backside-Illuminated High-Saturation Power Partially Depleted Absorber Photodetectors," *IEEE J. of Quantum Ele.*, V. 40, no. 9, pp. 1321-25, 2004.
4. K. J. Williams, D. A. Tulchinsky, J. B. Boos, D. Park, P. G. Goetz, and W. S. Rabinovich, "High-Current Photodetectors as Efficient and High-Power RF Output Stages," *IEEE Microwave Photonics Conference*, MWP 2006.
5. H Pan, A Beling, J E Bowers, and J.C Campbell, "High-power high-linearity flip-chip bonded modified uni-traveling carrier photodiode," *Optics Express*, V. 19, No. 19, 18 Nov 2011.

6. K. J. Williams, L. T. Nichols, and R. D. Esman, "Photodetector nonlinearity limitations on a high-dynamic range 3 GHz fiber optic link," *IEEE J. of Lightwave Tech.,* V. 16, No. 2, pp. 192-199, 1998.

7. J. D. McKinney, K. Colladay and K. J. Williams, "Linearization of Phase-Modulated Analog Optical Links Employing Interferometric Demodulation," *IEEE J. of Lightwave Tech.,* V. 27, No. 9, pp. 1212-1220, May 2009.

8. A. S. Hastings, D. A. Tulchinsky, and K. J. Williams, "Photodetector Nonlinearities Due to Voltage-Dependent Responsivity," *IEEE Photonics Tech. Lett.,* V. 21, No. 21, pp. 1642-1644, Nov 2009.

9. P.S. Devgan, V.J. Urick, J.D. McKinney, and K.J. Williams, "Cascaded Noise Penalty for Amplified Long-Haul Analog Fiber-Optic Links," *IEEE Trans. Microw. Theory Tech.,* vol. 55, no. 9, p. 1973-77, Sept 2007.

Lateral Scalability of Inverted p-down InAlAs/InGaAs Avalanche Photodiode

Masahiro Nada, Haruki Yokoyama, Yoshifumi Muramoto, Tadao Ishibashi and Satoshi Kodama

NTT Photonics Laboratories, NTT Corporation
3-1, Morinosato Wakamiya, Atsugi City, Kanagawa Pref., 243-0918 Japan
Author e-mail address: nada.masahiro@lab.ntt.co.jp

Abstract — **This work investigated the field confinement behavior and junction size scaling for an inverted p-down avalanche photodiode, which requires no selective diffusion or ion−implantation and regrowth techniques. With this structure, the small spread of electric field of only 1 μm is obtained. The maximum 3-dB bandwidth of 23 GHz at the multiplication factor of 4.5 is obtained with the diameter of 20 μm, with responsivity as high as 0.91 A/W. These results indicate that the structure is favorable for high-speed and high-sensitivity operation.**

Index Terms — **Avalanche breakdown, High-speed electronics, Optical receivers, Optical fiber communication, Optoelectronic devices, Photodiodes, III-V semiconductor materials.**

I. INTRODUCTION

Avalanche photodiodes (APDs) are key devices for improving receiver sensitivity in optical fiber communications systems typically using a channel rate of 10-Gbit/s, and are still important for systems with bit rates beyond 10 Gbit/s, such as 25 Gbit/s.

To increase the sensitivity of APD receivers, a basic requirement is high unity-gain responsivity of the APD. Otherwise, an APD receiver loses sensitivity even if the multiplication gain is high. However, due to the trade-off between responsivity and bandwidth, the reduced absorption layer thickness generally leads to lower responsivity. An evanescent-coupled waveguide, in which a thinner absorption layer is allowed, is one approach to overcome the trade-off [1, 2].

From the application point of view, however, an APD structure with vertical illumination has important advantages over the evanescent-coupled design, such as easier fabrication, simpler optical coupling, and better device stability with a guard-ring structure. To realize vertically illuminated high-speed APDs with high responsivity, we previously proposed an inverted p-down structure [3, 4]. This structure is fabricated without Zn-diffusion or ion-implantation (which is commonly used to fabricate conventional vertically illuminated APDs [5-7]) and thus suitable for the scaling of junction diameters needed for high-speed operations.

In this study, we investigated the scalability of the inverted p-down APD by analyzing the I-V and C-V characteristics as well as the optical frequency response. We found that the equivalent lateral spread of the active APD area surrounding the top n-contact mesa periphery was as small as 1 μm. Furthermore, we found sufficient f_{3dB} and multiplied-responsivity for one with sufficiently large diameter for easy optical coupling to a fiber. This performance potentially meets the requirements for 25-Gbit/s operation. This performance potentially meets the requirements for 25-Gbit/s operation.

II. DEVICE STRUCTURE

Figure 1 shows a schematic cross section of the inverted p-down APD. The p-contact, p- and undoped InGaAs absorption, p-field control, InAlAs avalanche, n-field control, undoped edge-field buffer, and n-contact layers are grown on semi-insulating InP substrate by MOCVD. The thicknesses of the avalanche and total absorption layers are 100 nm and 1 μm, respectively. The absorption layer has a hybrid structure consisting of the p- and undoped absorption layers, which is called the MIC design [8]. The n-contact layer, which defines the active area, is formed in a mesa with a diameter ranging from 5 to 30 μm, and a second mesa contains layers down to p-InGaAs absorption layer. The width of the terrace on the second mesa surrounding the n-contact mesa is typically constant. Although the edge field inherently arises beneath the n-contact mesa periphery due to the shape of the n-contact mesa, the edge breakdown can be avoided by optimizing the thickness of edge-field buffer layer. The fabricated APD has a

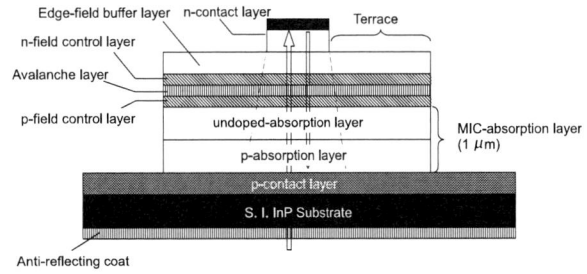

Fig. 1. Schematic cross-sectional view of the proposed inverted p-down APD.

back-illuminated structure, so the backside of the InP substrate is coated with an anti-reflection materials. A reflecting mirror is formed on the n-type contact mesa.

III. APD CHARACTERISTICS

In order to confirm that the edge-breakdown is prevented by the inverted p-down structure and that the multiplication factor (M) rises smoothly against the applied voltage, we carried out I-V measurements under illumination and dark conditions. Figure 2 shows the I-V characteristics and voltage dependence of M measured for a fabricated APD. The diameter of the n-type contact mesa (nominal active area) is 20 um, which provides a sufficient area for optical coupling. The onset voltage (V_{on}) and the breakdown voltage (V_b) are 14.5 and 26 V, respectively. M = 10 is achieved at 22 V. From the fitting of the increase of photocurrent assuming the electric field dependence of local ionization coefficient of InAlAs [9], the responsivity at unit gain was estimated to be 0.91 A/W at the wavelength of 1550 nm. As expected, the M rises smoothly up to 40, indicating that the edge-breakdown does not occur.

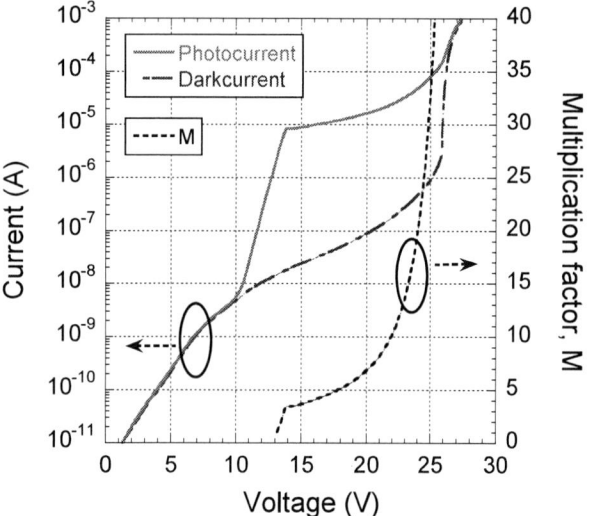

Fig. 2. I-V characteristics and voltage dependence of multiplication factor (M).

IV. LATERAL SCALING CHARACTERISTICS

Due to lateral electric field spread, the effective active area of the present APD structure can be larger than that of the n-contact mesa. Figure 3 shows the n-contact mesa area dependence of dark current at a voltage of 2.0 V. The inset shows the I-V characteristics of each device. In the I-V characteristics, the dark currents increase rapidly at voltages up to 10.0 V and tend to increase gradually at voltages above

that. Judging from our previous report [3], it seems that 10.0 V is the voltage where the main component of the dark current switches from surface-leakage to tunneling current due to the complete depletion of n-field control layers.

The dark currents at 2.0 V here are seen to depend mostly on the square root of the n-contact mesa area. Thus, the main component of 2.0-V dark current is attributed to the surface-leakage current. Furthermore, the surface leakage current is low enough (surface leakage current density of 250 fA/μm), suggesting that the electric field is sufficiently confined into the device.

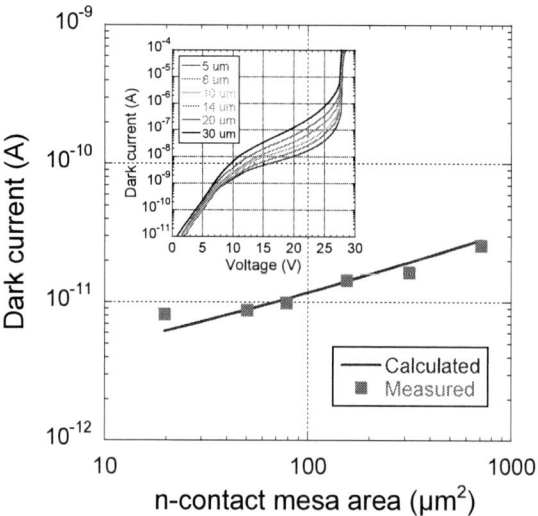

Fig. 3. Small-mesa area dependence of dark current at a voltage of 2.0 V.
Inset: Dark I-V characteristics (n-contact mesa diameter: 5- 30 μm).

Figure 4 shows the n-contact mesa area dependence of dark current at 22 V (M = 10), where the dark current is dominated by the tunneling current in the present structure. The dark currents increase almost proportionally to the n-contact mesa area. The areas of the second mesa do not proportionally increase because the terrace widths are constant. This indicates that the active area of the APD is tightly defined not by the total mesa area, i. e., the second mesa area, but by the n-contact mesa. In order to estimate how the electric field spreads to the terrace region from the n-contact mesa area, fitting of the dark current was done assuming the spread of the electric field of 0, 1, and 2 μm. These calculated lines are shown in the Fig. 4, too. The spread of the electric field of 1 μm well explains the measured data.

Changes in device capacitance with n-contact mesa area are also investigated by measuring C-V characteristics, which are shown in figure 5. Like the dark currents in Fig. 4, the capacitances also proportionally increase with n-contact mesa area. The fitting line with an area spread of 1 μm indicates good agreement with measured data. The I-V and C-V results confirm that the n-contact mesa can tightly confine the electric field and define the active area in the p-down inverted structure. Since the lateral spread of the active area from the n-contact mesa into the second mesa is only 1 μm, this APD structure is favorable for junction scaling.

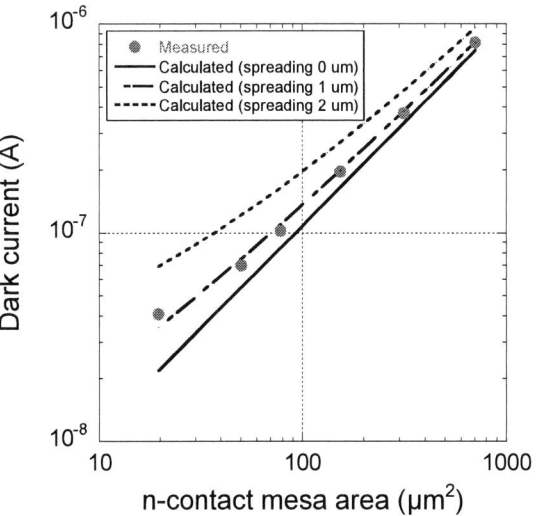

Fig. 4. Dark current variation versus n-contact mesa area at 22 V.

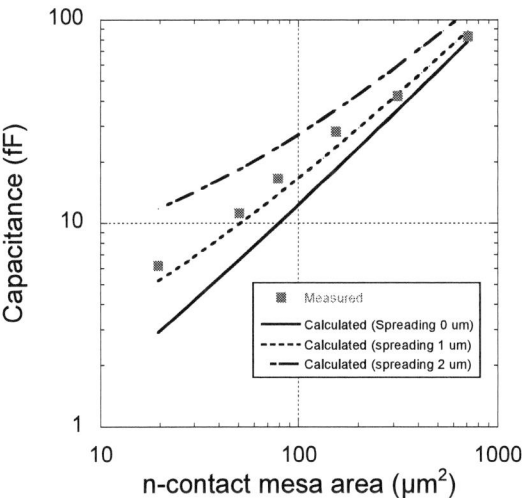

Fig. 5. Device capacitance variation versus n-contact mesa area at 22 V.

Figure 6 shows the change in f_{3dB} with M for a fabricated APD with an n-contact mesa diameter of 20 μm, which is

sufficiently large for easy optical coupling. The inset shows the O/E response against the frequency at M = 10, where f_{3dB} is 18.5 GHz. At this condition, the multiplied responsivity reaches 9.1 A/W. These values are sufficient for high-speed, high-sensitivity operations beyond 10 Gbit/s, such as 25 Gbit/s. The maximum f_{3dB} of as large as 23 GHz (at M = 4.5), is observed in spite of the thick absorption layer of 1 μm, which is typically used in 10-Gbit/s APDs. At M > 15, the measured f_{3dB} inversely decreases against M. From this trend line, the gain-bandwidth product (GBP) is estimated to be 235 GHz, which is comparable to previously reported ones [4.6] for the InAlAs-avalanche layer thickness of 100 nm.

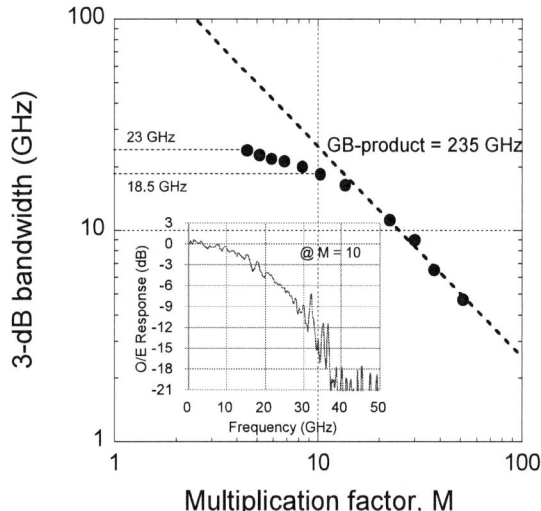

Fig. 6. 3-dB bandwidth against multiplication factor (M) of the fabricated APD.
Inset: The OE response.

The most important concern in scaling the inverted p-down structure is how the f_{3dB} changes with the n-contact mesa area. Figure 7 shows the measured and calculated n-contact mesa area dependence of f_{3dB} at M = 6 for mesa diameters ranging from 5 to 30 μm. Generally speaking, the f_{3dB} of APD is dominated by carrier transit time, CR, and GBP limitations. We used the measured capacitances (Fig. 5) and estimated GBP of 235 GHz (Fig. 6) in the calculated ones. In the measured ones (•), though the change is gradual, the f_{3dB} increases when the n-type contact mesa area is reduced. By ignoring the carrier transit time, the f_{3dB} is expected to increase up to 40 GHz with decreasing area of the n-contact mesa (■). On the other hand, assuming the intrinsic response (carrier transit time limitation), which includes the carrier transit times of absorption, edge-field buffer, field control, and avalanche layers (■), the calculated ones well explain the change in the measured f_{3dB} for a wide range of n-contact mesa area. Then the f_{3dB} saturates at smaller mesa areas due to the carrier transit time. These results indicate that our APD can achieve further high-speed operation by relaxing the carrier transit time limitation; that is, they indicate sufficient scalability for APD.

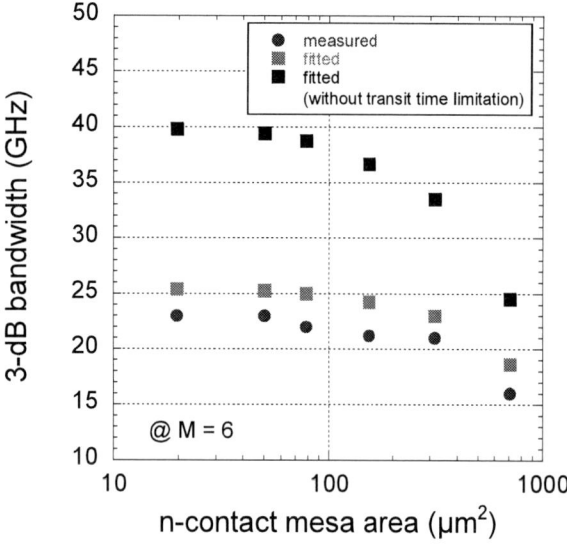

Fig. 7. Measured and calculated n-contact mesa dependence of 3-dB bandwidths.

V. SUMMARY

We have investigated the lateral scalability of an inverted p-down APD. From the n-contact mesa area dependence of dark current and device capacitance, we found that the n-contact mesa can tightly define the APD active area. The spread of the electric field is estimated to be only 1 μm, which allows the scaling of junction size. A device with 20-μm n-contact mesa size yielded a sufficiently high f_{3dB} of 18.5 GHz (at M = 10) and large unity gain responsivity of 0.91 A/W, which shows a good potential for high-sensitivity 25-Gbit/s applications. The measured variation of f_{3dB} with n-contact mesa size (5-30 μm in diameter) is well explained by taking into account device capacitance and transit-time-limited f_{3dB}. All of these observations show that the inverted p-down APD structure has excellent lateral scalability.

VI. ACKNOWLEDGEMENT

The authors thank S. Ando and K. Honda for device fabrication. The authors are grateful with T. Akeyoshi and T. Enoki for their continuous encouragement.

REFERENCES

[1] N. Yasuoka, H. Kuwatsuka, A. Kuramata, T.Uchida, Y. Yoneda and S. Nakai, "High-speed and high-efficiency InP/InGaAs waveguide avalanche photodiodes for 40 Gbit/s transmission systems," *Proc. Optical Fiber Communication Conf. (OFC)*, Paper TuM2, 2002.
[2] S. Shimizu, K. Shiba, T. Nakata, K. Kasahara and K. Makita, "40 Gbit/s waveguide avalanche photodiode with p-type absorption layer and thin InAlAs multiplication layer," *Electron. Lett.*, vol. 43, pp. 476, 2007.
[3] M. Nada, Y. Muramoto, N. Shigekawa, H. Yokoyama, T. Ishibashi and S. Kodama, "Inverted p-down avalanche photodiode with low-high-low field profile," *Jpn. J. Appl. Phys.*, vol. 51, pp. 02BG03, 2012.
[4] M. Nada, Y. Muramoto, H. Yokoyama, T. Ishibashi and S. Kodama, "InAlAs APD with high multiplied responsivity-bandwidth product (MR-bandwidth product) of 168 A/W·GHz for 25 Gbit/s high-speed operations," *Electron. Lett.*, vol.46, pp397-399, 2012.
[5] E. Yagyu, E. Ishimura, M. Nakaji, T. Aoyagi, and Y. Tokuda, "Simple planar structure for high-performance AlInAs avalanche photodiodes," *IEEE Photonics Technol. Lett.* vol. 18, pp. 1264, 2006.
[6] M. Lahrichi, G. Glastre, E. Derouin, D. Carpentier, N. Lagay, J. Decobert, and M. Achouche, "240-GHz Gain-Bandwidth Product Back-Side Illuminated AlInAs Avalanche Photodiodes," *IEEE Photonics Technol.* Lett. vol. 22, pp.1373, 2010.
[7] Y. Hirota, Y. Muramoto, T. Takeshita, T. Ito, H. Ito, S. Ando, and T.Ishibashi, "Reliable non-Zn-diffused InP/InGaAs avalanche photodiode with buried n-InP layer operated by electron injection mode," *Electron. Lett.* vol. 40, pp. 1378, 2004.
[8] Y. Muramoto and T. Ishibashi, "InP/InGaAs pin photodiode structure maximising bandwidth and efficiency," *Electron. Lett.*, vol. 39, pp. 1749, 2003.
[9] Y. L. Goh, D. J. Massey, A. R. J. Marshall, J. S. Ng, C. H. Tan, W. K. Ng, G. J. Rees, M. Hopkinson, J. P. R. David and S. K. Jones, "Avalanche Multiplication in InAlAs," *IEEE Trans. Electron Devices*, vol. 54, pp. 11, 2007.

Phase Characterization of Intermodulation Distortion in High-Linearity Photodiodes

Yang Fu, Huapu Pan, Andreas Beling, Joe Campbell

Department of Electrical Engineering
University of Virginia
Charlottesville, VA, United States
yf4f@virginia.edu

Abstract — **We present a simple method to obtain both magnitude and phase information of intermodulation distortion (IMD) in photodiodes. The IMD3 of the high-linearity InGaAs/InP photodiode exhibits 180 degree phase changes around its zero crossings.**

I. Introduction

Nonlinear distortion in photodiodes is one of the major factors that limit the performance of photonic systems in terms of spurious-free dynamic range (SFDR), phase noise and efficiency [1, 2]. Two-tone and three-tone intermodulation distortion (IMD) measurements are commonly used for photodiode nonlinearity characterization. Over 40 dBm of output third-order intercept point (OIP3) have been reported in waveguide photodiodes [3, 4] and the OIP3 of surface-normal photodiodes exceeds 50 dBm [5].

Conventional two-tone and three-tone measurements only provide the magnitude information of IMDs. More insight into the nonlinear distortion phenomenon can be gained from the phase of IMDs, which also benefits the development of linearization techniques [2, 6]. To the best of our knowledge, there is no report on the phase of IMDs in photodiodes in the published literature. Several techniques exist to measure IMD phase for RF amplifiers [7, 8].

In this paper, we demonstrate a simple method based on the feed-forward concept to measure the relative phase of IMDs in photodiodes. The method is broadband and only requires minor modifications of the conventional three-tone setup [3, 5]. The results of IMD phase measurement for high-linearity modified uni-traveling carrier (MUTC) photodiode are presented and analyzed as to their implications for nonlinearity study.

II. Experiment

A. Measurement Setup

The IMD phase measurement setup is shown in Figure 1. It can be viewed as the combination of a two-tone setup and a third "IMD cancellation" branch. The light sources are three DFB lasers at the wavelengths of 1544nm, 1545nm and 1546 nm. The three signal generators and the spectrum analyzer share a common 10 MHz reference frequency. The signal

generator in the IMD cancellation branch is set at the frequency of the IMD tone being measured. The phase and magnitude of the signal in this branch are controlled by a tunable optical delay line and an attenuator. The optical signals from the three branches are combined together in the coupler before entering the EDFA for amplification. The RF output from the photodiode is directed into the spectrum analyzer through AC coupling.

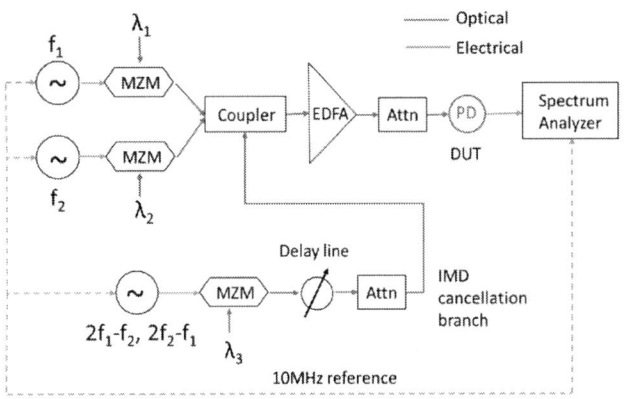

Figure 1 The IMD phase measurement setup

B. System Operation

The IMD magnitude is determined the same way as for a two-tone setup while the light signal in the third branch is blocked. The phase measurement is accomplished through cancellation of the IMD tone in the photodiode's output spectrum by adjusting the phase and magnitude of the signal in the "IMD cancellation" branch.

The maximum phase error due to imperfect cancellation can be expressed as

$$\varphi_{max} = \sin^{-1}\left(10^{-\frac{\alpha}{20}}\right) \qquad (1)$$

where α is the IMD power cancellation in dB. 35 dB of cancellation is needed to reduce the phase error below 1.0 degree. Closer matching of the IMD and the cancellation tone power will relax this requirement. The minimum cancellation for the following measurements is larger than 20 dB, which results in a maximum phase error of 5.7 degree. The power level of the signal generators and EDFA are kept constant during measurements to avoid introducing any phase errors.

III. RESULTS AND DISCUSSION

A. The Device under Test

The device under test is a 40 μm-diameter InGaAs/InP MUTC photodiode with heavily-doped absorber [5]. Its OIP3 value is over 50 dBm at low frequencies (< 3 GHz) and remains above 45 dBm at 20 GHz. The MUTC photodiode has 0.4 A/W responsivity and 3dB bandwidth of 13 GHz at -6 V bias voltage.

B. OIP3 and the Relative Phase of IMD3

For the measurements reported here the frequencies of the fundamental tones were 3.3111 GHz and 3.3207 GHz, respectively. A constant modulation depth of 55% was maintained during the test. The results are shown in Figure 2. Only the results for the upper IMD3 product (IMD3U) are included for sake of clarity. Similar peaking behavior of OIP3 has been observed in other high-linearity photodiodes [4, 9] and can be explained by a model based on nonlinear responsivity [5, 10].

The most striking feature in the IMD3 phase plot is the existence of 180 degree transitions at the photocurrent levels where OIP3 peaks, although the second transition at larger photocurrent is not as well defined as the first one. The first sharp 180 phase transitions may imply major changes in nonlinear processes such as switching from gain expansion into compression [8]. The more gradual phase transition around the second OIP3 peak could indicate competing nonlinearities with different phase characteristics instead of a single dominating mechanism. Photodiode linearity can be improved by taking advantage of these phase changes. For example, biasing multiple photodiodes at appropriate photocurrents/voltages and combining their outputs so that the IMD3 products cancel each other. Further study is needed to establish the connection between IMD phase change and underlying physical mechanisms.

IV. CONCLUSIONS

We have developed a simple method to characterize the relative phase of IMD in photodiodes. The unique 180 degree phase transitions found in MUTC photodiode holds great potential for future research into nonlinear distortion phenomenon.

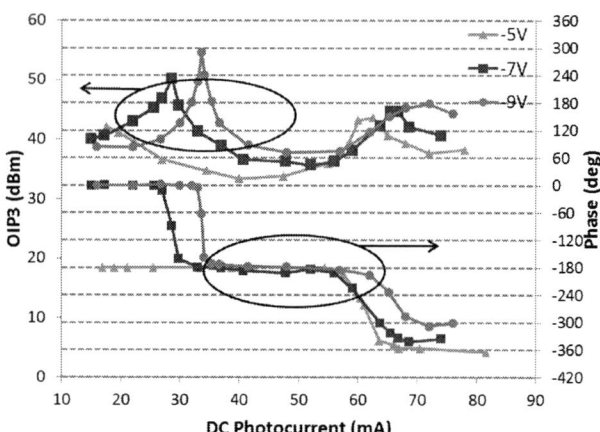

Figure 2 The measured OIP3 and relative phase of the upper IMD3 product (IMD3U) for MUTC photodiode at different dc photocurrent and bias voltages

ACKNOWLEDGMENT

This work is sponsored by the DARPA TROPHY program.

REFERENCES

[1] C. H. Cox, III, et al., "Limits on the performance of RF-over-fiber links and their impact on device design," Microwave Theory and Techniques, IEEE Transactions on, vol. 54, pp. 906-920, 2006.

[2] V. J. Urick, et al., "Long-Haul Analog Photonics," Lightwave Technology, Journal of, vol. 29, pp. 1182-1205, 2011.

[3] A. Ramaswamy, et al., "Measurement of intermodulation distortion in high-linearity photodiodes," Opt. Express, vol. 18, pp. 2317-2324, 2010.

[4] S. M. Madison, et al., "Third-order intermodulation distortion characterization of variable confinement slab-coupled optical waveguide photodiodes," in Photonics Conference (PHO), 2011 IEEE, 2011, pp. 23-24.

[5] P. Huapu, et al., "Characterization of High-Linearity Modified Uni-Traveling Carrier Photodiodes Using Three-Tone and Bias Modulation Techniques," Lightwave Technology, Journal of, vol. 28, pp. 1316-1322, 2010.

[6] J. McKinney, et al., "Linearization of Phase-Modulated Analog Optical Links Employing Interferometric Demodulation," J. Lightwave Technol., vol. 27, pp. 1212-1220, 2009.

[7] A. Walker, et al., "Simple, Broadband Relative Phase Measurement of Intermodulation Products," presented at the ARFTG Conference, 65th, Long Beach, California, 2005.

[8] K. A. Remley, et al., "Simplifying and interpreting two-tone measurements," Microwave Theory and Techniques, IEEE Transactions on, vol. 52, pp. 2576-2584, 2004.

[9] M. Chtioui, et al., "High Responsivity and High Power UTC and MUTC GaInAs-InP Photodiodes," Photonics Technology Letters, IEEE, vol. 24, pp. 318-320, 2012.

[10] F. Yang, et al., "Characterizing and Modeling Nonlinear Intermodulation Distortions in Modified Uni-Traveling Carrier Photodiodes," Quantum Electronics, IEEE Journal of, vol. 47, pp. 1312-1319, 2011.

High speed AlInGaAs/InGaAs quantum well waveguide photodiode for wavelengths around 2 microns

Hua Yang[1], Nan Ye[1], Marina Manganaro[1], Agnieszka Gocalinska[1], Kevin Thomas[1], Emanuele Pelucchi[1],

Brendan Roycroft[1], Frank H. Peters[1, 2] and Brian Corbett[1]

1 Tyndall National Institute, Lee Maltings, Prospect Row, Cork, Ireland;

2 Department of Physics, University College Cork, Cork, Ireland

Abstract — A high speed waveguide photodiode fabricated with AlInGaAs/InGaAs multiple quantum wells for 2 micron wavelength detection is reported. The fabricated photodiode shows a photoresponsivity of 0.3 A/W at around 2μm wavelength and a 3 dB bandwidth around 7 GHz.

Index Terms — Photodiode, 2μm wavelength, high speed, AlInGaAs/InGaAs, quantum well, optical communication

I. INTRODUCTION

Optical fiber communication networks require continually increasing bandwidth to meet the exponentially growing demands on the transmission capacity [1, 2]. With the impending saturation of the 1550 nm wavelength band there is a need to open up new spectral regions [3]. Photonic crystal fibers offer the potential of transmission capacity increases of an order of magnitude with multiple-input-multiple-output (MIMO) operation of the multi-mode fiber while a further order of magnitude capacity increase can be achieved through the ultra-low loss and nonlinearity offered by multi-mode photonic band gap fiber [4,5]. These benefits are predicted to be realized in a new transmission wavelength low loss window around 2.0 μm. Therefore an entire suite of optical components needs to be developed to exploit this possibility. A single mode laser at 2 μm has been successfully achieved on an InP substrate for the transmitter part [6, 7]. A photodiode at detection wavelength around 2 μm is another indispensable component for the receiver, which has to be developed. In this paper, we present the design and fabrication of a high speed waveguide photodiode with AlInGaAs/InGaAs quantum wells on InP substrate designed for 2.0 μm wavelength.

II. DESIGN AND FABRICATION

A. Material design and growth

To develop the photodiode for the 2 μm wavelength, the first step is to obtain material with a proper bandgap. $In_xGa_{1-x}As$, lattice-matched to InP with x=0.53, has been one of the most

TABLE I
EPITAXIAL LAYER STRUCTURE

Layer No.	Material	Thick (nm)	PL (nm)	Doping (cm-3)
11	InGaAs	100		P+/ 1 x 10^{19}
10	InP	1700		P /1 x 10^{18}
9	InGaAsP	10	1300	P/ 1 x 10^{18}
8	InP	80		P/ 5 x 10^{17}
7	InP	200		
6	AlInGaAs	80	1240	
5	InGaAs QW	9	>2010	
4	AlInGaAs	10	1240	
3	InGaAs QW	9	>2010	
2	AlInGaAs	150	1240	
1	InP	1200		N /8 x 10^{17}
Subs	InP			Semi-insulating

popular materials used for photodetectors in current communication systems with a high responsivity in a wide wavelength range which covers between 1310 nm and 1550 nm but cuts off at a wavelength of 1680 nm [8,9]. However, this material system is still promising to extend to 2 μm wavelengths by increasing the In composition in the $In_xGa_{1-x}As$ compound to around 80%. At the same time, this introduces the large strain in the material because of the lattice mismatch with the InP substrate. Thus, it is a challenge for the epitaxial growth to get a high quality crystal by properly controlling the strain and growth conditions. In our designed structure, the intrinsic core consisting of two 9-nm-thick $In_{0.82}Ga_{0.18}As$ compressively highly strained quantum wells separated by a 10-nm-thick AlInGaAs (PL=1240nm) barrier layers and sandwiched by 80 nm and 110 nm thick undoped AlInGaAs layers which are lattice-matched to InP on top and bottom respectively were optimized to achieve the 2 μm wavelength. In addition, another key design consideration is free carrier absorption which is significant for holes at these long wavelengths. For this reason the waveguide mode was partially offset from the p-side of the junction and the p-doping concentration was reduced close to the active layer. The designed epitaxial material structure is shown in Table 1.

In our experiments, the metal organic vapor phase epitaxy (MOVPE) method was used to grow the designed material on (100) semi-insulating InP substrate. Zn and Si were used as p and n type dopants respectively. The composition of the compounds as well as the optimized growth condition to control the strain was found and the material was successfully grown with high crystal quality with the desired bandgap wavelength. Fig. 1 presents the simulated and measured x-ray diffraction (XRD) spectrum and we can see the peaks of measured XRD spectrum fit very well with the simulated spectrum under ideal condition which indicates the lattice mismatch in the quantum wells has been managed very well and the grown material has the desired quality as designed. Fig. 2 shows the measured photo-luminescence (PL) spectrum of the grown material, and the target PL peak wavelength of 2100 nm has been successfully obtained. The good quality of the crystal and the proper bandgap of material set a good foundation for the final device's performance.

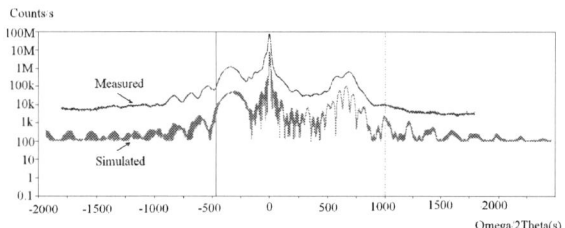

Fig.1 Simulated and measured XRD spectrum of the grown wafer

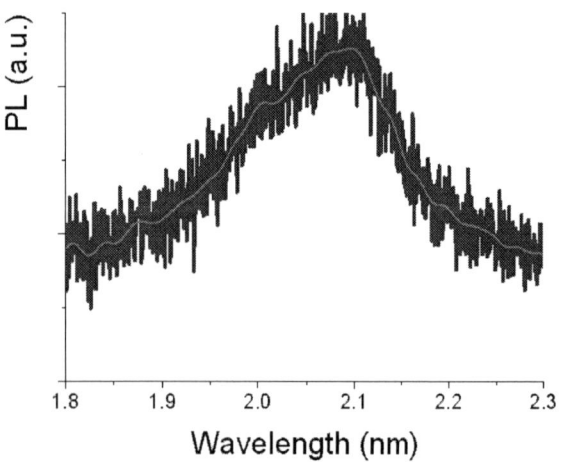

Fig.2 Measured PL spectrum of the grown wafer

B. Device design and Fabrication

In our device design, an edge coupled waveguide structure was adopted instead of the top illuminated mesa structure for achieving both high speed and high efficiency simultaneously. This is accomplished by independently optimizing the length and thickness of the absorber, which can be done because the incident direction of light and current flowing direction in waveguide PD differ from each other. It is especially significant with only two quantum wells in a very thin absorption layer. In addition, the waveguide PD also has the benefit for monolithic integration with other conventional waveguide optoelectronic devices based on InP substrates such as laser diodes, semiconductor optical amplifiers, and electro-absorption modulators. In order to achieve high speed which is determined by the RC time constant as well as the carrier transit time for the photodiode, a small intrinsic area is preferred for low capacitance. A deep etched ridge through the quantum wells layers was adopted in the devices. At the same time, coplanar ground-signal (GS) electrodes with benzocyclobutene (BCB) polymer underneath for planarization and passivation, as well as semi-insulating InP substrate were designed for high speed by minimizing the parasitic capacitance from the electrode pads. The proposed configuration of PD device is shown in Fig. 3.

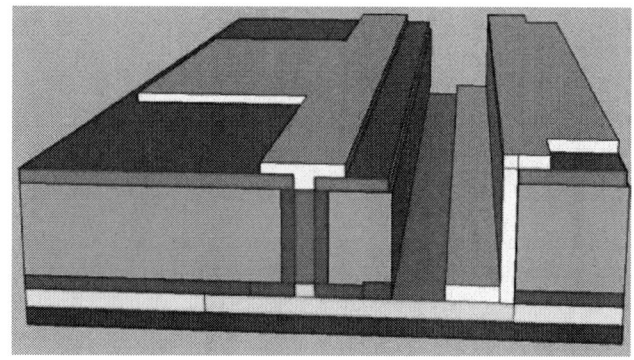

Fig.3 Structural configuration of the proposed photodiode

During device fabrication, the deep ridge waveguide was formed by room temperature dry etching with $Cl_2/CH_4/H_2$ in an Oxford 100 ICP system where a 500 nm thick SiO_2 layer was used as the etching mask. Following deposition of a 150 nm thick SiNx passivation layer BCB 4024-35 was spin-coated and cured at 250 degrees for one hour, followed with etching back of the BCB in a SF_6/O_2 ICP process until a planarizes ridge is obtained. After that, 150 nm thick SiN_x was deposited on top of BCB to smooth the etched surface and increase the adhesion with metal electrodes. Next, ohmic contact windows for p and n were opened on SiN_x and BCB by dry etching with SF_6/O_2 ICP and finally 300 nm thick TiAu was evaporated for both p and n contact metal and pads using electron beam evaporator. The top view microscope image of

978-1-4673-1725-2/12 $31.00 © 2012 IEEE

the fabricated PD array and the SEM picture of the cross section of waveguide PD are shown in Fig. 4 and Fig. 5, respectively.

Fig.4 Top view of fabricated photodiode array

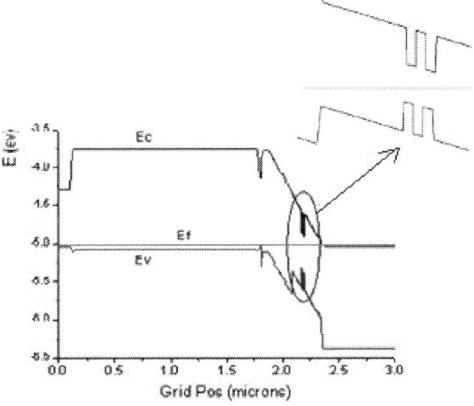

Fig.5 SEM picture of cross section of fabricated photodiode

III. MEASUREMENTS AND DISCUSSION

The fabricated devices were cleaved into different lengths for measurements. At first, the DC performance of the fabricated PD was characterized. Light with a wavelength at around 2 microns was coupled into the waveguide of the PD via lens ended fiber while the PD was reverse-biased under

different voltages applied through high speed GS probes. The generated photocurrent at a certain incident optical power was collected and then calculated into photoresponsivity. Fig. 6 shows the photoresponsivity of PD changing with the bias voltage. It can be seen for a 300 μm long device with a 3μm wide absorption region, the photoresponsivity linearly increases with the bias voltage and goes to the maximum of 0.3 A/W after 3 V reverse bias. The low responsivity of the device under less than 3 V bias voltage could result from the large band offset between the barriers and quantum wells because when the light is absorbed in the wells, the photo-generated carriers, especially the holes will be confined in the quantum wells and need be swept out with a high electric field for the larger band discontinuities. There is also a big valence band discontinuity between the AlInGaAs layer and the InP layer above the quantum wells which will need a large electric field for the holes to overcome the potential step. This can be seen from the simulated band diagram of the material shown in Fig. 7.

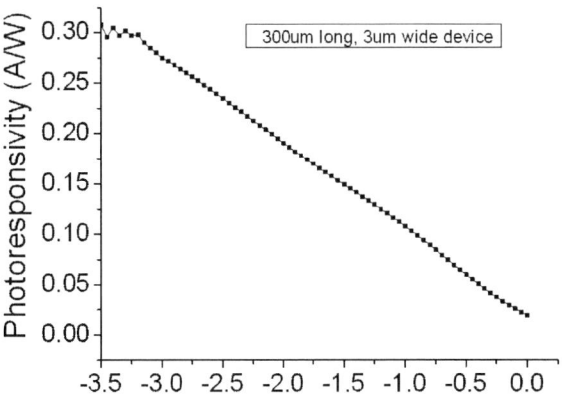

Fig.6 Measured responsivity of photodiode vs bias voltage

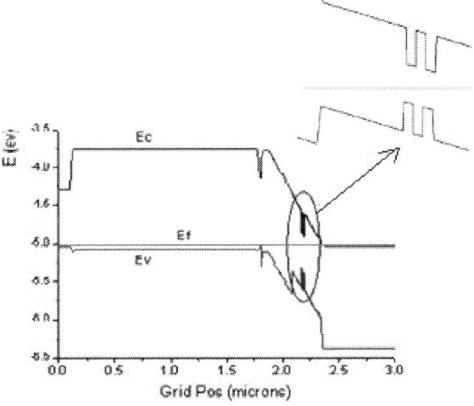

Fig.7 Simulated band diagram of the material

The small signal response of the PD was also measured with our high speed measurement set-up. A modulated light signal with -3 dBm optical power was coupled into the waveguide of PD which is biased with high speed GS probe. The measured 3 dB bandwidth of the 300μm long device with a 3μm wide absorption region under different voltage is shown in Fig.8.

From the measurement results, the small signal 3dB bandwidth of the device can go up to around 7 GHz with a reverse bias of more than 5 volts. This can be potentially used for data transfer in10 Gb/s systems.

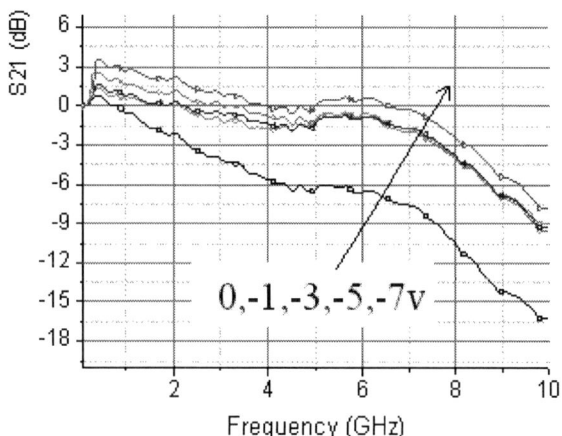

Fig.8 Measured small signal frequency response of the photodiode as a function of bias voltage.

IV. CONCLUSION

An AlInGaAs/InGaAs quantum well waveguide photodiode design to operate at 2 microns wavelengths was successfully designed and fabricated. The fabricated device shows efficient light absorption at the 2 μm wavelength and demonstrates high speed performance in the 7 GHz range, which is promising for use in future 2 μm wavelength optical communication systems.

ACKNOWLEDGEMENT

This work was supported by EU project FP7-ICT under grant 258033 (MODE-GAP) and also partly based on the work supported by Science Foundation Ireland under grant 07/SRC/I1173 (PiFAS).

REFERENCES

[1] Y.Miyamoto and S.Suzuki, "Advanced optical modulation and multiplexing technologies for high-capacity OTN based on 100Gb/s channel and beyond," IEEE Commun. Mag., vol.48, no.3, pp. S65-S72, March 2010

[2] A.Sano, T.Kobayashi, E. Yoshida, Y.Miyamoto,"Ultra-high capacity optical transmission technologies for 100Tbit/s optical transport networks," IEICE trans. Commun., vol. E94-B, no.2, February, 2011

[3] http://www.modegap.eu/

[4] P. Roberts, F. Couny, H. Sabert, B. Mangan, D. Williams, L. Farr, M. Mason, A. Tomlinson, T. Birks, J. Knight, and P. St. J. Russell, "Ultimate low loss of hollow-core photonic crystal fibres" , Optics Express., vol. 13, pp. 236-244, Jan 2005.

[5] H.S. Chen, H.P.A. van den Boom, A.M.J. Koonen, "30-Gb/s 3 x 3 Optical mode group-division-multiplexing system with optimized joint detection," IEEE Photon. Techno. Lett., vol.23, no.18, pp.1283-1285, September 2011

[6] R. Phelan, J. O'Carroll, D. Byrne, C. Herbert, J. Somers, and B. Kelly, "InGaAs/InP Multiple Quantum-Well Discrete-Mode Laser Diode Emitting at 2 μm", IEEE Photon. Tech. Lett., vol 24 pp. 649 – 651 2012

[7] C. Lauer, et al., "80 °C continuous-wave operation of 2.01-μm wavelength InGaAlAs-InP vertical-cavity surface-emitting lasers," IEEE Photon.Technol. Lett., vol. 16, no. 10, pp. 2209–2211, Oct. 2004.

[8] Y.H. Huang, C.C. Yang, T. Peng, F.Y. Cheng, M.C. Wu, Y.T.Tsai, C.L.Ho, I.M. Liu, C.C. Hong, and C.C. Lin, "10-Gb/s InGaAs p-i-n photodiodes with wide spectral range and enhanced visible spectral response," IEEE Photon.Technol. Lett., vol. 19, no. 5, pp. 339–341,March 2007

[9] S. Kagawa, K. Inoue, I. Ogawa, Y. Takada, and T. Shibata, "Wide-wavelength InGaAs/InP PIN photodiodes sensitive from 0.7 to 1.6 μm," Jpn. J. Appl. Phys., vol. 28, no. 10, pp. 1843–1846, 1989.

MOCVD BASED ZINC DIFFUSION PROCESS FOR PLANAR InP/InGaAs AVALANCHE PHOTODIODE FABRICATION

O.J. Pitts, M. Hisko, W. Benyon, A.J. SpringThorpe

Canadian Photonics Fabrication Centre
National Research Council, Ottawa, Ontario K1A 0R6, Canada

Abstract — We investigated the diffusion of zinc into InP/InGaAs avalanche photodiode structures using dimethylzinc in an MOCVD reactor. Diffusion profiles were measured by secondary ion mass spectrometry and compared with cleaved cross-sections imaged by scanning electron microscopy, in order to accurately target the diffusion depth for device fabrication. The dependence of the diffusion depth on the diffusion temperature, partial pressures of dimethylzinc and phosphine, and diffusion time is reported. Diffused devices exhibit, in some cases, a step increase in dark current at or near the punch-through voltage. We show that the dark current above the punch-through voltage is proportional to the junction area and originates in the bulk of the material. The dependence of this bulk dark current contribution on the diffusion process parameters has been studied in detail, and a reduction of three orders of magnitude was achieved.

Index Terms — Diffusion processes, photodiodes, epitaxial layers, dark current.

I. INTRODUCTION

The diffusion of Zn into InP is a critical process step in the fabrication of high quality planar InP/InGaAs avalanche photodiodes (APDs). Diffusion using the sealed ampoule technique or spin-on films suffers from a lack of reproducibility and uniformity. In particular, APD fabrication requires precise control of the diffusion depth, in order to accurately define the multiplication width. Diffusion in a metalorganic chemical vapor deposition (MOCVD) reactor using dimethylzinc (DMZn) or diethylzinc (DEZn) has been shown to result in excellent uniformity and reproducibility [1-5]. Advantages of the MOCVD diffusion technique are the ability to independently control the wafer temperature, diffusant source flow and phosphine (PH₃) overpressure.

Previously published dark current curves for diffused APDs show, in some cases, a step-like increase in the dark current at the punch-through voltage [6]. It has been suggested that step increases in dark current in photodiodes may be attributable to mid-gap electron traps [7]. In this work, we report the dependence of the diffusion depth on process parameters and achieve excellent uniformity of the multiplication width across 3" wafers. We show that the magnitude of the dark current above the punch-through voltage is strongly influenced by the diffusion conditions and study its dependence on the process parameters.

II. EXPERIMENT

The epitaxial structures used for this study were grown on n-type InP substrates by MOCVD and consist of a Si-doped buffer layer, undoped InGaAs absorption layer, InGaAsP grading layer, Si-doped InP charge sheet layer, and undoped InP cap layer, as shown in Fig. 1. PECVD-deposited silicon nitride was used as the diffusion mask and was patterned by standard photolithography. For each wafer, a single Zn diffusion was performed using DMZn in the MOCVD reactor. The resulting devices do not have guard rings or a shaped diffusion front [8] and are thus not optimized for preventing edge breakdown, but allow the diffusion depth and dark current to be investigated as a function of process conditions. The wafer temperature during diffusion, phosphine overpressure and DMZn source flow were varied. The reactor pressure was kept constant at 100 Torr. The temperatures reported are wafer surface temperatures measured by emissivity-corrected pyrometry.

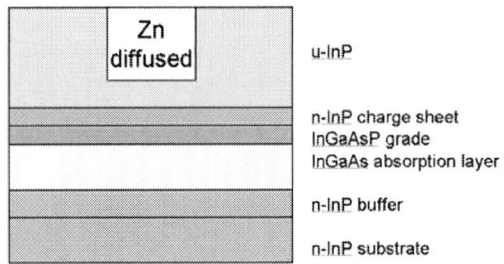

Fig. 1.   Epitaxial structure of the APD device

Current-voltage characteristics of diffused wafers were measured by probing the Zn-diffused surface without metallization. A subset of the wafers was cleaved for scanning electron microscope (SEM) imaging of the cleaved cross-section. A large diffused area was included in the diffusion mask to allow for secondary ion mass spectrometry (SIMS) profiling. Additional data were obtained from blanket-diffused wafers without a diffusion mask.

III. RESULTS AND DISCUSSION

Fig. 2 shows a SEM image of a cross-section of a Zn diffused area. The contrast between the p-type and n-type

978-1-4673-1725-2/12 $31.00 © 2012 IEEE

material is sufficient to image the diffusion front without any delineation etch, and a slight contrast change is also visible at the top of the InGaAsP grading layer. In order to calibrate the SEM magnification scale, the thickness of the top InP layers (marker "2") is measured from the image. The measured diffusion depth (marker "1") is then scaled based on the InP layer thickness, determined from the growth time and *in situ* optical reflectivity measurement of the growth rate [9].

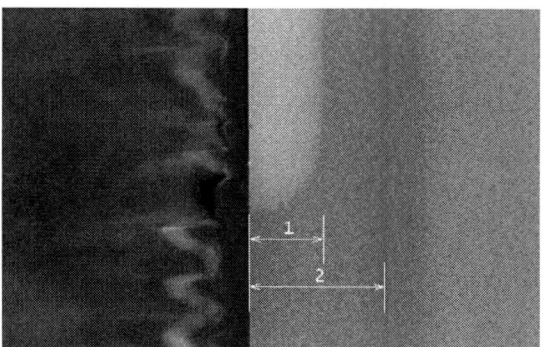

Fig. 2. SEM image of a cross-section of a diffused device. Marker 1 represents the depth of the diffusion front, 1.7 μm and Marker 2, the thickness of the top InP layers, 3.2 μm.

A SIMS profile of the Zn concentration measured in a ~1 cm^2 diffused area on the same wafer is shown in Fig. 3. The Zn concentration shows an abrupt diffusion front, and the point at which the Zn concentration reaches the instrumental background level of ~1×10^{16} cm-3 is used as the measurement of the diffusion depth. For the diffusion conditions used for this wafer, with a wafer temperature of 470 °C and with a DMZn partial pressure of 14 mTorr, the diffusion depth measured by SIMS agrees with the SEM measurement. However, at higher temperatures we have observed an increase in the diffusion depth for small device areas, associated with lateral transport of Zn across the mask [10].

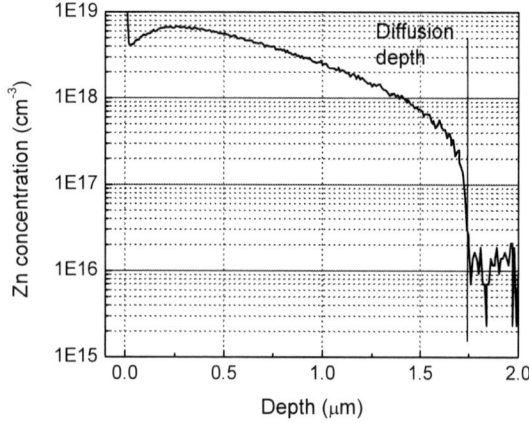

Fig. 3. SIMS Zn concentration profile of a diffused wafer.

Fig. 4 shows the dependence of the diffusion depth on diffusion time. The data shown are from wafers diffused at a temperature of 520 °C with a DMZn partial pressure of 14 mTorr, and the diffusion depth values are obtained from SIMS measurements. The data are a good fit to the theoretically expected linear dependence on the square root of the diffusion time. The best-fit regression line shows only a small offset, ~3 sec., on the time axis at zero diffusion depth.

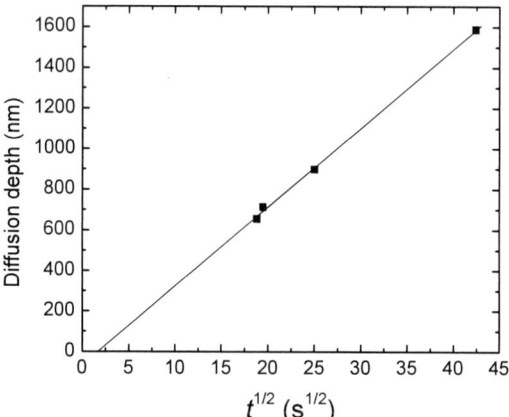

Fig. 4. Dependence of the diffusion depth on the square root of the diffusion time.

Fig. 5 shows dark current and photocurrent curves for devices with a junction area of 1.5×10^{-3} cm^2 from four wafers processed with different diffusion conditions. The photocurrent curves were obtained under white light illumination using a microscope light. The diffusion times were adjusted to account for different diffusion rates at the three different temperatures. The photocurrent increases at the punch-through voltage, which is near 25 V for all four wafers. The diffusion times for these wafers were adjusted to obtain a multiplication width of 0.7 μm. All, except curve (d), exhibit a step increase in the dark current at a reverse bias ~10 V greater than the punch-through voltage.

In order to measure the bulk-related dark current contribution for the wafer diffused at 535 °C and 1.8 Torr partial pressure (p.p.) of PH$_3$, the silicon nitride diffusion mask was removed. Fig. 6 shows the dark current and photocurrent curves for this wafer, for devices with a junction area of 1.5×10^{-3} cm^2, with the mask present (a) and after removing the silicon nitride mask (b). The surface leakage current is sufficiently reduced after removing the mask to allow the bulk-related step in the dark current to be measured.

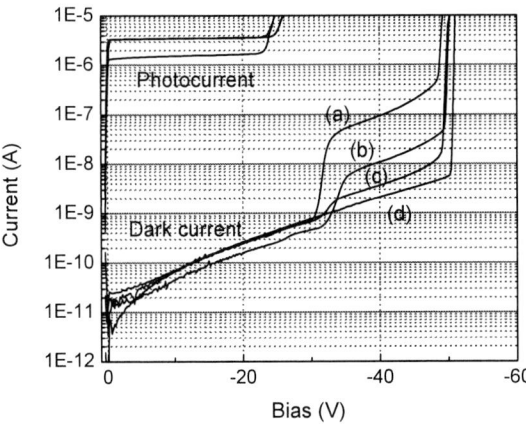

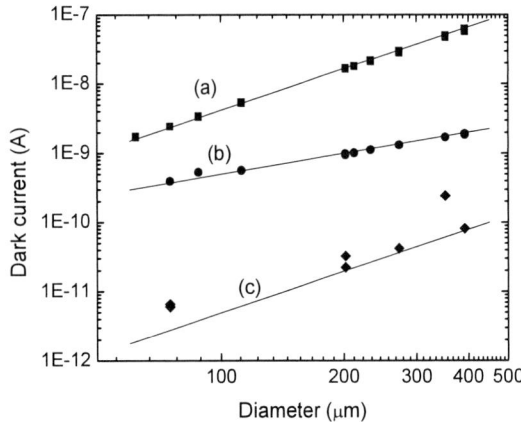

Fig. 5. Dark current and photocurrent for identical epitaxial structures diffused using four different conditions: (a) 470 °C, 1.0 Torr p.p. PH₃, (b) 520 °C, 1.0 Torr p.p. PH₃, (c) 520 °C, 1.8 Torr p.p. PH₃, (d) 535 °C, 1.8 Torr p.p. PH₃

Fig. 7. Dark current as a function of device diameter: Diffusion at 470 °C, 1.0 Torr p.p. PH₃ (a); Diffusion at 535 °C, 1.8 Torr p.p. PH₃ – before removing diffusion mask (b); after removing the diffusion mask (c).

Fig. 7 shows the variation in dark current as a function of device diameter, at a reverse bias 5 V above the step in the dark current. For the wafer diffused at 470 °C and 1.0 Torr p.p. PH₃, line (a) fits the data with a slope of 2 – the current is proportional to the area, indicating a bulk-related origin. For the wafer diffused at 535 °C and 1.8 Torr p.p. PH₃, line (b) with a slope of 1 fits the current measured before removing the diffusion mask. In this case, surface leakage current is the dominant current contribution. Line (c) is fitted to the data measured after removing the diffusion mask. Although there is more scatter in the data measured on the bare surface, the larger diameter devices are a good fit to the line with a slope of 2, indicating a dominantly bulk-related dark current. The bulk-related dark current contribution for the wafer diffused at 535 °C, 1.8 Torr p.p. PH₃ is three orders of magnitude lower than for the wafer diffused at 470 °C, 1.0 Torr p.p. PH₃.

To evaluate the uniformity of the multiplication width across a 3" diameter wafer, current-voltage curves were measured at 2.5 mm intervals in a line across the center of the wafer diffused at 535 °C and PH₃ partial pressure of 1.8 Torr. These measurements were made before removing the silicon nitride diffusion mask. Fig. 8 shows the variation in breakdown voltage (defined as the dark current reaching 10^{-6} A) and punch-through voltage (defined as the photocurrent reaching 10^{-6} A). The position labeled as 0 mm corresponds to the wafer center, and negative values are in the direction of the center of the susceptor, which holds three wafers. In the positive direction, the measurements end ~12 mm from the edge of the wafer due to the presence of a large diffusion area for profiling measurements near the edge. The breakdown voltage shows excellent uniformity with a standard deviation of 0.7 V, while the punch-through voltage shows slightly greater variation, with a standard deviation of 1.5 V. Both parameters exhibit a slight trend to higher bias voltage toward the outer edge of the susceptor.

In addition to the temperature and PH₃ partial pressure variations shown in Figs. 5 and 6, the DMZn partial pressure was varied for two of the conditions. Table 1 lists the wafer temperature, the PH₃ and DMZn partial pressures, as well as the ratio of diffusion depth to the square root of the diffusion time and the dark current density at a reverse bias 5 V greater than the bias corresponding to the step in the dark current. For the wafer diffused at 535 °C, the current density listed corresponds to the dark current measured after removing the silicon nitride mask.

The ratio of diffusion depth to the square root of the diffusion time gives an indication of the relative rates of diffusion under the different conditions. Keeping other variables fixed, the rate is slightly lower at 520 °C than at 470 °C, and considerably reduced at 535 °C, consistent with the

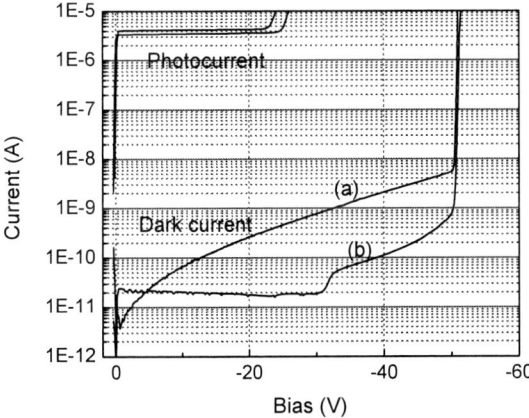

Fig. 6. Dark current and photocurrent for wafer diffused at 535 °C, 1.8 Torr p.p. PH₃: (a) with diffusion mask present, (b) after removing the diffusion mask.

trend we have reported in previous work [10]. The PH_3 partial pressure does not appear to have a significant effect on the rate of diffusion. The rate of diffusion increases with DMZn partial pressure at 520 ºC but is independent of the DMZn partial pressure at 470 ºC. This finding is consistent with the results of ref. [10], which show that diffusion at the lower temperature takes place from a solid ZnP_x film.

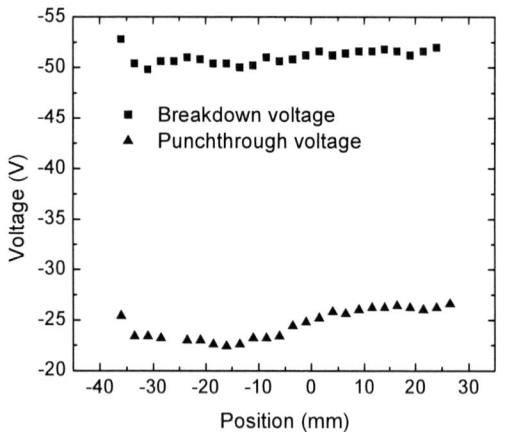

Fig. 8. Variation of breakdown voltage and punch-through voltage across wafer.

TABLE 1

SUMMARY OF DIFFUSION CONDITIONS AND RESULTS

Temperature	DMZn p.p.	PH_3 p.p.	$d / t^{1/2}$	I_D
(°C)	(mTorr)	(Torr)	($\mu m / s^{1/2}$)	(A / cm^2)
470	14	1.00	3.56E-02	5.2E-05
470	26	1.00	3.66E-02	1.3E-04
520	14	1.00	3.31E-02	8.0E-06
520	14	1.80	3.36E-02	2.4E-06
520	26	1.00	4.66E-02	6.5E-6
535	14	1.80	2.38E-02	6.9E-8

The dark current density decreases as the temperature or PH_3 partial pressure increase. Since PH_3 is only partially cracked at the temperatures used for diffusion, the dependence on temperature may in fact be due to an increase in the effective phosphorus overpressure as the temperature increases. Compared to the strong dependence on temperature and PH_3 partial pressure, the dark current density shows a comparatively weak dependence on the DMZn partial pressure, increasing as the DMZn partial pressure is increased. The strong dependence of the dark current density on diffusion conditions suggests that the step in the dark current is associated with defects formed during the diffusion process. Deep levels in heavily Zn-doped InP have been attributed to complexes of Zn and a P vacancy or vacancies [11, 12]. The dependence of the step in the dark current on PH_3 partial pressure and temperature suggests that the bulk contribution to

the dark current may be related to the formation of P vacancies at the surface and their migration into the bulk.

IV. CONCLUSIONS

We studied the diffusion of zinc into APD structures using DMZn as the source in an MOCVD reactor. The diffusion depth showed the expected dependence proportional to the square root of the diffusion time. The dark current above the punch-through voltage was shown to decrease strongly with increasing temperature and PH_3 flow, suggesting that Zn – P vacancy complexes play a significant role in determining the dark current.

ACKNOWLEDGEMENTS

The authors wish to thank Simona Moisa for SIMS measurements.

REFERENCES

[1] C. A. Hampel et al., J. Vac. Sci. Technol. A **22** (3), 916 (2004).
[2] A. van Geelen, et al., J. Cryst. Growth **195** (1998) 79.
[3] M. Wada, M. Seko, K. Sakakibara, Y. Sekiguchi, J. J. Appl. Phys. **28** (1989) L1700.
[4] J. Wisser, M. Glade, H.J. Schmidt, K. Heime, J. Appl. Phys. **71** (1992) 3234.
[5] N. Carr, J. Thompson, 7th European Workshop on Metal Organic Vapour Phase Epitaxy and Related Growth Techniques, Berlin, 8-11 June 1997 p. A1.
[6] L. E. Tarof, D. G. Knight, K. E. Fox, C. J. Miner, N. Puetz, H. B. Kim, Appl. Phys., Lett. **57** (7), 670 (1990).
[7] P. Philippe, P. Poulain, K. Kazmierski, B. de Cremoux, J. Appl. Phys. **59** (5), 1771 (1986).
[8] Y. Liu et al., Appl. Phys. Lett. **53** (14), 1311 (1988)
[9] P. Wolfram, E. Steimetz, W. Ebert, N. Grote, J.-T. Zettler, J. Cryst. Growth **272** (2004) 118
[10] O. J. Pitts, W. Benyon, D. Goodchild, A. J. SpringThorpe, J. Cryst. Growth **352** (2012) 249–252
[11] J. Slotte et al., Phys. Rev. B **67**, 115209 (2003)
[12] G. J. van Gurp, P. R Boudewijn, M. N. C. Kempeners, and D. L. A. Tjaden, J. Appl. Phys. **61** (5), 1846 (1987).

450 GHz Amplifier MMIC in 50 nm Metamorphic HEMT Technology

A. LEUTHER, A. TESSMANN, H. MASSLER, R. AIDAM, M. SCHLECHTWEG, O. AMBACHER

Fraunhofer Institute for Applied Solid State Physics (IAF),
Tullastrasse 72, D-79108 Freiburg, Germany
e-mail: arnulf.leuther@iaf.fraunhofer.de

Abstract — We present a passivated 50 nm gate length metamorphic high electron mobility transistor (mHEMT) technology optimized for the successful fabrication of submillimeter-wave MMICs. A BCB based planarization process is used for placing a second 450 nm wide gate head, which is defined by optical lithography, on top of a 50 nm e-beam written T-gate. Due to the very low intrinsic resistances of the realized mHEMT devices an extrinsic maximum transconduction $g_{m,max}$ of 2100 mS/mm was achieved together with an maximum drain current $I_{D,max}$ of 1300 mA/mm. Furthermore, transit frequencies f_T and f_{max} of 370 and 670 GHz were extrapolated. The f_{max} extrapolation is based on measured S-parameters up to 220 GHz and compared with the small signal model used for circuit design on the 50 nm mHEMT process. The presented transistor technology was used to fabricate a four-stage common source amplifier circuit in grounded coplanar waveguide topology demonstrating a linear gain of 13 dB at 450 GHz. Assuming matching losses of 1.5 dB per stage within the MMIC the measured circuit gain of 3.3 dB per stage is in good agreement with the 4.6 dB transistor gain predicted by the small signal model.

Index Terms — metamorphic high electron mobility transistor (mHEMT), InGaAs, submillimeter-wave amplifier, submillimeter monolithic microwave integrated circuit (S-MMIC).

I. INTRODUCTION

The coverage of the submillimeter-wave frequency regime of the electromagnetic spectrum with monolithic integrated HEMT devices and circuits was enabled by the reduction of the transistor gate length and by increasing the Indium content in the semiconductor heterostructures. Single transistors with a cut-off frequency of 688 GHz and MMICs with small signal gain up to 670 GHz [1, 2] based on InP HEMT devices have been presented. Submillimeter-wave frequency circuits find applications in advanced radar, communication and spectroscopic systems and enable compact and cost effective solutions.

To date, all published HEMT MMICs above 300 GHz were realized in sub 50 nm gate length technologies [3]. This paper presents for the first time a submillimeter-wave amplifier circuit using transistors with 50 nm gate length. The four-stage low-noise amplifier MMIC achieved a small

signal gain of 13 dB at 450 GHz, enabled by an optimized cross section of the 50 nm gate length metamorphic HEMT technology. The advantage of using larger gate length technologies is their typically higher robustness and process yield.

II. TECHNOLOGY

The metamorphic transistor heterostructure is grown by MBE on 100 mm semi-insulating GaAs wafers using a linear graded InAlGaAs buffer layer. The electrons are confined in a composite $In_{0.8}Ga_{0.2}As/In_{0.53}Ga_{0.47}As$ channel with $In_{0.52}Al_{0.48}As$ barriers. The layer sequence is capped with a saturation doped $In_{0.53}Ga_{0.47}As$ layer for improved ohmic contacts. Hall measurements on heterostructures without cap doping reveal an electron channel density of 4.2×10^{12} cm^{-2} together with a carrier mobility of 11,800 cm^2/Vs.

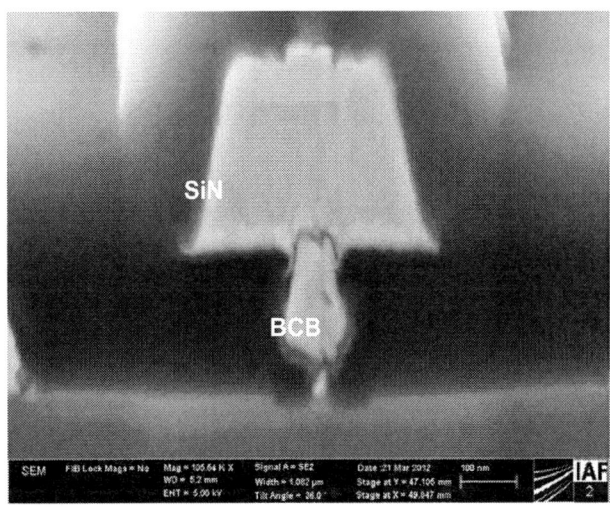

Fig. 1: SEM cross section of the passivated 50 nm gate with optically defined gate head. The cut through the gate finger was done by using a focused ion beam (FIB).

The 50 nm gate is defined by an electron beam lithography process using a four layer PMMA resist stack. After the wet chemical recess etching in a succinic based solution the wafers are metallized with a PtTiPtAu metal layer by electron beam evaporation. The head of the T-gate has a width of only 170 nm. This gate is encapsulated in BCB (ε_r= 2.65) to reduce parasitic gate capacitances compared to a SiN passivation. Due to the planarization of BCB during annealing the BCB thickness on top of the gate is reduced compared to the rest of the wafer. Subsequently, the entire wafer is dry etched in an ICP tool to uncover the gate tip. By using the first interconnection layer an electron beam evaporated second gate head with a total width of 450 nm is placed on top of the e-beam gate as shown in Fig. 1 [4]. Hereby, the gate line resistance is drastically reduced from 1400 to 280 Ω/mm which is necessary to achieve very high f_{max} values as well as a very low minimum noise figure F_{min}. A further advantage of the optical defined gate head compared to the gate head defined by e-beam writing is the larger distance to the semiconductor surface and a considerable reduced parasitic capacitance of the head.

The developed MMIC process also includes a 250 nm thick CVD deposited SiN layer used for device passivation and MIM capacitors, NiCr thin film resistors and a plated Au-layer in air bridge technology. After front side processing the wafers are glue bonded in vacuum to a 100 mm carrier wafer. The 100 mm GaAs wafers are thinned down to 50 µm thickness. Through substrate vias with 20 µm front side diameter are dry etched using a 12 µm thick resist mask. For the back side metallization and to realize the contact through the substrate vias to the front side a 2.7 µm thick plated Au layer is used.

III. EXPERIMENTAL RESULTS

Table 1
Electrical DC and RF parameters of the 50 nm mHEMT technology.

R_C (Ωmm)	0.04
R_S (Ωmm)	0.13
$I_{D,max}$ (mA/mm)	1300
V_{BD} (V)	3.0
$g_{m,max}$ (mS/mm)	2100
f_T (GHz)	370
f_{max} (GHz)	~ 670

The electrical DC and RF device parameters are listed in Tab. I. The output characteristics of a typical mHEMT are presented in Fig. 2. Due to the low intrinsic device resistances the R_{on} at V_G = 0.2 V was only 0.38 Ωmm. In

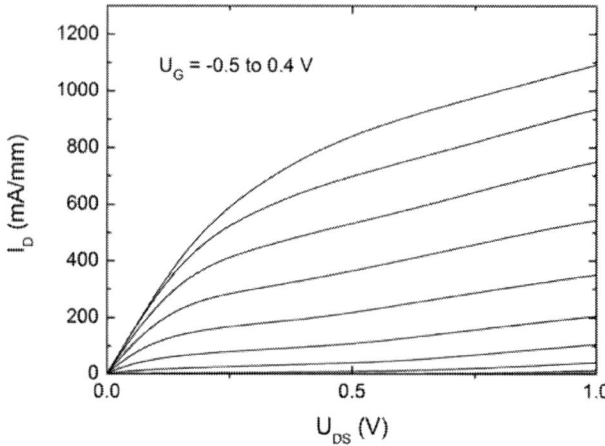

Fig. 2: Output characteristics of the 50 nm mHEMT for gate voltages varying from U_G = -0.5 V to +0.4V.

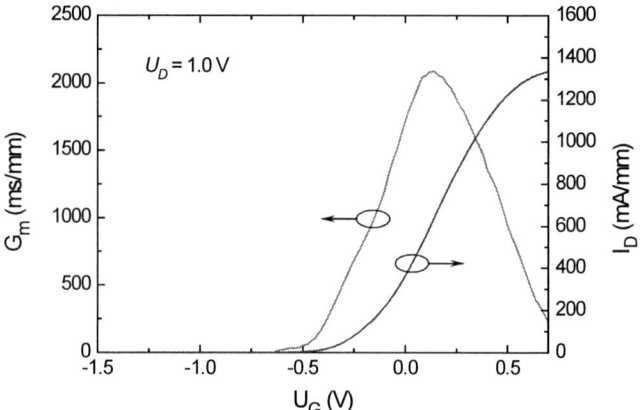

Fig. 3: Transfer characteristics of the 50 nm mHEMT for a drain voltage U_D= 1.0 V. A maximum transconduction $g_{m,max}$ = 2100 mS/mm was achieved.

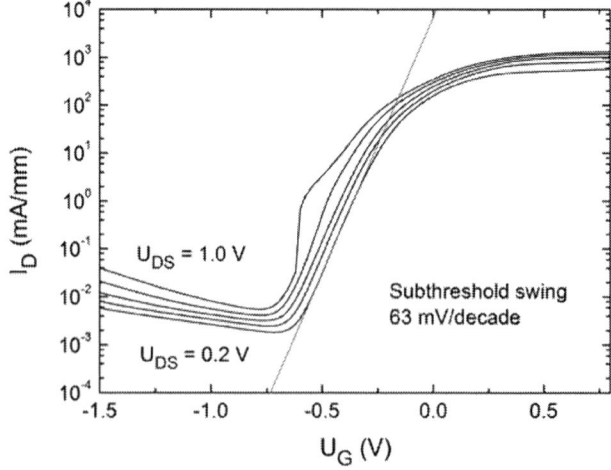

Fig. 4: For the drain current an I_{on}/I_{off} -ratio of 3×10^5 and a subthreshold swing of 63 mV/decade was measured.

Fig. 3 the transfer characteristic of the transistor is shown. The maximum extrinsic transconductance $g_{m,max} = 2100$ mS/mm was measured for a gate voltage of $V_G = 200$ mV. As a consequence of the low R_{on} the maximum drain current $I_{D,max}$ was as high as 1300 mA/mm. The transistor showed an I_{on}/I_{off}-ratio of 3×10^5 and a subthreshold swing of only 63 mV/decade (Fig. 4). To get a low pinch-off drain current, the gate leakage current had to be minimized by improving the homogeneity of the lateral recess etching width. The very low subthreshold swing value was achieved by reducing the barrier thickness and by using a Pt-gate sinking process. For V_{DS} values larger than 0.6 V a shoulder in the drain current curve appeared in the subthreshold region which might be caused by electron-hole pair generation.

In Fig. 5 RF-measurements of *MSG* and h_{21} up to 220 GHz are shown for a 2×15 μm gate width device. Based on the h_{21} data an f_T of 370 GHz was extrapolated. The Fraunhofer IAF small signal model (red line) developed for the mHEMT technology [5] predicts an f_{max} of 670 GHz, which is in good agreement with the MSG measurements up to 220 GHz.

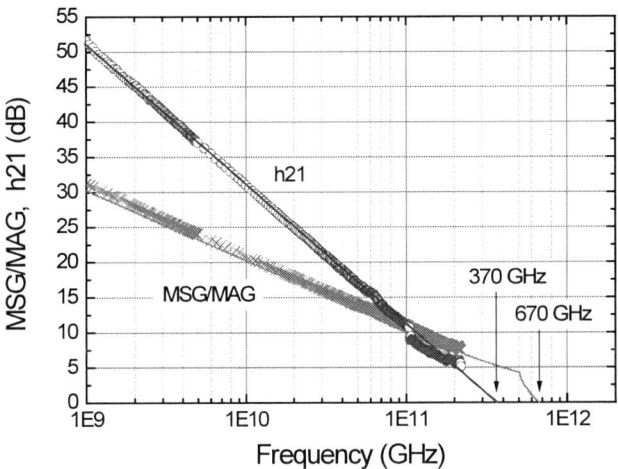

Fig. 5: Measured *MSG* and h_{21} of a 2×15 μm mHEMT. The extrapolation of h_{21} (blue line) results in a transit frequency $f_T = 370$ GHz. The red line shows the *MSG/MAG* characteristic of the small signal model. The *MSG* measurements up to 220 GHz are in good agreement with the small signal model used for circuit design.

Target of the process optimization was to achieve high circuit gain in the submillimeter-wave frequency regime. To verify the high frequency capability of the 50 nm technology, a four-stage submillimeter-wave mHEMT amplifier circuit was used demonstrating reasonable bandwidth and high small-signal gain in the WR-2.2 waveguide band (325 to 500 GHz). This amplifier was originally designed for the IAF 35 nm technology [6].

The schematic diagram of a single amplifier stage is shown in Fig. 6. The utilized transistors were realized in conventional common-source configuration and have a gate width of 2×5 μm, each. Special care was taken in the design of the RF shunt capacitance networks to ensure low-frequency stability of the submillimeter-wave circuit. Due to the compact coplanar circuit layout, the die size is only 0.37×0.63 mm^2. A chip photo of the MMIC is shown in Fig. 7.

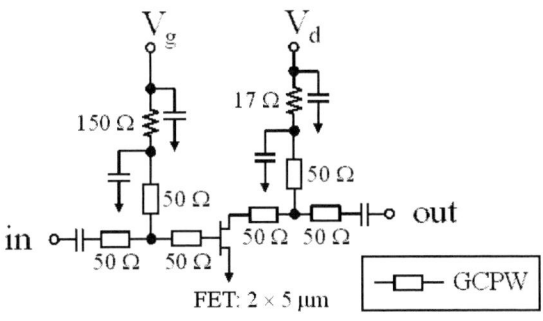

Fig. 6: Schematic diagram of a single 50 nm mHEMT amplifier stage (GCPW: grounded coplanar waveguide).

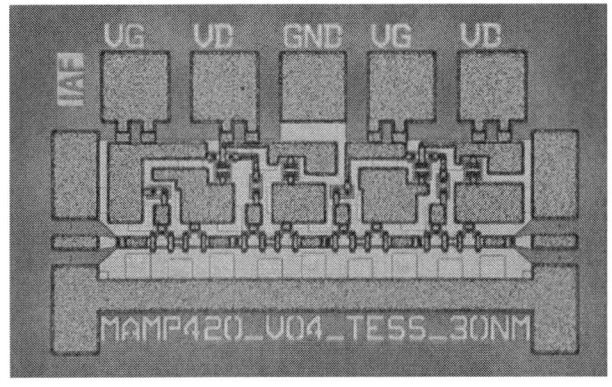

Fig. 7: Chip photo of the four-stage 450 GHz amplifier MMIC.

On-wafer S-parameter measurements were performed using an Agilent 8510C VNA system with an 85105A submillimeter controller, two Oleson WR-2.2 T/R frequency extension modules and two Cascade Infinity 500 RF-probes with a pitch of 60 μm. For an LRL-type calibration at the probe tip, a Cascade 138-356 calibration substrate was chosen.

At 450 GHz the amplifier demonstrates a small-signal gain S_{21} of more than 13 dB by applying a drain voltage of $V_D = 1$ V and a gate voltage $V_G = 0.2$ V (Fig. 8). The corresponding drain and gate currents at this bias point are $I_D = 23.7$ mA and $I_G = 5$ μA. The input and output return loss S_{11} and S_{22} at 450 GHz is better than -6 dB. To the

978-1-4673-1725-2/12 $31.00 © 2012 IEEE

authors knowledge this is the highest published operating frequency of any amplifier realized in 50 nm gate length HEMT technology. Assuming matching losses of 1.5 dB per stage within the MMIC the measured 3.3 dB circuit gain per stage is in excellent agreement with the 4.6 dB transistor gain predicted by the small signal model.

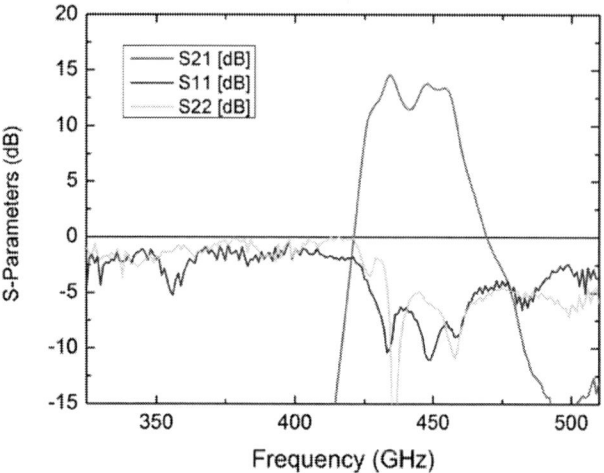

Fig. 8: On-wafer measured S-parameters of the 450 GHz amplifier MMIC. A linear gain of 13 dB was measured at 450 GHz.

IV. CONCLUSION

A 50 nm gate length mHEMT technology optimized for the fabrication of submillimeter-wave MMICs is presented. To combine low gate line resistance with drastically reduced parasitic gate capacitances, a second 450 nm wide optical defined gate head is placed on top of a 50 nm T-gate defined by e-beam lithography. A low gate-to-channel separation in combination with very low source resistance of $R_S = 0.13$ Ωmm and $R_{on} = 0.38$ Ωmm enable an extrinsic $g_{m,max}$ of 2100 mS/mm together with an $I_{D,max}$ of 1300 mA/mm. An f_{max} of 670 GHz is predicted by the small signal model which is in good agreement with measurement data up to 220 GHz. The optimized transistor technology was used to fabricate a four-stage common source amplifier circuit in grounded coplanar waveguide topology demonstrating a linear gain of 13 dB at 450 GHz. For the first time a submillimeter-wave amplifier could be demonstrated in a 50 nm gate length HEMT technology.

ACKNOWLEDGEMENT

The authors would like to thank their colleagues from the Fraunhofer IAF epitaxy, technology and high frequency department for their excellent contributions during epitaxial growth, wafer processing and characterization. This work was supported by the German Federal Ministry of Defence (BMVg) and the Bundeswehr Technical Center for Informa-

tion Technology and Electronics (WTD81) in the framework of the TERAMOSS program.

REFERENCES

[1] Dae-Hyun Kim, Berinder Brar and Jesús A. del Alamo, "fT = 688 GHz and fmax = 800 GHz in Lg = 40 nm In0.7Ga0.3As MHEMTs with gm_max > 2.7 mS/µm," *IEDM 2011*, p. 319 , December.

[2] William Deal, X. B. Mei, Kevin M. K. H. Leong, Vesna Radisic, S. Sarkozy, and Richard Lai, "THz Monolithic Integrated Circuits Using InP High Electron Mobility Transistors", *IEEE Tans. Terahertz Science and Technology*, vol. 1, no. 1, p. 25, September 2011.

[3] Lorene A. Samoska, " An Overview of Solid-State Integrated Circuit Amplifiers in the Submillimeter-Wave and THz Regime", *IEEE Tans. Terahertz Science and Technology*, vol. 1, no. 1, p. 9, September 2011

[4] Derek Smith, Gilles Dambrine, Jean-Claude Orlhac , "Industrial MHEMT Technologies for 80 - 220 GHz Applications", *3rd European Microwave Integrated Circuits Conference 2008*, p. 214, October 2008.

[5] Seelmann-Eggebert, M.; Schäfer, F.; Leuther, A.; Massler, H., "A versatile and cryogenic mHEMT-model including noise", : *IEEE MTT-S 2010 International Microwave Symposium*, p. 501-504, May 2010

[6] Tessmann A. et al., "Metamorphic HEMT MMICs and Modules Operating Between 300 and 500 GHz", *IEEE Journal of Solid State Circuits*, vol. 46, No.10, p. 2193, October 2011.

100nm-gate InAlAs/InGaAs HEMTs on plastic flexible substrate with high cut-off frequencies

Jinshan. SHI, N. Wichmann, Y. Roelens and S. Bollaert

Institute of Electronics, Microelectronics and Nanotechnology Technology, UMR-CNRS 8520, University of Lille, Villeneuve d'Ascq, 59652, France

jinshan.shi@ed.univ-lille1.fr

Abstract

This paper reports the transfer of 100nm-gate length high electron mobility transistors onto plastic flexible substrate. The layers of transistors are grown epitaxially on indium phosphide (InP) bulk substrate. The transfer of these transistors onto polyimide substrate is realized by an adhesive bonding technique. High cut-off frequencies f_T =120GHz, f_{max} =280GHz are demonstrated. These microwave performances are comparable with results obtained on 100nm-gate HEMT on rigid substrate (HEMT-RS), which provides a strong possibility to integrate high-frequency communication systems and high-speed processing applications into the flexible device in the near future.

I. Introduction

Owing to the combined advantages of bendability, toughness and light weight, flexible electronic technologies have been attracting considerable attention over the past decade. Up to now, a variety of technologies for these promising applications have been reported (e.g. organic thin film transistors, polycrystalline silicon as well as amorphous silicon) [1]-[4]. However, the main problem is that their low effective carrier mobilities definitely limit their performances in the higher frequency range applications. In order to reach better electrical performance, the use of the high electron mobility of heterostructure III-V semiconductor materials is proposed in this paper. We integrate InAlAs/InGaAs high electron mobility transistor (HEMT) onto the polyimide substrate (Kapton) and we embed it into thin-film micro-strip (TFMS) lines. Besides, both their electrical static and dynamic performances are reported.

II. Device structure and fabrication process

A. Device structure

Lattice-matched InAlAs-InGaAs HEMT layers are grown on an InP substrate using a solid source molecular beam epitaxy. The epitaxial structure of the actives layers is standard and is composed of a 20nm N^+ InGaAs(cap layer, $6\times10^{18}cm^{-3}$), a 12nm InAlAs (barrier), a Silicon δ−doping($5\times10^{12}cm^{-2}$), a 5nm InAlAs (spacer), a 15nm InGaAs (channel) and a 200nm InAlAs (buffer layer). Below of the buffer layer, a 20-nm InP layer is inserted, which allows to avoid a partially depletion of the channel layer due to the surface potential effect after the adhesive bonding process execution. Finally, in order to facilitate the transfer on flexible substrate, a 200nm InGaAs etch-stop layer is added. Fig. 1 shows the cross section of the HEMT layer structure before transferring onto polyimide substrate.

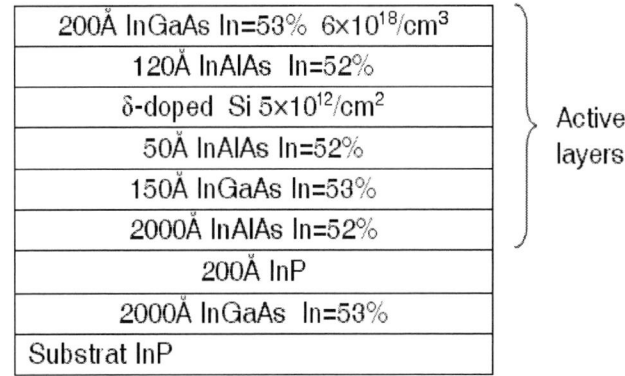

Fig. 1. The HEMT layer structure

B. Fabrication Process

Fig. 2a to Fig.2d show the main steps of the HEMT-FS fabrication process.

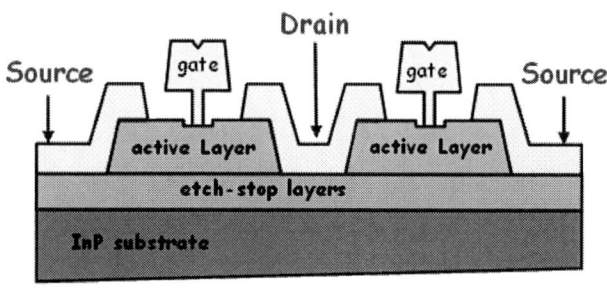

Fig. 2a. InAlAs/InGaAs HEMT fabrication (Lg=100nm)

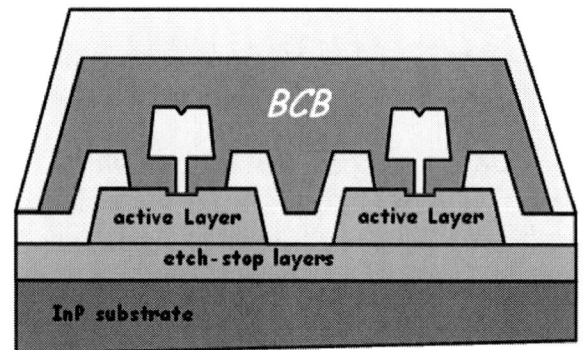

Fig. 2b. TFMS access fabrication

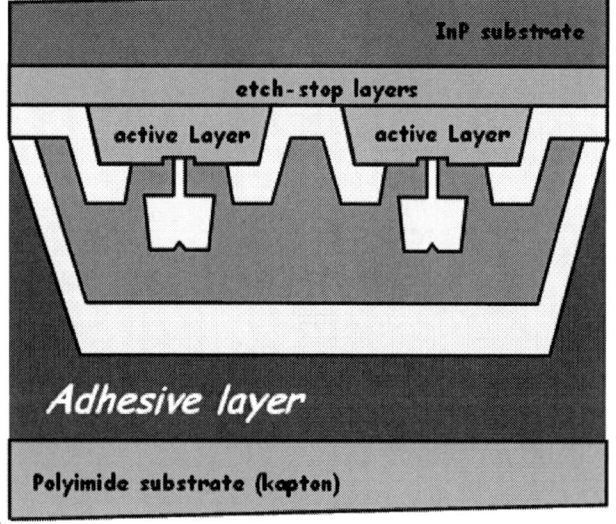

Fig. 2c. Transfer on flexible substrate

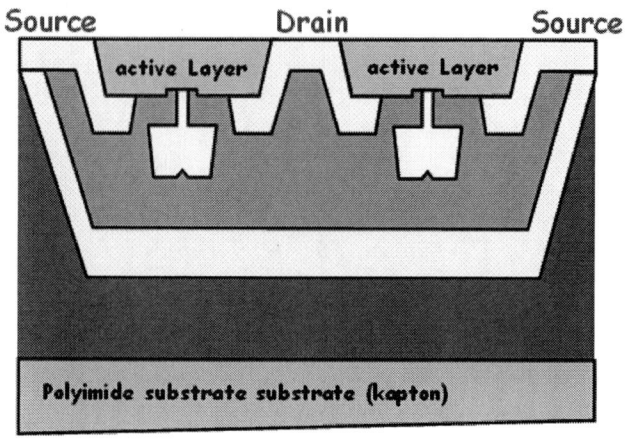

Fig. 2d. InP substrate and etch-stop layers removal

In the first main step, typical 100nm-gate HEMTs are fabricated (Fig. 2a). Ohmic contacts are realized by e-beam lithography and Ni/Ge/Au/Ni/Au evaporation with anneal at 295°C during 20s in a nitrogen atmosphere. Mesa isolation is achieved by wet etching using orthophosphoric acid based solution and an optical lithography. This wet etching is stopped in InGaAs etch-stop layer. Then, the bonding pads are deposited by the metallization of Ti/Au/Ti. Gate-recess process is carried out by wet chemical etching with a succinic acid based solution and Ti/Pt/Au/Ti gate metal is evaporated and lifted off. Note that at the end of this step, devices are not isolated on the wafer due to the conductive InGaAs etch-stop layer. So, no electrical characterization can be performed before the etching of InGaAs etch-stop layer. In the second step, TFMS access lines are defined as illustrated in Fig. 2b: Benzocyclobutene polymer island (BCB) is defined on the HEMT structures by a photolithography process. Another metal evaporation of Ti/Au/Ti is done to connect the two sides of the source bonding pads (ground plane). Next, in the third step, adhesive bonding process is performed using SU-8 as a bonding agent, supplied by Micro-CHEM (Fig. 2c). SU-8 is spun onto the HEMTs components and also on the polyimide substrate (Kapton). Then, we press them together in the vacuum ambiance (80°C, 30 min). Thanks to the ultraviolet transparency of polyimide film, we make SU-8 cross-linked by UV exposure. Finally, in the last main step, the InP substrate is selectively removed by a hydrochloric acid based solution and InGaAs etch-stop layer is selectively etched by a orthophosphoric acid based solution (Fig. 2d). This wet-etching process enables to stop accurately on InP passivation layer. Fig. 3 is the optical image of our HEMT-FS, showing an obvious mechanical bendability. In addition, we have fabricated standard HEMT on rigid substrate (HEMT-RS) on another wafer for comparison. Note that HEMT-RS do have the same epitaxial active layer as HEMT-FS and have identical

978-1-4673-1725-2/12 $31.00 © 2012 IEEE   234

device geometry with physical gate length of 100nm and gate width of 2×50 µm.

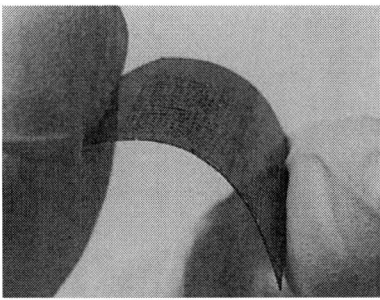

Fig. 3. Optical image of HEMT on flexible substrate

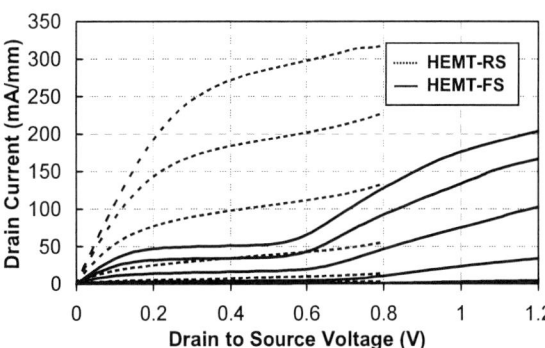

Fig. 4. Output characteristics of the HEMT on rigid substrate (HEMT-RS) and on flexible substrate (HEMT-FS). The topgate bias for HEMT-FS is 0.2V and for HEMT-RS is 0.1V, both gate steps are -0.1V.

III. Characterization

Hall measurements are performed at room temperature on HEMT-FS active layers which present a hall mobility of $7400 cm^2 V^{-1} s^{-1}$ and a hall density of $5.4×10^{12} cm^{-2}$ while HEMT-RS has a hall mobility and a hall density of $6900 cm^2 V^{-1} s^{-1}$ and $6,8×10^{12}$ cm^{-2} respectively. On wafer static and dynamic measurements are carried out at room temperature. For HEMT-FS, the value of maximum drain current Id_{MAX} equals to 210mA/mm at the gate to source voltage V_{GS}=0.2V. But for HEMT-RS, this value exceeds 300mA/mm at V_{GS}=0.1V (Fig. 4). Besides, according to the transfer characteristics (Fig. 5), the maximum extrinsic g_m of HEMT-FS is 800mS/mm at the drain-to-source voltage V_{DS}=1.2V. By contrast, this value is about 1000mS/mm at V_{DS}=0.8V on rigid substrate. Moreover, the threshold voltage is estimated to be -0.25V on HEMT-FS, while it reaches about -0.5V on HEMT-RS. The reasonable explanation for all these electrical static performance differences mentioned above is that the HEMT-FS are inverted by the transfer as shown in Fig.2c. After the final InGaAs etch-stop layer wet etching, the InP passivation layer whose surface is exposed to air, does not totally prevent a partial depletion of channel due to Fermi level pinning effect at the InP surface. Thus, a large depletion in the channel layer leads to the presence of kink effect phenomenon when impact ionization occurs. The kink effect can be clearly observed on output characteristics of HEMT-FS (Fig. 4) at V_{DS} around 0.7V. This phenomenon in HEMTs has been widely reported in the literature [5]-[6].

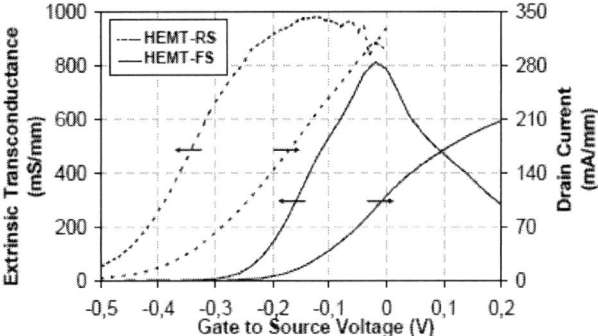

Fig. 5. Transfer characteristics at V_{DS}=0.8V (HEMT-RS) and V_{DS}=1.2V (HEMT-FS)

The S-parameters of the 100-nm-gate HEMT-FS and 100-nm-gate HEMT-RS are measured on wafer using a vector network analyzer. Fig. 6 and Fig 7 show the extrinsic current gain $/H_{21}/^2$ and the Mason's unilateral power gain Ug versus frequency for both devices. The HEMT-FS exhibits a f_T of 120 GHz and a f_{MAX} of 280 GHz. For HEMT-RS, cut-off frequencies f_T and f_{MAX} are 203 GHz and 215 GHz respectively. Despite of slight difference in terms of cut-off frequencies particularly due to the kink effect, RF performances obtained for both devices are very similar and seem to be not affected by adhesive bonding technique on flexible substrate.

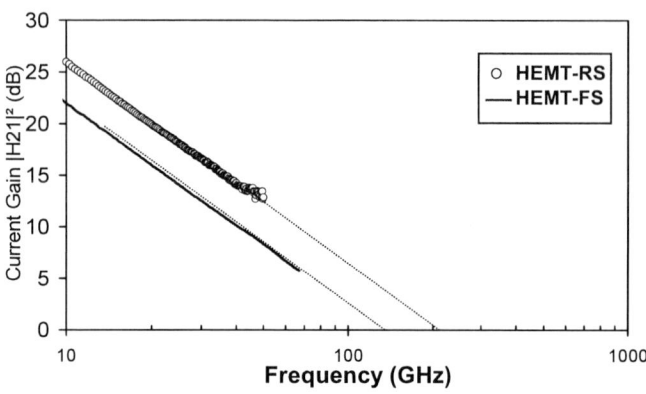

Fig. 6. Current gain $|H_{21}|^2$ as a function of frequency measured at V_{DS}=0.8V (HEMT-RS), V_{DS}=1.2V (HEMT-FS).

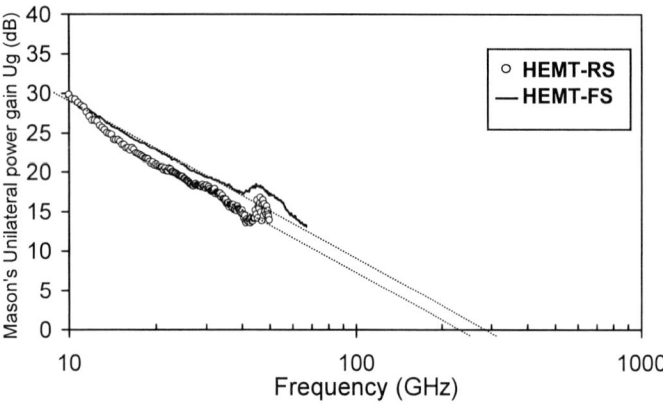

Fig. 7. The Mason's unilateral power gain U_g as a function of frequency measured at V_{DS}=0.8V (HEMT-RS), V_{DS}=1.2V (HEMT-FS).

All the comparisons of the electric static and dynamic performance are listed in Table. 1

Device:Lg=100nm	Rigid	Flexible
Id_{MAX} (mA/mm)	316	210
Gm_{DCMAX} (mS/mm)	1000	800
F_T (GHz)	203	120
F_{MAX} (GHz)	215	280

Table. 1   Comparison between the parameters of HEMT-RS and HEMT-FS

IV. Conclusion

In summary, this paper introduces a feasible method for transferring conventional HEMTs onto the flexible substrate. With this process, 100-nm-gate $In_{0.52}Al_{0.48}As/In_{0.53}Ga_{0.47}As$ HEMTs have been transferred onto polyimide film and electrically characterized in static and dynamic regime. Despite of the kink effect phenomenon on these devices due to a partial channel depletion, high cut-off frequencies (f_T =120 GHz and f_{MAX} =280 GHz) have been achieved. These performances are comparable to the ones obtained on 100nm-gate HEMT on rigid substrate. These good characteristics develop the potential to integrate high-frequency communication systems and other high-speed applications into the flexible devices. In our future work, to suppress the kink effect and further enhance the electrical performance of the device, we need to optimize the technological fabrication process and/or epitaxial layer for lowering the Fermi level surface pinning effect of HEMT-FS.

References

[1] U. Haas, H. Gold, A. Haase, G. Jakopic and B.Stadlober, "Submicron pentacene-based organic thin film transistors on flexible substrates", *Applied Physics Letters,* vol. 91, 043511, 2000.

[2] A.Z. Kattamis, R. J. Holmes, I. –C. Cheng, K. Long, J.C. Sturm, and S. Wagner, "High Mobility Nanocrystalline Silicon Transistors on Clear Plastic Substrates", *IEEE Electron Device Letters*, vol. 27 pp.49-53, January 2006.

[3] K. H. Cherenack, A. Z. Kattamis, B. Hekmatshoar, J. C. Sturm and S. Wagner, "Amorphous-Silicon Thin-Film Transistors Fabricated at 300°C on a Free-Standing Foil Substrate of Clear Plastic",*IEEE Electron Device Letters*, vol. 28 pp.1004-1006, November 2007.

[4] S. Saxena, D. C. Kim, J. H. Park, and J. Jang, "Polycrystalline Silicon Thin-Film Transistor", *IEEE Electron Device Letters*, vol.31 pp.1242-1244, November 2010.

[5] T. Akazaki, H. Takayangai and T. Enoki, "Kink Effect in an InAs-Inserted-Channel InAlAs/InGaAs Inverted HEMT at Low Temperature", *IEEE Electron Device Letters*, vol.17 pp.378-380, July 1996.

[6] R. T. Webster, S. Wu and A. F. M. Anwar, "Impact Ionization in InAlAs/InGaAs/InAlAs HEMT's", *IEEE Electron Device Letters*, vol. 21 pp.193-195, May 2000.

Analysis of Performances of InSb HEMTs
Using Quantum-Corrected Monte Carlo Simulation

Jun Sato[1], Yutaro Nagai[1], Shinsuke Hara[1], Hiroki I. Fujishiro[1], Akira Endoh[2], and Issei Watanabe[2]

[1]*Tokyo University of Science, 2641 Yamazaki, Noda, Chiba 278-8510,* [2]*National Institute of Information and Communication Technology, 4-2-1 Nukui-Kitamachi, Koganei, Tokyo 184-8795, Japan*

Abstract — The strained band structures of InSb are calculated, and then the DC and RF performances of the InSb HEMTs are analyzed by using the quantum-corrected Monte Carlo (MC) simulation. These are also compared with the InAs HEMTs. Although the compressive strain applied to the channel increases the electron effective mass, m^*, the InSb HEMTs still show the higher current drivability and the higher f_T than the InAs HEMTs from the lower V_{ds}. However, the severe impact ionization occurs from the lower V_{ds} owing to the smaller impact ionization threshold energy, E_{th}, although the compressive strain increases it. This restricts the InSb HEMTs within the low V_{ds} applications.

Index Terms — InSb HEMT, quantum-corrected Monte Carlo simulation, strain, delay time.

I. INTRODUCTION

The narrow band gap semiconductor of InSb has been attracted much attention as a promising channel material for future low power and high frequency applications, because of its extremely high electron mobility of 77,000 cm^2/Vs that has come from the smallest electron effective mass in III-V semiconductors [1]. So far, a few InSb HEMTs have been fabricated and reported [2]. Owing mainly to the immature epitaxial technique, however, their performances has still remained in their early stages, *e. g.*, the highest cutoff frequency, f_T, reported has been 305 GHz, which has been much lower than the world record of 688 GHz in the InGaAs HEMTs [2, 3]. The purpose of this paper is to investigate the potential of the InSb HEMTs by means of the quantum-corrected Monte Carlo (MC) simulation. Because of the small band gap energy of InSb, it has showed the serious impact ionization under a high electric field [4]. When it have been used as a channel, the strain has also been likely to accompany it owing to its lattice mismatch with the substrate. The strain has modified the electron effective mass and also the impact ionization frequency [4]-[6]. In this work, the DC and RF performances of the InSb HEMTs are calculated considering the strain effects. These are compared with those of the InAs HEMTs and discussed.

II. SIMULATION METHOD

A. Band calculation with strain

The channel is assumed to be on (001) substrate. The biaxial strain is applied to the channel when it is lattice-mismatched with the substrate [4, 5]. The strained band structure of the channel is calculated by the empirical pseudopotential method with the rigid ion approximation [4, 5]. The impact ionization threshold energy, E_{th}, that is minimum energy of the primary electron required for the impact ionization, is estimated by searching a pair of electrons and hole that conserve their energy and momentum [4, 5].

B. Quantum-corrected MC simulation

The device characteristics are calculated by means of the quantum-corrected MC simulation where the quantum mechanical and degeneracy effects are considered [7]-[9]. The three-valleys (Γ, L, X) non-parabolic spherical conduction band model is assumed for each material. The band parameters used in the MC simulation are extracted from the calculated band structures. The scattering mechanisms that are taken into account are acoustic, polar, non-polar optical phonons, ionized impurity, alloy and impact ionization. The scattering probability for the impact ionization is calculated using the Keldysh formula [10].

C. Delay time analysis

The delay time distribution, $\tau(x)$, is calculated along the channel by using the changes of the electron velocity and density caused by the small gate voltage, V_{gs}, perturbation [8, 9, 11]. Then, f_T is calculated as $1/2\pi\tau_{total}$, where τ_{total} is the total delay time integrating $\tau(x)$ from the source to the drain.

III. DEVICE MODELS

Fig. 1 shows the schematic illustration of the device geometry. The gate length, L_g, is 30 nm. Four types of the devices with different layer structures are simulated. Devices A and B have the InSb channels and devices C and D have the InAs channels. The barrier/buffer are Al$_{0.15}$In$_{0.85}$Sb in the device A, Al$_{0.25}$In$_{0.75}$Sb in the device B, Al$_{0.48}$In$_{0.52}$As in the device C and AlSb in the device D. The donor sheet density of the δ doping layer is constant as 2.0×10^{12} cm^{-2}. Owing to the lattice mismatch between the channel and the buffer/barrier, the compressive strain is applied to the channels in the devices A, B and C, or the tensile strain in the device D. The strain ratio, $\varepsilon_\parallel$, of the channels is -0.8, -1.3, -3.0 or $+1.3$ % in the devices

978-1-4673-1725-2/12 $31.00 © 2012 IEEE

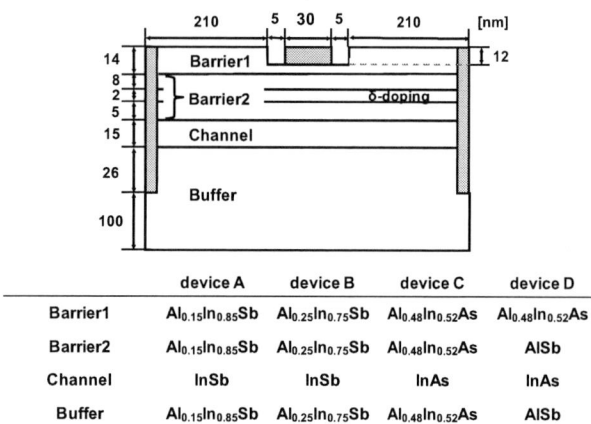

	device A	device B	device C	device D
Barrier1	Al$_{0.15}$In$_{0.85}$Sb	Al$_{0.25}$In$_{0.75}$Sb	Al$_{0.48}$In$_{0.52}$As	Al$_{0.48}$In$_{0.52}$As
Barrier2	Al$_{0.15}$In$_{0.85}$Sb	Al$_{0.25}$In$_{0.75}$Sb	Al$_{0.48}$In$_{0.52}$As	AlSb
Channel	InSb	InSb	InAs	InAs
Buffer	Al$_{0.15}$In$_{0.85}$Sb	Al$_{0.25}$In$_{0.75}$Sb	Al$_{0.48}$In$_{0.52}$As	AlSb
Strain ratio, $\varepsilon_\parallel$ [%]	-0.8	-1.3	-3.0	+1.3

Fig. 1. Schematic illustration of device geometry. Four types of devices with different layer structures are simulated.

A, B, C and D, respectively [4]. It is noted that the device C is unreal and just for comparison, because its channel thickness exceeds the critical value [5].

IV. SIMULATION RESULTS

A. Effective mass and impact ionization threshold energy

Fig. 2 shows the strained and unstrained conduction band structures of InSb (a) and InAs (b), where the insets show the enlarged Γ minima. $\varepsilon_\parallel$ is varied as −1.3, −0.8, 0 % for InSb (a) and −3.0, 0, +1.3 % for InAs (b), according to $\varepsilon_\parallel$ in the devices A - D. The electron effective mass, m^*, and E_{th} are calculated using them along the fundamental axes. Fig. 3 show the dependence of m^* (a) and E_{th} (b) on $\varepsilon_\parallel$ for InSb and InAs. In these figures, $\varepsilon_\parallel$ in the devices A - D is marked. m^* and E_{th} in InSb are smaller than those in InAs. In each material, the compressive strain makes m^* and E_{th} larger and the tensile strain makes m^* and E_{th} smaller [4, 5]. We note that the increase of E_{th} suppresses the incidence of the impact ionization [4]-[6].

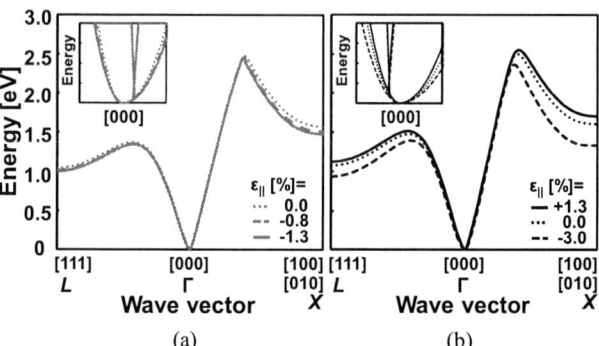

(a)   (b)

Fig. 2 Strained and unstrained conduction band structures of (a) InSb and (b) InAs. Insets show enlarged Γ minima.

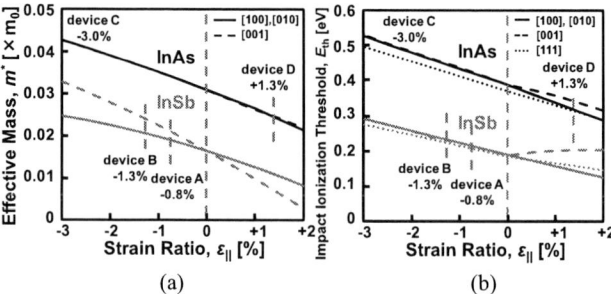

(a)   (b)

Fig. 3 Dependence of (a) effective mass, m^*, and (b) impact ionization threshold energy, E_{th}, on strain ratio, $\varepsilon_\parallel$, for InSb and InAs.

B. DC characteristics

Figs. 4 (a) - (d) show the drain current, I_{ds}, versus drain voltage, V_{ds}, characteristics for the devices A - D, where the numbers of the impact ionization events in the devices are also shown by the read lines. In these figures, the characteristics calculated ignoring the strain are shown by the dashed lines for comparison. The devices A and B with the InSb channels show the preferable I_{ds} - V_{ds} characteristics with the lower knee voltages of less than 0.2 V to the devices C and D with the InAs channels. It is also seen in the devices A, B and C, I_{ds} decreases from those without the strain, which is attributed to the increase of m^* by the compressive strain. Meanwhile, in the

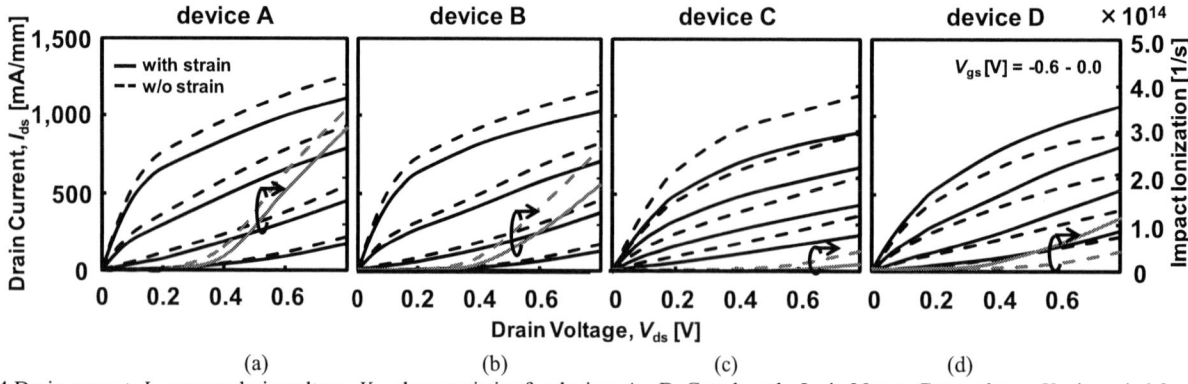

(a)   (b)   (c)   (d)

Fig. 4 Drain current, I_{ds}, versus drain voltage, V_{ds}, characteristics for devices A - D. Gate length, L_g, is 30 nm. Gate voltage, V_{gs}, is varied from −0.6 to 0.0 V. Numbers of impact ionization events are also shown by red lines. Dashed lines denote characteristics calculated ignoring strain.

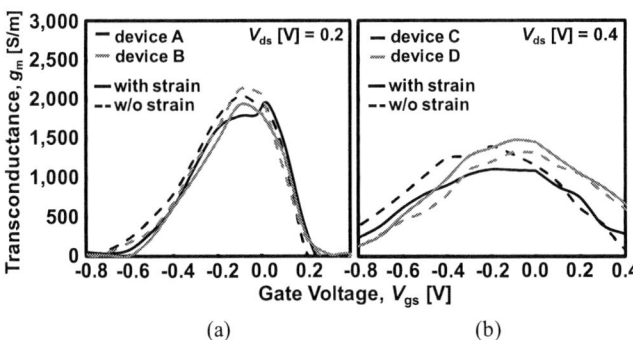

(a)　　　　　　　　　　(b)

Fig. 5 Transconductance, g_m, versus V_{gs} characteristics for (a) devices A and B and (b) devices C and D.

device D, I_{ds} increases contrarily, which is attributed to the decrease of m^* by the tensile strain. Fig. 5 shows the transconductance, g_m, versus V_{gs} characteristics for the devices A and B (a) and the devices C and D (b). V_{ds} is set to be 0.2 V for the devices A and B and 0.4 V for the devices C and D, as I_{ds} gets saturated. The devices A and B show the larger g_m than the devices C and D, indicating their superior current drivability. In the device A with the $Al_{0.15}In_{0.85}Sb$ barrier/buffer,

the electrons distribute into the barrier as V_{gs} increases, because of the shallower quantum well (QW) channel. Then, the g_m - V_{gs} curve is flattened and has the second peak at V_{gs} of about 0 V. In the device B with the $Al_{0.25}In_{0.75}Sb$ barrier/buffer, the deeper QW channel makes the two dimensional electron gas (2DEG) density larger, and the electrons no longer distribute into the barrier. Therefore, although the larger $\varepsilon_{\parallel}$ increases m^* more, the g_m - V_{gs} curve shows the larger single peak.

On the other hand, as seen in Fig. 4, the devices A and B show the severe impact ionization from the lower V_{ds} of about 0.3 V than the devices C and D, which is because of the smaller E_{th} in InSb than that in InAs. In the devices A, B, and C, the impact ionization becomes less pronounced as compared with those without the strain, which is attributed to the increase of E_{th} by the compressive strain. Meanwhile, in the device D, the impact ionization becomes more pronounced, which is attributed to the decrease of E_{th} by the tensile strain. In the device B, the quantum level in the deeper QW channel becomes higher and the larger $\varepsilon_{\parallel}$ increases E_{th} more. Both of them suppress the impact ionization somewhat.

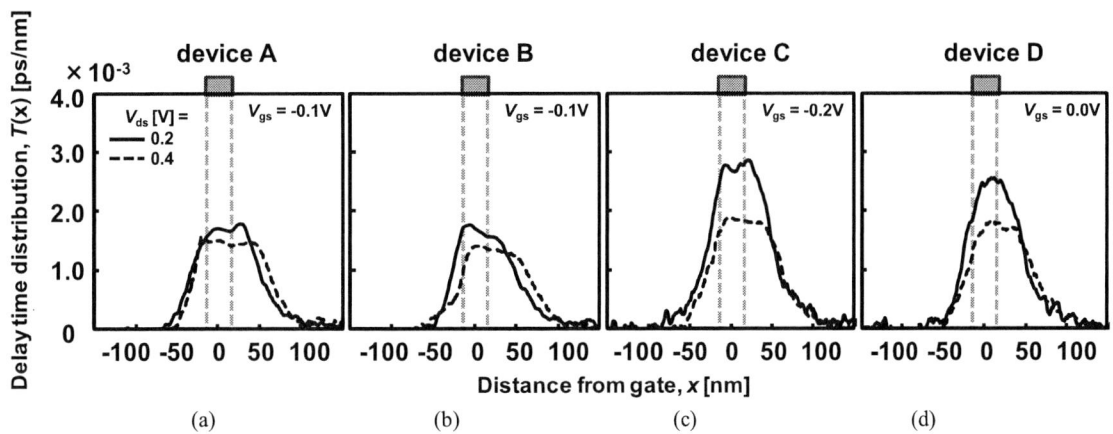

(a)　　　　(b)　　　　(c)　　　　(d)

Fig. 6 Delay time distribution, $\tau(x)$, along channel for devices A - D. L_g, is 30 nm. Strain is considered. Solid and dashed lines denote those at V_{ds} of 0.2 and 0.4 V, respectively. V_{gs} is biased as cutoff frequency, f_T, becomes maximum.

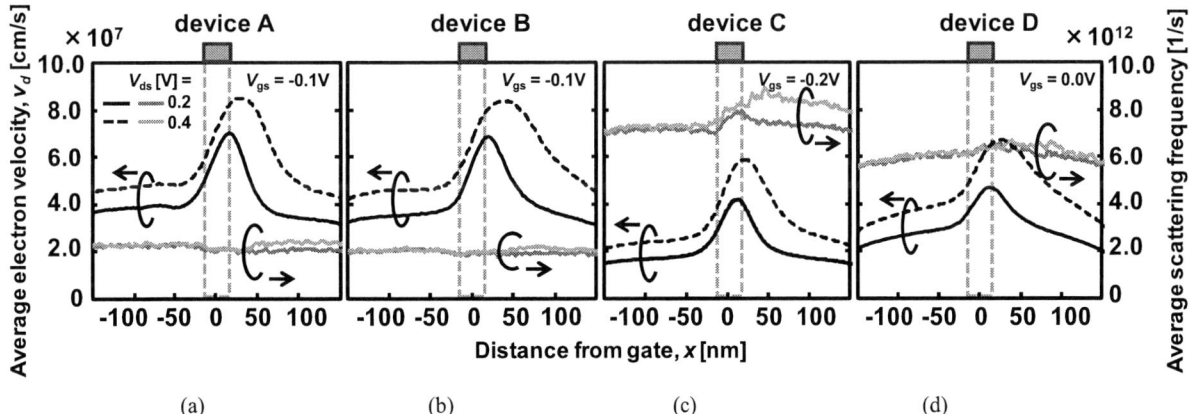

(a)　　　　(b)　　　　(c)　　　　(d)

Fig. 7 Distribution of average electron velocity in channel, v_d, for devices A - D. Average scattering frequencies in channels are also shown.

978-1-4673-1725-2/12 $31.00 © 2012 IEEE　　　　239

C. RF characteristics

Figs. 6 (a) - (d) show $\tau(x)$ for the devices A - D, where the solid and dashed lines denote those at V_{ds} of 0.2 and 0.4 V, respectively. V_{gs} is biased as f_T becomes maximum. The strain is considered. Figs. 7 (a) - (d) show the distribution of the average electron velocity in the channel, v_d, for the devices A - D, where the average scattering frequencies in the channels are also shown. In the devices A and B, the peaks of v_d under the gate reach 7.0×10^7 and 6.9×10^7 cm/s, respectively, even at V_{ds} of 0.2 V. These values exceed v_d in the steady state, indicating the remarkable velocity overshoot [4]: this is attributed to the smaller m^* and the smaller scattering frequency in the InSb channel. $\tau(x)$ distributes mainly under the gate and has the tails toward both sides [8, 9]. $\tau(x)$ in the devices A and B is smaller than those in the devices C and D at V_{ds} of 0.2 V, which is because of the larger v_d. As increasing V_{ds} from 0.2 to 0.4 V, τ (x) still more decreases under the gate, according to the increase of v_d. However, $\tau(x)$ also extends toward the drain, according to the extension of the high v_d region. Because of the smaller m^*, v_d in the device A is slightly larger than the device B. However, $\tau(x)$ in the device A is larger than the device B, which is because of the electron distribution into the barrier. On the other hand, in the devices C and D, $\tau(x)$ decreases considerably as increasing V_{ds} from 0.2 to 0.4 V, which is because of the delayed increase of v_d under the gate.

We calculate τ_{total}, and also divide it into the source delay, τ_s, the gate delay, τ_g, and the drain delay, τ_d [8, 9]. Fig. 8 shows the dependence of τ_s, τ_g, τ_d and f_T on V_{ds} for the devices A and B (a) and the devices C and D (b). The devices A and B show the sufficiently higher f_T even at V_{ds} of 0.2 V, owing to the smaller τ_{total}. While τ_g decreases as increasing V_{ds}, τ_d increases contrarily. Then, f_T tends to saturate with V_{ds}. On the other hand, the devices C and D still show the lower f_T at V_{ds} of 0.2 V, owing to the larger τ_{total}. However, τ_g decreases considerably with V_{ds}, which results in the decrease of τ_{total}. Then, f_T increases with V_{ds}. The estimated f_T in the devices A and B at V_{ds} of 0.2 V are 1,040 and 1,160 GHz, respectively. Meanwhile, those in the devices C and D at V_{ds} of 0.4 V are 900 and 1,020 GHz, respectively.

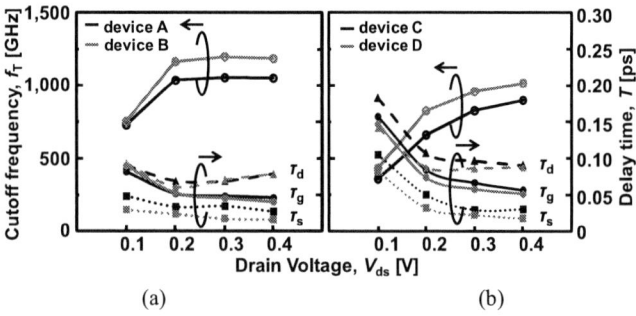

Fig. 8 Dependence of source delay, τ_s, gate delay, τ_g, drain delay, τ_d, and f_T on V_{ds} for (a) devices A and B and (b) devices C and D.

V. CONCLUSIONS

The strained band structures of InSb have been calculated, and then the DC and RF performances of the InSb HEMTs have been analyzed by using the quantum-corrected MC simulation. These have also been compared with the InAs HEMTs. Although the compressive strain applied to the channel has increased m^*, the InSb HEMTs have still showed the higher current drivability and the higher f_T than the InAs HEMTs from the lower V_{ds}. However, the severe impact ionization has occurred from the lower V_{ds} owing to the smaller E_{th}, although the compressive strain has increased it. This has restricted the InSb HEMTs within the low V_{ds} applications.

ACKNOWLEDGEMENT

This work is supported partly by Advanced Device Laboratories, Tokyo University of Science, and Grant-in-Aid for Scientific Research (C) (22560346).

REFERENCES

[1] U.P. Gomes, *et al.*, "Prospects of III-Vs for Logic Applications," *Journal of Nano- and Electronic Physics. Dig.*, vol. 4, no. 2, pp. 02009-1 - 02009-5, 2012.

[2] S. Datta *et al.*, "85nm Gate Length Enhancement and Depletion mode InSb Quantum Well Transistors for Ultra High Speed and Very Low Power Digital Logic Applications," *2005 IEEE IEDM Tech. Dig.*, pp. 763-766, 2005.

[3] Dae-Hyun Kim *et al.*, "f_T = 688 GHz and f_{max} = 800 GHz in L_g = 40 nm $In_{0.7}Ga_{0.3}As$ MHEMTs with g_{m_max} > 2.7 mS/μm," *2011 IEEE IEDM. Tech. Dig.*, pp. 13.6.1-13.6.4, 2011.

[4] H. Nishino *et al.*, "Monte Carlo Study of Strain Effect on High Field Electron Transport in InAs and InSb," *2010 IPRM Proceedings*, pp. 156-159. 2010.

[5] F. Machida *et al.*, "Strain Effects on Performances in InAs HEMTs," *2011 IPRM Proceedings*, pp. 437-440. 2011.

[6] S. Hara *et al.*, "Quantum-Corrected Monte Carlo Simulation of InSb HEMTs Considering Strain Effects," *2011 TWHM Abstracts*, p. 41, 2011.

[7] H. I. Fujishiro *et al.*, "Quantum Corrected Monte Carlo Analysis of Scaling Behavior of Nano-Scale InGaAs High Electron Mobility Transistors," *2007 ISCS Abstracts*, pp. 2795-2798, 2007.

[8] T.Takegishi *el al.*, "Theoretical Study on Performance Limit of Cutoff Frequency in Nano-Scale InAs HEMTs Based on Quantum-Corrected Monte Carlo Method," *2010 IEICE Trans. Electron.*, vol. E93-C, no. 8, pp. 1258-1265, 2010.

[9] T.Takegishi *el al.*, "Theoretical Study of Performance Limits in Nano-Scale InAs HEMTs Based on Quantum-Corrected Monte Carlo Method," *2009 IPRM Proceedings*, pp. 124-127. 2009.

[10] D.C.Herbert *et al.*, "Self-Consistent 2-D Monte Carlo Simulations of InSb APD," *IEEE Trans. Electron Devices.*, vol. 52, no. 10, pp. 2175-2181. 2005.

[11] Y. Kwon *et al.*, "Delay Time Analysis of Submicron InP-Based HEMT's," *IEEE Trans. Electron Devices.*, vol. 43, no. 2, pp. 228-237. 1996.

Optimized InP HEMTs for low noise at cryogenic temperatures

H. Rodilla, J. Schleeh, P.-Å. Nilsson and J. Grahn

GigaHertz Centre, Department of Microtechnology and Nanoscience (MC2), Chalmers University of Technology, SE-41296 Göteborg, Sweden

Abstract — **Experimental results and Monte Carlo simulations have been analyzed and compared at 300 K and 77 K for 130 nm gate-length InP HEMTs optimized for cryogenic 4-8 GHz low-noise amplifiers. The good agreement observed between simulations and experimental data for DC and small signal equivalent circuit parameters validates the simulation model. Compared to 300 K, an increase of 17% in the simulated mean electron velocity under the gate was observed at low drain current (100 mA/mm) when operating the device at 77 K. In addition, a better electron confinement in the channel was noted. The observations are consistent with an increase of the slope of the transconductance versus gate bias with reduced temperature. The high transconductance at low drain current is crucial for low noise operation of the InP HEMT at low temperature.**

Index Terms — **InP high electron mobility transistors, Cryogenic temperatures, Monte Carlo simulations, low noise.**

I. INTRODUCTION

InP HEMTs have been considered as the best devices for obtaining the lowest noise figure in cryogenic low-noise amplifiers (LNAs) working at or below 77 K. Despite being a well-known device structure, few studies are dedicated to the understanding and improvement of the device noise performance at cryogenic temperature. Indeed, it is well known that noise optimization at room temperature does not necessarily result in noise optimization at cryogenic temperatures [1].

In this work, we have used Monte Carlo (MC) simulations together with experimental results to analyze and understand the improvement in the InP HEMT noise performance at cryogenic temperatures. The studied 130 nm gate InP HEMTs have demonstrated a record average noise temperature of 1.6 K when integrated in a 4-8 GHz LNA at 10 K [2,3]. Simulation results have been compared to experimental data in order to validate our simulation method and increase the understanding of the device physics.

II. DEVICE DESCRIPTION AND SIMULATION MODEL

130 nm gate length InGaAs/InAlAs HEMTs have been fabricated on an InP substrate. The active region of the epitaxial structure was based on an $In_{0.52}Al_{0.48}As$ buffer followed by a 15 nm $In_{0.65}Ga_{0.35}As$ channel, a 3 nm $In_{0.52}Al_{0.48}As$ spacer, a Si delta-doping of 5×10^{12} cm^{-2}, an 11 nm $In_{0.52}Al_{0.48}As$ barrier and a 20 nm heavily doped (5×10^{19} cm^{-3}) $In_{0.53}Ga_{0.47}As$ cap layer. The fabrication process has been described earlier

[2]. STEM images of the final device provided accurate dimensions of the HEMT cross-section including epitaxial structure. Fabricated devices exhibited a source to drain distance of 1.2 μm and a gate recess width of 310 nm.

DC and small signal microwave measurements have been performed at room temperature and 77 K. Normally, cryogenic HEMT devices are operated at 4-15 K. In this study, an operational temperature of 77 K has been selected since quantum effects play a smaller role at 77 K than at 4 K which simplifies the simulations. Moreover, the experimental results did not show a significant difference between 4 K and 77 K in DC and RF performance of the InP HEMT.

A two-dimensional MC simulator consistently coupled with a Poisson solver was used [4,5]. MC simulations have been performed both at room temperature and 77 K. This model has previously been proven as a powerful method for simulation of InP HEMTs at room temperature [5]. In this work, the simulation model has been extended further in order to simulate at device temperatures down to 77 K.

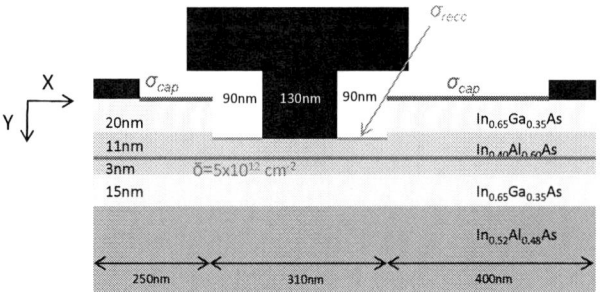

Fig. 1. Simulated InP HEMT.

For the MC simulations we have used the dimensions of the fabricated InP HEMT obtained from STEM pictures (Fig.1). As the influence of the doping in the cap layer is almost negligible in the intrinsic performance of the device, the doping level of the simulated cap layer has been reduced to 1×10^{12} cm^{-2} in order to decrease the simulation time. Ohmic source-recess distance has also been decreased for reducing simulation time as the omitted part can be considered as a contribution to the external source resistance R_s [4]. A

978-1-4673-1725-2/12 $31.00 © 2012 IEEE

negative surface charge of $\sigma_{cap}/e = -5 \times 10^{12}$ cm^{-2} has been included in the top of the cap layer in order to account for the surface charges appearing in the semiconductor–passivation interface [4]. This negative surface charge depends on the fabrication process in the case of the recess floor and we have considered $\sigma_{reces}/e = -2.2 \times 10^{12}$ cm^{-2}. Both values have been selected in order to reproduce the experimental Hall sheet electron density n_s in the channel. The same values of the surface charges was considered at 300 K and 77 K as experimental Hall n_s in the channel has been demonstrated constant with temperature.

III. RESULTS

When comparing 300 K static experimental results with 77 K, an improvement of the slope of the transconductance g_m (Fig. 2) is observed at the lower temperature. This increase in the slope of g_m at low drain current is crucial for low noise performance; in this low current regime the optimal bias point for low noise operation is normally located [3]. Also a shift of 0.1 V in the threshold voltage V_{th} has been experimentally observed when cooling down the devices from 300 K to 77 K (Fig. 2).

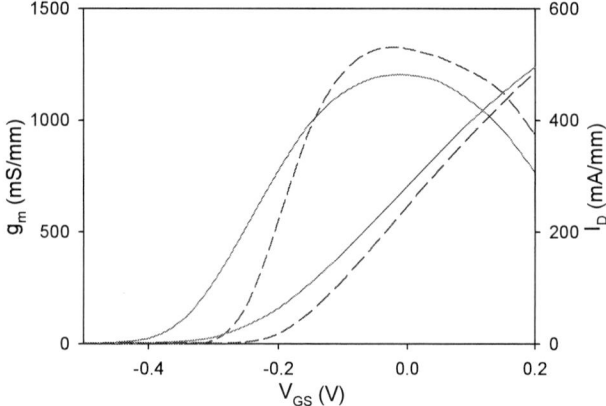

Fig. 2 Experimental results for g_m and I_D versus V_{GS}. Curves shown for 300 K (red solid lines) and 77 K (blue dashed lines), both for V_{DS}=0.6 V.

Static experimental results have been compared with MC simulations for both, 300 K and 77 K. For comparing intrinsic MC results with the extrinsic experimental data, external drain and source resistances, R_s and R_d and the Schottky barrier, V_{Sch} have been included in the simulations [5]. We have considered values of 0.1 Ω mm and 0.06 Ω mm for R_s and R_d at 300 K and 0.05 Ω mm and 0.03 Ω mm at 77 K. This reduction of the external resistances is in good agreement with the experimental results [2]. For the Schottky barrier we have used values of 0.35 V at 300 K and 0.38 V at 77 K.

A good agreement between simulated and experimental results has been observed for DC characteristics at 300 K

(Fig. 3) and 77 K (Fig. 4). Some discrepancies appear for high drain current levels, above 300 mA/mm, where surface traps play an important role, mainly at 77 K. The HEMTs for this study is intended for lower drain current levels below 150 mA/mm which is the optimum bias region for lowest noise operation in the amplifier. In this portion of the I-V curve, we observe a very good agreement between simulation results and experimental data for both 300 K and 77 K.

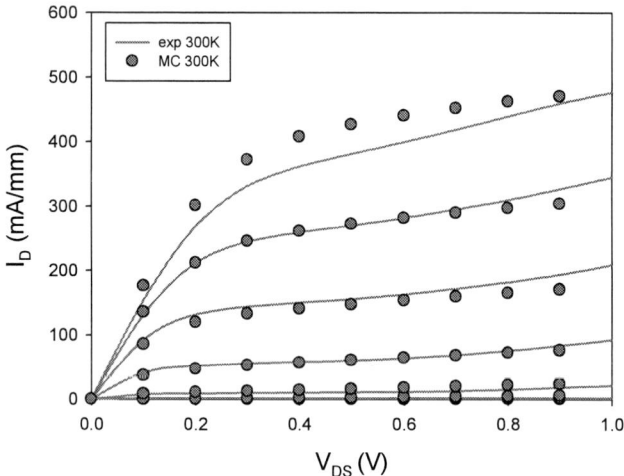

Fig. 3. Comparison between experimental results (solid lines) and MC (circles) I_D vs V_{DS} with V_{GS} from 0.1 V to -0.4 V in steps of 0.1 V at 300 K.

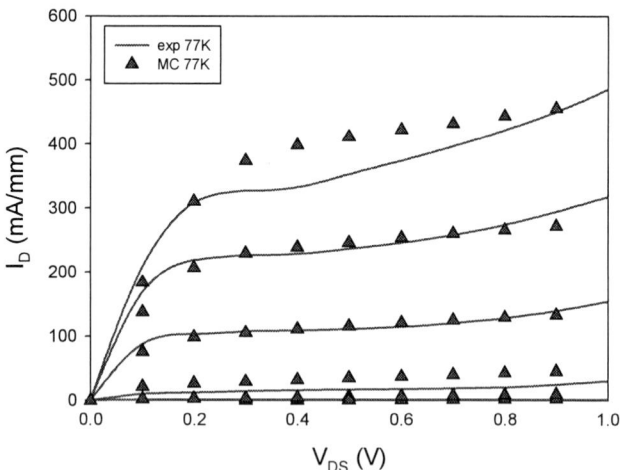

Fig. 4. Comparison between experimental results (solid lines) and MC (triangles) I_D vs V_{DS} with V_{GS} from 0.1 V to -0.4 V in steps of 0.1 V at 77 K.

After the DC comparison between experimental and MC results, small signal microwave measurements have been done and compared with MC simulations for 300 K and 77 K at

V_{DS}=0.6 V. Simulated small signal equivalent circuit (SSEC) parameters have been obtained by Fourier analysis of the response of the transistor current to voltage steps applied at the gate and drain contacts which leads us to the admittance parameters and subsequently SSEC parameters [6].

As we are interested in the low noise performance of the device, low current g_m and C_{gs} have been analyzed at 300 K and 77 K. An increase in the intrinsic transconductance of around 400 mS/mm at I_D=100 mA/mm is observed when cooling the HEMTs from 300 K to 77 K. Fig. 5 shows the good agreement between the experimental measurements and MC results for g_m at both, 300 K and 77 K.

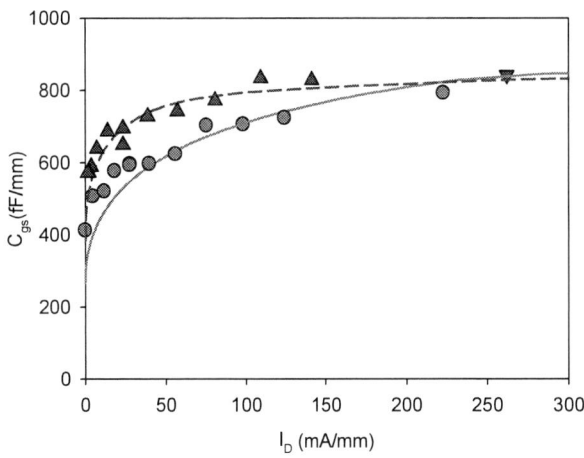

Fig. 6. Experimental (lines) and MC (symbols) results of C_{gs} as a function of I_D at 300 K (red solid line, circles) and 77 K (blue dashed line, triangles) for V_{DS}=0.6 V.

Figure 7 shows the simulated mean electron velocity along the InGaAs channel in the region under the gate at V_D=0.6 V and I_D=100 mA/mm. The expected increase in the mean electron velocity at the end of the gate is observed. This is due to the increase of the electric field at the end of the gate. Compared to 300 K, an increase in the mean electron velocity of 17% at 77 K is observed at the end of the gate.

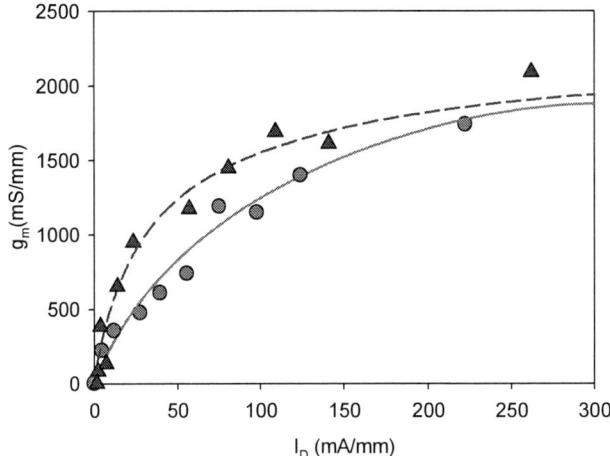

Fig. 5. Experimental (lines) and MC (symbols) results of g_m as function of I_D at 300 K (red solid line, circles) and 77 K (blue dashed line, triangles) for V_{DS}=0.6 V.

Together with the increase in the transconductance, an enhancement in gate-source capacitance C_{gs} is observed upon cooling from 300 K to 77 K. The C_{gs} increases almost 100 fF/mm at 25 mA/mm. (Fig. 6). Again, agreement between experiments and simulations are very good. No significant difference between 300 K and 77 K in neither experiments nor simulations is observed in C_{gd} and C_{ds} (not shown here).

The good agreement achieved for the SSEC parameters together with the DC agreement at both 300 K and 77 K validates our simulation model.

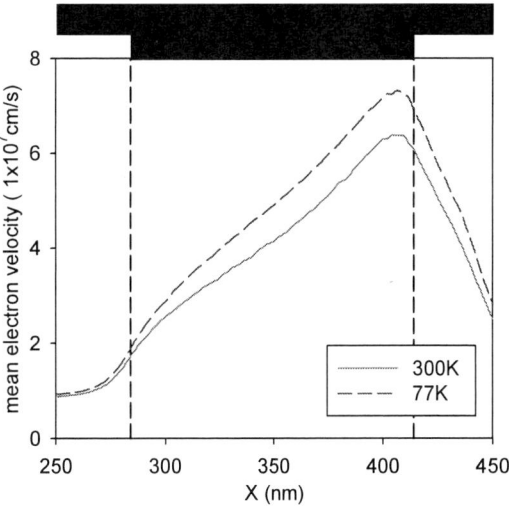

Fig. 7. Simulated mean electron velocity under the gate along the channel (X = horizontal direction) at 300 K (red solid lines) and 77 K (blue dashed line) for V_{DS}=0.6 V and I_D=100 mA/mm. Black vertical dashed lines indicate gate limits.

Compared to 300 K, a better confinement of the simulated electrons in the upper part of the channel has been observed at 77 K (Fig. 8). This is due to the decrease of the thermal energy of the electrons in the cooled HEMT.

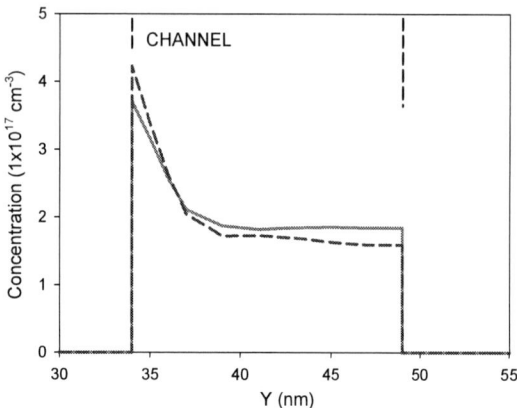

Fig.8. Simulated electron concentration profile under the gate along the Y direction at 300 K (red continuous line) and 77 K (blue dashed line) for V_{DS}=0.6 V and I_D=250 mA/mm. At 77 K, an increased electron concentration closer to the gate is seen.

The improvement in the electron confinement together with the increase in the electron velocity at low temperatures explains the increase of the slope of the transconductance at low drain current when cooling the InP HEMT. As a result, a reduction in the device noise temperature will be observed upon cooling of the devices [1].

III. CONCLUSIONS

A 130 nm gate length InGaAs/InAlAs HEMT optimized for cryogenic temperature has been presented and analyzed by comparing experimental data with Monte Carlo simulations at 300 K and 77 K. A very good agreement between experimental data and simulation results has been achieved in DC and small signal equivalent circuit parameters showing an increase of g_m and C_{gs} at low drain bias at cryogenic temperatures together with a shift in the threshold voltage. The increase of g_m in the low drain current region has been attributed to an increase in the mean electron velocity at the end of the gate region together with an improvement of the electron confinement in the upper part of the channel.

ACKNOWLEDGEMENT

This research has been developed in the GigaHertz Centre in a joint research project financed by the Swedish Governmental Agency of Innovation Systems (VINNOVA), Chalmers University of Technology, Omnisys Instruments AB, Wasa Millimeter Wave, Low-Noise Factory and SP Technical Research Institute of Sweden.

REFERENCES

[1] M. W. Pospieszalski, "Modeling of noise parameters of MESFETs and MODFETs and their frequency and temperature dependence," *IEEE Trans. Microwave Theory & Tech.*, vol. 37, no. 9, pp. 1340-1350, September 1989.

[2] J. Schleeh, G. Alestig, J. Halonen, A. Malmros, B. Nilsson, P-Å Nilsson, J. P. Starski, N. Wadefalk, H. Zirath, J. Grahn. "Ultra-low power cryogenic InP HEMT with minimum noise temperature of 1 K at 6 GHz," *IEEE Electr. Devices. Lett.*, vol. 33, no. 5,pp. 664-666, May 2012.

[3] J. Schleeh, N. Wadefalk, P-Å Nilsson, J. P. Starski, G. Alestig, J. Halonen, B. Nilsson, A. Malmros, H. Zirath, J. Grahn. "Cryogenic 0.5-13 GHz low noise amplifier with 3 K midband noise temperature," to be published in Proceedings of *IEEE MTT-S International Microwave Symposium IMS, 2012*.

[4] J. Mateos, T. González, D. Pardo, V. Hoel, Happy H and A. Cappy. "Improved Monte Carlo algorithm for the simulation of -doped AlInAs/GaInAs HEMT's" *IEEE Trans. Electr. Devices.*, vol. 47, no.1, pp. 250- 253, January 2000.

[5] J. Mateos, T. González, D. Pardo, V. Hoël, and A. Cappy. "Monte Carlo Simulator for the design Optimization of Low-Noise HEMTs," *IEEE Trans. Electr. Devices.*, vol. 47, no. 10, pp. 1950-1956, October 2000

[6] J. Mateos, T. González, D. Pardo, S Bollaert, T Parenty and A. Cappy. "Design optimization of AlInAs-GaInAs HEMTs for high-frequency applications," *IEEE Trans. Electr. Devices.*, vol. 51, no. 8, pp. 1228-1233, August 2004.

Metal-Organic Vapor-Phase Epitaxy Growth of InP-Based HEMT Structures with InGaAs/InAs Composite Channel

Hiroki Sugiyama, Takuya Hoshi, Haruki Yokoyama, and Hideaki Matsuzaki

NTT Photonics Laboratories, NTT Corporation, 3-1, Morinosato Wakamiya, Atsugi, Kanagawa243-0198, Japan

Abstract — **This paper reports the metal-organic vapor-phase epitaxy (MOVPE) growth of InP-based high electron mobility transistor (HEMT) structures with an InGaAs/InAs composite channel (CC). By optimizing the low-temperature growth conditions of the InGaAs/InAs CC, we obtained high-quality epiwafers with high electron mobility and a high-selectivity InP recess-etching-stopper layer. The mobilities exceed 18,000 cm²/Vs, which is comparable to the highest mobility ever reported for InP-based HEMT structures grown by molecular beam epitaxy (MBE). To our knowledge, this is the first report of MOVPE growth of InP-based HEMTs with an InGaAs/InAs CC.**

Index Terms — **MOVPE, HEMT, InGaAs, InAs, composite channel, mobility**

I. INTRODUCTION

The excellent high-frequency and low-noise characteristics of InP-based high electron mobility transistors (HEMTs) are beneficial for their use in millimeter-wave integrated circuits (MMICs) and have been demonstrated in various technical fields, such as broadband wireless communications [1]. The operation frequency of InP-based-HEMT ICs has recently been extended to the sub-millimeter range [2]. For the development of such high-performance HEMTs, the enhancement of channel electron mobility is essential. The use of a pseudomorphic InGaAs/InAs composite channel (CC) structure is quite effective for obtaining high mobility [3], and various ultrahigh-speed HEMTs with a InGaAs/InAs CC have been reported [2, 4, 5].

The InGaAs/InAs CC-HEMT epiwafers have been solely grown by molecular beam epitaxy (MBE) because the low growth temperature (T_g) in the MBE process is advantageous for suppressing three-dimensional growth and lattice relaxation of the highly strained InAs layer. In contrast, high T_g of typical metal-organic vapor-phase epitaxy (MOVPE) is not suitable for the growth of a highly strained InAs layer. In our previous study, we demonstrated MOVPE growth of lattice-matched (LM) InGaAs/In$_{0.8}$Ga$_{0.2}$As CC HEMT structures without lowering the conventional T_g [6]. A further increase of channel In content is strongly required in the MOVPE epiwafers. Recently, MOVPE-grown InGaAs/InAs multiple-quantum-well (MQW) lasers have been demonstrated by significantly lowering the MQW T_g [7]. However, to our knowledge, there have been no reports on the MOVPE growth of InGaAs/InAs CC HEMT structures.

In this paper, we report the MOVPE growth of InGaAs/InAs CC HEMT structures. By optimizing growth conditions, we managed to grow CC HEMT structures with room-temperature (RT) electron mobilities of over 18,000 cm²/Vs. The highest mobility reached 19,000 cm²/Vs at the sheet carrier concentration (Ns) of 2.8x10¹² cm⁻².

II. EXPERIMENT

Samples were grown on 3-inch semi-insulating InP substrates in a low-pressure vertical-flow reactor. Precursors were triethylgallium (TEG), trimethylindium (TMI), trimethylaluminum (TMA) for the group-III elements, arsine (AsH₃) and phosphine (PH₃) for the group-V elements, and disilane (Si₂H₆) for Si dopant. Figure 1 shows the layer structure of CC HEMT structures consisting of an LM InAlAs buffer, LM-InGaAs/InAs/LM-InGaAs CC, LM InAlAs spacer and barrier, InP wet-recess-etching stopper, and InGaAs/InAlAs ohmic-contact layer. Delta-doped Si was inserted between the spacer and barrier. As shown in Fig. 1, the T_g of the layers except for the InAs and upper InGaAs channel was over 600 °C and the same as that for conventional InP-based HEMT epiwafers [8].

The quality of the epitaxial layers was evaluated by X-ray diffraction (XRD), transmission electron microscopy (TEM) and atomic force microscopy (AFM). The van der Pauw

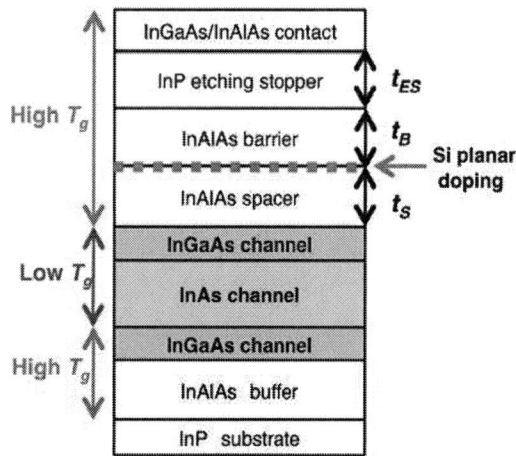

Fig. 1. Layer structure of InGaAs/InAs CC HEMT. The combinations of the layer thickness of the etching stopper, barrier, and spacer ($t_{ES}/t_B/t_S$) are 5/15/10 and 3/5/4 nm in this work.

978-1-4673-1725-2/12 $31.00 © 2012 IEEE

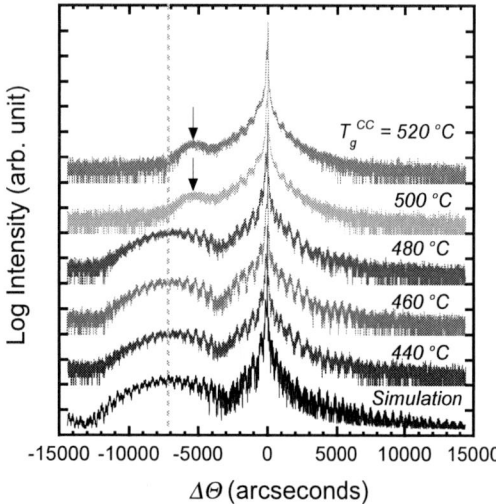

Fig. 2. Experimental and simulated (004) XRD patterns of InGaAs/InAs CC HEMT structures. The CC layers were grown at different T_g^{CC}'s.

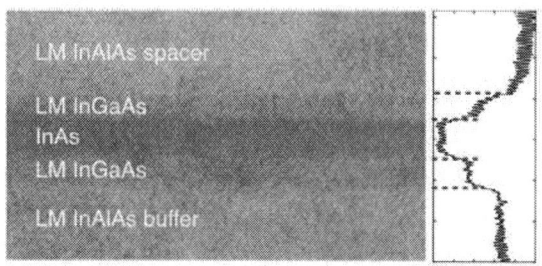

Fig. 3. Typical cross-sectional TEM image of InGaAs/ InAs CC region in HEMT structure. The average contrast profile of the image is on the right. The CC was grown at 440 ˚C.

method was used to measure Ns and mobility of channel electrons at RT. For the Ns and mobility measurements, the contact layer of samples was removed by selective wet etching before measurements [9].

III. RESULTS AND DISCUSSIONS

A. Structural characterization of InGaAs/InAs CC HEMT structures

The influence of the low CC T_g (T_g^{CC}) on epilayer quality was investigated by XRD measurements of HEMT structures. Here, we adopted a 9-nm-thick CC layer, which includes 3-nm-thick InAs [5]. Figure 2 shows examples of experimental and simulated (004) XRD patterns of CC HEMT wafers. When the CC layers were grown at 480˚C or lower, broad peaks originating from the pseudomorphic InAs channel were observed in the lower angle region, in good agreement with simulation. When the CC was grown at 500 ˚C or higher, the XRD pattern indicates the relaxation of InAs. The relaxation behavior of the InAs layer was similar to that

Fig. 4. AFM images (a)-(c) and (d)-(f) correspond to the evolution of surface morphology during InGaAs /InAs CC HEMT growth, where the CC layers were grown at 440 and 480 ˚C, respectively. (a) and (d) LM-InGaAs upper channel on 3-nm InAs before growth interruption. (b) and (e) LM-InGaAs upper channel after growth interruption. (c) and (f) The top of 3-nm-thick InP recess-etching stopper on 5-nm barrier/4-nm spacer after removing contact layer by wet etching.

reported for the MOVPE growth of InP-based InGaAs/InAs MQW lasers [7].

Figure 3 shows a typical cross-sectional TEM of the CC region in the HEMT structure grown at 440 ˚C, indicating that the 3-nm-thick InAs layer was grown coherently between LM InGaAs. We also attempted to increase InAs thickness and obtained 4-nm-thick InAs as described in the next section. Unintentional incorporation of silicon and carbon in the low-T_g grown layers was not detected by secondary ion mass spectroscopy.

The optimization of T_g^{CC} is also crucial for obtaining smooth heterointerfaces in HEMT layer structures. The AFM images in Fig. 4 summarize the evolution of surface morphology during the growth of InGaAs/InAs CC HEMT structures. The AFM images of as-grown LM InGaAs on 3-nm InAs in Figs. 4(a) and (d) correspond to the top of the upper InGaAs channel. Here, the samples were prepared by interrupting the growth sequence at the end of low-T_g CC growth. In both images, two-dimensional islands cover the sample surfaces. The smaller size of the islands in Fig. 4(a) indicates suppressed adatom migration at the lower T_g^{CC} of 440 °C. The root-mean-square roughness (*RMS*) and the maximum peak-to-valley height (R_{max}) in Fig. 4(a) are 0.13 and 1.1 nm, respectively, and smaller than those in Fig. 4(d) of 0.34 and 2.8 nm. In the growth sequence of CC HEMT structures, the low-T_g CC growth is followed by an interruption for substrate-temperature elevation and stabilization. This growth interruption causes surface smoothing. In the case of lower-T_g CC growth at 440 °C, the surface morphology changed from that in Fig. 4(a) to that in Fig. 4(b), where an almost regular monolayer (ML) step array was formed. In contrast, the surface with multiple steps and irregular terraces was observed as shown in Fig. 4(e) after the CC growth at 480 °C. The difference in surface-smoothing behavior is attributed to the initial two-dimensional-island sizes and surface roughness shown in Fig. 4(a) and (d). During the growth interruption, the smaller islands in Fig. 4(a) should decay faster and generate high-concentration adatoms, which migrate and form a regular step array [10]. In contrast, the larger islands in Fig. 4(d) should be comparatively stable and decay more slowly [11]. In this case, the adatom concentration near step edge tends to be lower than the equilibrium concentration during growth interruption, which should cause local step bunching or the formation of multiple steps [12]. The rougher surface might also enhance the formation of multiple steps. The details of the InGaAs-surface smoothing behavior will be discussed elsewhere [13].

The smooth channel/spacer heterointerface in Fig. 4(b) was inherited by InP wet-etching stopper as shown in Fig. 4(c) after the growth of the 4-nm spacer and 5-nm barrier layers. Here, the AFM image was taken after the contact layer of the HEMT structure had been removed by selective wet etching. The 3-nm-thick InP etching stopper with an almost regular ML-stepped surface provided excellent etching selectivity suitable for practical device fabrication. In contrast, the etching selectivity of the stopper layer became poor when the growth sequence followed the CC growth at comparatively higher T_g of 480 °C. The InP stopper surface showed multiple steps and hillock-like defects as shown in Fig. 4(f). The origin of hillock like defects is not clarified yet. However, such poor surface morphology is attributed to the multiple-stepped surface in Fig. 4(e). From the above results, the effect of T_g of CC on both the InAs channel itself and the quality of heterointerfaces in upper layers should be carefully investigated for the optimization of the growth sequence of CC HEMT structures.

B. Electron mobility in InGaAs/InAs CC HEMT structures

Figure 5 summarizes the RT Ns and mobility of various InP-based HEMT structures obtained in this work, our

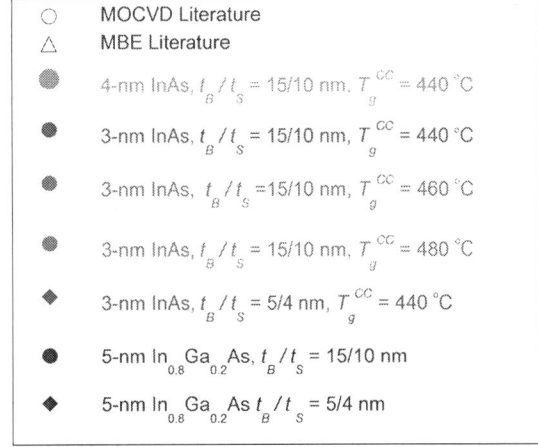

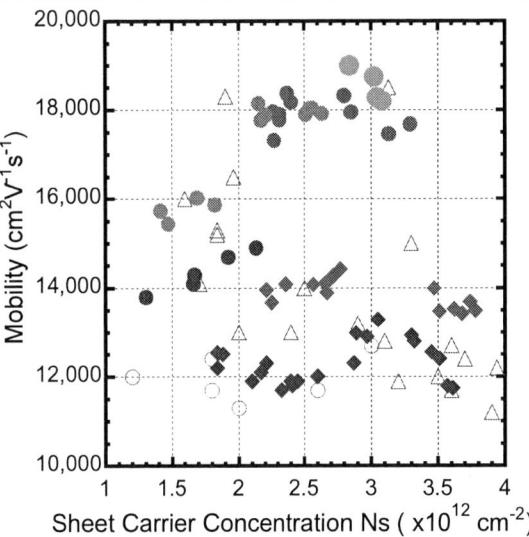

Fig. 5. Ns vs. mobility of various InP-based HEMT structures in this work and our previous work [6], compared with the literature data [14-33].

previous work [6], and in the literature [2, 14-33]. Here, the literature data includes both MBE- and MOVPE-grown InP-based HEMTs with an InAs or In-rich InGaAs channel. In our present and previous work, the InAlAs spacer thicknesses (t_s) of the samples were 10 and 4 nm for thicker (t_B = 15 nm) and thinner (t_B = 5 nm) barrier samples, respectively. The 10-nm-spacer structure was used to evaluate the effectiveness of InAs CC and demonstrate high mobility. The availability of our InGaAs/InAs CC for ultrahigh-speed transistors was also examined in the 4-nm-spacer structures. Compared with the previous data of InGaAs/In$_{0.8}$Ga$_{0.2}$As CC HEMT structures [6], the mobility is effectively enhanced by using the InGaAs/InAs CC in both 10 and 4-nm spacer structures. As shown in Fig. 5, high mobilities of over 18,000 cm²/Vs were obtained in the 10-nm-spacer structures at the Ns of around 2.2 to 3×10¹² cm⁻². The highest mobility reached 19,000 cm²/Vs at the Ns of

2.8×10^{12} cm^{-2} in the 4-nm-thick InAs channel. Compared with the record of 18,500 cm^2/Vs reported for MBE-grown InP-based HEMT structures [25], these mobilities are almost the same or even higher. The mobility in 10-nm-spacer structures in Fig. 5 does not seem to depend on T_g^{CC}. However, recent low-temperature (LT) measurements revealed that the T_g^{CC} dependence of LT mobility seems to reflect the anisotropy of interface roughness shown in Fig. 4. The details of LT measurements will be described elsewhere [13]. The samples with a 4-nm-spacer structure also exhibited high mobilities of around 14,000 cm^2/Vs at the Ns of around 2.2 to 3.5×10^{12} cm^{-2}. These results indicate that our MOVPE-grown InGaAs/InAs CC is favorable for the fabrication of ultrahigh-speed HEMT ICs and for boosting their operation speed toward the terahertz range.

IV. CONCLUSION

InP-based HEMT structures with InGaAs/InAs were successfully grown by MOVPE. The optimization of LT-growth conditions of InGaAs/InAs CC layers was critical for obtaining high-quality HEMT structures with smooth and abrupt heterointerfaces. The samples exhibited high electron mobilities of over 18,000 cm^2/Vs, which are some of the highest ever reported in MOVPE-grown InP-based HEMT structures.

ACKNOWLEDGEMENT

The authors thank Minoru Ida, Tomoyuki Akeyoshi, and Takatomo Enoki for their support and encouragement throughout this work. This work was partly supported by the R&D program on "multi-tens-gigabit wireless communication technology at subterahertz frequencies" of the Ministry of Internal Affairs and Communications, Japan.

REFERENCES

[1] T. Kosugi, A. Hirata, T. Nagatsuma, and Y. Kado, *IEEE Microwave Magazine*, vol. 10, no. 2, pp. 68-76, April 2009.

[2] W. Deal, X. B. Mei, K. M. K. H. Leong, V. Radisic, S. Sarkozy, and R. Lai, *IEEE Trans. Terahertz Sci. Technol.*, vol. 1, no. 1, pp. 25-32, September 2011.

[3] T. Akazaki, K. Arai, T. Enoki, and Y. Ishii, *IEEE Electron Device Lett.*, vol. 13, no. 6, pp. 325-327, June 1992.

[4] D. -H. Kim and J. A. del Alamo, *IEEE Electron Device Lett.*, vol. 31, no. 8, pp. 806-808, August 2010.

[5] H. Matsuzaki, T. Maruyama, T. Kosugi, H. Takahashi, M. Tokumitsu, and T. Enoki, *IEEE Trans. Electron Devices*, vol. 54, no. 3, pp. 378-384, March 2007.

[6] H. Sugiyama, H. Matsuzaki, H. Yokoyama, and T. Enoki, *Proc. 2010 Int. Conf. Indium Phosphide and Related Materials*, pp. 477-480, May/June 2010.

[7] T. Sato, M. Mitsuhara, K. Kakitsuka, T. Fujisawa, and Y. Kondo, *IEEE J. Sel. Top. Quantum Electron.*, vol. 14, no. 4, pp. 992-997, July/August 2008.

[8] H. Yokoyama, H. Sugiyama, Y. Oda, K. Watanabe, and T. Kobayashi, *Jpn. J. Appl. Phys.* vol. 42, pp. 4909-4912, 2003.

[9] K. Watanabe and H. Yokoyama, *Appl. Phys. Lett.*, vol. 76, pp. 973-975, 2000.

[10] V. L. Alperovich, I. O. Akhundov, N. S. Rudaya, D. V. Sheglov, E. E. Rodyakina, A. V. Latyshev, and A. S. Terekhov, *Appl. Phys. Lett.*, vol. 94, p. 101908, 2009.

[11] H. Hibino, C. –W. Hu, T. Ogino, and I. S. T. Tsong, *Phys. Rev. B.* vol. 63, p. 245402, 2001.

[12] M. Shinohara and N. Inoue, *Appl. Phys. Lett.*, vol. 66, pp. 1936-1938, 1995.

[13] H. Sugiyama, H. Irie, T. Hoshi, H. Yokoyama, and H. Matsuzaki, to be submitted.

[14] M. Kamada, T. Kobayashi, H. Ishikawa, Y. Mori, K. Kaneko, and C. Kojima, *Electron. Lett.,* vol. 23, no. 6, pp. 297-298, March 1987.

[15] Y. Mori and M. Kamada, *J. Crystal Growth*, vol. 93, pp. 892-899, 1988.

[16] A. Chin and T. Y. Chang, *J. Vac. Sci. Technol. B.,* vol. 8(2), pp. 364-366, March/April 1990.

[17] Y. Sugiyama, Y. Takeuchi, and M. Tacano, *J. Crystal Growth*, vol. 115, pp. 509-514, 1991.

[18] A. S. Brown, L. D. Nguyen, R. A. Metzger, A. E. Schmitz, and J. A. Henige, *J. Vac. Sci. Technol. B.,* vol. 10(2), pp. 1017-1019, March/April 1992.

[19] L. D. Nguyen, A. S. Brown, M. A. Thompson, and L. M. Jelloian, *IEEE Trans. Electron Devices,* vol. 39, no. 9, pp. 2007-2014, September 1992.

[20] K. B. Chough, C. Caneau, W-P. Hong, and J. –I. Song, *IEEE Electron Device Lett.*, vol. 15, no. 1, pp. 33-35, January 1994.

[21] M. Wojtowicz, R. Lai, D. C. Streit, G. I. Ng, T. R. Block, K. L. Tan, P. H. Liu, A. K. Freudenthal, and R. M. Dia, *IEEE Electron Device Lett.*, vol. 15, no. 11, pp. 477-479, November 1994.

[22] T. Akazaki, J. Nitta, H. Takayanagi, T. Enoki, and K. Arai, *Appl. Phys. Lett.*, vol. 65, pp. 1263-1265, 1994.

[23] T. Nakayama, H. Miyamoto, E. Oishi, and N. Samoto, *J. Electronic Materials*, vol. 25, no. 4, pp. 555-558, 1996.

[24] D. Xu, H. G. Heiß, S. A. Kraus, M. Sexl, G. Böhm, G. Tränkle, G. Weimann, and G. Abstreiter, *IEEE Trans. Electron Devices*, vol. 45, no. 1, pp. 21-30, January 1998.

[25] T. Nakayama and H. Miyamoto, *J. Crystal Growth*, Vol. 201/202, pp. 782-785, 1999.

[26] X. Letartre, P. Rojo-Romeo, J, Tardy, M. Bejar, M. Gendry, M. A. Py, M. Beck, H. J. Bühlmann, L. Ren, C. Villar, A. Sanz-Hervas, J. J. Serrano, J. M. Blanco, M. Aguilar, O. Marty, V. Souliére, and Y. Monteil, *Jpn, J. Appl. Phys.*, vol. 38, no. 2B, pp. 1169-1173 February 1999.

[27] D. Xu, J. Osaka, Y. Umeda, T. Suemitsu, Y. Yamane, and Y. Ishii, *IEEE Electron Device Lett.*, vol. 20, no. 3, pp. 109-112, March 1999.

[28] W. Z. Cai, Z. M. Wang, and D. L. Miller, *J. Vac. Sci. Technol. B.,* vol. 18(3), pp. 1633-1637, May/June 2000.

[29] M. S. Goorsky, R. Sandhu, R. Hsing, M. Naidenkova, M. Wojtowicz, T. P. Chin, T. R. Block, and D. C. Streit, *J. Vac. Sci. Technol. B.,* vol. 18(3), pp. 1658-1662, May/June 2000.

[30] T. Tanaka, K. Tokudome, and Y. Miyamoto, *Jpn, J. Appl. Phys.*, vol. 42, no. 8B, pp. L993-L995, August 2003.

[31] Y.-W. Chen, W.-C. Hsu, R.-T. Hsu, Y.–H. Wu, and Y.-J. Chen, *Solid-State Electron.*, vol. 48, pp. 119-124, 2003.

[32] D. -H. Kim, J. A. del Alamo, J.-H. Lee, and K.-S. Seo, *IEEE Trans. Electron Devices,* vol. 54, no. 10, pp. 2606-2613, October 2007.

[33] D. -H. Kim and J. A. del Alamo, *IEEE Electron Device Lett.*, vol. 29, no. 8, pp. 830-833, August 2008.

Selective-area growth InP-based nanowires and their optical properties

Junichi Motohisa

Graduate School of Information Science and Technology, Hokkaido University,
North 14, West 9, Sapporo 060-0814, Email: motohisa@ist.hokudai.ac.jp

Abstract—I describe our study on the growth of InP-based nanowires (NW) by selective-area metal-organic vapor-phase epitaxy (SA-MOVPE) and their optical properties. Vertically-aligned InP NWs were grown on InP (111)A oriented substrates, and they exhibited transition of crystal structures depending on the growth conditions. Laterally or vertically heterostructured NWs were also grown by SA-MOVPE based on InP NWs, and their optical properties were studied by using μ-PL measurement. InP NWs with a vertical pn-junction were also grown and was applied to NW light-emitting diodes (LEDs).

I. INTRODUCTION

Semiconductor nanowires (NWs) are attracting great deal of interest as new building blocks for nano-photonic and nano-electronic devices. Among them, InP-based NWs are especially promising for the application of photonic devices as well as electron devices, because they have a direct band gap compatible with optical fiber telecommunication bands. To date, field-effect transistors [1], photodetectors [2], light-emitting devices [1], [3], [4], and solar cells [5], [6] have been demonstrated using InP NWs. Various types of heterostructures, such as quantum dots (QDs) [3], [7], [8] or core-shell structures [9] have also been realized on the basis of InP NWs.

For the growth of NWs, the most commonly used approach utilizes vapor-liquid-solid (VLS) mechanism [10], [11], in which nanometer-sized "whiskers" are grown with a help of metal catalyst (typically gold). This method is simple, but has turned out to be very effective to realize various kinds of NWs and NW-based devices. However, one of the disadvantages is incorporation of metal catalysts, which could form deep levels and become non-radiative recombination centers. The other problem is their randomness in nature, that is, the position of NWs are determined by the position of metal catalysts, which are in most case random. To tackle the latter problem, attempts to define the pattern for metal catalysts are reported [12]. We have been reporting on an alternative method for the growth of III-V semiconductor NWs, that is, selective-area metalorganic vapor-phase epitaxy (SA-MOVPE) [13]. In this method, the substrate for the growth is partially covered with SiO$_2$ mask having an array of circular openings. By appropriately selecting the substrate orientation as well as growth conditions for MOVPE growth, vertically-aligned NW are grown selectively at the mask openings. This method does not rely on metal catalysts, and NWs are grown on the predetermined position of mask openings. Furthermore, it is possible to form III-V NWs directly on Si substrates.

In this report, we focus on the growth of InP-based NWs on InP substrates. We describe the optical properties of InP NWs with and without heterostructures, and application to light-emitting diodes.

II. GROWTH OF InP NANOWIRES

InP NWs were grown by SA-MOVPE on InP substrates partially covered with SiO$_2$, using trimethylindium (TMIn) and tertiarybutylphosphine (TBP) for source materials. Vertically-aligned InP NWs with hexagonal cross sections were grown on InP (111)A-oriented substrates, while SA-MOVPE of InP on (111)B substrates resulted in the formation of tri-pod like structures [14], [15]. Based on these results, we conclude that preferential growth direction of InP NWs is in the [111]A direction. Furthermore, we have found the crystal structure exhibits transition from wurtzite (WZ) to zincblende (ZB) depending on the growth conditions [16], [17]. Figure 1 summarizes the results of SA-MOVPE carried out different growth temperature T_G and V/III ratio of material supply. The NWs with negligible tapering were obtained at two distinctive growth conditions, that is lower V/III ratio and higher T_G (condition-B), and higher V/III ratio and lower T_G (condition-C). It is noted that the orientation of the hexagon is different between two types of NWs, and the cross sectional size is lager for NWs grown in condition-C. Transmission electron microscopy (TEM, Fig. 2) study revealed that the crystal structure of the NWs grown under condition-B and C was WZ and ZB, respectively. Furthermore, at intermediate growth conditions (higher T_G and V/III, condition-D), tapered NWs were formed and contained considerable mixing of ZB and WZ crystal structures [17].

The growth-condition dependent transition from ZB to WZ in InP was explained by the converge of P at the topmost surface of InP (111)A [16]. More specifically, at high temperature and low V/III, P-coverage is low and it favors WZ stacking in order to assist binding of adsorbed In atoms. When the temperature is lowered and V/III is raised, the surface has more P and it makes the energetic preference smaller between ZB and WZ stacking, resulting in the mixing of crystal structures. Degree of the tapering is quantitatively explained by the degree of mixing of two crystal phase [17]. In theoretical point of view, InP has the largest ionicity among conventional (non-nitride based) III-V semiconductors, and this favors WZ stacking more for smaller NW diameter [18]. In addition, formation energy of nuclei with WZ stacking is

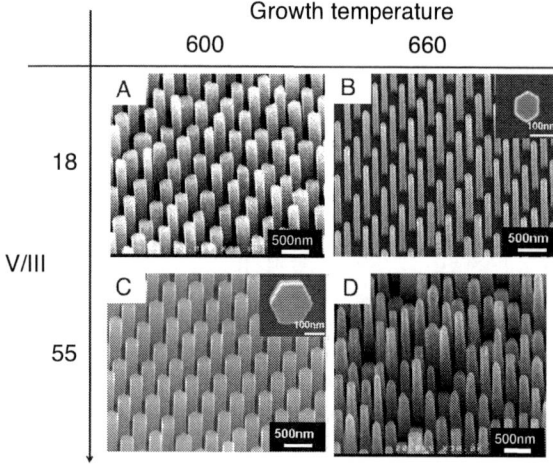

Fig. 1. Results of SA-MOVPE of InP on InP (111)A substrates under various growth temperature and V/III ratio.

the smallest at the edge of NWs [19], [20]. Furthermore, recent *ab-initio* calculation also indicates that formation of WZ phase is favored for high temperature and large V/III ratio conditions [21], which is fairly consistent with our experimental results.

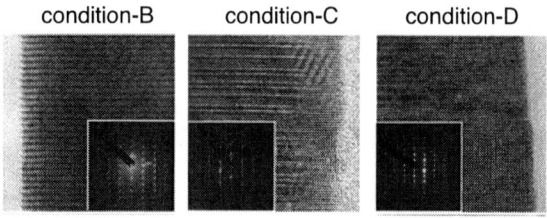

Fig. 2. Cross-sectional TEM image of InP NW grown under growth condition B, C, and D.

III. OPTICAL PROPERTIES OF InP NANOWIRES

We studied the optical properties of InP NW by μ-PL measurement at low temperature, and it was found that the transition energy of WZ-InP NW was located around 1.49~1.50 eV [22] at $T = 4$ K, and was larger than that of ZB-InP. This indicates that WZ-InP has larger band gap energy than ZB-InP, and temperature dependent PL study suggested that the it was lager by about 88 meV. This value is consistent with the theoretically estimated ones (~ 84 meV) [23]. We also compared the PL of WZ, ZB, and mixed NWs and the results are shown in Fig. 3. As we have just mentioned, the PL peak energy is higher in WZ-InP NWs than in ZB-InP NWs. Specifically, the mixed NWs exhibit a PL peak between these two type of NWs. This also can be explained by the mixing of crystal structures and formation of type-II quantum wells as follows. As predicted in [23], band alignment of ZB and WZ InP forms a staggered type-II band alignment along with the 84-meV change in bandgap which we have just mentioned. The electrons are strongly confined in the

ZB regions with 129-meV higher WZ barriers. We estimated that the average thickness of ZB segment to be 1.7 nm TEM observation. Transition between electrons confined in the ZB segment and less confined holes in WZ segment gives the transition energy of of about 1.48-1.49eV, consistent with the experimental observation.

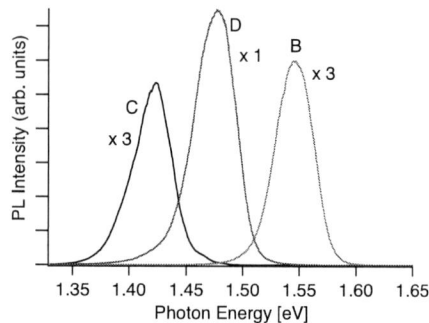

Fig. 3. PL spectra of InP NWs grown under conditions B, C, and D.

We also carried out μ-PL measurement of single NWs. For this, grown InP NWs were mechanically detached from the substrate, and were dispersed onto the SiO_2-coated Si substrate. Two types of room-temperature PL spectra are shown in Fig.4(a). In type-A NWs, the edges were tapered and showed normal PL. On the other hand, we observed a multiple peaks in NWs with sharp edges. These peaks are originated from Fabry-Perot resonance in NWs, because the peaks and their spacings $\Delta\lambda$ can be explained by $\Delta\lambda = \lambda^2/2Ln^*$, where L is the NW length and n^* is the group index given by $n^* = n - dn/\lambda/d$, and n is the refractive index of InP (see. Fig.4(b)). In addition, emission at the edge of NW (see inset of Fig. 4(b)) strongly shows the existence of waveguiding effect in type-B NWs.

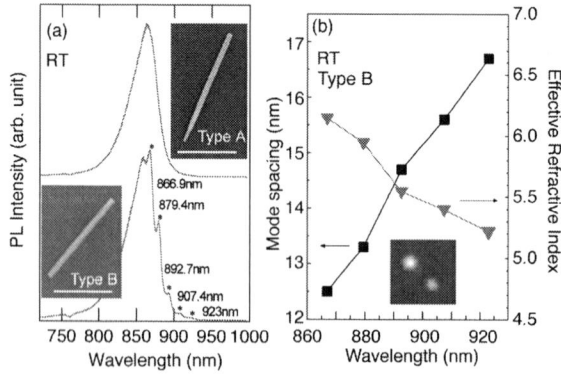

Fig. 4. (a) Room-temperature PL spectra of two types of InP NWs, whose SEM images are respectively shown in the inset. Scale bar, $3\mu m$. (b) Plot of the peak spacing and effective refractive index obtained from the results of type B NW. Inset shows an emission image.

IV. HETEROSTRUCTURED NANOWIRES

We have reported on two types of heterostructured InP-based NWs and their optical properties. The first one is core-multishell (CMS) NWs [9] shown in Fig. 5(a). Here, we first grew InP core NWs close to the condition-B. Then we grew InAs at 400°C, and InP was successively grown with the growth condition close to C. Thin layers of InAs are formed around the core InP, and are sandwiched by InP outer shell. This kind of core-multishell structure is possible since the growth rate in the vertical and lateral direction can be controlled by the growth condition, which is explained in the previous section. Figure 5(b) shows a result of a μ-PL measurement, exhibiting multiple peaks originating from InAs layer. This result indicates that InAs quantum wells, whose thickness is order of one or few monolayers and have atomically flat heterointerfaces, are formed on the sidewalls of InP NWs. More detailed investigation on the optical properties of this CMS-NWs have been carried out in collaboration with Prof. Masumoto, and are reported in [24], [25]

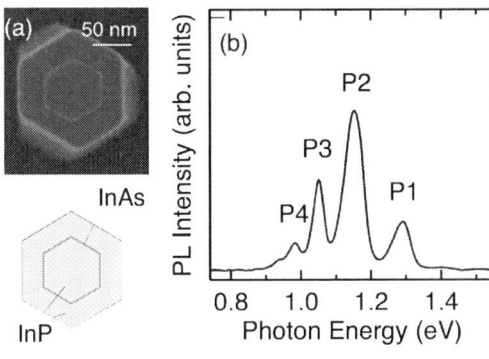

Fig. 5. (a) Cross-sectional SEM image of an InP-based core-multishell NW and its schematic illustration. (b) PL spectrum of core-multishell NWs.

Next, we show the results of the growth of InP/InAsP/InP heterostructures in Fig. 6(a) These NWs exhibit larger diameter at the top, and it is observed after the growth of second InP. For this reason, we think the heterostructure shown in Fig.6(b) is formed in these NWs. Figure 6 (c) shows PL spectra from a single NW and its dependence on excitation intensity. We got very sharp emissions of X and XX at 1.227 eV and 1.232 eV, respectively, The full width of half maximum of X was $400\mu eV$ at the excitation power density of 1 W/cm^2, as shown in Fig. 6(c). Figure 6 (d) represents excitation power dependence of X and XX. The peak X increases linearly as excitation power density, and XX shows squared dependence. Therefore, we conclude X and XX are emissions from exciton and biexcitons in a single InAsP QD embedded in NWs, respectively. It is noted that binding energy of biexciton is negative ($E_X - E_{XX} = -4.5$meV), This is presumably due to the strong Coulomb interaction between the holes of biexciton. Such negative binding of biexcitons are reported in Stranski-Krastanow InAs QDs, and indication of the formation of very small QDs [26]. Nature of QDs in NWs have also

been manifested by interference spectroscopy and by the single photon emission [27]–[29].

interaction between the holes in the QDs.

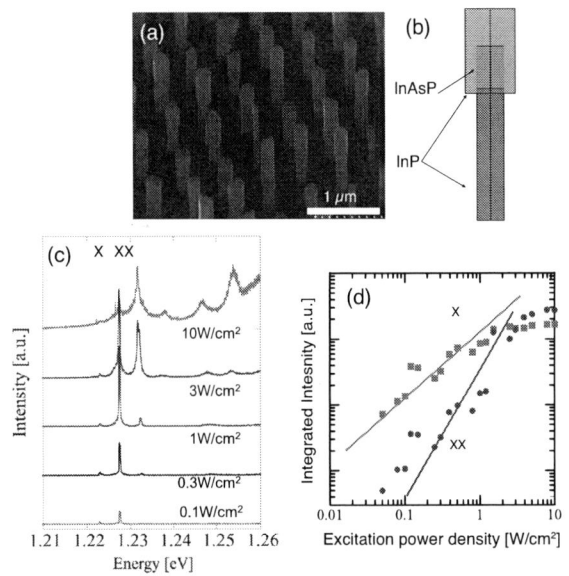

Fig. 6. (a) SEM image of InP/InAsP/InP NWs and (b) their schematic cross-sectional illustration. (c) PL spectra of the NW and their excitation intensity dependence. (d) Excitation power dependence of integrated peak intensity for X and XX.

V. NANOWIRE LIGHT-EMITTING DIODE

Finally, we describe the fabrication of LEDs using InP NWs. For this, we first grew InP NWs with axial pn junction by SA-MOVPE on p-type InP (111)A substrates. SiH$_4$ and diethylzinc (DEZn) were used for n- and p-type dopant sources, respectively. After the growth, the space between NWs was filled with polymer resin (benzocyclobutene, BCB) as a transparent electrical insulator by spin coating. The overlaid excess resin was removed by reactive ion etching (RIE) using CF$_4$ and O$_2$ to expose the top portion of NWs. Next, a transparent indium tin oxide (ITO) film (thickness 300 mm) electrode was first sputtered onto a NW array and patterned using photolithography and wet etching by HCl Finally, the backside electrode on the substrate was formed by 200-mm-thick Au-Zn. The whole sample was annealed in N$_2$ ambient for 15 min at 400 °C. A cross sectional SEM image of NW-LED is shown in Fig.7(a)

Figure 7(b) shows a typical result of the current-light output (I-L) and current-voltage (I-V) characteristics of the fabricated NW-LEDs. We can see fairly linear I-L characteristics and clear rectifying behavior with very small reverse bias leakage in I–V characteristics. The turn-on voltage was about 1.4 V, which is consistent with the band gap of WZ-InP. Reasonably uniform EL from 100 μm^2 regions, where the NW array was defined, was also confirmed, as shown in the inset of Fig. 7(c). Figure 7(c) shows EL spectra of the same LED up to a bias voltage of 3.5 V. The EL peak was located at

857 mm (~1.45 eV), slightly longer than the peak wavelength of the PL. The peak position shows no shift up to $I \sim 20$ mA. EL and fairly linear I-L characteristics were also confirmed in many of the fabricated devices.

By designing the pitch of the NW array, emission from individual NWs was confirmed. As we already described in the previous section, InP NWs containing single QDs are feasible, and it is expected that the emission at the optical fiber telecommunication bands are possible by controlling the alloy composition of InAsP. Therefore, Thus, our results are promising for realizing current-driven single-photon sources operating in optical fiber telecommunication bands using InP-based NWs.

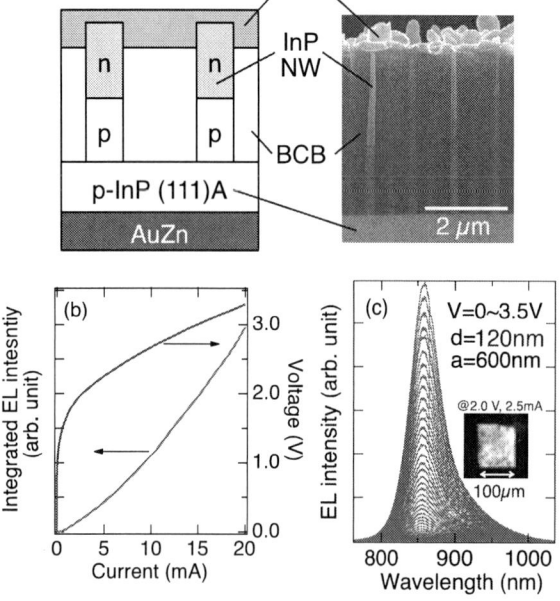

Fig. 7. (a) Schematic and SEM image of the cross section of an InP-NW-based LED. (b) Current-light output (I-L, red line) and current-voltage (I-V, blue line) characteristics of a typical NW-LED. (c) Emission spectra of LEDs measured up to a bias voltage of 3.5 V. Inset shows an emission image of a NW-LED at bias voltage of 2 V.

ACKNOWLEDGMENT

This work is done in collaboration with the group of Professor Takashi Fukui, Professor Kenji Hiruma, Professor Shinjiro Hara, Professor Masumoto, Professor Nomuara, Professor Ikuo Suemune, Dr. Val Zwiller. We thank Dr. Sasakura, Dr. Premila Mohan, Dr. Katsuhiro Tomioka, Dr. Ying Ding, Dr. M. Van Kouwen, Dr. Hajime Goto, Keitaro Ikejiri, Yasunori Kobayashi, Yusuke Kitauchi, Sandars Dorenbos, and Masatoshi Yoshimura and Yoshinori Kohashi for their experimental support. The work is partly financially supported by a Grant-in-Aid for Scientific Research, supported by Ministry of Education, Culture, Sports, Science and Technology, Japan.

REFERENCES

[1] X. Duan, Y. Huang, Y. Cui, J. Wang, and C. M. Lieber, *Nature*, vol. 409, pp. 66–69, 2001.

[2] J. Wang, M. S. Gudiksen, X. Duan, Y. Cui, and C. M. Lieber, *Science*, vol. 293, pp. 1455–1457, 2001.

[3] E. E. Minot, F. Kelkensberg, M. van Kouwen, A. J. van Dam, L. P. Kouwenhoven, V. Zwiller, M. T. Borgström, O. Wunnicke, M. A. Verheijen, and E. P. A. M. Bakkers, *Nano Lett.*, vol. 7, pp. 367–371, 2007.

[4] S. Maeda, K. Tomioka, S. Hara, and J. Motohisa, *Jpn. J. Appl. Phys.*, vol. 51, no. 2, p. 02BN03, Feb 2012.

[5] H. Goto, K. Nosaki, K. Tomioka, S. Hara, K. Hiruma, J. Motohisa, and T. Fukui, *Appl. Phys. Express*, vol. 2, no. 3, p. 035004, Feb 2009.

[6] M. Heurlin, P. Wickert, S. Fält, M. Borgström, K. Deppert, L. Samuelson, and M. Magnusson, *Nano Lett.*, vol. 11, no. 5, pp. 2028–2031, 2011.

[7] V. Zwiller, N. Akopian, M. van Weert, M. van Kouwen, U. Perinetti, L. P. Kouwenhoven, R. Algra, J. G. Rivas, E. Bakkers, G. Patriarche, L. Liu, J.-C. Harmand, Y. Kobayashi, and J. Motohisa, *C. R. Physique*, vol. 9, pp. 804–815, Jan 2008.

[8] Y. Kobayashi, J. Motohisa, K. Tomioka, S. Hara, K. Hiruma, and T. Fukui, *Proceedings of 2010 International Conference on Indium Phosphide & Related Materials (IPRM)*, June 2010.

[9] P. Mohan, J. Motohisa, and T. Fukui, *Appl. Phys. Lett.*, vol. 88, p. 133105, 2006.

[10] R. S. Wagner and W. C. Ellis, *Appl. Phys. Lett.*, vol. 4, no. 5, pp. 89–90, 1964.

[11] K. Hiruma, M. Yazawa, T. Katsuyama, K. Ogawa, K. Haraguchi, M. Koguchi, and H. Kakibayashi, *J. Appl. Phys.*, vol. 77, no. 2, pp. 447–462, Apr 1995.

[12] T. Mårtensson, P. Carlberg, M. Borgström, L. Montelius, W. Seifert, and L. Samuelson, *Nano Lett.*, vol. 4, no. 4, pp. 699–702, 2004.

[13] J. Motohisa, J. Noborisaka, J. Takeda, M. Inari, and T. Fukui, *J. Cryst. Growth*, vol. 272, no. 1-4, pp. 180–185, 2004.

[14] M. Inari, J. Takeda, J. Motohisa, and T. Fukui, *Physica E*, vol. 21, no. 2-4, pp. 620–624, 2004.

[15] P. Mohan, J. Motohisa, and T. Fukui, *Nanotechnology*, vol. 16, no. 12, p. 2903, 2005.

[16] Y. Kitauchi, Y. Kobayashi, K. Tomioka, S. Hara, K. Hiruma, T. Fukui, and J. Motohisa, *Nano Lett.*, vol. 10, no. 5, pp. 1699–1703, 2010.

[17] K. Ikejiri, Y. Kitauchi, K. Tomioka, J. Motohisa, and T. Fukui, and wurtzite crystal phase mixing and transition in indium phosphide nanowires," *Nano Lett.*, vol. 11, no. 10, pp. 4314–4318; *Nano Lett.*, vol. 12, no. 1, pp. 524–525, 2012.

[18] T. Akiyama, K. Sano, K. Nakamura, and T. Ito, *Jpn. J. Appl. Phys.*, vol. 45, no. 8/11, p. 275, 2006.

[19] F. Glas, J.-C. Harmand, and G. Patriarche, *Phys. Rev. Lett.*, vol. 99, no. 14, 146101, 2007.

[20] T. Yamashita, T. Akiyama, K. Nakamura, and T. Ito, *Jpn. J. Appl. Phys.*, vol. 50, no. 5, 055001, May 2011.

[21] T. Yamashita, "Theorical study on the shape and the crystal structure of inp nanowires (in japanese)," *PhD Thesis (Mie University)*, March 2012.

[22] Y. Kobayashi, M. Fukui, J. Motohisa, and T. Fukui, *Physica E*, vol. 40, no. 6, pp. 2204–2206, 2008.

[23] M. Murayama and T. Nakayama, *Phys. Rev.*, vol. B 49, pp. 4710–4724, 1994.

[24] B. Pal, K. Goto, Y. Masumoto, P. Mohan, J. Motohisa, and T. Fukui, *Appl. Phys. Lett.*, vol. 93, no. 7, 073105, 2008.

[25] Y. Masumoto, Y. Hirata, P. Mohan, J. Motohisa, and T. Fukui, *Applied Physics Letters*, vol. 98, no. 21, 211902, 2012.

[26] D. Sarkar, H. van der Meulen, J. Calleja, J. M. Becker, R. J. Haug, and K. Pierz, *J. Appl. Phys.*, vol. 100, 023109, Jan 2006.

[27] H. Sasakura, H. Kumano, I. Suemune, J. Motohisa, Y. Kobayashi, M. Kouwen, K. Tomioka, T. Fukui, N. Akopian, and V. Zwiller, *J. Phys.: Conf. Series*, vol. 193, p. 012132, 2009.

[28] S. Dorenbos, H. Sasakura, M. van Kouwen, N. Akopian, S. Adachi, N. Namekata, M. Jo, J. Motohisa, Y. Kobayashi, and K. Tomioka, *Appl. Phys. Lett.*, vol. 97, p. 171106, 2010.

[29] H. Sasakura, C. Hermannstädter, S. Dorenbos, N. Akopian, M. van Kouwen, J. Motohisa, Y. Kobayashi, H. Kumano, K. Kondo, and K. Tomioka, *Phys. Rev. B*, vol. 85, no. 7, p. 075324, 2012.

Single GaAs nanowire photovoltaic devices under very high power illumination

A. Lysov, C. Gutsche, W. Prost, F.-J. Tegude

Center for Nanointegration, Solid-State Electronics Dep., University of Duisburg-Essen, Duisburg, Germany, werner.prost@uni-due.de

Abstract — GaAs nanowire pn-diodes were used for single nanowire radial and axial photovoltaic device fabrication. In this study the performance under very high illumination power is measured and analyzed in terms open circuit voltage, and efficiency. It is found that GaAs nanowire photovoltaic devices can be operated under very high power of up to 120 W/cm² (λ = 532 nm) without detectable degradation. The short circuit current increases directly proportional to the illumination power while the open circuit voltage steadily increases up to the highest illumination. This study shows that both axial and radial nanowire pn-junctions are suitable for sun light conversion with very high concentration ratio (> 1,000).

Index Terms — nanowire, solar cells, high power illumination..

I. INTRODUCTION

The enhanced light collection efficiency as well as the highly reduced material consumption with respect to conventional thin film devices makes nanowires quite attractive for the development of photovoltaic devices [1-3]. For nanoscaled light absorption a high absorption coefficient is important that is provided by direct band gap III-V compound semiconductors. The high material costs along with efforts in nanoscaled patterning assume that concentrator cells with very high illumination power are recommended. In this study the performance of such an arrangement is studied for single nanowire photovoltaic elements.

II. EXPERIMENTAL

Both, axial [4] and radial [5] pn-junctions are grown by vapor-liquid-solid MOVPE growth on (111)B GaAs substrates. The axial doping modulation was realized by subsequent switching of doping precursors without any growth interruptions. A p-doping with mean hole concentration of 2×10^{19} cm^{-3} was achieved with zinc. The n-part of the diode was doped with tin and had a mean electron concentration of 1×10^{18} cm^{-3}. The carrier concentrations were determined by a mobility vs. carrier concentration model, which is discussed in [6] in more detail.

Radial pn-diodes were formed from an n-GaAs core surrounded by an intermediate lattice matched GaInP and an outer p-GaAs shell. An intermediate 30 nm thick nominally undoped GaInP shell is inserted to allow selective wet etching of the outer p-GaAs [5].

The as-grown structures were transferred to special pre-patterned carriers and finally contacted by electron beam lithography and lift-off technique. No passivation layer or dielectric coverage for enhanced photo absorption is applied.

Power dependent photocurrent measurements were performed at room temperature, steady state, and without any cooling. High power optical excitation was achieved by a homogenous laser illumination with λ= 532 nm focused by a microscope objective to a spot of 20 µm diameter. The conversion efficiency was determined under standard AM 1.5 G conditions, provided by a solar simulator operating with a 100 mWcm^{-2} (1 sun) illumination intensity.

III. PHOTOVOLTAIC ANALYSIS

A. Dark current analysis

Figure 1 shows SEM micrographs of the contacted axial (a) and radial (b) nanowire pn-junction diodes. The n-type part of the nanowire is on the right hand side. In case of the radial device the p-shell is selectively etched prior to contact formation. The p-contact for both devices is on the left hand side and is provided by Ti/Au metallization that do not require alloying.

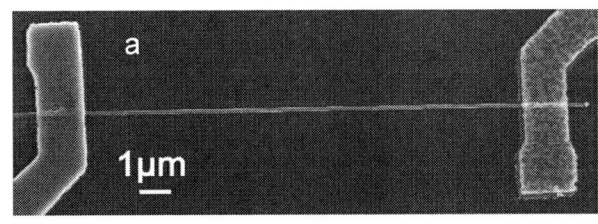
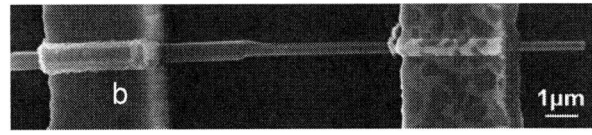

Fig. 1. SEM micrographs of (a) n-GaAs/p-GaAs axial pn junction and (b) n-GaAs/GaInP/p-GaAs radial pn-junction.

Measured and simulated I-V curves of both axial and radial nanowire pn-diode in forward direction are presented in Fig. 2. The measured data represent a typical pn-diode I-V characteristic. At low bias ($V_{D,ext}$ < 0.8 V) the current log(I_D)

978-1-4673-1725-2/12 $31.00 © 2012 IEEE

linearly increases proving for the basic set-up of the device model the applicability of the Shockley equation:

$$I_D = I_0 \cdot (e^{\frac{V_D}{n \cdot V_T}} - 1) \tag{1}$$

with the reverse current I_0, the ideality factor n, and the thermal voltage V_T. At high currents ($I_D > 0.1$ µA) the series resistance R_S results in additional voltage drop $I_D R_S$.

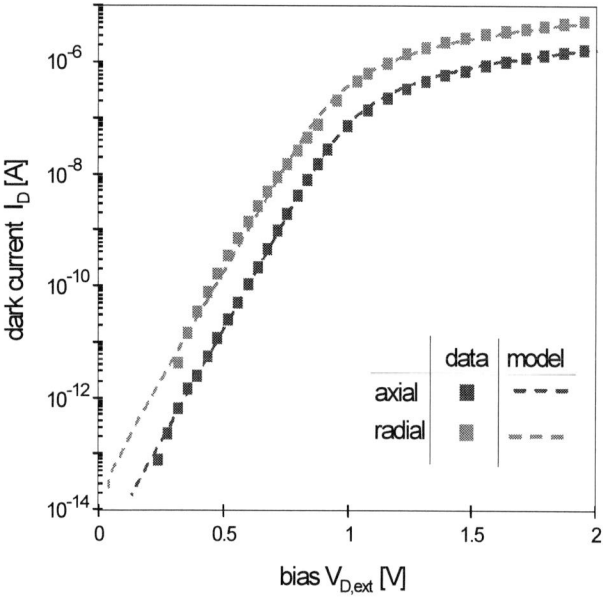

Fig. 2. Dark I-V current analysis of axial and radial nanowire diodes in forward direction at ambient temperature. Squares represent measured data, and the broken lines fitted data using the model parameters in table I.

TABLE I
DARK I-V MODEL DATA

model parameter	I_0 [fA]	J_0 [µA/cm²]	n	R_S [kΩ]
axial	1.8	51.6	2.11	520
radial	26.0	0.87	2.12	180

Using the Shockley equation the nanowire pn-diode I-V characteristic may be described with three model parameter such as reverse current I_0, the ideality factor n, and the additional series resistance R_S. Based on the extracted model parameters in table I an excellent agreement with the measured I-V characteristic is found (cf. Fig. 2) for both axial and radial devices.

The series resistance R_S in axial nanowire devices is dominated by the p-n transition region, only. Due to the Au-seeded VLS growth mode the dissolved dopant atoms in the Au particle cause a long transition from p-GaAs:Zn to n-GaAs:Sn with strong compensation effects. The long transition area is providing a separating electric field that might be favorable for the photovoltaic performance but increases the series to $R_S = 520$ kΩ in this case.

The determination of the total series resistance of the radial device (cf. Fig. 1b) sums up the contribution from the outer p-shell, the inner n-core, and the additional n-contact part. The carrier concentration in the p-GaAs:C shell has been determined to $N_A = 1 \cdot 10^{18}$ cm^{-3}. The contribution of the distributed series resistance of the shell can be estimated as follows:

$$R_{S,p} = \frac{1}{q N_A \mu_A} \cdot \frac{L/3}{\left[\pi \left(r_0 - d_{spc,p} \right)^2 - \pi \left(r_{Au} + d_i + d_p \right)^2 \right]} \tag{2}$$

with shell length $L = 8$ µm, shell outer radius $r_0 = 150$ nm, n-core radius $r_{Au} = 50$ nm, space charge length at the shell surface $d_{spc,p} = 26$ nm, and space charge length towards the core $d_{c,p} = 32$ nm (valid for $N_D = 1 \cdot 10^{18}$ cm^{-3}), and the thickness of the undoped GaInP shell $d_i = 20$ nm. The hole mobility μ_A depends on the carrier concentration and is assumed here to $\mu_A = 167$ cm²/Vs [6] resulting in the contribution of the shell to the total resistance to $R_{S,p} = 49$ kΩ. In order to fit to the total series resistance of $R_S = 180$ kΩ (cf. table I), an electron concentration in the core of $N_D = 1 \cdot 10^{18}$ cm^{-3} is calculated. After subtraction of the series resistance the intrinsic ideality factor n is extracted. In both cases the ideality factor is about $n \approx 2.1$ attributed to a non-abrupt pn-junction caused by (i) the transition area in axial devices, and (ii) the i-GaInP intershell in radial devices. The reverse saturation current I_0 in Shockley equation is given by

$$I_0 = q A n_i^2 \left(\frac{D_n}{N_A L_n} + \frac{D_p}{N_D L_p} \right) \tag{3}$$

with the junction area A, the elementary charge q, the intrinsic carrier concentration n_i, the diffusion coefficients for n- and p D_n, D_p, and the diffusion L_n, L_p. However, eq. (3) results in reverse saturation current that is several orders of magnitude lower than those extracted in table 1. This effect is attributed to a somewhat degraded minority carrier transport in nanoscaled devices [5] causing an excessive reverse saturation current.

B. AM 1.5 G measurements

The photovoltaic behavior of axial and radial nanowire diodes is investigated at low irradiation with a 1-sun simulator (AM 1.5 G, 0.1 W/cm²), and under high intensity monochromatic radiation using a focused laser beam. The I-V characteristics of the devices under AM 1.5 G irradiation is depicted in figure 3. The short circuit currents of 479 pA and 88 pA were measured for radial and axial nanowire diodes under standard illumination. Due to the larger area of the pn-junction radial diodes demonstrate under AM 1.5 G illuminatation a 5.4 times higher short circuit currents than axial devices.

Under open circuit condition the solar generated current is directed internally through the pn-junction causing the external voltage $V = V_{OC}$ and with eq. 1 holds:

$$V_{OC} = nV_T \cdot \ln(\frac{I_S}{I_0}) \qquad (4)$$

The open circuit voltage of the devices is $V_{OC,radial} = 0.5$ V and $V_{OC,axial} = 0.6$ V. Despite the large band gap of 1.42 (1.35) eV in GaAs (InP) based solar cells a $V_{OC} = 0.3...0.6$ V, only, is routinely observed [1-5]. A strong improvement is mandatory requiring an optimization of minority carrier transport by surface treatment and device layout optimization. The slope of the I-V curves under illumination in the vicinity of $V = 0$ results into a parallel resistance R_P:

$$R_P = \frac{\Delta V}{\Delta I}\bigg|_{V=0} \qquad (5)$$

Under low intensity illumination the parallel resistance can no longer be ignored and is measured to $R_{P,radial} = 10.2$ GΩ while the axial device exhibits an $R_{P,axial} = 34$ GΩ. The energy-conversion efficiency of the devices is $\eta_{radial} = 4.7$ % [5] and $\eta_{axial} = 9$ % [4].

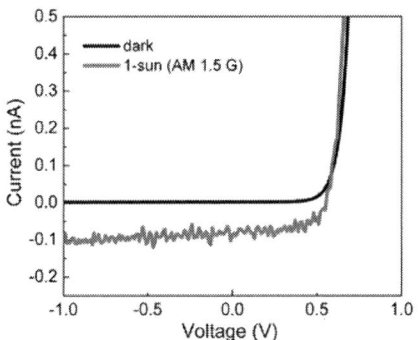

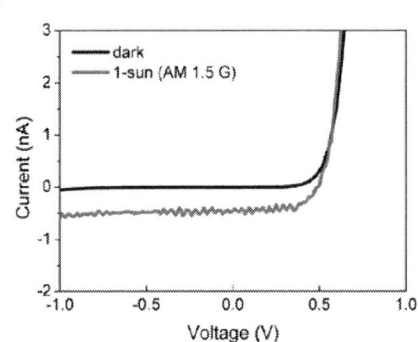

Fig. 3. Dark und 1-sun illuminated I-V curves of axial (a) and radial (b) nanowire diodes.

C. High-intensity, monochromatic irradiation

The high-intensity photovoltaic properties are tested with preliminary focused laser monochromatic ($\lambda = 532$ nm) irradiation with an intensity in the range of 6 W/cm² < P_{opt} < 120 W/cm² that might be roughly compared to a sun

concentration ratio of 60 to 1,200. In Fig. 4 the open circuit voltage V_{OC} is given versus irradiation power density P_{opt} while the whole device is illuminated. The linear increase of the open circuit voltage V_{OC} versus P_{opt} in a semi-log plot is to be expected according to eq. 4. The different slope and the cross-over of the open circuit voltage V_{OC} evolution of radial and axial devices was analyzed with the model given in the inset of Fig. 4. Best agreement to experimental data was obtained with the model parameters in table II. We found that the generated solar current I_S is linearly proportional to the illumination power density P_{opt}. The according conversion factors α were extracted from experimental intensity dependent photocurrent measurements and are given in table II for both devices. This approach enables to model the evolution of the open circuit voltage versus illumination power density P_{opt} given in Fig. 4. It is found that the decay of open circuit voltage for radial devices is caused by the parallel resistance R_P along with a higher reverse saturation current density and ideality factor. On the contrary, the axial devices improve under illumination both in terms of ideality factor and reverse saturation current attributed to a saturation of surface states.

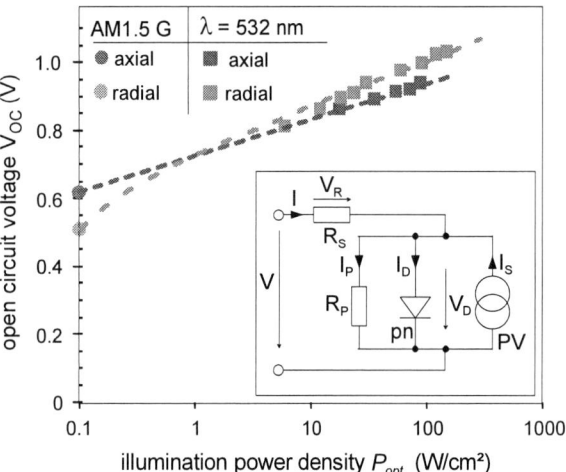

Fig. 4. Measured open circuit voltage of axial and co-axial single nanowire photovoltaic elements at various illumination density ($\lambda = 532$ nm or AM 1.5 G). The inset shows a simple equivalent circuit of the element.

TABLE II
ILLUMINATED I-V MODEL DATA

model parameter	α [nA/Wcm^{-2}]	J_0 [μA/cm^{-2}]	n	R_P [GΩ]
axial	0.89	4.7	1.8	34.0
radial	6	6.7	2.6	10.2

The results of the short circuit current versus illumination power density are given in Fig. 5. The experimental short circuit current data are directly proportional to P_{opt} (m = 1, Fig. 5a) indicating that no parasitic effect is degrading the performance even at very high irradiation power for both axial and radial devices, respectively. The incident power to solar current conversion factor α has been determined to $\alpha_{axial} = 0.89$ nA/Wcm^{-2} and to $\alpha_{radial} = 6$ nA/Wcm^{-2}. The radial devices demonstrate under high illumination power a 6.7 times higher photocurrent due to the larger absorption volume in the vicinity of the pn-junction. The impact of the series resistance may become evident at high photocurrent resulting in a substantial voltage drop V_R (cf. inset in Fig. 4) that may degrade the available output voltage.

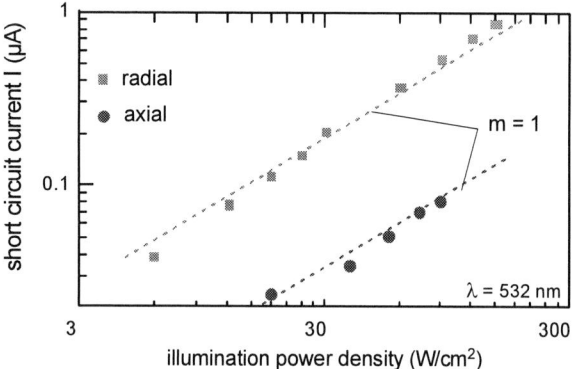

Fig. 5. Experimental short circuit current data of both axial and co-axial devices under various illumination power ($\lambda = 532$ nm).

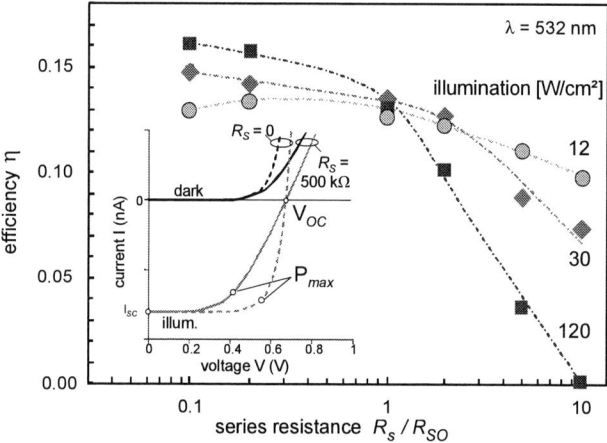

Fig. 6. Modeled efficiency of axial single nanowire photovoltaic elements as a function of the model parameter series resistance R_S normalized to the experimental resistance R_{SO}.

We have performed a sensitivity study on the impact of the series resistance based on the model in the inset Fig. 4 and model parameters extracted from experimental axial devices while R_P is neglected. The modeled efficiency η of the axial device for various input powers is given in Fig. 6. The data for $R_{SO}/R_S = 1$ represent the modeled efficiency η for the realized resistance $R_S = R_{SO}$ while the sensitivity to a modification of the series resistance is modeled in the in the range $0.1 < R_S/R_{S,O} < 10$. A possible reduction of the series resistance to 1/10 has almost no impact on the fill factor and the efficiency

up to 30 W/cm² while a further increase of the series resistance at high illumination powers would degrade the performance drastically.

IV SUMMARY

The VLS grown GaAs nanowire devices can be precisely modelled with very few parameters based on the Shockley equation. Single GaAs nanowire photovoltaic elements offer even without any optimization and dielectric surface coverage an energy conversion of 4.7 % (radial device) and 9 % (axial device). The highly n- and p-doped area provides sufficiently low series resistance. Therefore, it could be shown that the performance is limited neither by series nor by the parallel resistance but by a low open circuit voltage. Based on these achievement single GaAs nanowire devices are suitable for very high power illumination in excess of 1,000 suns. Further studies are necessary to improve the energy conversion yield in terms of device geometry, doping densities, and especially by improving the minority carrier transport. The latter aspect is all import to improve nanoscaled photovoltaic elements exhibiting higher surface induced carrier recombination mechanism.

ACKNOWLEDGEMENT

The contributions from Thorsten Wierzkowski, Franziska Maculewicz, Sarah Blumenthal, Christian Blumberg, Audrey Nekam, and Ebru Tuncel are gratefully acknowledged. This work has been partly supported by the Deutsche Forschungsgemeinschaft within the Forschergruppe FOR 1616, by the EuropeanUnion within the Ziel2 project "Nanowire solar cells and light emitters (Nasol)", and by the bmbf within the project "Nano-III-V PIN".

IV. REFERENCES

[1] B. Tian, X. Zheng, T.J. Kempa, Y. Fang, N. Yu, G. Yu, J. Huang, C.M. Lieber; "Coaxial silicon nanowires as solar cells and nanoelectronic power sources", *Nature* 449, 885, 2007.

[2] C. Colombo, M. Heiß, M. Grätzel, A. Fontcuberta i Morral; "Gallium arsenide p-i-n radial structures for photovoltaic applications", *Appl. Phys. Lett.* 94, 173108, 2009.

[3] H. Goto, K. Nosaki, K. Tomioka, S. Hara, K. Hiruma, J. Motohisa, T. Fukui; "Growth of Core–Shell InP Nanowires for Photovoltaic Application by Selective-Area Metal Organic Vapor Phase Epitaxy", *Appl. Phys. Express* 2, 035004, 2009.

[4] A Lysov, S Vinaji, M Offer, C Gutsche, I Regolin, W Mertin, W Prost, G Bacher, F-J Tegude; "Spatially resolved optoelectronic performance of axial GaAs nanowire pn-diodes", *Nano Res.* 4, 987-995, 2011.

[5] C Gutsche, A Lysov, D Braam, I Regolin, G Keller, Z-A Li, M Geller, M Spasova, W Prost, F-J Tegude, *Adv. Funct. Mater.* 2012, 22, 929–936.

[6] C. Gutsche, I. Regolin, K. Blekker, A. Lysov, W. Prost, F.-J. Tegude;" Controllable p-type doping of GaAs nanowires during vapor-liquid-solid growth", *J. Appl. Phys.*, 105, 024305, 2009.

Radial InP/InAsP Quantum Wells with High Arsenic Compositions on Wurtzite-InP Nanowires in the 1.3-μm Region

Kenichi Kawaguchi[1], Yoshiaki Nakata[1], Mitsuru Ekawa[1], Tsuyoshi Yamamoto[1], and Yasuhiko Arakawa[2]

[1]*Fujitsu Laboratories Ltd., Morinosato-Wakamiya 10-1, Atsugi, 243-0197 Japan*
[2]*INQIE and IIS, The University of Tokyo, 4-6-1 Komaba, Meguro-ku, Tokyo 153-8505, Japan*

Abstract — **Wurtzite (WZ) InP nanowires (NWs) with radial InP/InAsP quantum wells (QWs) having an arsenic composition in the rage of 0.43-0.60 were grown using metalorganic vapor phase epitaxy, and their optical properties were investigated. These InAsP QW layers with a high arsenic content and WZ crystal phase were successfully grown using WZ-InP NWs whose crystalline structure was controlled by sulphur doping. Photoluminescence (PL) of individual NWs with radial InP/InAsP QWs was clearly observed at room temperature. The PL wavelengths were successfully controlled by adjusting the radial QW thickness and arsenic composition of InAsP, and emissions in the 1.3-μm region were demonstrated.**

Index Terms — **InP nanowire, wurtzite crystal, radial InP/InAsP quantum well.**

I. INTRODUCTION

III-V compound semiconductor nanowires (NWs) have attracted much attention as building blocks for optoelectronic devices with small sizes because of their distinctive shape [1]-[8]. In particular, the NWs composed of InP-based materials that have been well developed for telecom devices are promising candidates for applications in the near-infrared wavelength region. Introducing heterostructure is crucial for functionalizing the NWs such that they can be applied to actual devices. Techniques for fabricating heterostructures in the axial direction have been widely investigated [1], [3]-[5]. In addition, demand for the development of heterostructures formed in the radial direction is increasing, since these structures are expected to have the advantage of providing a larger active region on the basis of high aspect ratios of these NWs [2], [6]-[8]. Recently, an attempt to fabricate radial $InAs_{0.2}P_{0.8}/InP$ quantum wells (QWs) with emission wavelengths of approximately 1 μm using InP NW cores with a wurtzite (WZ) crystal phase was reported [8]. Since InP NWs have a poly-type issue of WZ and zinc-blende (ZB), controlling crystal phase is important for forming high-quality NW heterostructures. From a practical point of view, the use of InAsP with higher arsenic compositions is desirable as this affords strong carrier confinements suitable for room-temperature operating devices and long wavelength emissions including the 1.3-1.5 μm telecom band region. Also, studying the effects of varying the arsenic content in InAsP structures is necessary to achieve a fundamental understanding of the crystal growth of radial NW heterostructures.

In this work, we investigated the growth of WZ-InP NWs with radial InAsP QWs having a high arsenic composition (0.43-0.60) using metalorganic vapor-phase epitaxy (MOVPE). The optical properties of these novel WZ-NW heterostructures were also investigated.

II. EXPERIMENTAL PROCEDURES

The MOVPE growth of NWs with radial QWs was performed on InP(111)B substrates. A schematic description of the procedure for fabrication of the radial NW heterostructures is shown in Fig. 1. Au colloidal particles with the diameter of 100 nm were used as catalysts for the vapor-liquid-solid NW growth [9], [10]. 3-μm-long InP NWs were grown at 400°C using $(CH_3)_3In$ (TMIn) and PH_3 as precursors. Sulphur doping with H_2S was applied in order to enhance formation of the WZ crystal phase in the NWs [11], [12]. The supply amounts of the sources were set to 0.217, 155, and 0.5 ccm for TMIn, PH_3, and H_2S, respectively. During the InP NW growth, HCl (0.08 ccm) was also introduced to suppress NW tapering [13]. After growth of the InP NWs was complete, the Au particles were removed using a wet chemical etch process, which suppressed axial growth during the subsequent radial growth process. Next, 100-nm-thick undoped InP buffer layers, single InAsP QWs, and 40-nm-thick InP capping layers were radially grown at 530°C. AsH_3 was used as the precursor of As. The radial InAsP QW thickness was varied from 1 to 8 nm by adjusting the growth time. The nominal arsenic composition was varied from 0.43 to 0.60 and was verified for multiple InAsP/InP QWs grown on InP(001) substrates. An InP shell structure without an InAsP layer was grown as a reference. The crystalline structures of the NWs were investigated using x-ray diffraction (XRD) measurements for samples with NW ensemble standing on the InP substrates. Photoluminescence (PL) measurements of individual NWs dispersed on Si substrates were performed using a green laser (532 nm) as the excitation light source.

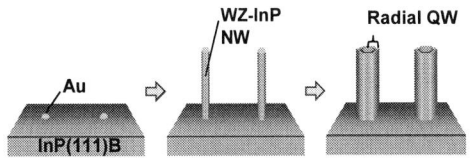

Fig. 1. Procedure for fabrication of InP NWs with radial InAsP/InP QWs.

III. RESULTS AND DISCUSSION

An SEM image of the InP NW cores is shown in Fig. 2 (a). The NW core shape was well-defined by the Au catalytic particles. The diameter at the top of the NW was approximately 120 nm, and the degree of tapering, excluding the root segment, was approximately 30%. An SEM image of InP NWs with radial InAs$_{0.43}$P$_{0.57}$/InP QWs is shown in Fig. 2 (b). Growth in the radial direction was clearly observed for the radial QW sample while axial growth was obviously suppressed. Tapering was less than 5%. Thus, straight and smooth NWs with a radial QW were obtained.

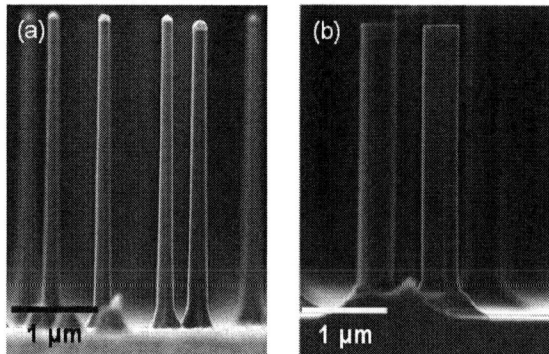

Fig. 2. SEM images of (a) InP NW cores and (b) InP NWs with radial InAs$_{0.43}$P$_{0.57}$/InP QWs.

In order to evaluate the vertical lattice constants of the NWs, ω-2θ scans were taken around the ZB-InP(111) Bragg reflection condition. Figure 3 shows XRD spectra of InP NW samples with varying radial InAs$_{0.43}$P$_{0.57}$ QW thickness. The samples gave two peaks, which were the diffractions from ZB-InP substrates and NWs. Compared with the ZB-InP substrate, the sample of InP NWs with an InP shell (0-nm InAsP) showed a 0.37% larger vertical lattice constant, thus indicating that the InP NWs have a WZ crystal phase [14].

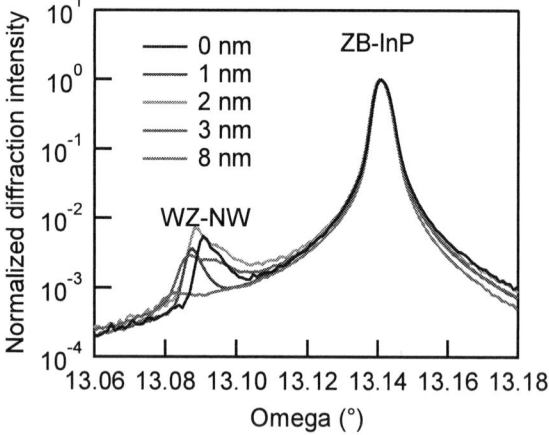

Fig. 3. XRD spectra of NW samples with varying radial InAs$_{0.43}$P$_{0.57}$ QW thickness.

As the radial QW thickness was increased, the NW diffraction peak shifted toward lower angles, indicating an increase in the average vertical lattice constant of the NWs. An increase of 0.060% was observed for the NW sample with an 8-nm-thick InAsP QW. The change in the vertical lattice constant indicates that the InP NW cores were under elastic tensile strain, which was related to the small volume of the InP NWs [8].

Figure 4 shows XRD spectra of NW samples with varying arsenic composition of the radial InAsP QW with the thickness of 2 nm. These samples also showed NW diffractions around the WZ crystal regions. As the arsenic composition was increased, the NW peak shifted slightly toward a lower angle corresponding to the increase in the lattice constant of 0.037%, indicating that increased amount of arsenic in the QW layers affected the elastic strain in the NW heterostructures. For QW samples with a higher arsenic content, additional spectral features were indicated between the substrate and NW peaks. Further investigation is necessary to identifying the origin of the diffraction profile.

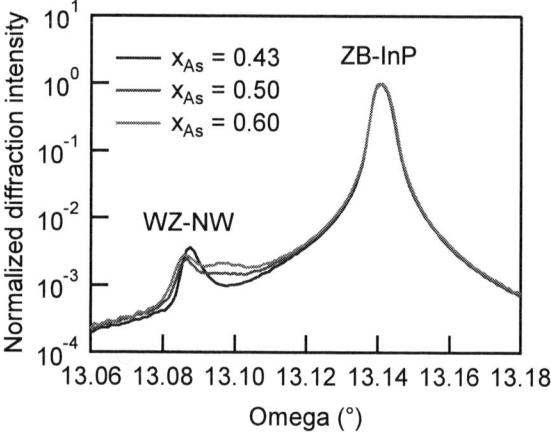

Fig. 4. XRD spectra of NW samples with varying arsenic composition of the radial InAsP QW with the thickness of 2 nm.

Room-temperature PL spectra from single NWs with varying InAs$_{0.43}$P$_{0.57}$ QW thickness are shown in Fig. 5. The excitation power density was estimated to be 0.5 kW/cm^2. The InP NW with an InP shell showed an emission at approximately 870 nm, which was 60-nm shorter than the emission wavelength of ZB-InP, indicating a WZ crystal structure [15], [16]. ZB-related luminescence was found to be well-suppressed. A PL measurement of naked InP NW cores was also attempted, but the degree of luminescence from these structures was not sufficient for qualification. This is expected to result from suppression of radiative recombination by the high-density sulphur dopants. The result indicates that the light emission from the NW with an InP shell came from the undoped shell itself. As the QW thickness was increased from 1 to 8 nm, the PL wavelength was redshifted from 1087 to 1287 nm, whereas the PL intensity was maintained. The wavelength shift was due to the change in quantum confinements imposed by the radial QW thickness. The FWHM of the PL peaks for the NWs with QWs was approximately 100 meV. The observed peak broadening is considered to result primarily from alloy fluctuation in the InAsP structures with a high arsenic content, which is often observed with InAsP films grown on planar InP substrates.

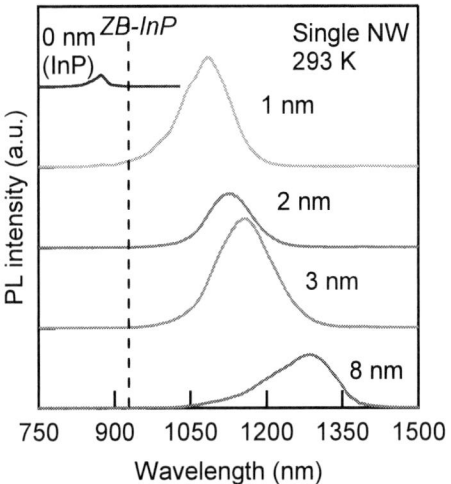

Fig. 5. Room-temperature PL spectra of single NWs with varying InAs$_{0.43}$P$_{0.57}$ QW width.

The change in the PL peak energies of the radial QWs was also analyzed. Figure 6 shows the PL peak energy shifts associated with variation in the QW thickness. Calculated values using the one-particle model are also plotted (broken line). In the calculation, the WZ-InAsP bandgap energy estimated with the bowing parameter of 0.23 was used [17]. Regarding the effective masses, the theoretical values for InP and InAs with WZ crystal phase were adopted [18], and the values for WZ-InAsP were estimated by linear interpolation.

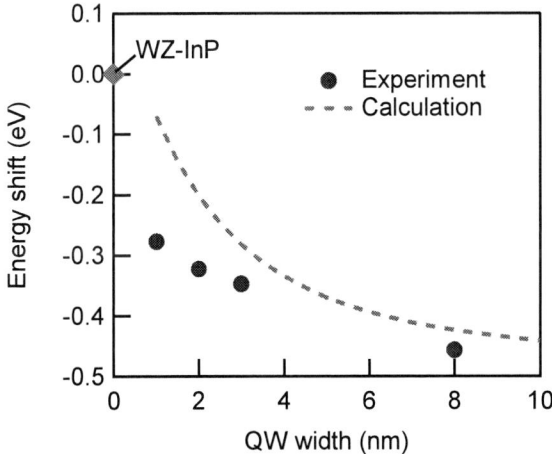

Fig. 6. PL peak energy shifts of WZ-InP NWs with radial InAs$_{0.43}$P$_{0.57}$ QWs as a function of QW width.

The calculated results showed a shift in energy (0.4 eV) when the QW thickness was changed from 1 nm to 8 nm, similar to the experimental findings. The change in the experimental values is, however, rapid in the narrow QW region, which might indicate that the actual effective masses are heavier and/or the effective QW thickness is greater owing to a contribution from the facet corners of the NW sidewalls.

Controlling the emission wavelengths by adjusting the arsenic content of the radial InAsP QW layers was then investigated. PL spectra of single NWs with 2-nm-thick InAsP QWs are shown in Fig. 7. As the arsenic composition was increased from 0.43 to 0.60, the PL wavelength increased from 1129 to 1315 nm. Owing to the thinness of layers, highly arsenic-enriched InAsP materials with large strains were successfully grown while maintaining crystalline quality.

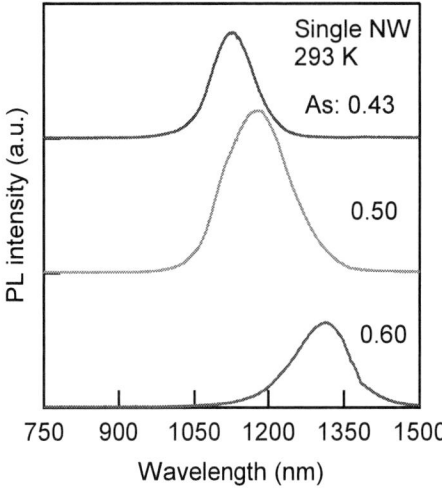

Fig. 7. Room-temperature PL spectra of single NWs with 2-nm-thick InAsP QWs having different As compositions.

The PL peak energy shifts for the radial QWs on the InP NWs were compared with the changes in the bulk InAsP bandgap energy. Figure 8 shows the relationship between the peak energy shift and arsenic composition. The offset was set to zero for the arsenic composition of 0.43. Bulk InAsP bandgap energy shifts are plotted with a broken line. The NW PL shifts were well fitted to the bulk bandgap energy shifts, indicating that control of the PL wavelength was successful. Also, it was confirmed that the energy shifts were comparable to those of the ZB-InAsP/InP MQWs on InP(001).

Thus, light emissions in the 1.3-μm region were obtained using InAsP materials as radial QW layers. These results indicate that the radial QW thickness and arsenic composition of InAsP alloy layers on the WZ-InP NW sidewalls can be controlled in the same manner used in the conventional film growth on planar ZB substrates.

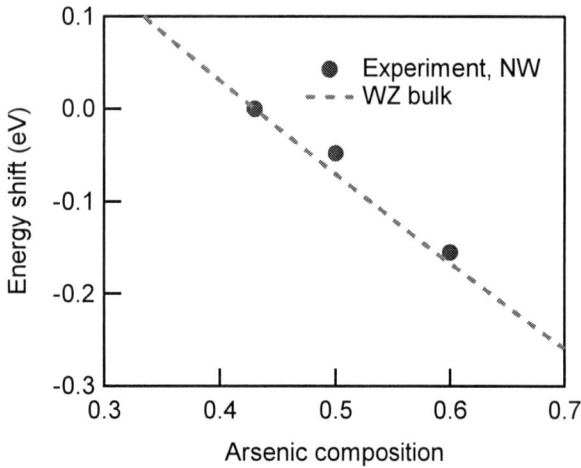

Fig. 8. PL peak energy shifts of WZ-InP NWs with 2-nm-thick InAsP QWs as a function of As composition.

IV. SUMMARY

In summary, we realized high-quality InP NWs with radial InAsP/InP QWs having a high arsenic content on S-doped WZ-InP NW cores and observed PL from single NWs at room temperature. We showed that the PL wavelength was controllable by adjusting the radial QW thickness and arsenic composition, and demonstrated emissions in the 1.3-µm region. These NWs are promising for nanometer-scale near-infrared light emitting devices.

ACKNOWLEDGMENT

We thank Prof. Lars Samuelson and Dr. Magnus T. Borgström from Lund University for fruitful discussions on this subject. This work was supported by Project for Developing Innovation Systems of the Ministry of Education, Culture, Sports, Science and Technology (MEXT), Japan.

REFERENCES

[1] M. T. Borgström, V. Zwiller, E. Müller, and A. Imamoglu, "Optically Bright Quantum Dots in Single Nanowires", *Nano Lett.*, vol. 5, no. 7 pp. 1439–1443, 2005.

[2] P. Mohan, J. Motohisa, and T. Fukui, "Fabrication of InP/InAs/InP core-multishell heterostructure nanowires by selective area metalorganic vapor phase epitaxy", *Appl. Phys. Lett.*, vol. 88, no. 13, 133105 (3pp), 2006.

[3] A. Fuhrer, L. E. Frloberg, J. N. Pedersen, M. W. Larsson, A. Wacker, M.-E. Pistol, and L. Samuelson, " Few Electron Double Quantum Dots in InAs/InP Nanowire Heterostructures", *Nano Lett.*, vol. 7, no. 2, pp. 243–246, 2007.

[4] M. Tchernycheva, G. E. Cirlin, G. Patriarche, L. Travers, V. Zwiller, U. Perinetti, and J.-C. Harmand, "Growth and Characterization of InP Nanowires with InAsP Insertions", *Nano Lett.*, vol. 7 , no. 6, pp. 1500-1504, 2007.

[5] E. D. Minot, F. Kelkensberg, M. van Kouwen, J. A. van Dam, L. P. Kouwenhoven, V. Zwiller, M. T. Borgström, O. Wunnicke, M. A. Verheijen, and E. P. A. M. Bakkers, "Single Quantum Dot Nanowire LEDs", *Nano Lett.* vol. 7, no. 2, pp. 367-371, 2007.

[6] C. P. T Svensson, T. Mårtensson, J. Trägårdh, C. Larsson, M. Rask, D. Hessman, L. Samuelson, and J. Ohlsson, " Monolithic GaAs/InGaP nanowire light emitting diodes on silicon", *Nanotechnology*, vol. 19, no. 30, 305201 (6 pp), 2008.

[7] K. Tomioka, J. Motohisa, S. Hara, K. Hiruma, and T. Fukui, "GaAs/AlGaAs Core Multishell Nanowire-Based Light-Emitting Diodes on Si", *Nano Lett.*, vol. 10, no. 5, pp. 1639–1644, 2010.

[8] K. Kawaguchi, M. Heurlin, D. Lindgren, M. T. Borgström, and L. Samuelson, "MOVPE growth and optical properties of wurtzite InP nanowires with radial InP/InAsP quantum wells," *23rd International Conference on Indium Phosphide and Related Materials*, Berlin, Germany, May 22-26, 2011.

[9] J. H. Woodruff, J. B. Ratchford, I. A. Goldthorpe, P. C. McIntyre, and C. E.D. Chidsey, "Vertically Oriented Germanium Nanowires Grown from Gold Colloids on Silicon Substrates and Subsequent Gold Removal," *Nano Lett.*, vol. 7, no. 6, pp. 1637–1642, 2007.

[10] H. J. Joyce, Q. Gao, H. H. Tan, C. Jagadish, Y. Kim, J. Zou, L. M. Smith, H. E. Jackson, J. M. Yarrison-Rice, P. Parkinson, M. B. Johnston. "III–V semiconductor nanowires for optoelectronic device applications," *Progress in Quantum Electronics*, vol. 35, no. 2-3, pp. 23–75, 2011.

[11] M. H. M. van Weert, A. Helman, W. van den Einden, R. E. Algra, M. A. Verheijen, M. T. Borgström, G. Immink, J. J. Kelly, L. P. Kouwenhoven, and E. P. A. M. Bakkers, " Zinc Incorporation via the Vapor-Liquid-Solid Mechanism into InP Nanowires," *J. Am. Chem. Soc.*, vol. 131, no. 13, pp. 4578-4579, 2009.

[12] G. L. Tuin, M. T. Borgström, J. Trägårdh, M. Ek, L. R. Wallenberg, L. Samuelson and M.-E. Pistol, "Valence band splitting in wurtzite InP nanowires observed by photoluminescence and photoluminescence excitation spectroscopy," *Nano Res.*, vol. 4, no. 2, pp. 159-163, 2011.

[13] M. T. Borgström, J. Wallentin, J. Trägårdh, P. Ramvall, M. Ek, L. R. Wallenberg, L. Samuelson, and K. Deppert, "In Situ Etching for Total Control Over Axial and Radial Nanowire Growth", *Nano Res.*, vol. 4, no. 2, pp. 159–163, 2011.

[14] D. Kriegner, E. Wintersberger, K. Kawaguchi, J. Wallentin, M. T. Borgström, and J. Stangl, "Unit cell parameters of wurtzite InP nanowires determined by x-ray diffraction," *Nanotechnol.*, vol. 22, no. 42, 425704 (7pp), 2011.

[15] M. Mattila, T. Hakkarainen, M. Mulot, H. Lipsanen, "Crystal-structure-dependent photoluminescence from InP nanowires," *Nanotechnology*, vol. 17, no. 6, pp.1580–1583, 2006.

[16] A. Mishra, L. V. Titova, T. B. Hoang, H. E. Jackson, L. M. Smith, J. M. Yarrison-Rice, Y. Kim, H. J. Joyce, Q. Gao, H. H. Tan, C. Jagadish,"Polarization and temperature dependence of photoluminescence from zincblende and wurtzite InP nanowires," *Appl. Phys. Lett.*, vol. 91, no. 26, 263104, 2007.

[17] J. A. Van Vechten and T. K. Bergstresser, "Electronic structures of semiconductor alloys," *Phys. Rev. B*, vol. 1, no. 8, pp. 3351-3358, 1970.

[18] A. De and Craig E. Pryor, "Predicted band structures of III-V semiconductors in the wurtzite phase," *Phys. Rev. B*, vol. 81, no. 15, pp. 155210-1-13, 2010.

Site-controlled growth of InP/InGaP quantum dots

V. Baumann, F. Stumpf, S. Kremling, T. Steinl, A. Forchel, C. Schneider, S. Höfling, and M. Kamp

Technische Physik and Wilhelm Conrad Röntgen Research Center for Complex Material Systems
Universität Würzburg, Am Hubland, 97074 Würzburg, Germany

Abstract

We report on site-controlled growth of InP/InGaP quantum dots (QDs) on GaAs substrates. Shallow nanoholes etched into a InGaP layer are used as nucleation sites for the QDs. Optimized growth conditions and the use of strain mediated nucleation allow us to realize QD arrays with excellent long range ordering on hole pitches as large as 1.25 μm. Single QD lines with an average linewidth of 553 μeV and best values below 200 μeV are observed. Second-order photon-autocorrelation measurements show clear single photon emission with $g(2)(0)=0.13\pm0.01$.

I. INTRODUCTION

During the last decades, a steady progress in gaining control over quantum dot (QD) properties has allowed scientists to carry out break-through experiments in the field of semiconductor quantum optics. The most commonly used dots are (Ga)InAs/GaAs QDs with transition energies in the near infrared. Another possibility are InP quantum dots with an emission in the visible (red) part of the spectrum [1]. This offers several advantages, for example a higher detection efficiencies of commercial Si-based avalanche single photon detectors and the compatibility with free-space quantum key distribution [2, 3]. Since the performance of many single QD based devices critically depends on the QD position in e.g. a microresonator, technologies allowing for the integration of single QDs at the optimum position are highly desirable. In fact, various techniques for site-controlled QD growth in various material systems like SiGe QDs [4, 5] and In(Ga)As QDs on GaAs [5-9] and InP substrates [10,11] have been reported. Although efforts have been made to investigate lateral ordering mechanisms of Stranski-Krastanov QDs in the InP/InGaP material system [12-14], site-controlled growth resulting in accurately ordered QD-arrays has so far been elusive.

Our work is devoted to establish a reliable and robust site-controlled growth technique of InP QDs, with QD distances in the order of typical photonic resonator device dimensions. For this purpose, we investigated the site-controlled deposition of InP QDs on a lithographically patterned InGaP layer, which was grown lattice matched on GaAs by gas-source molecular beam epitaxy

II. QUANTUM DOT GROWTH

The fabrication technology for the site-controlled growth is based on a combination of lithographic patterning of nucleation centers and subsequent gas-source molecular beam epitaxy (GSMBE). Two different growth schemes were used, which we refer to as "single layer dots" and "strain-coupled dots". The corresponding layer structures are illustrated in fig. 1 (a) and (b).

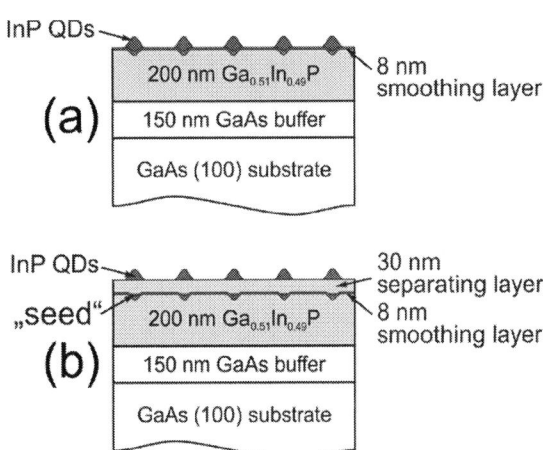

Fig. 1: Sample structure (a) „single layer dots". (b) „strain-coupled dots".

The nucleation layer was prepared by the growth of a GaAs buffer on a (100) GaAs substrate, followed by a 200 nm thick layer of lattice-matched InGaP. In a next step, mesas with dimensions of 300 μm × 300 μm and marker structures were defined by optical lithography and wet chemical etching. The resulting etch depth was 900-1000 nm. The samples were then coated with Polymethyl-methacrylat (PMMA) and arrays of nanoholes arranged in a square pattern with varying nanohole period were defined on top of the mesa structures by electron beam lithography, aligned to the previously fabricated markers. The pattern was then transferred into the semiconductor by reactive ion etching in an Ar/Cl₂ plasma. The initial depth of the nanoholes was determined to be 15-20 nm, with diameters of about 50-80 nm. The PMMA resist was removed with pyrrolidone in an ultrasonic bath and the samples were cleaned with H_2SO_4 and H_2O. After a chemical oxide removal with NH_4OH the samples were immediately transferred into the MBE system to reduce further surface oxidation.

In a following thermal cleaning step, the samples were heated up to 300 °C to remove surface contaminants. The samples were then exposed to a flux of thermally cracked phosphine gas at a substrate temperature of 375°C to further clean the surface by the hydrogen impinging on it. If the temperature of this cleaning step is too high, the hole pattern on the surface deteriorates by material migration on the GaInP surface, as illustrated in fig. 2. Fig. 2a and b show atomic force microscopy (AFM) images of samples which were heated to a temperature of 300 °C (fig. 2a) and 400 °C (fig. 2b) for 60 minutes. Fig. 2c shows a comparison of line profiles extracted from the AFM data across representative nanoholes.

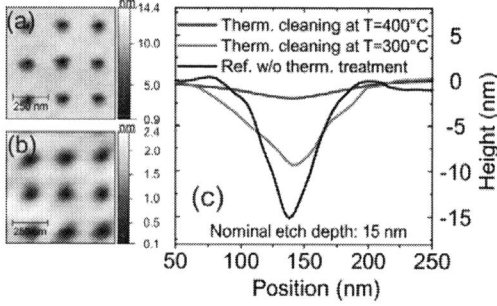

Fig. 2: AFM micrographs of the sample surface after thermal cleaning for 60 min. at 300 °C (a) and 400 °C (b). Surface profiles extracted across nanoholes on a reference sample without thermal cleaning, after 60 min. at 300 °C and after 60 min. at 400 °C.

Heating the samples up to 300 °C for 60 minutes reduces the nanohole depth by 5 nm compared to original depth measured immediately after etching. A temperature of 400 °C results in very shallow patterns with only 2 nm depth. The substrate temperature during the cleaning step was therefore fixed at 375°C, which offers a good compromise between structural integrity of the nano-holes and cleaning efficiency.

Growth was initialized by deposition of a thin layer of 8 nm $Ga_{0.51}In_{0.49}P$ to smooth the surface. For samples with a single layer of dots (fig. 1a), an amount of 1.5-1.6 monolayers (ML) of InP was deposited on the GaInP surface. The deposition was performed in a number of deposition cycles separated by a short growth interruption. After deposition of the QDs, the samples were either transferred out of the MBE system for structural characterization of the surface, or capped with 70 nm of $Ga_{0.51}In_{0.49}P$ for optical studies. For samples with the strain coupled dots (fig. 1b) a reduced amount of 1.3-1.4 ML of InP was deposited on the first 8 nm thick smoothing layer. We refer to this InP layer as the seeding layer, it is not sufficient for nucleation of obvious QDs, but partly fills the nanoholes. The resulting strain around these seeds allows for

the application of vertical stacking techniques in the following $Ga_{0.51}In_{0.49}P$ separating layer. On top of the GaInP separating layer with a thickness of 30-50 nm, we deposited 1.5-1.6 ML of InP, resulting in the formation of dots which nucleate aligned to the nanohole pattern of the seeding layer. The InP deposition for the seeding layer as well as for the QD layer was carried out by means of a cycled deposition with short growth interruptions between the deposition cycles.

A. Single layer layout quantum dots

Different growth parameters, such as substrate temperature, phosphor supply and material amount have been varied to optimize the yield of site-controlled QDs and to reduce the number of randomly grown QDs in between the desired nucleation sites. Applying a cycled deposition technique, we observed a strong influence of the number of deposition cycles on the QD nucleation. Fig. 3 shows AFM micrographs of samples with a single layer of InP QDs. A total of 1.6 ML was deposited on the sample in 8, 16 and 32 cycles. For a deposition with 8 cycles (fig. 3a), a lateral ordering is barely observable. The area density of randomly grown QDs is about $9.2 \cdot 10^9$ cm^{-2} and several large QDs can be observed around each nucleation site. If the InP is deposited in 16 cycles (fig. 3b), the area density of randomly grown QDs is strongly reduced by one order of magnitude to $8.1 \cdot 10^8$ cm^{-2}. The holes are now occupied by larger QDs, with a trend to the formation of QD molecules.

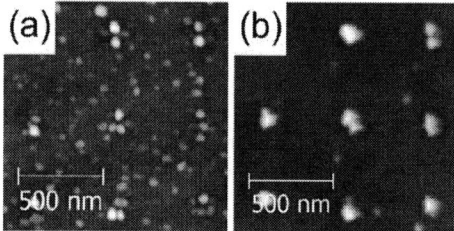

Fig. 3: QD growth by cyclic deposition of 1.6 ML InP using (a) 8 cycles and (b) 16 cycles.

For 32 deposition cycles, the formation of randomly grown QDs is further reduced and almost every nanohole is occupied by a single QD, as shown in fig. 4.

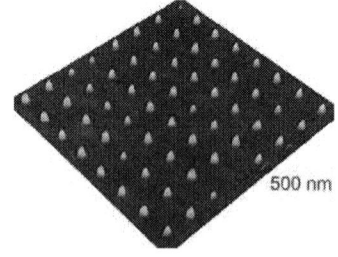

Fig. 4: AFM-micrograph (3D-representation) of site-controlled InP QDs with 500 nm pitch. Excellent suppression of the formation of interstitial QDs was achieved by optimization of the growth conditions.

B. Strain coupled dots

The growth of quantum dot using the strain-coupled scheme offers several advantages over the single layer dot approach described in the previous section: First of all, the separation of the initial seeding layer and the actual QDs allows the use of different growth conditions (and especially different substrate temperatures) for the two layers. As discussed above, the seeding layer has to be deposited at moderate temperatures to prevent degradation of the nanoholes. The actual QD deposition can be performed at higher temperatures due to the fact that the seed is preserved by the separating layer. The lateral alignment of the QDs is then established less by surface irregularities presented by the nanoholes than by means of a strain coupling to the seeding layer. In addition, the QDs are spatially separated from carriers trapped at fabrication-induced impurities or crystal defects at the etched surface which can induce severe emission linewidth broadening due to spectral diffusion.

Fig. 5 shows AFM micrographs of QDs grown using strain-mediated nucleation. The period of the nanohole pattern was varied between 500 nm and 2 μm. The seeding layer was grown at 375 °C by means of cyclic deposition of 1.4 ML InP at a growth rate of 26 nm/h. The substrate temperature was then raised by 20 °C after the growth of a 30 nm $Ga_{0.49}In_{0.51}P$ separation layer. The quantum dot growth was performed at 395 °C, using a cyclic deposition of 1.6 ML InP as for the single layer dots.

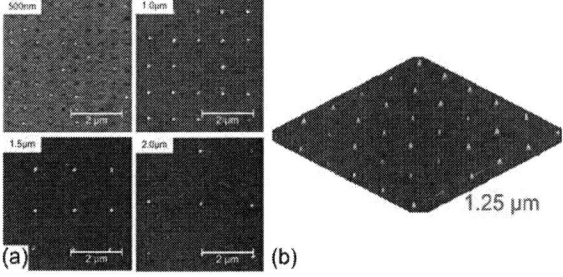

Fig. 5: (a) AFM micrographs of site-controlled quantum dots grown in the strain-coupled mode for different nanohole periods ranging from 500 nm to 2 μm. (b) AFM micrograph of a QD array with 1.25 μm period.

In the first array with 500 nm period, numerous nanoholes are not occupied by QDs or only a formation of rather small QDs at the edge of a nanohole is observable, which we attribute to the fact that the net amount of InP being present on the surface is insufficient for nucleation of QDs at the area density of the predefined nucleation sites. The number of unoccupied nucleation sites is much smaller for the arry with 1μm period, since the four times more material is now availabe for the QD formation at each site. A clear benefit of the strain coupled nucleation is the reduction of randomly nucleating interstitial QDs by a factor of two, allowing the fabrication of quantum

dot arrays with periods up to 1.25 μm (see fig. 5b) that contain almost no insterticial dots. Above 1.75 μm the number of QDs randomly nucleating in between the sites significantly increases.

III. OPTICAL CHARACTERIZATION

The ensemble emission of the QDs is centered around 670 nm (1.85 eV) and has a width of approximately 50 meV. The excellent long range ordering and high suppression of interstitial island formation facilitated probing of individual QDs by micro-photoluminescence (PL) spectroscopy. Single QD lines with an average linewidth of 553 μeV and best values below 200 μeV were observed in a wavelength range of about 670 nm (1.85eV).

To proof the principle suitability of our site-controlled QDs as quantum emitters, we have performed second-order auto-correlation measurements under pulsed excitation. In this experiment, the QD was excited with a pulsed laser at a wavelength of 375 nm and a pulse frequency of 80 MHz. Fig. 6 shows the corresponding time integrated photoluminescence PL spectrum. Due to the low area density of our QDs, only one dominant emission feature of a site controlled QD is visible in the entire spectral range. The autocorrelation histogram shown in the inset of fig. 6 was recorded using a free space Hanbury Brown and Twiss setup. The strong suppression of a coincidence peak around τ=0 proves the quantum light emission of the QD. By analyzing the area under below the peaks, we extract a value of $g^{(2)}(\tau=0)=0.13\pm0.01$, which demonstrates the feasibility to use capability of our structures to these dots as emitters be exploited in triggered single photon sources.

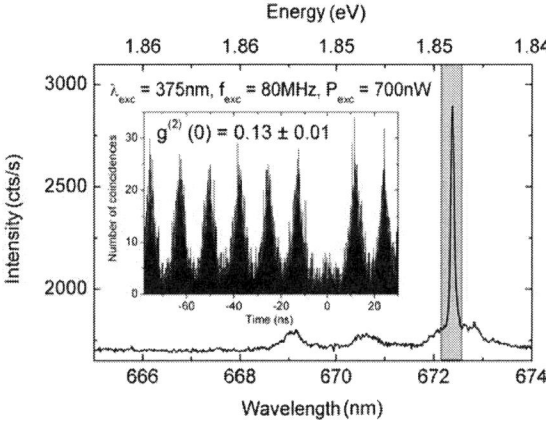

Fig. 6: Micro-photoluminescence spectrum showing a single excitonic emission line from a site-controlled QD. The inset shows pulsed second order photon-autocorrelation measurements on this positioned QD revealing with a value $g^{(2)}(0)=0.13\pm0.01$

IV. CONCLUSION

We reported routes for the reliable growth of site-controlled growth of InP QDs on GaAs substrates. GaInP layers patterned with shallow nanoholes were used to define the nucleation position of the InP QDs. Good lateral ordering of the QDs grown directly on the patterned surface was achieved for periods up to 750 nm. Limitations due to thermal degradation of the surface pattern above 375 °C present a major challenge for site-controlled growth on larger pitches. A strain-coupled growth technique was applied to overcome this difficulties, allowing us to enhance surface migration via the substrate temperature and achieve an excellent long range ordering for periods up to 1.25 μm, which is particularly promising for a deterministic single QD integration in micron-sized microresonator devices. It also facilitates probing of individual QDs by micro-PL spectroscopy, which revealed single QD emission lines. We consider the establishment of a growth routine for positioned InP QDs as an important step towards integration of single InP/GaInP quantum dots into photonic devices and a progress upon the scalable realization of efficient sources of triggered single photons in the visible red spectral range.

Acknowledgment

The authors would like to thank and M. Wagenbrenner for technical assistance. Financial support by the Federal German Ministry of Education and Research (BMBF) within the project 'QPENS' is gratefully acknowledged.

REFERENCES

[1] W.M. Schulz, R. Roßbach, M. Reischle, G.J. Beirne, M. Bommer, M. Jetter, and P. Michler, Phys. Rev. B **79**, 035329 (2009)

[2] R. J. Hughes, J. E. Nordholt, D. Derkacs, C. G. Peterson, New J. Phys. **4**, 43 (2002).

[3] J. L. Duligall, M. S. Godfrey, K. A. Harrison, W. J. Munro, J. G. Rarity, New J. Phys. 8, 249 (2006).

[4] J. Stangl, V. Holy, G. Bauer, Rev. Mod. Phys. **76**, 725 (2004)

[5] O.G. Schmidt, Lateral Alignment of Epitaxial Quantum Dots, edited by O. G. Schmidt, Springer, Berlin, 2007

[6] T. Ishikawa, S. Kohmoto, K. Asakawa, Appl. Phys. Lett. **73**, 1712 (1998)

[7] L. O. Mereni, V. Dimastrodonato, R. J. Young, E. Pelucchi, Appl. Phys. Lett. **94**, 223121 (2009)

[8] M.H. Baier, E. Pelucchi, E. Kapon, S. Varoutsis, M. Gallart, I. Robert-Philip, and I. Abram, Appl. Phys. Lett. **84**, 648 (2004)

[9] C. Schneider, M. Strauß, T. Sünner, A. Huggenberger, D. Wiener, S. Reitzenstein, M. Kamp, S. Höfling, and A. Forchel, Appl. Phys. Lett. **92**, 183101 (2008)

[10] H. Z. Song T. Usuki, S. Hirose, K. Takemoto, Y. Nakata, N. Yokoyama, Y. Sakuma, Appl. Phys. Lett. **86**, 113118 (2005)

[11] D. Dalacu, M. E. Reimer, S. Fréderick, D. Kim, J. Lapointe, P. J. Poole, G. C. Aers, R. L. Williams, W. Ross McKinnon, M. Korkusinski, and P. Hawrylak, Laser Photonics Rev. **4**, 283 (2009)

[12] R. Roßbach, W.M. Schulz, M. Reischle, G.J. Beirne, M. Jetter, P. Michler, J. Cryst. Growth **298**, 595 (2007)

[13] A. Ugur, F. Hatami, W. T. Masselink, A. N. Vamivakas, L. Lombez, M. Atatüre, Appl. Phys. Lett. **93**, 143111 (2008)

[14] R. Rödel, A. Bauer, S. Kremling, S. Reitzenstein, S. Höfling, M. Kamp, L. Worschech A. Forchel, Nanotechnology **23**, 015605 (2012)

Catalyst Design for Native Oxide Based Selective Area

InP Nanowire Growth

Yonatan Calahorra, Yaakov Greenberg, Shimon Cohen and Dan Ritter

Department of Electrical Engineering, Technion - Israel Institute of Technology,

Haifa 32000, Israel

Email: yon.calahorra@gmail.com

Abstract—E-beam lithography based nanowire catalysts are defined by a two dimensional parameter space, spanned by metallization thickness and resist pinhole diameter. We report that native oxide based selective area nanowire growth allowed reducing the metallization thickness of catalysts down to 1/20 of the resist pinhole diameter, without thermal catalyst splitting; contrary to native oxide free nanowire growth, where catalyst splitting is a limiting effect. This parameter space allows growing similar-diameter nanowires, by two different parameter sets. In one such case, nanowires of about 50 nm grew at considerably different rates determined by the metallization thickness; indicating that at given conditions, nanowire diameter does not solely determine nanowire growth rate.

I. INTRODUCTION

A. Background

Understanding nanowire (NW) growth physics is crucial for successful realization of NW based devices. Two distinct practices for NW growth exist - catalyst assisted and catalyst free; this report deals with e-beam defined catalyst assisted NW growth.

Three approaches exist for growth catalyst realization (sorted here by increasing process complexity): *i)* dewetting of a continuous metallic layer; *ii)* dispersing prepared nanoparticles; *iii)* e-beam lithography (EBL) - all shown schematically in Fig. 1. The thermal induced dewetting of a pre-deposited metallic layer, is the simplest method; however, poor control over NW location and catalyst size is obtained. Using nanoparticles as catalysts increases process complexity but allows greater control of catalyst size; essentially, the only major geometrical parameter in this method is the nanoparticle diameter (assuming it is sphere-like).

The EBL method consists of a lithography step, where a pinhole opening in the resist layer is defined, and a metallization step, where the evaporated thickness is determined. In cost and efficiency terms, EBL catalyst definition is costly compared to definition by metallic layer dewetting or by nanoparticles. However, it offers precise control over the catalyst location,

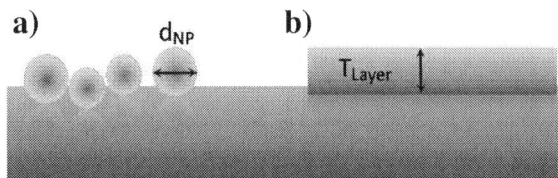

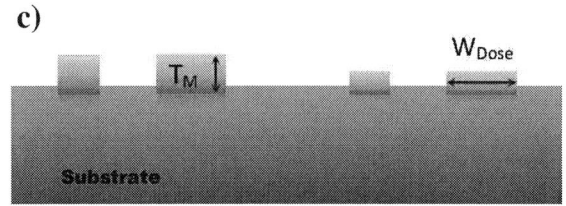

Fig. 1. Nanowire catalyst preparation approaches, and characterizing parameters: a) nanoparticles - characterized by their diameter; b) continuous metallic layer - characterized by its thickness; c) EBL - characterized by resist pinhole diameter (W_{Dose} - a function of the e-beam dosage), and metallization thickness (T_M).

and lower particle diameter variance; therefore it is favorable for systematic NW growth study, and for applications which require location control.

There is no inherent relation between the pinhole diameter and the metallization thickness in EBL, and together they define a two-dimensional catalyst design space. As shown in Fig. 1c, different pinholes can be defined in a single lithography step, and similar lithography products can be subjected to different metallization processes. This fundamental attribute allows realizing a wide range of thickness to diameter ratios for e-beam defined catalysts, and examination of subsequent effects on NW growth. This design space is not available for the other catalyst realization methods; EBL therefore adds another NW growth parameter, to an existing long list (growth system, temperature, materials, precursors, flows, dopants).

B. Catalyst Shape and Size Effects

Nanowire properties are intimately related to catalyst properties; primarily, the catalyst diameter directly determines the NW diameter[1] [1]. Moreover, nanowire diameter usually relates inversely to NW growth rate, *i.e.*, nanoparticles of small diameter catalyze thin NWs, which grow faster than thicker NWs catalyzed by larger nanoparticles, within the same growth [1], [2]. Recently, the shape of a nanoparticle based catalyst was found to effect NW growth rate, where faceted gold catalysts yielded longer NWs than obtained from spherically shaped catalysts, in similar surface density and nominal diameter [3]. Regarding e-beam defined catalysts, a metallization thickness to diameter ratio of 1/3 was found to translate pinhole diameter to NW diameter; ratios much smaller caused catalyst splitting following thermal annealing [4]. Dalacu *et al.* succeeded growing NWs from catalysts with a thickness-diameter ratio smaller than 1/3, utilizing SiO_2 based selective area NW growth. In this method, catalysts are deposited inside SiO_2 openings which prevent catalyst splitting [5].

Recently we have reported the growth of InP nanowires on an InP(111)B substrate in a selective area manner, based upon the wafer's native oxide. Unlike the common growth procedure, native oxide is not thermally desorbed prior to NW growth, and subsequently acts as a selective area mask which inhibits bulk growth in between NWs [6]. Growth conditions in this method greatly differ from the common, oxide-free, NW growth conditions. In this publication, we show that our method enables NW growth from catalysts with a small thickness-diameter ratio, and that the metallization thickness effects NW growth non-trivially.

II. EXPERIMENTAL

A. Nanowire Growth

Nanowires were grown in a metal organic molecular beam epitaxy system for 40 minutes, at 405 °C, with TMIn flux corresponding to a planar growth of 25 nm and PH_3 flow of 4 sccm. A pre-growth thermal treatment at 455 °C, which does not remove the surface oxide, was employed. The two experimental parameters were e-beam dot dose (pinhole diameter estimated at 35-45 nm), and the gold metallization thickness (2-25 nm).

---

[1]This relation may differ in initial stages of growth, or change due to the Gibbs-Thomson effect, but is observed in a wide range of growth methods for NWs of about 25 nm and above in diameter.

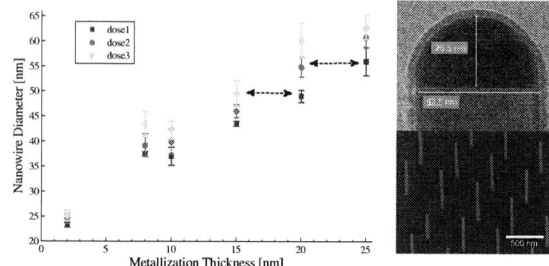

Fig. 2. Nanowire diameter vs. metallization thickness for three e-beam dot dose levels, showing that similar NW diameters were obtained with different parameter sets (two such cases are indicated by dashed arrows). Blue squares, red circles, and green diamonds correspond to 0.7, 0.9, and 1.2 dose factors multiplying the nominal dot dose (0.0045 pC). To the right, upper inset shows TEM image (at the [112] zone axis) of the catalyst of a NW originating from a 20 nm metallization; lower inset shows a NW array SEM image, which is used to measure NW length.

B. Nanowire Characterization

Nanowire and catalyst geometry were studied by scanning electron microscopy (SEM, Hitachi S4700) with a 30° tilted view, and transmission electron microscopy (TEM, FEI Tecnai G2 T20).

III. RESULTS & DISCUSSION

A. Nanowire Diameter

Figure 2 shows NW diameter vs. catalyst metallization thickness, for different e-beam dot dose levels (different W_{Dose}). NW diameter increased together with either metallization thickness or pinhole diameter (dose), as expected due the enlarged catalyst volume, implying a substantial effect of both parameters. As indicated by horizontal dashed arrows, several thickness-dose combinations may result in similar NW diameters; such cases are explained either by formation of completely identical catalysts, or by formation of catalysts with a similar catalyst-substrate interfacial diameter.

The resist pinhole diameter is not known a priori; by comparing the gold volume in the catalyst to the evaporated volume, it can by calculated by the following

$$V_{catalyst} \cdot \eta_{Au} = \pi \left(\frac{W_{Dose}}{2} \right)^2 \cdot T_M \qquad (1)$$

with η_{Au}, the gold associated part of the catalyst volume ($V_{catalyst}$), and the evaporated gold volume on the right-hand side of the equation.

The inset of Fig. 2 shows a TEM image of a NW catalyst, with a total volume of about 38,000 nm^3 calculated for it; 60%

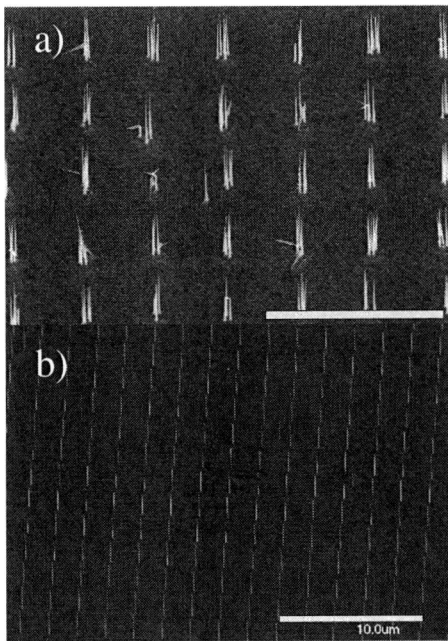

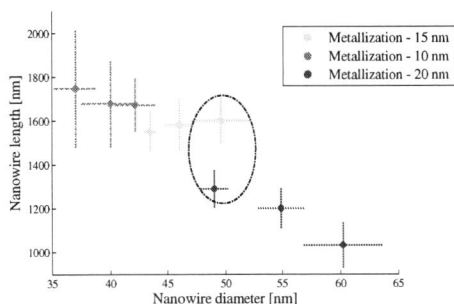

Fig. 4. Length vs. diameter for nine sets of NWs, showing that NWs of the same diameter, may grow at different rates (dashed ellipse). Blue (dash-dot), green (solid), and red (dash) sets correspond 20, 15, and 10 nm metallization thickness. Crosses correspond to standard deviation.

Fig. 3. Tilted SEM images of grown NWs: a) 15 nm catalyst metallization, surface oxide removal at 495 °C; b) 2 nm catalyst metallization, pre-growth heat treatment at 455 °C; scale bars are 10 nm.

of this volume is attributed to gold[2]. The 52.5 nm diameter is in the range measured for dose 2, therefore the 20 nm metallization yields a 40 nm pinhole diameter for this dose.

The pinhole diameter does not depend on metallization, therefore (according to Fig. 2) the 10 nm metallization set roughly translates the pinhole diameter to catalyst and NW diameter, with a thickness-diameter ratio of about 1/4. This finding also agrees with the rule of thumb reported by Martensson *et al.*, the favored 1/3 thickness-diameter ratio for EBL based NW catalysts. Thicker or thinner metallizations (compared to about 10 nm, for the above dose value), correspond to larger or smaller thickness-diameter ratios, which result in NWs thicker or thinner than the pinhole diameter.

Generally, NW growth from catalysts of a small thickness-diameter ratio is hindered by catalyst splitting; however, we report NW growth without catalyst splitting for thickness-diameter ratios as small as 1/20. Figure 3a shows NWs grown after a thermal treatment which removes the surface oxide (as indicated by the surface roughness); these NWs originate in catalysts of 15 nm metallization, and should not be subjected to such severe splitting, according to the 1/3 rule of thumb.

---

[2]*i*) assuming a hemispherical shape (radius of 26.25 nm) terminated by a cylinder (radius of 26.25 nm and length of 2.25 nm) corresponding to the total catalyst height of 28.5 nm; *ii*) considering that gold and indium atoms are similar in size, and that indium constitutes about 40% of the post growth catalyst [7].

Nevertheless, at each catalyst location 4-5 NWs grow. The NWs on Fig. 3b grew without surface oxide removal; these NWs originate in catalysts of 2 nm metallization, and still did not exhibit splitting.

We suggest that the thin native oxide layer surrounding the catalysts in our growth method, prevents catalyst splitting. Therefore, our growth method allows NW growth with diameters considerably smaller than the resist pinhole diameter (*e.g.*, 25 nm diameter in a 40 nm pinhole for a 2 nm metallization). In the report by Dalacu *et al.*, NWs were grown in similar conditions, and we believe a similar mechanism acted to keep thin catalysts intact.

B. Nanowire Growth Rate

We further examined the length of the grown NWs. Surprisingly, nanowires of similar diameters, originating from different metallization-dose sets, exhibited distinguishable growth rates. Figure 4 shows length vs. diameter of NWs originating from catalysts of different parameters (corresponding to data presented in Fig. 2). Only NWs grown simultaneously were used to compare NW length. The general trend, increasing length with reducing diameter, is maintained; however, a discontinuity in that relation exist. The dashed ellipse marks the length-diameter relations measured for NWs growing out of catalysts with metallization thickness of 20 nm (dose 1), and of 15 nm, (dose 3). The diameters measured were 49 ± 1.2, and 49.6 ± 2.6 nm, correspondingly. Nevertheless, although having the same diameter, these NWs measured 1292 ± 88 and 1601 ± 104 nm in length; varying by much more than the measurement standard deviation.

The origin of this result is not yet known, and further investigations regarding catalyst shape and composition, and NW crystalline structure and quality, are required to understand it.

Nevertheless, this result suggests that metallization thickness in our growth method is a physical growth parameter as well as a technological one.

IV. Conclusion

Unlike common growth methods, native oxide based selective area allows growth of NWs with a metallization thickness to diameter ratio as small as 1/20, without catalyst splitting; thus pushing the limits of e-beam lithography downwards. The design space spanned by the metallization thickness and dot dose level, allows obtaining NWs of similar diameters, using different parameter sets. In a specific case, this operation resulted in NWs of similar diameter growing at remarkably distinguishable rates - a result suggesting a non-trivial effect of catalyst geometry on growth.

References

[1] Kimberly A. Dick. A review of nanowire growth promoted by alloys and non-alloying elements with emphasis on Au-assisted III-V nanowires. *Progress in Crystal Growth and Characterization of Materials*, 54(3-4):138 – 173, 2008.

[2] L. E. Fröberg, W. Seifert, and J. Johansson. Diameter-dependent growth rate of inas nanowires. *Phys. Rev. B*, 76(15):153401, Oct 2007.

[3] Pin Ann Lin, Dong Liang, Samantha Reeves, Xuan P.A. Gao, and R. Mohan Sankaran. Shape-controlled au particles for inas nanowire growth. *Nano Letters*, 12(1):315–320, 2012.

[4] T Mårtensson, M Borgström, W Seifert. B J Ohlsson, and L Samuelson. Fabrication of individually seeded nanowire arrays by vapour-liquid-solid growth. *Nanotechnology*, 14(12):1255, 2003.

[5] Dan Dalacu, Khaled Mnaymneh, Xiaohua Wu, Jean Lapointe, Geof C. Aers, Philip J. Poole, and Robin L. Williams. Selective-area vapor-liquid-solid growth of tunable inasp quantum dots in nanowires. *Applied Physics Letters*, 98(25):251101, 2011.

[6] Yonatan Calahorra, Yaakov Greenberg, Shimon Cohen, and Dan Ritter. Native-oxide-based selective area growth of inp nanowires via metalorganic molecular beam epitaxy mediated by surface diffusion. *Nanotechnology*, 23(24):245603, 2012.

[7] Linus E. Fröberg, Brent A. Wacaser, Jakob B. Wagner, Sören Jeppesen, B. Jonas Ohlsson, Knut Deppert, and Lars Samuelson. Transients in the formation of nanowire heterostructures. *Nano Letters*, 8(11):3815–3818, 2008. PMID: 18811210.

High Speed VCSELs for Optical Interconnects

*A. Larsson, J.S. Gustavsson, Å. Haglund, J. Bengtsson, B. Kögel, P. Westbergh, R. Safaisini, E. Haglund,
K. Szczerba, M. Karlsson, and P.A. Andrekson*

*Photonics Laboratory, Department of Microtechnology and Nanoscience,
Chalmers University of Technology, SE-412 96 Göteborg, Sweden, anders.larsson@chalmers.se*

Abstract — **This paper presents an overview of our recent work on high speed, oxide confined, 850 nm vertical cavity surface emitting lasers (VCSELs). With proper active region and cavity designs, and techniques for reducing capacitance and thermal impedance, we have reached a modulation bandwidth of 23 GHz and demonstrated 40 Gbps transmission. Using an integrated mode filter for reducing the spectral width we have extended the reach on multimode fiber at 25 Gbps from 100 to 500 m. Improved link capacity was also demonstrated using a more spectrally efficient multi-level modulation format (4-PAM). Finally, a MEMS-technology for wafer scale integration of tunable high speed VCSELs was developed, enabling a tuning range of 24 nm, a 6 GHz modulation bandwidth, and 5 Gbps transmission.**

Index Terms — **Vertical cavity surface emitting lasers, modulation, dynamics, tuning, optical interconnections.**

I. INTRODUCTION

Vertical cavity surface emitting lasers (VCSELs) are being developed towards higher speed, higher operating temperature, and higher efficiency to satisfy demands for higher interconnect capacity, higher interconnect density, and reduced power consumption in e.g. data centers and high performance computing systems. Recent efforts on improving the performance of GaAs-based VCSELs at wavelengths in the range 850-1100 nm have enabled a modulation bandwidth in excess of 20 GHz [1-4], data rates at or above 40 Gbps [5-8], high speed operation at elevated temperatures [7], and have proved that an energy dissipation of less than 100 fJ/bit is achievable under direct current modulation [9].

Fig.1 summarizes the speed vs. temperature performance reported in the literature. While VCSELs at 850 nm (the wavelength in existing datacom standards) operate up to 40 Gbps at 25°C and up to 30 Gbps at 85°C, VCSELs at 980 nm operate at an even higher data rate at a given temperature. This suggests that the use of deeper and more strained InGaAs quantum wells (QWs) and the more extensive use of binary compounds favors speed. Fig.2 presents a summary of reported values for VCSEL energy dissipation at different data rates. Clearly, small aperture VCSELs can be very energy efficient (<100 fJ/bit) while the dissipated energy increases with increasing data rate as the VCSEL has to be biased at a higher current for sufficient bandwidth.

In what follows, we present an overview of recent work on high speed 850 nm VCSELs at Chalmers University of Technology. This includes fixed wavelength VCSELs as well as tunable VCSELs. The former has also been equipped with an integrated mode filter for extending reach and a multi-level modulation format has been used for extending reach and capacity.

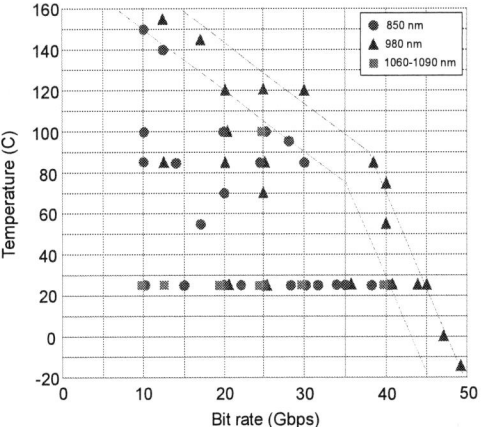

Fig.1 A compilation of data rates and operating (ambient) temperatures for GaAs-based VCSELs at different wavelengths from the literature. In most cases performance was proved by error-free transmission over multimode fiber. The lines indicate state-of-the-art for VCSELs at 850 (red) and 980 nm (blue).

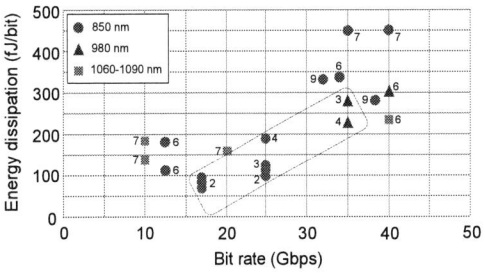

Fig.2 Energy dissipation per bit for GaAs-based VCSELs at different wavelengths compiled from the literature. The numbers next to the symbols show the diameter in μm's of the current/optical aperture. VCSELs within the box are single or quasi-single mode while the others are multimode. The ambient temperature is 25°C.

978-1-4673-1725-2/12 $31.00 © 2012 IEEE

II. HIGH SPEED VCSEL DESIGN AND PERFORMANCE

The design of our high speed fixed wavelength VCSEL is shown in Fig.3. It employs an active region with strained InGaAs/AlGaAs QWs and thin transport layers for high differential gain, low gain compression, and fast carrier transport and capture [10]. To minimize current induced self-heating, and thereby the impact of thermal effects on the modulation speed, we use graded interfaces and modulation doping in the distributed Bragg reflectors (DBRs) for low resistance and low free carrier absorption and a binary compound in most of the bottom DBR for low thermal impedance [11]. To reduce capacitance we use multiple oxide layers, with two deep oxide layers for current and optical confinement and an additional four shallow oxide layers for a further reduction of capacitance [12].

The cavity photon lifetime was found to have a significant impact on VCSEL performance, including the modulation response and bandwidth [4]. Therefore, the impact of the photon lifetime was investigated by systematically varying the reflectivity of the top DBR [13]. The reflectivity was fine-tuned by the thickness of the top DBR layer which sets the phase of the reflection at the semiconductor/air interface. With decreasing photon lifetime there is initially a rapid reduction of the damping of the modulation response, followed by a reduction of the resonance frequency. The net effect is a maximum bandwidth at an optimum photon lifetime.

The VCSEL slope efficiency and output power also showed a strong dependence on photon lifetime. A VCSEL with a 7 μm oxide aperture and the photon lifetime set to an optimum value for speed (3.3 ps) had a threshold current of 0.4 mA, a slope efficiency as high as 1.0 W/A, and a maximum output power of 6.2 mW.

Fig.4 shows the small signal modulation response for the same VCSEL. The bandwidth reaches a maximum value of 23 GHz, which is a record bandwidth for 850 nm VCSELs. Results from data transmission over OM3+ multimode fiber using the 7 μm aperture, 3.3 ps photon lifetime VCSEL are shown in Fig.5. In a back-to-back (BTB) configuration, 40 Gbps error-free transmission was achieved for the first time, while with a 100 m long fiber data could be transmitted up to 35 Gbps [6].

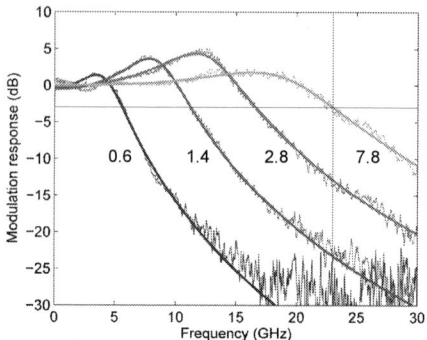

Fig.4 Small signal modulation response for the 3.3 ps photon lifetime, 7 μm oxide aperture VCSEL at different bias currents (in mA).

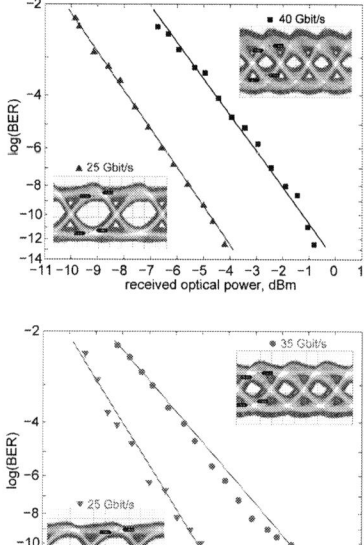

Fig.5 Results from transmission experiments. Upper: BTB transmission. Lower: transmission over 100 m OM3+ fiber.

III. HIGH SPEED VCSELs WITH INTEGRATED MODE FILTER

To enable longer distance transmission over multimode fiber, the spectral width of the VCSEL has to be reduced to mitigate effects of modal and chromatic fiber dispersion. Rather than reducing the size of the oxide aperture [14], which increases VCSEL resistance, current density, and internal temperature, we have integrated a mode filter to suppress higher order modes, thereby reducing the RMS spectral width. The concept is illustrated in Fig.6. The thickness of the top layer of the top DBR is increased such that an anti-phase reflection occurs at the semiconductor/air interface, thereby increasing the cavity loss. We then etch back to restore the in-

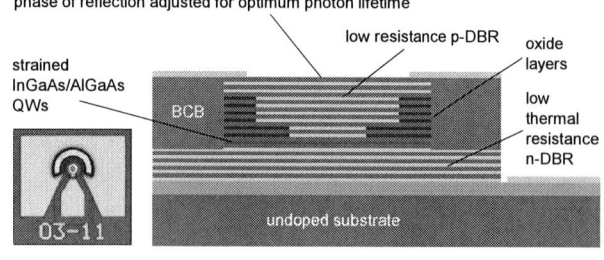

Fig.3 Cross-sectional view of the high speed VCSEL design.

phase reflection above the center of the aperture, creating a shallow surface relief which lowers the cavity loss for the lower order modes and suppresses higher order modes [15].

Spectral measurements (Fig.7) revealed a transition from multimode to quasi-single mode emission when applying the mode filter to a VCSEL with a 5 μm oxide aperture [16]. The RMS spectral width was reduced from 0.9 to 0.3 nm while the modulation bandwidth was reduced from 22 to 19 GHz. The reduction of bandwidth is due to an increase of the photon lifetime which leads to stronger damping of the modulation response. Transmission experiments showed that the maximum transmission distance at 25 Gbps could be extended from 100 to 500 m when applying the mode filter [16]. The results from the transmission experiments are shown in Fig.8.

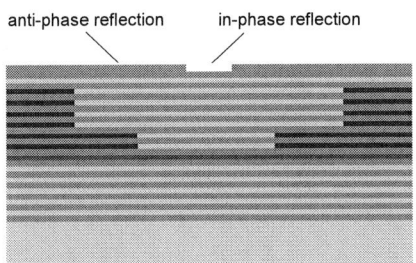

Fig.6   Cross-sectional view of a VCSEL with a shallow surface relief etched into the top layer of the top DBR for suppression of higher order modes.

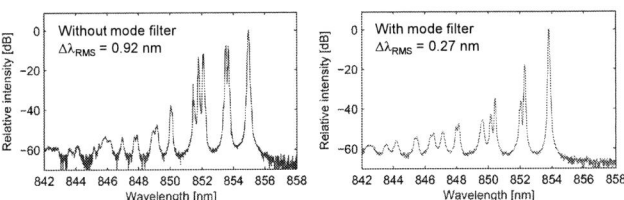

Fig.7   Emission spectra for VCSELs without (left) and with (right) an integrated mode filter.

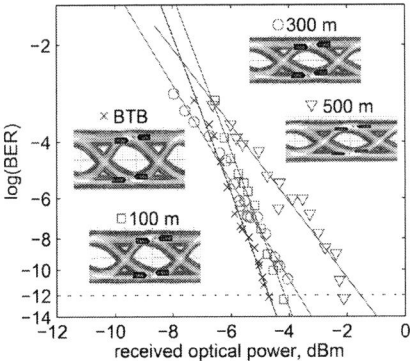

Fig.8   Results from 25 Gbps transmission experiments using a mode filtered high speed VCSEL and various lengths of OM3+ fiber.

IV. MULTILEVEL MODULATION

Multilevel modulation can be used to increase the data transfer capacity and/or reach of a single channel optical link with a given bandwidth, beyond what is possible with traditional on-off keying (OOK) modulation. This is due to the higher spectral efficiency of the multi-level modulation format.

We have applied a 4-PAM modulation format (4 intensity levels (symbols), with 2 bits per symbol) which offers lower complexity and better receiver sensitivity than subcarrier modulation [17]. Transmission experiments were carried out at a symbol rate of 15 Gbaud/s (equivalent to a bit rate of 30 Gbps). The link bandwidth was <10 GHz, limited primarily by the 12 GHz photoreceiver. The received eyes are shown in Fig.9 while the link performance is shown in Fig.10.

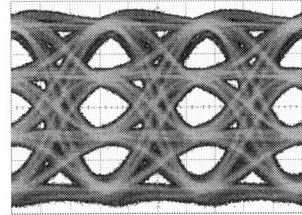 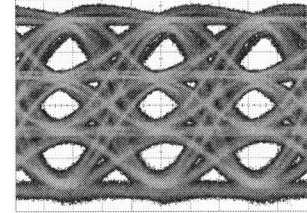

Fig.9   Received eyes at 15 Gbaud/s 4-PAM (30 Gbps) transmission. Left: BTB. Right: 200 m of OM3+ fiber.

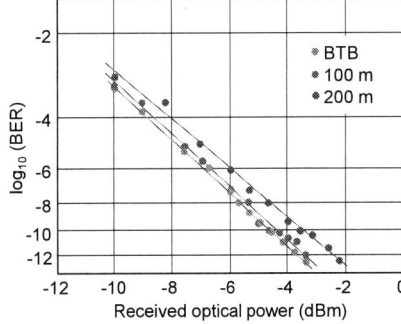

Fig.10 Results from 15 Gbaud/s 4-PAM (30 Gbps) transmission over various lengths of OM3+ fiber.

The results show that 4-PAM enables 30 Gbps transmission over 200 m of multimode fiber even with a link bandwidth <10 GHz. This would not be possible using OOK modulation. A higher link bandwidth and the use of an 8-PAM modulation format may eventually enable a single channel capacity approaching 100 Gbps.

V. HIGH SPEED TUNABLE VCSELs

Future optical interconnects in e.g. computing systems may use wavelength division multiplexing and high speed tunable optical sources for the interconnect network to be able to adapt to the irregular and time varying traffic patterns. This calls for the development of tunable high speed VCSELs.

Our tunable VCSEL (Fig.11) employs a "half-VCSEL" on top of which a spherical dielectric DBR is integrated using surface micromachining techniques [18]. The spherical DBR ensures single transverse mode operation and electro-thermal tuning is accomplished by passing a current through a resistive heating element deposited on the dielectric DBR. The "half-VCSEL" was designed for high speed modulation.

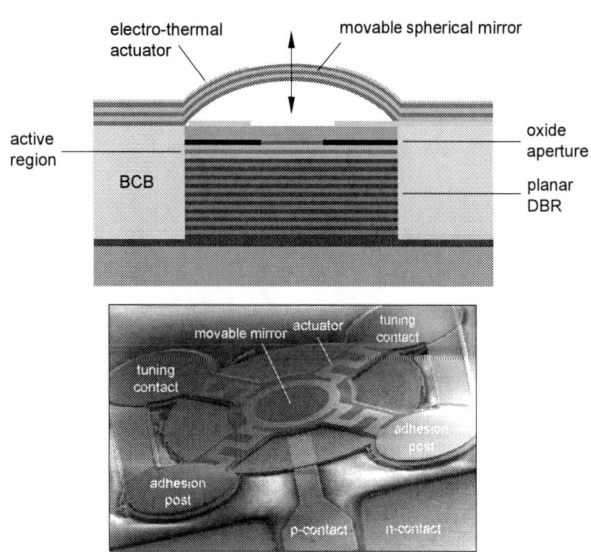

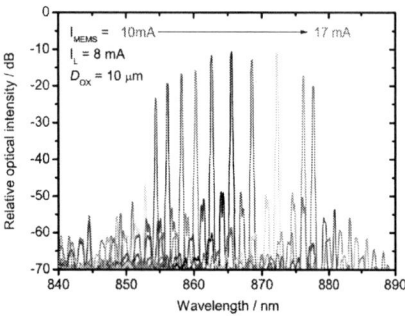

Fig.11 Upper: cross-sectional view of the tunable high speed VCSEL. Lower: SEM image of a fully functional MEMS-tunable VCSEL.

Fig.12 shows that the MEMS-tunable VCSEL has a mode-hop free tuning range of 24 nm. The maximum modulation bandwidth was 6 GHz and exceeds 3 GHz over a wavelength range of 16 nm. This has enabled transmission at 5 Gbps [19].

Fig.12 Emission spectra at a bias current of 8 mA and tuning currents from 10 to 17 mA.

VI. Acknowledgement

We acknowledge the collaborative efforts by the partners in the European projects VISIT and SUBTUNE and at the Department of Signals and Systems at Chalmers. Financial support was provided by the European Union and the Swedish Foundation for Strategic Research.

References

[1] K. Yashiki et al., "1.1 µm range high speed tunnel junction vertical cavity surface emitting lasers", *IEEE Photon. Techn. Lett.* 19, 1883 (2007).

[2] Y.C. Chang et al., "Efficient high data rate tapered oxide aperture vertical cavity surface emitting lasers", *IEEE J. Sel. Top. Quantum Electron.* 15, 704 (2009).

[3] A. Mutig et al., "Frequency response of large aperture oxide confined 850 nm vertical cavity surface emitting lasers", *Appl. Phys. Lett.* 95, 131101 (2009).

[4] P. Westbergh et al., "Speed enhancement of VCSELs by photon lifetime reduction", *Electron. Lett.* 46, 938 (2010).

[5] N. Suzuki et al., "High speed 1.1 µm range InGaAs-based VCSELs," *IEICE Trans. Electron.* E92-C, 942 (2009).

[6] P. Westbergh et al., "40 Gbit/s error-free operation of oxide-confined 850 nm VCSEL", *Electron. Lett.* 46, 1014 (2010).

[7] P. Wolf et al., "High performance 980 nm VCSELs for 12.5 Gbit/s data transmission at 155°C and 49 Gbit/s at -14°C, *Electron. Lett.* 48, 389 (2012).

[8] A.V. Rylyakov et al., "A 40 Gb/s VCSEL-based full optical link", *Proc. Optical Fiber Communication Conference*, paper OTh1E1 (2012).

[9] P. Moser et al., "Energy efficient VCSELs for green data and computer communication", *Proc. SPIE* 8276, 82760J (2011).

[10] S.B. Healy et al., "Active region design for high speed 850 nm VCSELs", *IEEE J. Quantum Electron.* 46, 506 (2010).

[11] P. Westbergh et al., "High speed, low current density 850 nm VCSELs", *IEEE J. Sel. Top. Quantum Electron.* 15, 694 (2008).

[12] A. Larsson et al., "High speed VCSELs for short reach communication", *Semicond. Sci. Techn.*, 26, 014017 (2011).

[13] P. Westbergh et al., "Impact of photon lifetime on high speed VCSEL performance", *IEEE J. Sel. Top. Quantum Electron.* 17, 1603 (2011).

[14] G. Fiol et al., "Multimode optical fiber communication at 25 Gbit/s over 300 m with small spectral width VCSELs", *Electron. Lett.* 47, 810 (2011).

[15] Å. Haglund et al., "Single fundamental mode output power exceeding 6 mW from VCSELs with a shallow surface relief", *IEEE Photon. Techn. Lett.* 16, 368 (2004).

[16] E. Haglund et al., "25 Gbit/s transmission over 500 m multimode fiber using 850 nm VCSEL with integrated mode filter", *Electron. Lett.* 48, 517 (2012).

[17] K. Szczerba et al., "30 Gbps 4-PAM transmission over 200 m of MMF using an 850 nm VCSEL", *Opt. Exp.* 19, B203 (2011).

[18] B. Kögel et al., "Integrated MEMS-tunable VCSELs using a self-aligned reflow process", *IEEE J. Quantum Electron.* 48, 144 (2012).

[19] B. Kögel et al., "Integrated MEMS-tunable VCSELs with high modulation bandwidth", *Electron. Lett.* 47, 764 (2011).

Slotted tunable laser with monolithic integrated mode coupler

James R. O'Callaghan[1,*], Brendan Roycroft[1], Wei-Hua Guo[2], Qiao-Yin Lu[2], Chris L. Daunt[1, 3], Kevin Thomas[1], Emanuele Pelucchi[1], John Donegan[2], Frank H. Peters[1, 3] and Brian Corbett[1]

[1]*Tyndall National Institute, Lee Maltings, Cork, Ireland*
[2]*School of Physics Trinity College Dublin*
[3]*Department of Physics University College Cork*

Abstract — **We present a tunable laser based on partially reflective slots etched into the active ridge of a laser the output of which is monolithically coupled into a lower waveguide using a tapered coupler. The laser demonstrates a tuning range of 2.8 THz covering 8 channels separated by 400 GHz. The coupling efficiency to the lower waveguide is up to 90%. This approach requires no epitaxial re-growth making it suitable for low cost photonic integration with additional functional devices such as an integrated IQ modulator.**

I. INTRODUCTION

InP based photonic integrated circuits (PICs) are a critical technology in providing the core functions of next-generation optical networks in order to meet increasing bandwidth demands [1]. The elimination of fiber coupling between components serves to both reduce cost and to increase reliability. The twin waveguide (TG) approach to photonic integration offers a route by which a variety of components of differing structure, function and bandgap can be integrated on the same wafer of a photonic integrated circuit using a single epitaxial growth [2]. The simplest TG structure has two vertically stacked coupled waveguides separated by a cladding layer, and is grown in a single epitaxial growth step. All the components and their integration are defined by post-growth processing. Typically, the upper waveguide is used for active devices with gain (e.g., lasers, semiconductor optical amplifiers), whereas the lower waveguide has a larger bandgap energy and is used for on-chip manipulation of light such as routing, splitting, phase modulation, etc. In this paper, we demonstrate the integration of a tunable laser with a passive lower waveguide. The tunable laser is based on the introduction of slots into the active ridge waveguide for the purposes of providing feedback thus eliminating the dependence of reflectance from at least one of the chip facets [3]. The passive waveguide in this paper has a similar effective index and mode profile to that of a Mach-Zehnder configuration as used for IQ modulation [4].

II. SIMULATION

A layer structure consisting of a top active waveguide and a lower passive waveguide was simulated using Fimmwave/ Fimmprop from Photon Design. Four depths were defined, the top of the material, the depth of the upper ridge which remains above the active waveguide, the depth of the n-contact just below the upper waveguide, and the lowest depth which is below the lower waveguide. This then requires three etches. A basic taper is used to start with, and then the dimensions are adjusted to maximise coupling from the upper waveguide active fundamental mode to the lower waveguide passive fundamental mode. Furthermore, the confinement factor of the upper waveguide can be varied by changing the thickness of the cladding layers, and the separation between the two waveguides can be varied by changing the thickness of an InP spacer layer which separates the waveguides. Values for refractive index of the various layers were taken from the literature. By systematically varying the structure parameters, including number of taper segments, a design was found of a coupler with a length of about 250 μm and greater than 95% coupling efficiency, assuming no absorption in the coupler. A mask was designed with an electrical contact on the coupler in order to bias it at transparency to overcome residual absorption. Figure 1 shows the side elevation and plan view of

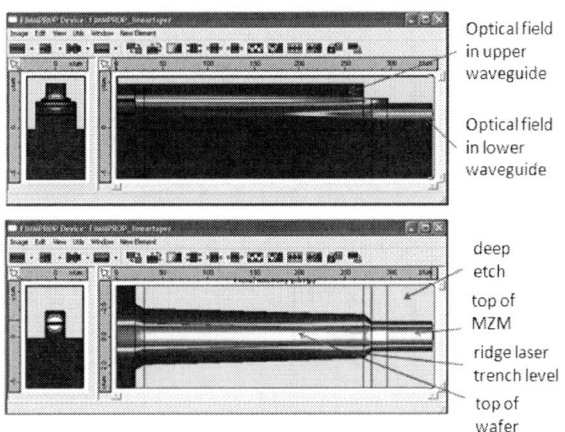

Fig. 1. Generic device design and simulated optical coupling between waveguides. An overall efficiency greater than 95% is calculated.

the optimised structure. The optical intensity can be seen to transfer from the upper to the lower waveguide whilst retaining fundamental mode operation.

978-1-4673-1725-2/12 $31.00 © 2012 IEEE

III. EPITAXIAL DESIGN

The layer structure is grown on an n-type InP substrate. An undoped 320 nm thick AlInGaAs slab waveguide is first grown. This acts as the lower waveguide, and the bandgap of 1330 nm makes it optically transparent at the laser wavelength. On this an n-doped 750 nm thick InP layer is grown to separate the lower waveguide from the laser waveguide. This layer also acts as an n-contact plane for the laser. The laser structure is then grown and consists of six 7.5 nm thick AlInGaAs quantum wells (λ=1520 nm) with 10 nm AlInGaAs barriers (λ=1270 nm). An InGaAsP etch stop layer is used to control the etch depths and the top InP cladding is then grown. This is then capped with a 50 nm thick InGaAsP and a 150 nm thick p+ InGaAs contact layer. Table 1 below shows the epitaxial structure in detail.

TABLE I

EPITAXIAL STRUCTURE OF INTEGRATED MODE COUPLER

	Lyr	Material	Thickness [nm]	Doping
p	25	InGaAs	150	1E19p (Zn)
p	24	Q1.3 InGaAsP	50	2-3e18p (Zn)
p	23	InP	1340	1E18p (Zn)
p	22	Q1.3 InGaAsP	20	1E18p (Zn)
p	21	InP	50	1e18p (Zn)
p	20	AlInAs	40	7E17p (Zn)
i	19	Q0.97 AlGaInAs	35	-
i	18	Q1.1 AlGaInAs	35	-
qw	6x6	Q1.5 AlGaInAs	7.5	-
b	5x7	Q1.27 AlGaInAs	10	-
i	4	Q1.1 AlGaInAs	80	-
n	3	InP	750	1-2e18n (Si)
i	2	Q1.27 AlGaInAs	320	-
i	1	InP	100	1-2e18n (Si)
n		InP	Substrate	n

IV. DEVICE DESIGN

The upper waveguide and laser is processed into a 2.5 μm ridge waveguide laser with etched slots for mode control [3]. The laser is contacted on three distinct sections so as to allow tuning of the emitted wavelength. The total length of the laser section is 1930 μm while the length of the tapered coupler section is 210 μm. The lower ridge is deeply etched through the passive waveguide, and the width tapers down to 1.2 μm by the end of the coupler. A schematic of the device is shown in Figure 2.

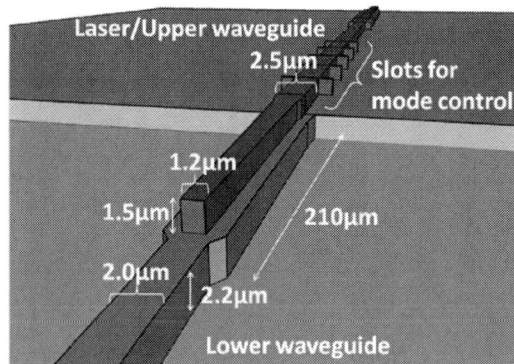

Fig. 2.   Schematic of a slotted laser and coupler to lower waveguide

V. DEVICE FABRICATION

The laser ridge is first defined then etched using Cl$_2$/CH$_4$/H$_2$ ICP etch in an Oxford 150 Plasmalab system. The ridge process is then finished with a brief HCl:H$_3$PO$_4$ wet etch to the InGaAsP etch stop layer. Figure 3 shows a SEM image of the laser ridge and an etched slot for mode control. The

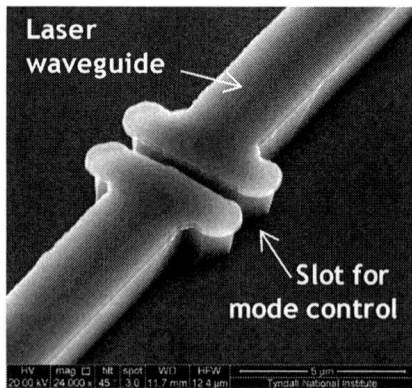

Fig. 3.   SEM image of ridge waveguide and mode control slot of tunable laser.

waveguide is orientated perpendicular to the <110> plane providing a slightly negatively sloped sidewall for the ridge section and a slightly positive slope for the mode slot region for mode control due to the crystallographic nature of the etch. The SiO$_2$ mask is then removed in a HF etch and a new layer of SiO$_2$ is deposited. This acts as the hard mask for formation of the lower waveguide. The lower waveguide is then etched using the same ICP chemistry to a depth of 2.2 µm. Figure 4 shows a SEM image of the coupling region between the upper (laser) and lower waveguide. The hard mask is again removed

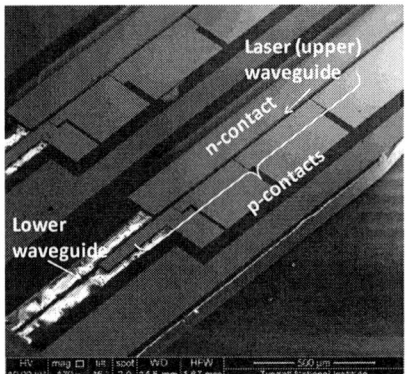

Fig. 4. SEM image of upper laser waveguide and coupling section to lower waveguide

and replaced and the n-contacts are defined and etched. A protective layer of SiO$_2$ is deposited via PECVD and the structure is planarised using BCB. After planarisation an opening to the n-contacts is etched in the BCB. The p- and n-metal contacts are then deposited. A SEM image of the completed devices is shown in Figure 5. The light coloring around the lower waveguide is due to charging of the BCB cladding on the lower ridge.

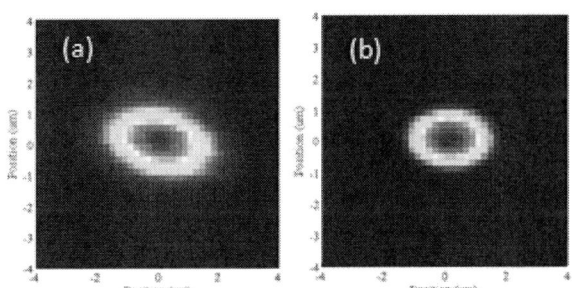

Fig. 5. SEM image of completed tunable laser with waveguide coupler.

VI. EXPERIMENTAL RESULTS

The laser can be controlled using the three tuning contacts. A fourth contact is used to bias the taper to transparency, as the laser QWs are absorbing along the coupler, but this contact is not required for control of the laser. The Vernier design of the laser produces discrete channels, in this case spaced at 400 GHz. The channels achieved from one of the finished devices are shown in Figure 6. The lasing threshold current of the

facetless laser was 25 mA per section for the three laser sections, and all channels could be reached by varying the section currents from 25 to 100 mA. In this figure, the current in the taper section was set at 30 mA. By varying the taper current the power can be equalized to a much greater extent.

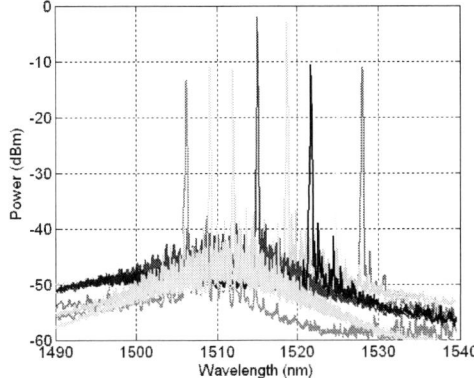

Fig. 6. Emission spectra spaced at 400 GHz covering 2.8 THz.

For a Mach-Zehnder modulator based on the quantum confined Stark effect (QCSE) [4] one way to achieve a low operational voltage is if the optical mode is tightly confined with a high mode overlap with the QWs. In the present structure the passive lower waveguide is designed to have greater optical confinement than the upper waveguide. The modal properties were measured by biasing the laser section and scanning a lens ended fiber across the output of the upper and lower waveguides. The focal plane was found by scanning the fiber in 3D. The result in Figure 7 shows that the mode has been successfully confined in the lower waveguide.

Fig. 7. Laser emission from a) upper and b) lower waveguide.

The coupling efficiency of the tapered region was measured in two ways: First by comparing the emitted power from the upper and lower waveguides and secondly by injecting light via lens-ended fibre from an external tunable laser and measuring Fabry-Perot transmission fringes. From the FP transmission spectrum, one can obtain both the waveguide loss and fibre coupling loss. The waveguide loss can be obtained from

$$\alpha = -\frac{1}{L}\ln\left(R\frac{1+\sqrt{1/r}}{1-\sqrt{1/r}}\right)$$

(1)

where L is the waveguide length, R is the power reflectivity of the facets and r is the ratio of the peak to the minimum transmitted power of the measured fringes. The fibre coupling loss can be measured by subtracting the peak fringe power from the fibre-to-fibre coupled power (assumed to be 100%) to get the total loss which consists of the waveguide and coupling losses. Subtracting the waveguide loss (calculated from exp(-αL)) gives the coupling loss. This can then be halved to get the coupling loss at each end of the device. The validity of this procedure was verified using a simulation of the FP fringes that one would obtain in an 'ideal' lossy waveguide, and it was found that the numbers extracted according to the above procedure matched the known values used in the simulation. By applying this procedure to test FP ridges of just the upper waveguide and just the lower waveguide, one can use the loss results to find the coupler loss in a full upper waveguide/coupler/lower waveguide structure, as the only unknown is the coupler loss.

VII. Conclusions

A tunable laser has been fabricated as an active waveguide, and the light coupled to a second monolithic passive waveguide. All waveguides were processed using standard photolithography with no re-growth necessary. The mode in the lower waveguide is compatible with a Mach-Zehnder modulator, allowing in principle fully integrated tunable laser-MZM devices, including IQ modulators, as needed for advanced modulation format data transmission. This work was supported by Science Foundation Ireland under Grant SFI 10/CE/I1853 CTVR II.

References

[1] L. Coldren, S. Nicholes, L. Johansson, S. Ristic, R. Guzzon, E. Norberg, and U. Krishnamachari, "High Performance InP-Based Photonic ICs—A Tutorial", *J. Lightw. Technol.,* vol. 29, no. 4, pp 554-570, 2011.

[2] P. V. Studenkov, M. R. Gokhale, and S. R. Forrest, "Efficient coupling in integrated twin waveguide lasers using waveguide tapers," *IEEE Photon. Tech. Lett.,* vol. 11, no. 9, pp. 1096-1098, September 1999.

[3] D. C. Byrne, J. P. Engelstaedter, W. Guo, Q. Lu, B. Corbett, B. Roycroft, J. O'Callaghan, F. H. Peters, and J. F. Donegan, "Discretely Tunable Semiconductor Lasers Suitable for Photonic Integration," *IEEE Sel. Topics Quant. Electron.* vol. 15, no. 3, pp. 482-487, 2009.

[4] J. O'Callaghan, B. Roycroft, W.-H. Guo, C. LL. M. Daunt, J. F. Donegan, F. H. Peters, B. Corbett, "Refractive index contributions to phase shifting in InP based 30 GHz bandwidth n-i-n Mach-Zehnder Modulators", IPRM, Berlin, Germany, 22-26 May, 2011

C-band Operation of Lateral-grating-assisted Lateral Co-directional Coupler Tunable Laser with High-mesa Buried Heterostructure

T. Suzuki[1], H. Arimoto[1], T. Kitatani[1], A. Takei[1], T. Taniguchi[1], K. Shinoda[1],
S. Tanaka[1], T. Ido[1], A. Nakamura[2], and K. Naoe[2]

[1]Central Research Laboratory, Hitachi, Ltd.
[2]Opnext Japan, Inc.

Abstract

To decrease the operational power consumption of a laser module, a lateral-grating-assisted lateral co-directional coupler (LGLC) tunable laser with a high-mesa buried heterostructure (BH) was fabricated. The LGLC laser demonstrated successful tunable laser operation in the C-band wavelength range of 37 nm with 40 mA injection current into an LGLC filter at 50°C.

I. Introduction

Tunable lasers with indium phosphide (InP) tunable reflectors and filters, for instance, sampled-grating distributed-Bragg-reflector (SG-DBR) lasers, super-structure-grating (SSG) DBR lasers, and grating-assisted co-directional coupler lasers with a rear-sampled grating reflector (GCSR), have been widely studied [1]-[3]. These tunable lasers have advantages (namely, small footprint and low power consumption) that make them useful for realizing tunable 10-Gigabit laser modules for metro networks.

SG-DBR lasers and SSG-DBR lasers are typical tunable lasers. In these lasers, two DBR-based semiconductor filters are integrated with the gain region. By exploiting the Vernier-effect of the two DBRs, these lasers have been reported to operate successfully over a 40 nm wide wavelength range [1][2]. To prevent unexpected "mode hop," however, they require complex algorithms for controlling the current injected into each DBR.

In comparison SG-DBR and SSG-DBR lasers, lasers such as a grating-assisted co-directional coupler laser with a rear-sampled grating reflector (GCSR) adopt a combination of a comb-shaped filter with high wavelength selectivity (such as a DBR) and a widely tunable filter with medium wavelength selectivity [3]; accordingly, they can provide a simpler current control procedure, because the lasing wavelength can be simply shifted in accordance with the selected wavelength of the widely tunable filter.

In our previous work, a tunable laser with a lateral-grating-assisted lateral co-directional coupler (LGLC) tunable filter was proposed, and it demonstrated lasing characteristics in a wide wavelength range, namely, over 65 nm [4][5]. Similar to GCSR lasers, the LGLC laser has a simple wavelength-selection mechanism. One of the key features of the LGLC laser, integrated with the LGLC filter monolithically, is that it

has no abrupt structure that causes unstable filter characteristics. Since the LGLC filter, composed of a grating-assisted co-directional coupler, is located in the same plane as the waveguides, a smooth mode conversion in the LGLC filter can be obtained.

The above-mentioned tunable lasers are generally composed of four sections, namely, a gain section, a phase section, and two tunable filters (or reflectors), and all sections are operated by current injections. To obtain a tunable laser operated with low power consumption, all sections are required to have high carrier-injection efficiency. Since lasers with a high-mesa buried heterostructure (BH) generally have a high confinement factor of both carrier and photon, a high-mesa BH is one solution for attaining low-power operation [6].

In this study, a LGLC tunable laser with a high-mesa BH was fabricated. The LGLC filter and LGLC tunable laser (with 4 μm thickness high-mesa BH) were successfully fabricated by deep etching and epitaxy processing for the first time. The laser demonstrated C-band operation, and its fundamental laser characteristics were compared with those of a LGLC laser with a previously reported planar BH (PBH) structure [4][5].

II. Structure and Fabrication

An overview of the LGLC laser-which consists of a LGLC filter, a phase-control section, a gain section, and a DBR-is shown in Fig. 1(a). Since the LGLC filter has a medium wavelength selectivity, the LGLC filter can select one of the multiple reflection peaks of the SG-DBR and provides single-mode operation. The phase-control section is used for fine wavelength tuning by current injection. A cross-sectional view of the LGLC filter with the high-mesa BH is shown in Fig. 1(b). The LGLC filter has two asymmetric cores. One is a high-refractive-index waveguide (HΔ-WG) comprised of a quaternary (Q)1.4 InGaAsP layer and a Q1.3 InGaAsP layer; the other is a low-refractive-index waveguide (LΔ-WG) comprised of a Q1.3 InGaAsP layer only. A long-period grating, which is formed by a mesa-width modulation of the HΔ-WG, is located laterally in plane with the WGs. Since the LΔ-WG is covered with a semi-insulated (SI)-InP, the current can be injected into the HΔ-WG only.

When the incident light to the LGLC filter is launched into the HΔ-WG, a phase-matched light (wavelength: λ_{mat}) is coupled to the LΔ-WG. On the other hand, a phase mismatched light (wavelength: λ_{mis}) cannot be coupled to the LΔ-WG; thus, it propagates through the HΔ-WG. λ_{mat} is given by

$$\lambda_{mat} = \Lambda(n_1 - n_2) \qquad (1)$$

where n_1 and n_2 are the effective indices of the 1st eigen mode (mainly confined in the HΔ-WG) and the 2nd eigen mode (mainly confined in the LΔ-WG), respectively. When a current is injected only to the core of the HΔ-WG, the peak of the transmittance wavelength of the LGLC filter can be tuned. In the LGLC filter, the high-mesa BH is expected to be effective in the case of both high carrier confinement for the HΔ-WG and high-isolation characteristic for the LΔ-WG.

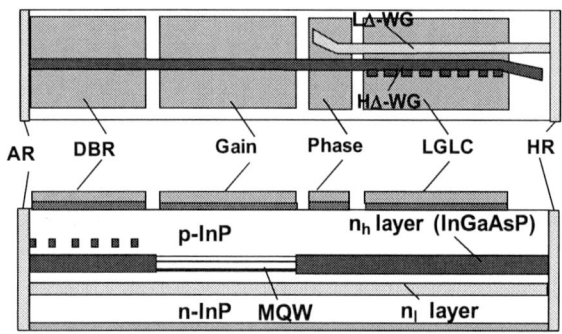

(a) Overview,

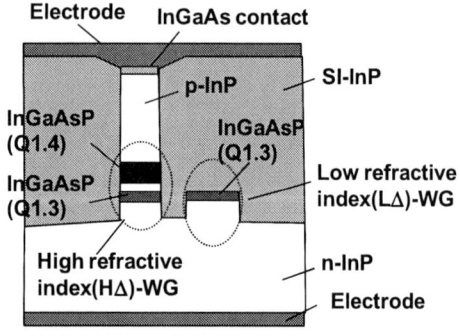

(b) Cross-sectional view of LGLC filter with high-mesa BH

Fig. 1. Schematic structure of LGLC laser

The fabrication procedure of the LGLC filter with high-

mesa BH structure is explained as follows. The twin InGaAsP layers, composed of a p-doped InGaAs contact layer, p-doped InP clad layer, Q1.4 InGaAsP core layer, and Q1.3 InGaAsP core layer, are grown by metal organic vapor phase epitaxy (MOVPE) on an n-doped InP substrate. Thicknesses of Q1.4 and Q1.3 layers are 300 and 150 nm, respectively. An InP layer, grown between the InGaAsP layers, is utilized as the etch-stopped layer. After the MOVPE growth, the waveguide structure included in the long-period grating of the HΔ-WG is formed by photolithography and dry etching. Next, the upper InGaAsP layer of the LΔ-WG is removed by photolithography and wet etching. After that, the LΔ-WG is buried with SI-InP so that current is only injected into the HΔ-WG. Finally, a metal is deposited on both the p-doped InGaAs layer and the n-doped InP substrate. To separate the HΔ-WG and LΔ-WG, the etching depth of the high-mesa structure is over 4 μm. A scanning electron microscope (SEM) photograph of the LGLC and phase section is shown in Fig. 2. This is the first time such a deep mesa of an asymmetrical co-directional coupler has been buried.

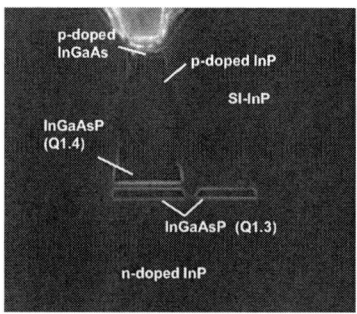

(a)　LGLC section

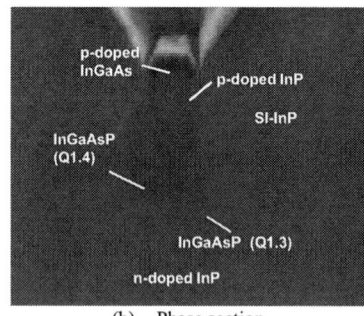

(b)　Phase section

Fig.2. SEM photograph of the LGLC laser

III. CHARACTERISTICS

To investigate a fundamental characteristic, namely, threshold current and slope efficiency, of the fabricated lasers with the high-mesa BH structure, simple Fabry-Perot (FP) lasers were evaluated. Figure 3 shows the light output power as a function of injection current density of the fabricated FP lasers with both high-mesa BH and PBH structures. These lasers have a 400 μm long cleaved facet and waveguide widths are around 1 to 1.4 μm. Operational temperature is 25°C. Similar to the PBH lasers, the fabricated high-mesa BH lasers exhibit low threshold current and high slope efficiency. Furthermore, in the high-current-density range, optical output power of the high-mesa BH lasers is higher than that of the PBH lasers. This result means that the high-mesa BH lasers have high-carrier-confinement characteristics.

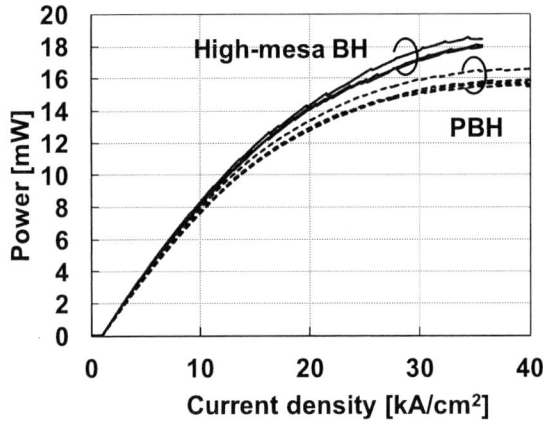

Fig. 3. Light output power as a function of current density in Fabry-Perot lasers with both high-mesa BH and planar BH (PBH). Operational temperature of lasers is 25°C.

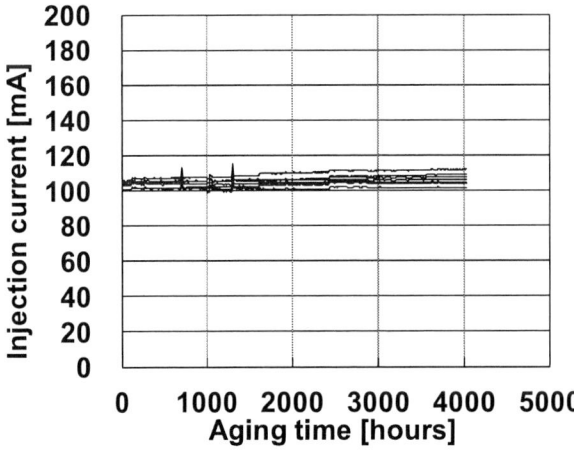

Fig. 4. Variation of injection current as a function of aging time. An auto power control (APC) aging test was carried out and under the following aging- test conditions: aging temperature of 85°C, and optical power of around 4 to 5 mW.

To assess the reliability of the high-mesa BH lasers, an aging test, namely, auto power control (APC), on the FP lasers (more than 10 samples) was carried out. The conditions of aging test are temperature of 85°C and optical power of around 4 to 5 mW, giving injected current density of about 25 kA/cm^2. Injection current is plotted as a function of aging time (4000 hours) in Fig. 4. Injection current of FP lasers to maintain constant optical power is increased during aging. Increment of injection current is, however, less than 5% at aging time of 4000 hours. If activation energy is assumed as 0.7 eV, the estimated lifetime for 50°C normal-use condition is over 5 years. This result proves the high reliability of the high-mesa BH FP lasers.

The LGLC filter characteristic of the LGLC laser is shown in Fig. 5. Vertical axis is a lasing wavelength of the LGLC laser, and horizontal axis is injection current of the LGLC section. Injection currents of the gain, phase, and DBR sections are constant, namely, 100, 0.5, and 5 mA, respectively. The measurement temperature was 50°C. Lasing wavelength was shifted toward short wavelength when injection current was increased. The reason that lasing wavelength is mode hopped about 6 nm is that it jumped from one reflection peak of the SG-DBR to the next. The wavelength shift of the high-mesa BH laser is about 37 nm after the injection current of LGLC section was increased from 0.7 to 40 mA. A C-band tunable laser operation at over 37 nm was demonstrated by the fabricated LGLC lasers. This wavelength shift is, however, less than that of the PBH laser, which has a wavelength shift of 37 nm produced by a LGLC current of only 10 mA. We suppose this small wavelength shift (and large injection current of the LGLC section) of the high-mesa BH laser is not caused by the device design itself and can be improved to the same value as the PBH laser.

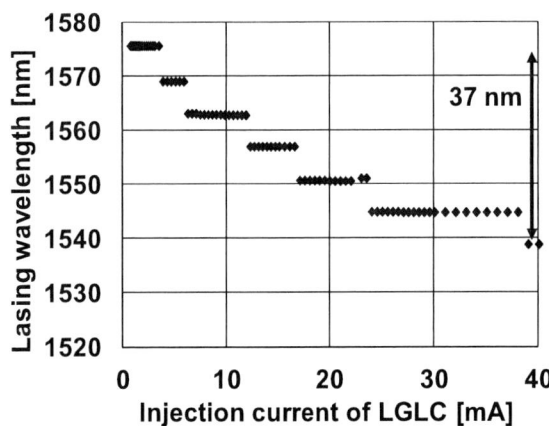

Fig. 5. Lasing wavelength of the LGLC tunable laser as a function of injection current of LGLC section. Injection currents of the gain, phase, and DBR sections are 100, 0.5, and 5 mA, respectively. Operational temperature is 50°C.

IV. SUMMARY

A lateral-grating-assisted lateral co-directional coupler (LGLC) tunable wavelength laser with 4 µm thick high-mesa buried hetero-structure (BH) was fabricated for the first time. Light output power of a Fabry-Perot (FP) laser with high-mesa BH was measured in both initial state and after aging test. The initial characteristics and reliability of high-mesa BH FP were measured. The fabricated LGLC lasers demonstrated C-band tunable laser operation with over 37 nm wavelength shift under a 50°C semi-cooled temperature condition.

REFERENCES

[1] V. Jayaraman, A. Mathur, L. A. Coldren, and P. D. Dapkus, "Extended Tuning Range in Sampled Grating DBR Lasers," *IEEE Photon. Technol. Lett.*, vol. 5, pp. 489-491, 1993.

[2] G. Sarlet, G. Morthier, and R. Baets, "Control of Widely Tunable SSG-DBR Lasers for Dense Wavelength Division Multiplexing," *IEEE J. Lightwave Technol.*, vol. 18, pp. 1128-1138, 2000.

[3] M. Oberg, S. Nilson, K. Streubel, J. Wallin, L. Backbom, and T. Klinga, "74 nm Wavelength Tuning Range of an InGaAsP/InP Vertical Grating Assisted Codirectional Coupler Laser with Rear Sampled Grating Reflector," *IEEE Photon. Technol. Lett.*, vol. 5, pp. 735-738, 1993.

[4] T. Suzuki, H. Arimoto, T. Kitatani, A. Takei, T. Taniguchi, K. Shinoda, S. Tanaka, and S. Tsuji, "Wide Tuning (65nm) Semi-Cooled (50°C) Operation of a Tunable Laser based on a Novel Widely Tunable Filter," *OFC'11*, OWD7, 2011.

[5] T. Suzuki, H. Arimoto, T. Kitatani, A. Takei, T. Taniguchi, K. Shinoda, J. Igrashi, A. Nakamura, K. Naoe, M. Okayasu, S. Tanaka, and S. Tsuji, "Tunable Laser Integrated with Lateral-Grating-Assisted Lateral Co-directional Coupler Filter," *IEEE Photon. Technol. Lett.*, vol. 23, No. 19, pp. 1391-1393, 2011.

[6] H. Arimoto, T. Kitatani, T. Tsuchiya, K. Shinoda, A. Takei, H. Uchiyama, M. Aoki, and S. Tsuji, "Wavelength-Tunable Short-Cavity DBR Laser Array With Active Distributed Bragg Reflector," *IEEE J. Lightwave Technol.*, vol. 24, No.11, pp. 4366-4371, 2006.

Multiple coherent outputs from single growth monolithically integrated injection locked tunable lasers

P. E. Morrissey [1,2], W. Cotter [1,2], J. O'Callaghan [1], H. Yang[1], B. Roycroft [1], D. Goulding [1], B. Corbett [1] and Frank H. Peters [1,2]

1. Tyndall National Institute, University College Cork, Ireland

2. Physics Department, University College Cork, Ireland

Abstract — we present a photonic integrated circuit (PIC) designed to create multiple coherent optical signals for the generation of coherent modulation formats, such as DPSK. A Multimode Interference Coupler (MMI) is monolithically integrated with single facet Slotted Fabry Perot (SFP) lasers at its input and output arms. In this work we show that light from a master SFP laser can be used to injection lock both output slave SFPs for use in the generation of coherent optical channels.

Index Terms — Coherent sources, fabrication and characterization, injection locked lasers, tunable lasers, waveguide devices.

I. INTRODUCTION

In order to satisfy the ever increasing demand placed upon current network communications, novel optical transmission systems and modulation formats have been developed. At higher transmission rates, On-Off keying modulation is now being replaced by multi-level modulation formats for increased robustness and spectral efficiency. Differential Phase Shift Keying (DPSK) and Differential Quadrature Phase Shift Keying (DQPSK) are two such modulation formats [1].

Typically, implementing these modulation formats requires the use of one or more Mach Zehnder modulators. However it has also been shown that these signals can be generated by a Photonic Integrated Circuit (PIC) employing electroabsorption modulators (EAMs) [2]. Recent work has demonstrated the monolithic integration of Slotted Fabry Perot Lasers (SFPs) [3] with SOAs and EAMs based upon a single epitaxial growth step. This enables the development of a single epitaxial grown, fully monolithic DPSK transmitter PIC with integrated laser sources.

A key requirement of such a PIC is the ability to manipulate 2 or more coherent channels independently in both amplitude and phase in order to generate the symbols for transmission. Previously, DQPSK signals have been created on a PIC by splitting and recombining the light from a single laser source [2]. This resulted in a 40dB loss coming from the fibre coupling and the optical splitters and combiners. To overcome this insertion loss problem, in this paper we examine the use of injection locked SFPs [4] to amplify the independent channels while retaining coherence. Both the master and slave SFP lasers are monolithically integrated on chip with an SOA and a

1 x 2 MMI splitter. Light from the master laser is split and used to injection lock the slave SFPs resulting in higher power, single mode operation from both the slave laser devices.

II. DEVICE DESIGN & TESTING

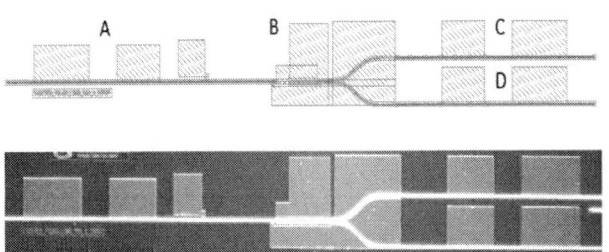

Fig. 1. Schematic of photonic integrated circuit (PIC) used for coherent output generation. A single facet laser (A) is integrated on chip with a 1 x 2 MMI splitter (B). The outputs from the MMI are coupled to two further integrated lasers (C and D). Hatched boxes represent metal contact pads.

A schematic of the device is shown in Figure 1 along with an image of the actual device. The structure was based on a standard 1550 nm laser material that uses 5 x AlInGaAs QWs in the active region on an InP substrate. A two section single facet SFP laser was designed on this material with the required optical feedback coming from the cleaved facet on one end, and etched slotted mirrors on the other end. The mirror section consisted of 7 slots, 1.7 µm deep with a slot width of 0.88 µm. A 2.5 µm wide ridge waveguide provided lateral optical confinement.

Three of these SFP lasers were monolithically integrated on-chip with pseudo-passive waveguide interconnects. Two acted as slave lasers (SL) in the PIC we were investigating with the third acting as the master laser (ML). The ML is shown at A in Figure 1 with two large contact pads for biasing the gain and mirror sections. At the mirror side of the laser, an isolated, 520µm straight waveguide section connected it to the input side of a 1 x 2 MMI indicated by B in Figure 1. The MMI was designed for optimal splitting between the two outputs. The

978-1-4673-1725-2/12 $31.00 © 2012 IEEE

optimal device had a width of 12.5 μm with a corresponding length of 195 μm. At the output of the MMI two constant curvature s-bends of radius 350 μm were used, to separate the outputs by 250 μm, for ease of testing.

The straight waveguide connecting the ML to the MMI, the MMI itself and the output bends are all inherently lossy near 1550 nm, the proposed operational wavelength of the PIC. To compensate for the losses that would otherwise exist, metal contacts were deposited on all these pseudo-passive sections which enable biasing of the sections to control gain and amplification of the optical signal. Applying a reverse bias also allowed these sections to act as photodiodes. At each output of the MMI, two identical single facet SFP lasers, indicated by C and D, in Figure 1, were also fabricated. These lasers acted as the slave lasers in the PIC.

III. DEVICE TESTING

The PIC was tested using a custom built chuck and probe station as shown in Figure 2. Eight probes in total were required to bias the 3 SFPs and common pseudo-passive waveguide sections. The SFPs were all driven individually with the SOA, MMI and MMI output bends driven collectively from a common current source. At the master SFP side of the PIC, a lensed fiber was positioned at the facet of the waveguide. At the slave laser side, a lensed fiber was also positioned. This lensed fiber could be moved between the slave laser outputs, which were located precisely 250 μm apart.

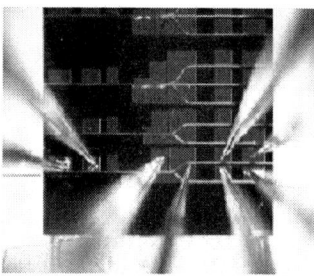

Fig. 1.   Experimental Setup showing probes required for testing of the PIC. Lensed fiber was positioned at the input and output of the chip which was used to examine the output from the master and slave lasers.

LIV curves of the SFPs were taken by fixing the current in the mirror section and varying the current in the gain region. The passive waveguide sections were reversed biased to -2 V and used as photodiodes to determine the light output from the mirror side of both the master and slave lasers. LIV curves of the commonly driven passive sections were taken with light output monitored through the reversed biased mirror sections of the SFPs. This was required to confirm that these sections

did not setup a lasing cavity themselves, which would interfere with the slave and master laser outputs.

Both slave lasers were driven slightly above the determined threshold of the SFP while the pseudo passive waveguide sections were biased to transparency. The master laser bias was varied until single mode operation was achieved and this was then injected into both slave lasers. Due to having just one lensed fiber at the SL side of the chip, it was only possible to look at the output from one SL at a time. The spectrum from each SL was examined separately on an Optical Spectrum Analyzer (OSA). Changing the bias on the gain section of the master or slave SFPs allowed the spectra from both lasers to be either red or blue shifted. This was done to bring the main mode of master laser into alignment with a resonant side mode of the slave laser. This resonant side mode of the slave laser was then injection locked using the output of the master laser. The locked output from the ML and two SL was recorded on the OSA with the traces shown in Fig. 4-5.

With both SL SFPs injection locked to the ML SFP, we investigated the coherence of the outputs of these SL SFPs with the ML. Since we were unable to examine both SL outputs at the same time, we examined whether each output was coherent with the master laser separately. The SL and ML outputs were each sent through a Mach Zehnder Modulated operated at the null and driven to generate a dual sideband suppressed carrier signal. The traces for both were examined on an electronic spectrum analyzer (ESA). The next step involved modulating the SL output and beating this with the unmodulated ML output. The beating of these signals was observed on the ESA and the coherence of the signals was determined.

IV. RESULTS

A.   LIVs of SFPs and Passive Waveguide Sections

The mirror section of each SFP on the device was biased to 30mA while the current in the gain section was varied. Based on the LIV curves we found that threshold for devices occurred at 40 mA. LIV curves for the commonly driven passive sections showed behavior similar to that of an LED; producing spontaneous emission but with no lasing occurring in the cavity. Based on these curves, the passive sections were biased to 50 mA which was sufficient to overcome absorption losses.

B.   Outputs from Free-Running Slave and Master Lasers

Initially, the SL and ML SFP laser outputs were observed with the passive waveguide sections left unbiased, thereby becoming totally absorbing. The ML was biased at 105 mA in the gain section and 30mA in mirror section which resulted in a single mode output at 1577 nm with an SMSR of ~30 dB as

978-1-4673-1725-2/12 $31.00 © 2012 IEEE   282

shown in Figure 4. The two SL SFPs were both operated above threshold with a gain bias of 45 mA and mirror bias of 30 mA. The resonant cavity modes for Slave Laser 1 when taken at the facet side of the laser on an OSA are shown in blue in Figure 5. Slave Laser 2 shows similar output spectra.

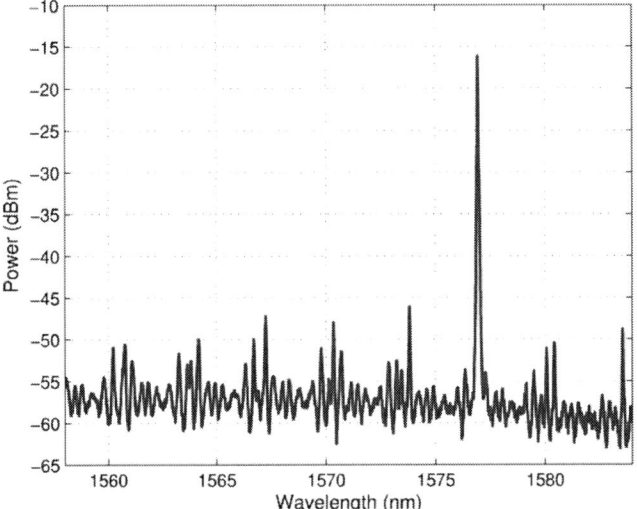

Fig. 4.   Output spectra of master laser. Lases at 1577 nm with an SMSR of 30 dB.

C. Outputs from Injection Locked Slave and Master Lasers

With the MMI and passive waveguide sections now biased at 50mA, light from the master later could be split through the MMI and injected into the mirror side of each SL. The currents of the master laser were chosen so that the main mode of the ML was aligned with a resonant mode of the free-running SL. With the passive sections now transparent, the SL SFP outputs became locked to the injected ML SFP. The outputs from both slave lasers were examined separately on an OSA, with the locked spectra from Slave Laser 1 shown as green in Figure 5.

Slave Laser 1 had a preferential lasing mode present at ~1565 nm when free-running. With light from the master injected into the slave laser cavity we see that the SL becomes injection locked, with the preferential mode being suppressed by 24 dB. The SL has been forced to lase at the master laser wavelength with an SMSR of ~30 dB. Figure 6 shows the overlapped output traces from the master and both slave lasers as measured on an OSA. We see that all other modes from the SL have been suppressed and that they lase strongly at the master wavelength. The power in each slave laser is ~ 9 dB lower than injected master signal. This drop in power arises from inherent MMI splitter losses, losses from going through the bends at the output of the MMI and also coupling losses at the output of the device.

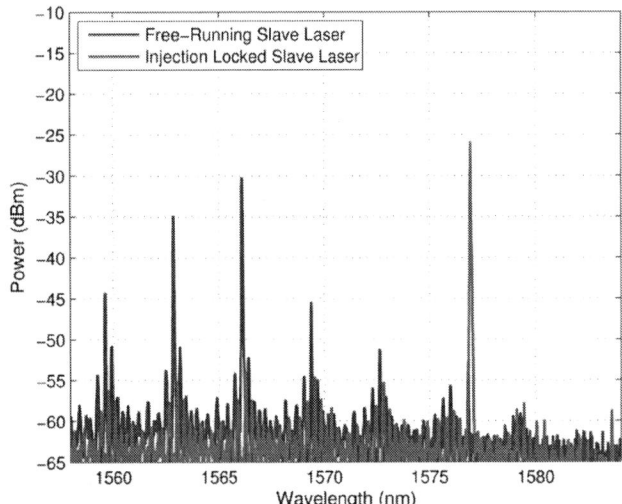

Fig. 5.   Output spectra of the slave laser without (blue) and with (green) injection of master laser. When injection locked the slave laser is forced to lase at the master laser wavelength, with other modes suppressed.

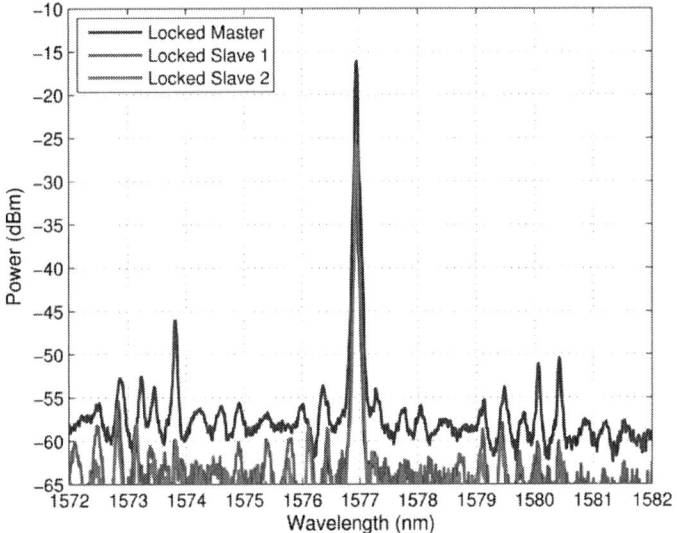

Fig. 6.   Outputs from the both locked slave lasers overlapped with the master laser output. Slave lasers have both been forced to lase at the master wavelength exactly.

D. Coherence of Slave Laser Outputs

With the SL locked to the ML, the SL output was taken and passed through a MZM operated at 10 GHz and driven at the null. Two sidebands 20 GHz apart were generated with the carrier suppressed. This modulated signal was passed through an amplifier then combined with the signal from the master

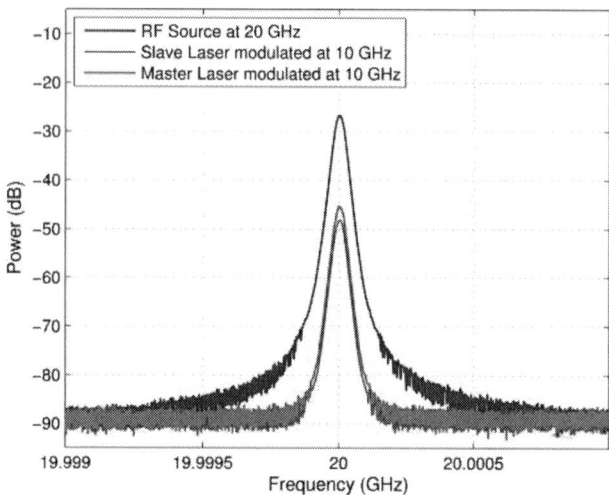

Fig. 7. ESA trace of the RF Source, Modulated Slave and Modulated Master Lasers.

laser. The beat note between these two signals was then observed on an ESA. Figure 7 shows the ESA trace of the RF source itself operated at 20 GHz. (Blue). Also shown is a trace of the 20 GHz tone generated from the beating between the sidebands of the 10 GHz modulated SL (Green) and the same for the modulated ML output (Red). The linewidth of the ESA signal generated for the modulated SL 1 was found to be ~30 kHz. This agrees well with the determined linewidth of the RF source from the ESA signal.

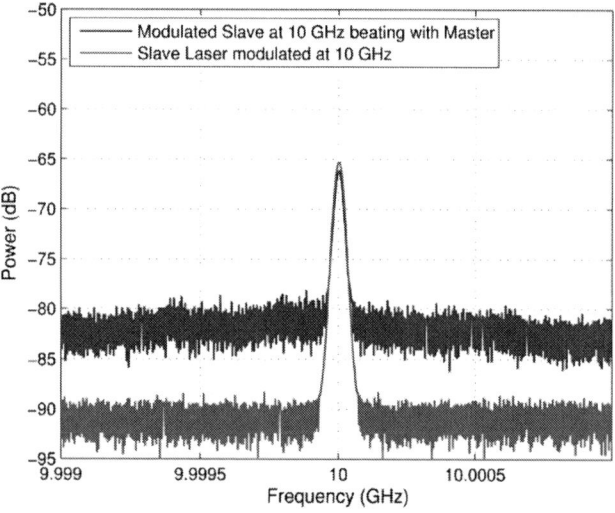

Fig. 8. ESA trace outputs at 10 GHz from beating master SFP with modulated slave SFP 1.

Figure 8 shows the trace of the signal near 10 GHz on the ESA for the modulated locked SL 1 beating with the unmodulated ML (Blue). Again, we also include the trace of just the modulated SL signal (Green) as observed on the ESA).

As seen from Figure 8, we indeed have 2 coherent signals mixing together when we beat the ML and the modulated SL 1 together. The linewidth of the ESA signal generated between the two remains at ~30 kHz, the observed linewidth from the ESA trace of the RF source. No broadening occurs and we see the noise floor increasing from the addition of both signals. The same result is obtained for the modulated SL 2 beating with the master. We have shown that our two SL SFPs are coherent with the ML, and by extension, coherent with each.

CONCLUSION

A novel photonic integrated circuit was designed to investigate the generation of multiple coherent outputs on chip. The output from a single facet SFP laser was split and coupled to two output SFP lasers which were biased above threshold. The output lasers were injection locked to the input master laser with their preferential free-running modes being suppressed. The locked lasers lased at the master wavelength and then were shown to each be coherent with the master laser. This provides a particularly useful means of generating coherent outputs on chips for use in advance modulation formats. Future work aims to investigate the generation of increased numbers of coherent output channels. This work was supported by the Science Foundation Ireland under Grant 07/SRC/I1173, under its CSET, Centre for Telecommunications Value-chain Research and under its Research Frontiers Programme.

REFERENCES

[1] M. Daikoku et al, "100 Gbit/s DQPSK transmission experiment without OTDM for 100 G Ethernet transport," presented at the Optical Fiber Comm. Conf., Anaheim, CA, 2006, Paper PDP36.

[2] C. R. Doerr et al, "Compact High-Speed InP DQPSK Modulator," IEEE Photonics Technology Letters, vol. 19, no. 15, pp. 1184–1186, 2007.

[3] D. C. Byrne, J. P. Engelstaedter, W. Guo, Q. Lu, B. Corbett, B. Roycroft, J. O'Callaghan, F. H. Peters, and J. F. Donegan, "Discretely Tunable Semiconductor Lasers Suitable for Photonic Integration," IEEE Sel. Topics Quant. Electron. vol. 15, no. 3, pp. 482-487, 2009.

[4] Roycroft, B.; Mondal, S.K.; Lambkin, P.; Engelstaedter, P.; Corbett, B.; Peters, F.H.; Smyth, F.; Barry, L.; Phelan, R.; Donegan, J.F.; Ellis, A.D.; , "Fast Switching Tunable Laser Sources for Wavelength Division Multiplexing in Passive Optical Access Networks," Indium Phosphide & Related Materials, 2007. IPRM '07. IEEE 19th International Conference on, vol., no., pp.606-609, 14-18 May 2007.

Low-Threshold Operation of LCI-Membrane-DFB Lasers with Be-doped GaInAs Contact Layer

Mitsuaki Futami[1], Takahiko Shindo[1], Kyohei Doi[1], Tomohiro Amemiya[1,2],
Nobuhiko Nishiyama[1], Shigehisa Arai[1,2]

[1]*Department of Electrical and Electronic Engineering, Tokyo Institute of Technology*
[2]*Quantum Nanoelectronics Research Center, Tokyo Institute of Technology*
2-12-1-S9-5 O-okayama, Meguro-ku, Tokyo 152-8552, Japan
E-Mail futami.m.aa@m.titech.ac.jp, arai@pe.titech.ac.jp

Abstract — One of the promising candidates to solve a problem of a performance limitation of LSI is replacing electrical global wirings by on chip optical interconnections. We proposed and realized lateral-current-injection (LCI) type membrane DFB lasers for this purpose. In this paper, we report a new type LCI membrane DFB laser by introducing Be-doped GaInAs contact layer to the initial wafer structure so as to make simple fabrication of p-contact. As the result, a threshold current of as low as 3.8 mA, which was much lower than the previously reported value of 11 mA, was obtained for the stripe width of 1.5 μm and the cavity length of 250 μm.

Index Terms — GaInAsP/InP, lateral current injection, membrane structure, quantum-well laser, semiconductor laser.

I. INTRODUCTION

The performance of the LSI will soon expected to confront the limitation due to ohmic heating, RC delay, power consumption, and crosstalk in the global wiring. Alternatively, photonic integrated circuits (PICs) in LSI are very intriguing approaches to address the problems confronted in the electrical global wiring [1], and an ultralow power consumption laser is strongly required for such optical interconnections. As a criterion for PICs to be successfully integrated into LSI, meanwhile, the acceptable optical pulse energy of the light source is set to be 100 fJ/bit or less [2]. To meet this strict requirement, micro-cavity lasers such as vertical-cavity surface emitting lasers VCSELs), microdisk lasers, and photonic crystal (PC) lasers have been reported as very low power consumption light sources [3]-[5]. Recently, very low pulse energy operations (less than 100 fJ/bit) of PC lasers [6],[7] and VCSEL [8] were demonstrated.

As a promising candidate for the light source, we proposed and demonstrated a GaInAsP/InP membrane distributed feedback (DFB) laser consisting of a thin semiconductor core layer sandwiched by low-index claddings such as air, benzocyclobutene (BCB), or SiO$_2$. The membrane structure produces a large refractive-index difference between the core layer and the cladding layers and supports strong optical confinement to the active region, leading to ultralow power consumption. In our previous report, low threshold (irradiated power: 0.34 mW) under room temperature continuous wave (RT-CW) optical pumping was successfully demonstrated [9],[10].

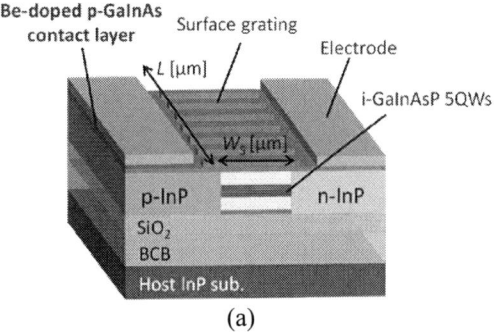

(a)

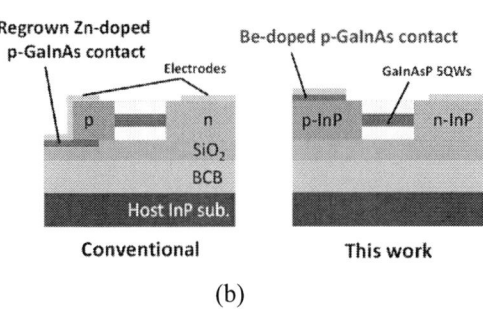

(b)

Fig. 1 Schematic structure of the LCI-membrane laser with Be-doped contact layer. (a) overall view of a device. (b) comparison of cross sectional view with that of a conventional LCI-membrane laser.

Toward an injection-type membrane laser, a lateral current injection (LCI) structure [11] was introduced and an injection-type GaInAsP/InP membrane DFB laser was demonstrated by using BCB adhesive bonding [12]. Recently, room-temperature pulsed operation with threshold current of 11 mA was realized for LCI-membrane-DFB lasers with InP surface grating [13], however, the threshold current was much higher than the expected value from the theory. In this research, new initial wafer structure consisting of a Be-doped GaInAs contact layer was introduced to achieve lower threshold operation of LCI-membrane-DFB lasers and to simplify the fabrication process.

II. DESIGN AND FABRICATION

One of the causes of the high threshold operation is degradation of optical property due to impurity diffusion.

978-1-4673-1725-2/12 $31.00 © 2012 IEEE

Table. 1 GaInAsP/InP epitaxial layer structure

Name	materials	Doping concentration	Thickness
Cap layer	InP	undoped	20 nm
OCL	InP	undoped	155 nm
QW(×5)	$Ga_{0.22}In_{0.78}As_{0.81}P_{0.19}$	undoped	6 nm
barrier(×6)	$Ga_{0.26}In_{0.74}As_{0.49}P_{0.51}$	undoped	10 nm
OCL	$Ga_{0.21}In_{0.79}As_{0.46}P_{0.54}$	undoped	155 nm
Etch stop layer	InP	undoped	50 nm
p contact	$Ga_{0.47}In_{0.53}As$	$8\times10^{18}/cm^3$	50 nm
Etch stop layer	InP	undoped	100 nm
Etch stop layer	$Ga_{0.47}In_{0.53}As$	undoped	300 nm

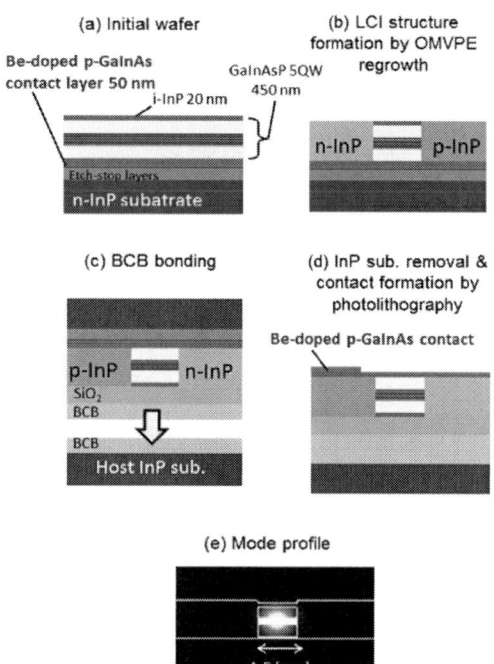

Fig. 2 (a)-(d) Fabrication process of the membrane laser with Be-doped p-GaInAs contact layer. (e) Mode profile of the designed structure for a W_S of 1.5 μm.

So far, Zn has been adopted as a p-type dopant of a GaInAs highly doped contact layer. However it has a large diffusion coefficient, leading to degradation of luminescence properties. In contrast, Be or C are well known as p-type dopants with low diffusion coefficient. Figure 1(a) shows the schematic structure of our membrane DFB laser with a Be-doped contact layer. Top and bottom cladding layers were composed of air ($n = 1$) and SiO$_2$ ($n = 1.45$), respectively. The core layer consisted of five 1% compressively-strained $Ga_{0.22}In_{0.78}As_{0.81}P_{0.19}$ quantum-wells (CS-5QWs, 6-nm-thick), 0.15% tensile-strained $Ga_{0.26}In_{0.74}As_{0.49}P_{0.51}$ barriers (10-nm-thick),

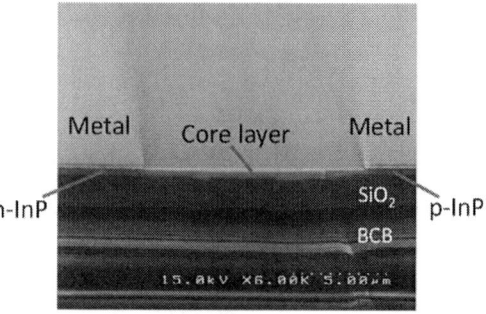

Fig. 3 Cross sectional SEM view of the fabricated device.

sandwiched by optical confinement layers (OCLs, 155-nm-thick for both side). The total thickness of the core layer was 450 nm including a 50-nm-thick InP cap layer for surface passivation. As can be seen from Fig. 1(b), an introduction of Be as a p-type dopant of GaInAs contact enables electrodes to easily access to the contact due to an absence of regrowth process of the contact layer, which can simplify the fabrication process of LCI-membrane lasers.

The device was fabricated as follows. GaInAsP/InP epitaxial layers with a Be-doped p-GaInAs contact layer, as shown in Table. 1, were grown on an n-InP substrate by gas-source molecular-beam epitaxy (GSMBE). The Be-doped contact layer was grown below the OCLs and the QWs, which was different from the previous initial wafer (Zn-doped contact was grown at the top of the wafer). The process flow is simply illustrated in Fig. 2(a)-(d). The LCI structure was fabricated by two-step organometallic vapor-phase epitaxy (OMVPE) selective area regrowth. First, a mesa stripe (7-μm-wide and 400-nm-high) was formed by reactive-ion-etching (RIE) with a SiO$_2$ mask, and n-InP ($N_D = 4 \times 10^{18}$ /cm^3) was selectively regrown on both sides of the mesa as a cladding layer. Next, one side of the cladding layer was etched, and p-InP ($N_A = 4 \times 10^{18}$ /cm^3) were regrown in the same way. Then, after depositing 1-μm-thick SiO$_2$ and 2-μm-thick BCB layers, the wafer was bonded upside down on an InP host substrate, and the BCB was hard-baked at 250°C for 1 hour under a N$_2$ atmosphere. Subsequently, the InP host substrate and etch-stop layers were removed by polishing and wet chemical etching. The top Be-doped GaInAs contact layer (the layer order was reversed by bonding) was then removed except for p-contact section, and Ti / Au electrodes were deposited on both p- and n-side electrodes. Finally DFB pattern was formed by electron-beam lithography (EBL) and CH$_4$/H$_2$ RIE on the InP cap layer. The depth of the InP surface grating was set to be 30 nm and corresponding index-coupling coefficient κ_i was estimated to be 150 cm^{-1}. The grating period and the equivalent refractive index for a device with a stripe width of 1.5 μm are 253.75 nm and 3.10, respectively. Figure 2(d) shows a fundamental mode profile of the designed LCI-membrane-DFB laser in case for stripe width W_S of 1.5 μm. This profile indicates that

978-1-4673-1725-2/12 $31.00 © 2012 IEEE   286

propagating light is well confined to the active region and the optical confinement factor was estimated to be 2%/well.

A scanning-electron-microscopic (SEM) image of fabricated device is shown in Fig. 3, where metal contact pads were formed by lift-off process with $(10 + W_S)$ μm wide mask and AZ5218 photoresist. As can be seen, metal contact pads were formed approximately 5μm away from the stripe edge.

III. EXPERIMENTAL RESULTS

Figure 4 shows the light output properties of an LCI-membrane-DFB laser with a Be-doped GaInAs contact layer (solid line) and a previously reported membrane laser from conventional initial wafer (dashed line). A low threshold operation of 3.8 mA and a differential quantum efficiency of 8.2%/facet were obtained under a RT-pulsed condition (1μs width and 1 kHz repetition) for a device with a cavity length of 250 μm and a stripe width of 1.5 μm. This corresponds to a threshold current density of 1.01 kA/cm^2 (203 A/cm^2/well) when the current is regarded as uniformly injected into the entire stripe region. This threshold is much lower and the output is higher than those of previously reported results [13].

The lasing spectrum was measured using a multimode fiber directly aligned to the cleaved facet. Figure 5 shows the lasing spectrum of a device at a bias current of $2I_{th}$, whose cavity length and stripe width are 250 μm and 1.5 μm, respectively. As can be seen, it showed multi-mode mode operation, not a single-mode operation under DFB mode, which may be attributed to a failure in the formation of the surface grating. From the measured resonant mode spacing of 1.32 nm, the effective refractive index of the waveguide n_{eff} is calculated to be 3.75 which is approximately 5% smaller than that of conventional (vertical current injection type) GaInAsP/InP lasers emitting at 1500-1600 nm wavelength range. Since the equivalent refractive index of this waveguide structure n_{eq} and its wavelength dispersion are calculated to be 3.03 and -0.41 μm^{-1}, respectively, n_{eff} is calculated to be 3.67, which is almost the same as that measured from the mode spacing.

Figure 6 shows a plot of the reciprocal of the measured differential quantum efficiency η_d as a function of the cavity length L and its liner approximation. An internal quantum efficiency η_i of 17% and waveguide loss α_{WG} of 5.9 cm^{-1} were obtained from the following relation,

$$\frac{1}{\eta_d} = \frac{1}{\eta_i}\left[1 + \frac{\alpha_{WG}L}{\ln(1/R)}\right]. \tag{1}$$

The internal quantum efficiency η_i was much lower than that of our conventional DH lasers ($\eta_i \sim 70\%$) and also that of LCI lasers prepared on semi-insulating InP substrates ($\eta_i \sim 40\%$) [15]. This is due to a large amount of carrier leakage in OCLs and carrier recombinations at dielectric-semiconductor interfaces, both of which originate from the LCI structure. Hence an effective threshold current for radiative recombinations is estimated to be 1/2 of measured values. Fortunately, we already reported some

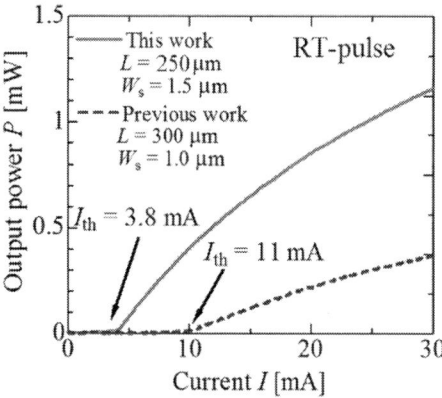

Fig. 4 Lasing characteristics of the membrane laser.

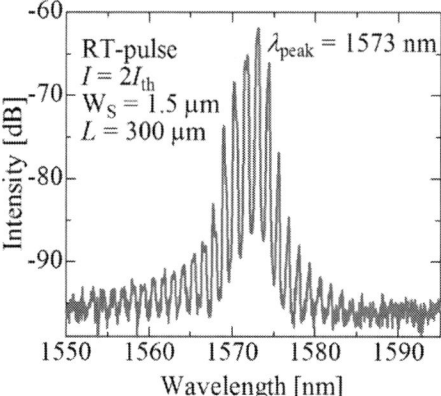

Fig. 5 Lasing spectrum of the membrane laser.

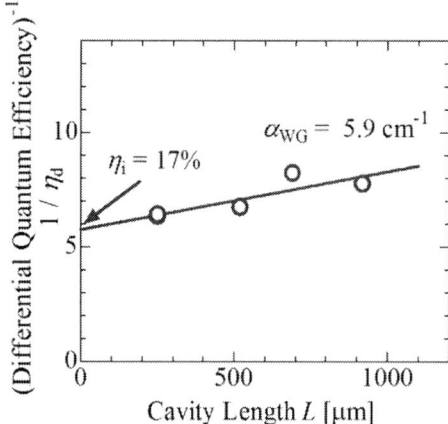

Fig. 6 Reciprocal of differential quantum efficiency as a function of the cavity length.

unique core structures suited for LCI-membrane lasers and can expect their highly-efficient operations with η_i of around 70% [16], [17], threshold current of 1/3-1/4 lower value can be expected by adopting these core structures.

By introducing a thinner core layer (~200 nm) and shorter cavity structure (~50 μm) as well as the matching of peak gain and the Bragg wavelengths, LCI-membrane-

DFB laser will achieve a single-mode operation together with ultralow threshold current.

IV. CONCLUSION

As a step toward a realization of an photonic integrated circuits in LSI, we investigated the LCI-membrane-DFB laser with wire Be-doped p-GaInAs contact layer. As the result, a threshold current of 3.8 mA and a differential quantum efficiency of 8%/facet were obtained for the cavity length of 250 μm and stripe width of 1.5 μm. In addition, an internal quantum efficiency and a waveguide loss were evaluated to be 17% and 5.9 cm^{-1}, respectively.

ACKNOWLEDGMENT

This research was financially supported by the Ministry of Education, Culture, Sports, Science and Technology (MEXT), Japan, the JSPS through Grants-in-Aid for Scientific Research (#24246061, #22360138, #21226010, #23760305, #10J08973) and also by the Ministry of Internal Affairs and communications through SCOPE, and the Council for Science and Technology Policy (CSTP), JSPS through FIRST program.

REFERENCES

[1] G. Chen, H. Chen, M. Haurylau, N. A. Nelson, D. H. Albonesi, P. M. Fauchet, and E. G. Friedman, "Predictions of CMOS compatible on-chip optical interconnect," *VLSI J.*, Vol. 40, No. 4, pp. 434-446, July 2007.

[2] D. A. B. Miller, "Device requirements of optical interconnects to silicon chips," *Proc. IEEE*, Vol. 97, No. 7, pp. 1166-1185, July 2009.

[3] M. Fujita, R. Ushigome, and T. Baba, "Continuous wave lasing in GaInAsP microdisk injection laser with threshold current of 40 μA," *Electron. Lett.*, Vol. 36, No. 9, pp. 790-791, Apr. 2000.

[4] N. Nishiyama, C. Caneau, B. Hall, G. Guryanov, M. H. Hu, X. S. Liu, M/=J. Li, R. Bhat, and C. E. Zah, "Long-wavelength vertical-cavity surface-emitting lasers on InP with lattice matched AlGaInAs-InP DBR grown by MOCVD," *IEEE J. Sel. Top. Quantum Electron.*, Vol. 11, No. 5, pp. 990-998, Sept./Oct. 2000.

[5] B. Ellis, M. Mayer, G. Shambat, T. Sarmiento, J. Harris, E. E. Haller, and J. Vuckovic, "Ultra-low threshold, electrically pumped quantum dot photonic crystal nanocavity laser," *Nat. Photonics*, Vol. 5, pp. 297-300, May 2011.

[6] S. Matsuo, A. Shinya, T. Kakitsuka, K. Nozaki, T. Segawa, T. Sato, Y. Kawaguchi, and M. Notomi, "High-speed ultracompact buried heterostructure photonic crystal laser with 13fJ of energy consumed per bit transmitted," *Nature Photon.* Vol. 4, No. 9, pp. 648-654, Aug. 2010.

[7] Shinji Matsuo, Koji Takeda, Tomonari Sato, Masaya Notomi, Akihiko Shinya, Kengo Nozaki, Hideaki Taniyama, Koichi Hasebe, and Takaaki Kakitsuka, "Room-temperature continuous-wave operation of lateral current injection wavelength-scale embedded active-region photonic-crystal laser," *Opt. Express*, Vol. 20, No. 4, pp. 3773-3780, Feb. 2012.

[8] P. Moser, W. Hofmann, P. Wolf, J. A. Lott, G. Larisch, A. Payusov, N. N. Ledentsov, and D. Bimberg, "81 fJ/bit energy-to-data ratio of 850 nm vertical-cavity surface-emitting lasers for optical interconnects," *Appl.*

Phys. Lett., Vol. 98, No. 23, pp. 231106-1-231106-3, June 2011.

[9] S. Sakamoto, H. Naitoh, M. Otake, Y. Nishimoto, S. Tamura, T. Maruyama, N. Nishiyama, and S. Arai, "Strongly index-coupled membrane BH-DFB lasers with surface corrugation grating," *IEEE J. Sel. Top. Quantum Electron.*, Vol. 13, No. 5, pp. 1135-1141, Sept./Oct. 2007.

[10] S. Sakamoto, H. Naitoh, M. Ohtake, Y. Nishimoto, T. Maruyama, N. Nishiyama, and S. Arai, "85 °C continuous-wave operation of GaInAsP/InP-membrane buried heterostructure distributed feedback lasers with polymer cladding layer," *Jpn. J. Appl. Phys.*, Vol. 46, No. 47, pp. L1155-L1157, Nov. 2007.

[11] K. Oe, Y. Noguchi, and C. Caneau, "GaInAsP lateral current injection lasers on semi-insulating substrates," *IEEE Photon. Technol. Lett.*, Vol. 6, No. 4, pp. 479–481, Apr. 1994.

[12] T. Okumura, T. Koguchi, H. Ito, N. Nishiyama, and S. Arai, "Injection-type GaInAsP/InP membrane buried heterostructure distributed feedback laser with wirelike active regions," *Appl. Phys. Express*, Vol. 4, No.4 pp. 042101-1-042101-3, Mar. 2011.

[13] T. Shindo, M. Futami, T. Okumura, R. Osabe, T. Kouguchi, T. Amemiya, N. Nobuhiko, and S. Arai., "Lasing Operation of lateral-current-injection membrane DFB laser with surface grating," *the 16th Opto-Electronics and Communications Conference (OECC2011)*, 6D3-7, July 2011.

[14] M. Futami, T. Shindo, T. Okumura, R. Osabe, T. Koguchi, T. Amemiya, N. Nobuhiko, and S. Arai., "Stripe width dependence of internal quantum efficiency and carrier injection delay in lateral current injection GaInAsP/InP lasers," *the 16th Opto-Electronics and Communications Conference (OECC2011)*, 7D2-2, July 2011.

[15] T. Okumura, H. Ito, D. Kondo, N. Nishiyama, and S. Arai, "Continuous wave operation of thin film lateral current injection lasers grown on semi-insulating InP substrate," *Jpn. J. Appl. Phys.*, Vol. 49, No.4 pp. 040205-1-040205-3, Apr. 2010.

[16] M. Futami, K. Shinno, T. Shindo, K. Doi, T. Amemiya, N. Nobuhiko, and S. Arai, "Improved quantum efficiency of GaInAsP/InP top air-clad lateral current injection lasers," *the 1st Optical Interconnects Conference (OI2012)*, TuB-3, May 2012.

[17] M. Futami, T. Shindo, T. Koguchi, K. Shinno, T. Amemiya, N. Nobuhiko, and S. Arai, "GaInAsP/InP lateral current injection laser with uniformly distributed quantum wells structure," *IEEE Photon. Technol. Lett.*, Vol. 24, No. 11, pp. 888–890, June 2012.

9781467317252